International Conference on Information Technology and Engineering Management (ITEM 2020)

Held online due to COVID-19

IOP Conference Series: Materials Science and Engineering
Volume 851

Batam, Indonesia
2 - 4 April 2020

ISBN: 978-1-7138-3614-8
ISSN: 1757-8981

Printed from e-media with permission by:

Curran Associates, Inc.
57 Morehouse Lane
Red Hook, NY 12571

Some format issues inherent in the e-media version may also appear in this print version.

Institute of Physics
Dirac House, Temple Back
Bristol BS1 6BE UK

Phone: 44 1 17 929 7481
Fax: 44 1 17 920 0979

techtracking@iop.org

Additional copies of this publication are available from:

Curran Associates, Inc.
57 Morehouse Lane
Red Hook, NY 12571 USA
Phone: 845-758-0400
Fax: 845-758-2633
Email: curran@proceedings.com
Web: www.proceedings.com

TABLE OF CONTENTS

IOP Conf. Series: Materials Science and Engineering **851** (2020) 011001 doi:10.1088/1757-899X/851/1/011001

PREFACE

Dear Participants,

Though we had a great start in 2020, neither I (nor probably any of you) imagined the difficulties of the situation we now find ourselves in caused by the outbreak of COVID-19. This year we decided to substitute physical presentation with video presentation due to this situation. The 2020 International Conference on Information Technology and Engineering Management (ITEM) proceeding is supported by **IOP Conference Series, Material Science and Engineering** and other organizations.

The 2020 International Conference on Information Technology and Engineering Management (ITEM), is going to focus on all the major aspects in the fields of information technology and engineering management. ITEM conference links researchers, scientists, engineers, students and practitioners from different branches of information technology and engineering management from around the world. ITEM provides opportunities from both academic as well as industry to discuss, share, exchange and to extend knowledge, view and experience in respective fields. ITEM will be instrumental in generating ideas and disseminating information to a greater extent than is possible through more conventional forums.

We would like to thank all the committees for their great support on organizing the conference. We also would like to thank all the reviewers for their great effort on reviewing the papers submitted to ITEM 2020. Special thanks to all the researchers and students who with their work and participate in the conference.

Ikhsan Siregar

Chair, ITEM 2020

IOP Conf. Series: Materials Science and Engineering **851** (2020) 011002 doi:10.1088/1757-899X/851/1/011002

Peer review statement

All papers published in this volume of *IOP Conference Series: Materials Science and Engineering* have been peer reviewed through processes administered by the proceedings Editors. Reviews were conducted by expert referees to the professional and scientific standards expected of a proceedings journal published by IOP Publishing.

IOP Conf. Series: Materials Science and Engineering **851** (2020) 012001 doi:10.1088/1757-899X/851/1/012001

Optimal Production Quantity in Rubber Industry

I Siregar[1] , I Rizkya[1], Rahmi M Sari[1], Khalida Syahputri[1]
[1]Department of Industrial Engineering, Faculty of Engineering, Universitas Sumatera Utara, Medan, Indonesia

E-mail: ikhsan.siregar@usu.ac.id

Abstract. The increasing demand for rubber makes rubber production companies continue to make improvements in production planning determination continuously in order to meet consumer demand. The largest rubber production company in Indonesia is still experiencing problems in its production activities. In some periods the company still often experiences production mismatches with consumer demand, especially in Centrifuged Latex products as the highest demand product. The incompatibility of the quantity of demands with the production amount results in the emergence of other costs and the company loses the chance to get maximum income. Consequently, it is essential to analyse the company making planning activities using the goal programming method. Based on the outcomes of data processing carried out, company profits with goal programming solutions of Rp. 43.778.032.320, which is an increase of 0,99% compared to production according to consumer demand. The percentage increase in profits is relatively small, this is shows that the production planning carried out by the company is optimal.

1. Introduction

The progress of a company can be seen from the increase in the quantity of consumer demands for the products produced. However, fluctuations in demand for the quantity of products from consumers can also cause problems in the management of resources which hamper the smooth process on the production floor. The smooth production process can be achieved by the right production planning activities by the company. This activity relates to what must be produced determination, how much is produced and what resources are needed to get the product planned. The goals of production planning can be concluded by way of many tails: deliver the volume and making to encounter arranged or predictable request, guarantee appropriate and positional accessibility of materials and mechanisms, provision of a steady flow of work through all departments, provide a balanced work between various departments involved in manufacture operation, make obtainable adequate manufacturing orders to allow good management, and management and archives and deliver passable information to capture letdown and postponement [1]. Production planning is a movement transaction with invention resolve, the quantity of making products, when the product is shaped, and the amount of labors wanted to become the creation resolve [2]. Production planning is a process of determination in the all level of manufacturing output to meet the planned level of sales and desired inventory. Production planning deals with volume, timeliness of completion, capacity utilization, and load planning determination. The production plan in this case must be coordinated with company planning. There are several types of production planning. Based on the time period, there will be long-period planning, medium-term planning, and short-term period planning. These three types of planning require different planning processes (also their input and output) at each other [3, 4].

Companies engaged in rubber product are still experiencing problems in the production process. The company is still experiencing obstacles in its production process. In some periods the company still often experiences production mismatches with consumer demand, especially in centrifuged latex products, which is the highest demand products. In some periods the company experienced inability to meet consumer demand and some periods experienced excess production. The excess production results in the emergence of excess stock which leads to the other costs such as storage costs and the company's goal to obtain maximum income is not achieved. Therefore companies need to do production planning in the optimal production quantity to meet consumer needs. Production planning in determination of optimal production quantity is done by several methods, one of them is the goal programming method. The goal programming method is the right method used in production planning because it has the potential to resolve conflicting aspects between elements in production planning, namely consumer, production, and manufacturing processes[5]. Goal Programming is one of a method that can solve the problem with more than one purposes. The objectives of Goal Programming method are to maximize profit of the company and minimize the distribution cost [6]. Goal programming is a branch of multi-objective optimization and a branch of multicriteria decision analysis, which is called optimization programming. Goal programming models are similar to the other linear programming models where linear programs only have one objective and goal programming have several objectives [7].

Research to overcome the problems of production planning has been done a lot before. The research was conducted by Wiwik Anggraeni in 2014 at a garment company that manufactures various types of products. Goal programming is used to determine the quantity of product combinations by paying attention to several objectives such as income maximization, production costs minimization and machine usage maximization. The result is production planning achieved by three objectives of production planning are to produce profits are greater than the profit target up to 4 times, production costs are smaller than expected production costs, machine usage is still below the engine of target usage and the planning production process is optimal because all of target deviation variables are 0 [8]. Other studies presented a research using goal programming to determine the tasks of the patrol administrators of manpower allocation. This method offerings a non-linear goal programming model for patrolmen allocation problem to road sections within a patrol area. The model is applied via a situation case of the East section of Visakhapatnam. The consequences of the model are treasured to the watch administrator for seeing departmental goals and importance structure, in totalling to obtainable historical data, in the task of patrol manpower for a assumed urban zone [9].

2. Methodology

The research was conducted at one of the industries in Medan city produced centrifuged latex. The object examined in this study is the production volume of centrifuged latex for 1 year. Research starts with analyse to observe and see the state of the industry. After observation, the topic and purpose of the study are determined according to the conditions on the production floor. Then data collection is needed to overcome the problems of production planning occurs. Data collection in the form of the quantity of product demands, production cost, lead time, usage and availability of raw materials and the other data. The data collected is then processed using the goal programming method. The stages of the work process in overcoming the production planning problem begins with forecasting demand. Forecasting is done in order to estimate the volume of rubber production in the next period. Calculation of forecasting is done by the decomposition method and consider the smallest of error estimation value. The next step is to do production volume planning using the goal programming method. Goal Programming, the most commonly used goal programming solution methods were introduced by Lee (1972) and Ignizio (1976) [10, 12] other method using neural network prediction [11]. This method starts by determination of decision variables. Decision variables must be determined before formulating the objective function and constraints. After the decision is obtained, then the objective function and the constraints for the value of the objectives to be achieved are determined.

IOP Conf. Series: Materials Science and Engineering **851** (2020) 012001 doi:10.1088/1757-899X/851/1/012001

After that, the achievement function is determined by the decision variables combination with the objectives and constraints function. The function of achievement is done by using LINDO software to control the optimal manufacture amount. With this software, the optimal production quantity will be obtained for centrifuged latex.

3. Result and Discussion

3.1. Forecasting of Product Demand

Predicting conducted goals to forecast production volume in the following period by means of the decomposition method. Forecasting consequences are designated based on the minimum error estimation. Forecasting consequences obtained by decomposition methods consider the smallest error approximation can be seen in Figure 1.

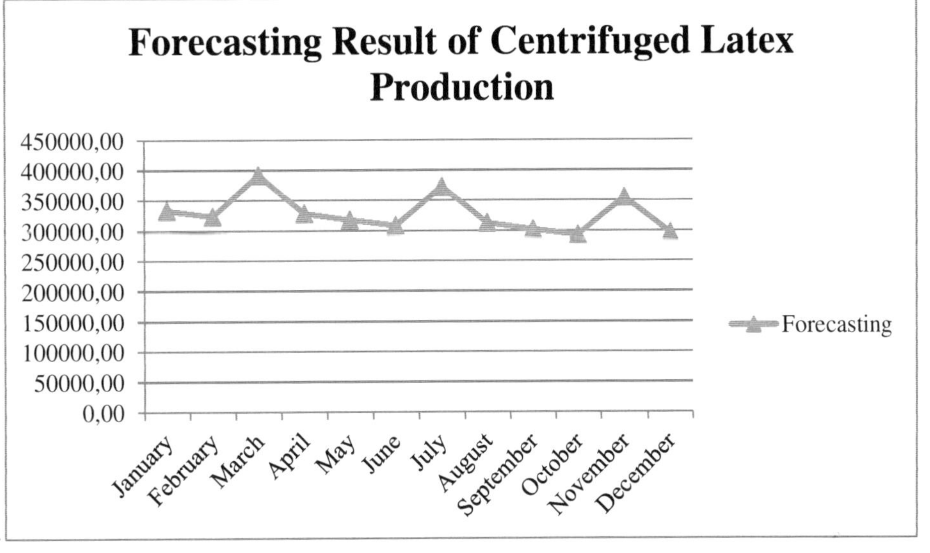

Figure 1. Forecasting Result of Centrifuged Latex Production for 1 Year

Based on the diagram overhead, it can be seen that the volume of request for centrifuged latex products has fluctuated every period. The largest quantity of demands is in March. The smallest number of demands is in October. These outcomes indicate that here is a fluctuation in product demand for separately period order and good production planning is needed so that all stresses in each period can be fulfilled.

3.2. Production Planning Using Goal Programming

3.2.1. Determination of Decision Variables

Determination of decision variables is the basis for making a decision model to get the solution. Decision variables are outputs will be optimized and it meet the criteria of subjective, constraints and priorities. Decision variables for production planning in the company are as follows:

X_1 = Production Quantity of Centrifuged Latex at Each Month (kg) (1)

3.2.2. Determination of Constraint Variables

Some variables in constraint functions formulation include:

1. Constraints of product lead time and availability of working hours for centrifuged latex product

Availability of working hours as a control function is used to see the relationship between production time and the quantity produced. Constraints function of product lead time and availability of working hours for centrifuged latex product in January is:

$$4X_1 \leq 2.401.200 \quad (2)$$

Formulation of Goal Programming method

$$4X_1 + d_1^- - d_1^+ = 2.401.200 \quad (3)$$

2. Constraints of the usage and availability of raw materials for centrifuged latex product
The usage and availability of raw materials as a function of constraints is to see the relationship between usage and availability of raw materials with the quantity of products produced. Constraints function of usage and availability of raw materials for centrifuged latex product in January is:

$$2X_1 \leq 900.000 \quad (4)$$

Formulation of Goal Programming method

$$2X_1 + d_2^- - d_2^+ = 900.000 \quad (5)$$

3.2.3. Determination of Objective Variables
The selection of objectives is based on information from company management and also based on conclusions taken from observations and data collection. The objective to be achieved is the demand meet for centrifuged latex, namely production maximization and profit maximization.

1. Production volume maximization
Forecasting result for the quantity of centrifuged latex production with a safety stock of 0,05% every month. Formulation of volume maximization in January is:

$$X_1 + 0{,}05X_1 \geq 339.803 \quad (6)$$

Formulation of Goal Programming method

$$X_1 + 0{,}05X_1 + d_3^- - d_3^+ = 339.803 \quad (7)$$

2. Profit maximization
Profit projections can be seen by taking into calculation the quantity of products and forecasting. Formulation of profit maximization in January is:

$$9.680X_1 \geq 3.289.283.360 \quad (8)$$

Formulation of Goal Programming method

$$9.680X_1 + d_4^- - d_4^+ = 3.289.283.36 \quad (9)$$

3.2.4. Determination of Achievement Functions
Achievement formulation for production planning optimization problems using Goal Programming is arranged based on the objectives to be achieved. The formulation of achievements for January is:

Min DA1 + DA2 + DB3 + DB4
SUBJECT TO
X1 + 0.05 X1 + DB3 – DA3 = 339803
9680X1 + DB4 –DA4 = 3289293040
4X1 + DB1 – DA1 = 2401200
2X1 + DB2 – DA2 = 900000
END

By using the formulation overhead, the objectives purpose and constraints purpose is calculated by means of the LINDO (Linear Interactive Discrete Optimizer) software to control the best creation quantity. Recapitulation of production planning for 12 months from Goal Programming achievements purpose by means of the LINDO software program shown in Table 1.

IOP Conf. Series: Materials Science and Engineering **851** (2020) 012001 doi:10.1088/1757-899X/851/1/012001

Table 1. Result of Optimal Production Quantity Using Goal Programming Method

Month	Optimal Production Quantity (Kg)
January	339.803
February	372.795
March	413.946
April	380.955
May	339.803
June	372.795
July	413.946
August	380.955
September	339.803
October	372.795
November	413.946
December	380.955

The results of the Goal Programming method by means of the help of LINDO software program provides information about optimal completion. Based on the optimal results obtained through Goal Programming in Table 1, comparison of the quantity in actual production and recommendation shown in Table 2.

Table 2. Comparison of Actual and Recommendation in Optimal Production Quantity

Month	Optimal Production Quantity Actual (Kg)	Optimal Production Quantity Recommendation (Kg)
January	401.779	339.803
February	348.632	372.795
March	214.960	413.946
April	242.335	380.955
May	334.980	339.803
June	403.186	372.795
July	375.098	413.946
August	386.534	380.955
September	322.398	339.803
October	307.764	372.795
November	400.352	413.946
December	285.771	380.955

Table above shows that the optimum quantity of centrifuged latex products must be produced every month using the goal programming method does not indicate the big difference from the quantity of production forecasting results are carried out using the company actual data.

4. Conclusion

Goal programming is used to control the best quantity of production and exploit the usage of fresh materials by resources utilization owned by the company. The optimum quantity of centrifuged latex products must be produced every month using the goal programming method does not indicate the big difference from the quantity of production forecasting results are carried out using the company actual data. The company's advantage with goal programming solutions is IDR 43.778.032.320, which only increased by 0,99% compared to production based to consumer demand. The percentage increase in profits is relatively small, this shows that the production planning carried out by the company is optimal.

References

[1] Patrick A O., and Emmanuel S 2010 *Leonardo Electronic Journal of Practices and Technologies* 1**6** 201-217
[2] S. Sinulingga, 2013 *Metode Penelitian* (Medan: USU Press)
[3] Supriyadi 2016 *SINERGI* **20** (2) 157-164
[4] V. Vincent, 2011 *Production Planning and Inventory Control Berdasarkan Pendekatan Sistem Terintegrasi MRP II dan JIT Menuju Manufakturing 21* (Jakarta: PT. Bumi Aksara)
[5] Fauziyah 2016 *Jurnal Matematika Mantik* 2 (1) 52-59
[6] Hidaen B, Jaya A I, Resnawati 2018 *Jurnal Ilmiah Matematika dan Terapan* 15 (1) 77-85
[7] Kesava R V, Sunitha S V, Raghunath Y R 2017 *Journal of Mathematics* 13 (2) 01-04
[8] Wiwik A, 2014 *Journal of Management Science* **3** (4) 52-61
[9] Venkata S K., Daniel G V, Suresh C 2015 *ELK International Journal of Patrol Engineering Research and Applications* **5** (10) 142-153
[10] Orumie., and Ukamaka Cynthia 2017 *Mathematics Applied in Science and Technology* **9** (1) 1-22
[11] Rahmat R F, Nurmawan S, Sembiring S, Syahputra M F, Fadli 2018 Adaptive neuro-fuzzy inference system for forecasting rubber milk production *IOP Conference Series: Materials Science and Engineering* **308**(1),012014
[12] Siregar I, Nasution A A, Prasetio A, and Fadillah K 2017 Analysis of production flow process with lean manufacturing approach. In *IOP Conference Series: Materials Science and Engineering* Vol. **237**, No. 1, p. 012002 IOP Publishing.

IOP Conf. Series: Materials Science and Engineering **851** (2020) 012002 doi:10.1088/1757-899X/851/1/012002

Determination of Joint Lot Size using the supply chain approach with Vendor Managed Inventory (VMI) Method at PT. XYZ

Abdillah Arif Nasution[1*], Erlina[1], Rujiman[1], and Chandra Situmeang[2]
[1]Department of Accounting, Universitas Sumatera Utara, Medan, Indonesia
[2]Universitas Negeri Medan

*Email: badinst@usu.ac.id

Abstract. PT. XYZ is a company producing animal feed products which is located in the Medan Region. This company is a branch of Z Group company based in Surabaya. In the process company business activities, both from the process of ordering raw materials to suppliers up to the sale of products to buyers, often happens planning mismatches in determining production lots so often cause overstock and stockout. This is based on the lack of integration each component in the supply chain. Each component is still using the traditional way that is by making each plan without coordinating with each other. This manufacturing company still needs to do improvements in the production planning system. The absence of the PPIC Department make planning in the production in the company becomes out of control. This study aims to improve the supply chain system by method VMI and determine the optimum combined lot size by the JELS method. The JELS calculation results show an optimal Q of 8,692 bags, with QM for 437,082 bags with a frequency of product shipments 50 times and the frequency of replenishment of raw materials 13 times.

1. Introduction

PT. XYZ is a company engaged in feed production live stock. The main product of the company is chicken feed, whereas as Supporting products are animal feed for pigs, cattle, goats, ducks, and fish. In the company's operational processes meet the demand with the Pull method demand system, which is a demand fulfillment system by considering external conditions or in other words demand is determined by consumers themselves. The advantage of this system is that the company can recognize changes that occurs in the external environment and can make policies for anticipating these changes. But in reality, the demand is must be met always uncertain and cause difficulties in determine the production lot size policy. Lack of information can be cause chaos in the supply chain.

ISA 105 (Mash) is the most abundant product and significantly produced by PT. XYZ compared to 35 other types of feed products and is make to stock so that sometimes it can cause build up product (over stock) or product shortage (stock out) when planning not accurate. There are other types of feed products that are made to order, such as feed products for cows, ducks, pigs, and so on.Supply chain system for layer 105 chicken (Mash) chicken feed products at PT. XYZ-Medan Mill can be seen in Figure 1.

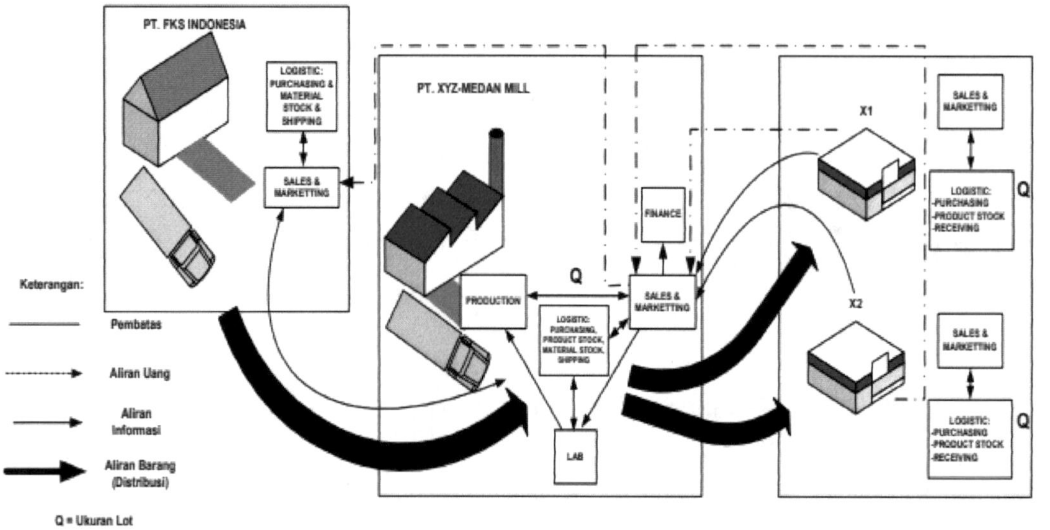

Figure 1. Existing Feed Supply Product System 105 ISA(Mash) at PT. XYZ-Medan Mill

Figure 1. shows the supply chain system of layer chicken feed products105 ISA (Mash). The raw material comes from an importer company, PT. FKS Indonesia, a single company importing corn raw materials for the region North Sumatra. The company meets the material requirements according to the arrival of imported materials from abroad and the request from the factory need imported raw materials. Receiving raw materials cannot be done directly at the time of order. The buyer must at leastwaited for a month because the product was imported from abroad the shipping process takes quite a long time. Therefore, parties feed producers must be able to plan exactly when the raw material ordered and available when needed so that the production process can run smoothly and can meet the amount of demand for feed products by consumers. The demand for feed products is done by Sri companiesWinarti (SW) and Bak Tjai (BT) through the sales department at PT. XYZ-Medan Mill which will then be processed in the production section. On Figure 1. It can also be seen that manufacturers and buyers of feed products determine each lot size for. This is why it is not integrated feed product supply chain system 105 ISA (Mash). Planning done often not in accordance with the reality that occurs in the field so causing the number of products produced is not optimum.

Based on the company's previous supply chain picture, it was also founddata on the number of requests, the amount of production, and the amount of supply of feed products 105 ISA (Mash) for the period September 2017 - August 2018 which can be seen in Table 1.

Table 1. Amount of Demand, Production Amount and Product Inventory Chicken Feed 105 ISA (Mash) PT. XYZ Medan

Period	Number of Requests (Bags)		Total Requests
	SW (Bags)	BT (Bags)	
September, 2017	21.355	9.426	30.781
October, 2017	25.241	9.076	34.317
November, 2017	23.628	11.662	35.290
December, 2017	21.175	11.166	32.341
January, 2018	24.946	11.538	36.484
February 2018	27.215	9.935	37.150
March,, 2018	26.844	11.039	37.883
April, 2018	24.249	10.775	35.024

May, 2018	26.185	10.788	36.973
June, 2018	28.695	12.287	40.982
July, 2018	30.234	11.513	41.747
August, 2018	27.469	10.602	38.071

Source : PT. XYZ

Table 1. shows that there are different differences andlead to product accumulation (over stock) in the period September 2017, October 2017, December 2017, April 2018, May 2018, and June 2018 as well product shortages (stock out) in the period November 2017, January 2018, February 2018, March 2018, July 2018 and August 2018. From observations at company, the occurrence of overstock and stock out is caused by a demand tend to change and the absence of special sections or departments for production and inventory planning in determining lot orders clearly between suppliers, companies and buyers. Companies and buyers respectively plan their own needs and supplies, so the difference planning often happens. Stacking of product inventory in the product warehouse will pose an expiration risk, because feed products are products perishable. But if the company cannot meet the number of products in accordance with the request of the buyer, it will lead to lost sales. Of course these things will cause risks, both cost and integrity between companies.

2. Literature Review

2.1. Supply Chain Concept

Supply chain may have often been heard. In Indonesian often found translation of the supply chain as "procurement chain or supply or supply of goods and services ". Basically supply chain is a network of organizations involving upstream and downstream relations (upstream-downstream), in different processes and activities, which generate tangible value in goods and services in the hands of users or final consumer. The different processes and activities involve relations between actors, from producers or suppliers, makers or processor, distributor or distributor to the user or consumer [1].

According to NyomanPujawan[2], Supply chain is a network companies that jointly work to create and deliver a product to the end user. Companies These usually include suppliers, factories, distributors, stores or retail, as well supporting companies such as logistics service companies.

The concept of supply chain is a new concept in looking at the probleminternal logistics. The old concept sees logistics more as an internal problem each company and its solution is focused on solving internally in their respective companies. In this new concept, a problem logistics is seen as a broader problem that stretches very long from basic materials to finished goods used by end consumers, which is a supply chain. Seeing this definition, you can it is said that the supply chain is a logistics network. In this supply chain is a supply chain, also called "model four step "(the four step model), which consists of elements, [1] that is:

a. Suppliers
b. Manufacturers
c. Distributors
d. Retailers

According to Indrajit [1], key factors are managed in supplychain includes:

a. Management of relationships with consumers.
b. Management of services to consumers.
c. Demand management system.
d. Manufacturing flow regulation system.
e. Raw material procurement system (Office Stationery).
f. Development of commercial products and processes.
g. Product return

In a supply chain there are usually 3 types of flow that must be managed.First is the flow of goods that flows from upstream to downstream. Examples are raw materials sent from suppliers to parts logistics. After the raw materials are finished producing, they are sent to the distributor, then to retailers or retailers, then to end users. The second is flow money and the like that flows from downstream to upstream. The third is flow information that can occur from upstream to downstream or vice versa.

Information about inventory of products that still exist in each division often needed by distributors and by agencies. Information about availability production capacity owned by the supplier is also often needed by agency. Information about the delivery status of raw materials is often needed by the sending and receiving agency. Shipping agencies must share information like this so stakeholders can monitor in the interest of more accurate planning. Figure 2 gives conceptual illustration of a supply chain.

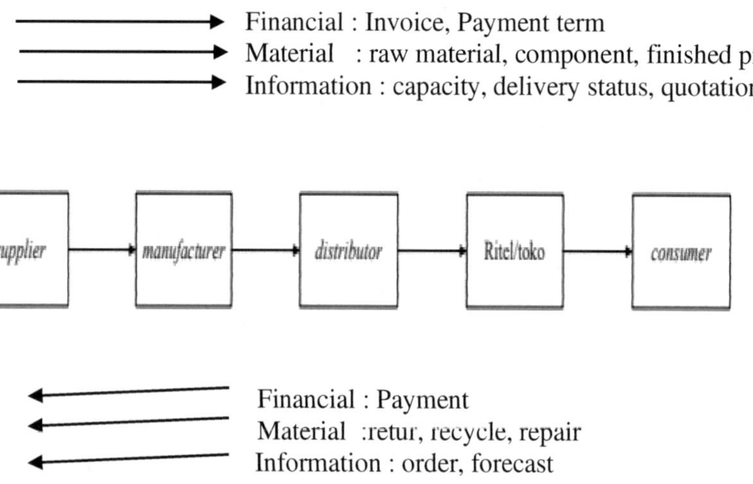

Figure 2. Simplification of Supply Chain Models and 3 Kinds of Flow Managed

2.2.Supply Chain Strategy
According to Chopra [3], supply chain strategies are needed to helpachievement of the desired agency goals in the agency strategy. To be able to Win the market competition, there are 3 stages that must be carried out including:
a. Institutions must understand the needs of consumers in each market segment.
b. Agencies must understand the designed supply chain.
c. The agency has the ability to change the structure of the supply chain if there is a mismatch between the supply chain design with the desire consumer.

2.3. Supply Chain Management (SCM)
The term SCM was first coined by Oliver & Weber in years1982. If the supply chain is the physical network, i.e. companies involved in supplying raw materials, producing goods, as well as send it to end users, SCM is a method, tool, or approach its management. But it should be stressed that SCM requires an approach or integrated method based on the spirit of collaboration [4][5].

3. Research Method
3.1. Type of Research
This type of research is action researchresearch). Action research is a research conducted for get practical findings / for decision making purposes operational (Sinulingga, 2011) [6]. Because the goal is for decision making operational to develop new skills or new approaches then This research contributes less to science

3.2. Object of research
The research object observed was the number of animal feed products 105ISA (Mash) is produced and will be sent to buyers who are partners company business namely X1 and X2 by considering defective products and backorder in the supply chain network with a single supplier, costs, as well inventory and information systems for implementing Supply Chain Management.

3.3. Variable Research

A. Independent variables, namely:

a. Number of requests, number of products ordered by customers per unit time

b. Message Fee, i.e. the amount of money that must be spent for support product ordering activities. In this case the cost of the message consists of the two indicators namely the cost of ordering raw materials by manufacturing and cost of ordering finished products by the customer.

c. Save Cost, i.e. the amount of money that must be sacrificed for support storage activities. In this case the cost of the message consists of several indicators namely the cost of storing raw materials, the cost of storing products so by manufacturing, the cost of storing the finished product by the customer.

B. Dependent variable, which is the optimum economic lot size (Joint Economic Lot Size).

3.4. Method of Collecting Data

In this study the data collection techniques used werein the form of:

1. Interview techniques, namely by conducting interviews with parties' company management to get the information needed in order achievement of research objectives.
2. Observation technique, which is to recapitulate stock level and demand in the current period of the product 105 ISA Mash.

4 Result and discussion

4.1 Supply Chain Analysis System Based on the Vendor Method Managed Inventory (VMI)

Supply chain systems with vendor managed inventory methods are usedto determine new policies by determining economically optimal lotsfor both parties. Then do a good integration of the suppliers, manufacturers and distributors will increasingly influencesmooth flow system. One of the supply chain integration policies is implementing the vendor managed inventory (VMI) method, which is policy all inventory planning is carried out by the producer company (PT. XYZ-Medan Mill). The customer receives input from the company manufacture. Robustness of information and trust integration is needed in this case. This policy is carried out as a basis for determining size optimum combined lot. As for integration with VMI in the company can be seen in the following figure 3.

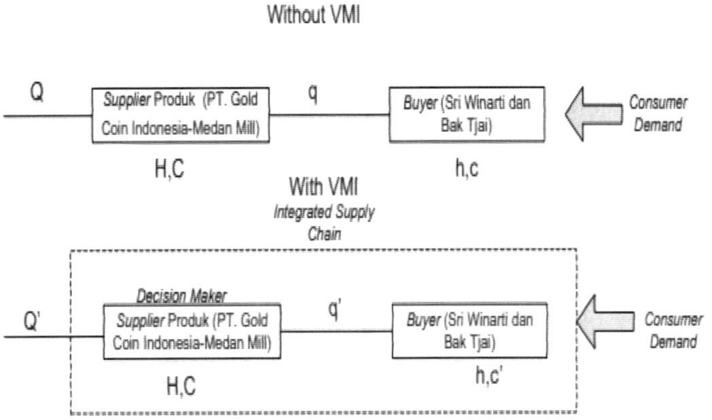

Figure 3.VMI Work Network Integration Model at PT. XYZ-Medan Mill

4.2 Analysis of the Difference Between Request Amount and Total Available

Based on the results of data processing, the recapitulation of the difference results was found between the number of requests and the total available is as follows.

Table 2. Difference Between Demand Amount, Amount of Production, andInventory of Chicken FeedProducts 105 ISA (Mash) PT. XYZ-Medan Mill

Period	Number of requests (Bags)		Total Request (Bags)	Rate Production (Bags)	Total Available (Bags)	Difference (Bags)
	SW (Bags)	BT (Bags)				
Sept 17	21.355	9.426	31.500	34.000	34.000	3.219
Oct 17	25.241	9.076	33.720	36.939	36.939	2.622
Nov 17	23.628	11.662	31.320	33.942	33.942	(1.348)
Dec 17	21.175	11.166	36.000	34.652	34.652	2.311
Jan 18	24.946	11.538	31.500	33.811	33.811	(2.673)
Feb 18	27.215	9.935	33.600	30.927	30.927	(6.223)
Mach 18	26.844	11.039	41.100	34.877	34.877	(3.006)
Apr 18	24.249	10.775	40.800	37.794	37.794	2.770
May 18	26.185	10.788	42.000	44.770	44.770	7.797
June 18	28.695	12.287	36.600	44.397	44.397	3.415
July 18	30.234	11.513	36.300	39.715	39.715	(2.032)
Aug 18	27.469	10.602	38.071	36.848	36.848	(1.223)
Total			437.043	433.320	442.672	

The table above shows that there are variations that cause product buildup (over stock) in the period September 2017, October 2017, December 2017, April 2018, May 2018, and June 2018 and product shortages (stock out) in the period November 2017, January 2018 , February 2018, March 2018, July 2018 and August 2018. Based on the discussion of these problems, it is necessary to have a good plan in deciding how much should be provided by the manufacturing company to meet the demand. Planning department or commonly referred to as PPIC Department is needed to control optimal production and be able to meet consumer needs. The PPIC department must create a new strategy in solving problems that occur and one of the strategies that can be usedin determining the number of lots is the application of the JELS method with the supply chain approach.

4.3 Analysis of Optimal Joint-Lot Size Iteration Calculation

Based on the results of data processing, recapitulation of iteration results from the optimal Joint-Lot size based on Total Cost is as follows:

Table 3. Recapitulation of Iteration results TC

Numb.	m	n	Q (Bags)	B (Bags)	QM (Bags)	TC (Rp)
1.	1	1	95.866	548	96.414	1.650.985.790
2.	1	2	51.860	593	104.313	1.589.532.704
3.	1	180	2.420	2.488	438.088	5.130.179.586
4.	2	1	134.851	770	135.621	2.119.221.625
5.	2	2	71.536	817	143.889	1.990.271.781
6.	2	162	2.688	2.487	437.943	4.809.015.958
7.	3	1	164.846	942	165.788	2.493.391.892
8.	3	2	86.819	992	174.630	2.318.373.721
9.	3	149	2.917	2.482	437.115	4.693.487.338
10.	4	1	190.163	1.086	191.249	2.813.874.350
11.	4	2	99.779	1.140	200.698	2,601.882.279

12.	4	138	3.149	2482	437.044	4.638.652.917
13.	5	1	212.483	1.214	213.697	3.098.614.109
14.	5	2	111.234	1.271	223.739	2.855.004.304
15.	5	128	3.398	2.484	437.428	4.611.745.586
16.	6	1	232.671	1.329	234.000	3.357.418.746
17.	6	2	121.614	1.389	244.617	3.085.720.252
18.	6	118	3.686	2.484	437.432	4.592.484.493
19.	7	1	251.242	1.435	252.677	1.540.731.090
20.	7	2	131.174	1.499	263.847	3.299.067.205
21.	7	108	4.026	2.483	437.291	4.576.831.991
22.	8	1	268.531	1.534	270.065	3.819.186.261
23.	8	2	140.082	1.600	281.764	3.498.494.011
24.	8	99	4.402	2.849	438.287	4.576.060.935
25.	9	1	284.773	1.627	286.400	4.028.963.001
26.	9	2	148.457	1.696	298.610	3.686.346.262
27.	9	89	4.893	2.487	437.964	4.564.687.901
28.	10	1	300.137	1.714	301.851	4.227.708.181
29.	10	2	156.384	1.786	314.554	3.864.467.349
30.	10	79	5.506	2.484	437.458	4.556.099.303
31.	11	1	314.751	1.798	316.549	4.416.973.558
32.	11	2	163.927	1.873	329.727	4.034.219.436
33.	11	70	6.225	2.489	438.239	4.560.065.687
34.	12	1	328.717	1.877	330.594	4.598.012.600
35.	12	2	171.138	1.955	344.231	4.196.690.774
36.	12	60	7.254	2.486	437.726	4.552.973.845
37.	13	1	342.113	1.954	344.067	4.771.791.548
38.	13	2	178.057	2.034	358.148	4.352.732.164
39.	13	50	8.692	2.482	437.082	4.548.249.599*
40.	14	1	355.004	2.028	357.032	4.939.131.544
41.	14	2	184.718	2.110	371.546	4.503.029.947
42.	14	41	10.615	2.486	437.701	4.556.791.385
43.	15	1	367.442	2.099	369.541	5.100.716.759
44.	15	2	191.146	2.183	384.475	4.648.221.147
45.	15	32	13.615	2.488	438.168	4.568.977.828

The optimal lot size calculation aims to determine the size of the lot of the shipment of raw materials from suppliers to the company or the calculation of the optimum lot from the company to the buyer based on the most optimum total cost. Based on the calculation after iterating for the delivery frequency m = 1 to m = 15 according to Table 3 we found a solution with a minimum TC when m = 13 and n = 50 with Rp. 4,548,249,599. It can be seen in table 3 that companies should order 13 times the raw material and ship the product 50 times to the buyer. One time ordering of raw materials and one delivery in one year is assessed because this feed product is a consumption product that has a proper period of consumption.

5.Conclussion

Based on the analysis of processing and discussion of data, several conclusions can be drawn, namely:

1. Based on the calculation of total cost using the Joint Economic Lot Sizing method, the optimum total cost is Rp. 4,548,249,599 with a shipping lot size to buyers of 8,692 bags, producing as many as QM = 437,082 bags and maximum product backorder, B = 2,483 bags and by scheduling the delivery of finished products to the buyer as much as 50 times with the frequency of ordering raw

materials 13 times. When compared with existing conditions, the application of this method is better because it can save costs by Rp. 1,239,128,918.
2. Stock out and over stock often occur in companies which indicate that there is a deviation between production capacity and the number of consumer requests.
3. Sensitivity analysis shows that the cost of ordering raw materials and disposal costs are the costs that have the most significant influence on changes in total cost.
4. Factors causing the problem are the lack of integration and cooperation between suppliers, companies and buyers in determining lots. The PPIC department is responsible to the manufacturing company of the to coordinate and determine a plan in the production process is important to be formed immediately.

Acknowledgments
Authors thank to Allah, lord of the universe who always bestows his grace along with guidance, so that the authors can finish the research.

References

[1] Richardus E I 2002 *Konsep Manajemen Supply Chain*. Jakarta :Gramedia.
[2] Pujawan I N 2005. *Supply Chain Management*, Edisi I. Surabaya: Penerbit Guna Widya.
[3] Cao M 2010. Supply chain collaborative advantage: a firm's perspective *International Journal of Production Economics*, vol. **128**, no. 1, pp. 358–367.
[4] Hosseini S K 2014. Optimizing An Integrated Vendor-Managed Inventory System for A Single-Vendor Two-Buyer Supply Chain with Determining Weighting Factor for Vendor's Ordering Cost. *International Journal of Production Economics*, vol. **153**, ,pe. 295–308.
[5] Alaqeel K, Shakkah M S, Rahmat R F, Alfageeh A, Budiarto R 2017 Organization Readiness and ERP Implementation in Albaha University *IOP Conference Series : Materials Science and Engineering* **190**(1), 0123049
[6] Sinulingga S 2011 *Metode Penelitian* . Medan : USU press

IOP Conf. Series: Materials Science and Engineering **851** (2020) 012003 doi:10.1088/1757-899X/851/1/012003

Application of Preventive Maintenance in Planning Maintenance of Tablet Printing Machines in Drug Companies

I Siregar[1], A A Nasution[2]
[1]Department of Industrial Engineering, Faculty of Engineering, Universitas Sumatera Utara, Medan, Indonesia
[2]Departement of Accounting, Faculty of Economics and Business, Universitas Sumatera Utara, Medan, Indonesia

Email : ikhsan.siregar@usu.ac.id

Abstract. Companies that implement the production process will always be faced with market competition. The process is done on the floor of the production required to produce a quality product and is able to win the competition. Machine condition greatly affects the results of the production process. A good machine will produce a quality product, otherwise the machine that his condition is not good will result in products that are not qualified. Companies engaged in the manufacture of the drug often have damage to the machine where the damage is mostly caused by engine component of a machine failure function so that the machine could not start the process when the operating damage to one of the machines at work, then the machine production process will be interrupted because it takes time to replace components of the machine. For that to be done, a step to improve the system of corrective maintenance which currently used the company so that the damage to the engine at the time of the working hours of the machine can be avoided. The purpose of this research is to get the engine maintenance schedule intervals by performing a critical component replacement tablet printing machine based on data obtained by the drug companies. Determining the schedule engine maintenance done by using Reliability Engineering. Of research results in retrieved that components are frequently damaged components namely Punch and Dies and the Pulley with the value of the respective machine reliability of 0.3429 and 0.4183. While the time Interval component replacement punch and dies the optimum is 791 hours per cycle of usage which means that the components of the punch and dies must be replaced at a time when the machine operates 791 hours. Whereas in the pulley component the optimum replacement time interval is 1695 hours per cycle of damage where the pulley component must be replaced when the engine is operating 1695 hours.

1. Introduction

The growing manufacturing industry has triggered companies to always strive to remain able to survive in the market competition that can arise between companies and similar products. The process carried out on the production floor is required to produce quality products and be able to win the competition. To produce a quality product, the company must be able to carry out every process so that high quality output can be produced [1]. The condition of the machine greatly affects the results of the production process. A good machine will produce a quality product, on the other hand a machine that is not in good condition will produce a quality product too. A good machine can be seen based on the level of reliability of the engine. Reliability is the possibility that a part of a machine or product can function properly in a certain time related to the level of confidence that an equipment or component will perform its function without experiencing problems or damage at the time of operating conditions that are still implemented in the time period [2, 3].

The level of machine reliability in producing a product depends on the maintenance carried out on the machine. Maintenance is a collection of actions executed on assets with the aim of retaining assets, or specified, a specified condition. The maintenance function is an essential part of any asset intensive organization, and needs to support the organization's business objectives [4]. Maintenance is an activity to maintain or maintain factory facilities or equipment and make repairs or adjustments or replacements needed so that there is a satisfactory state of production operations in accordance with what is planned [5]. Machine maintenance has an important role in determining the company's production activities with regard to the smooth and congestion of production, slowness, and production volume and production efficiency. The cessation of the operation of the engine will result in a constrained production process. Apart from that, the products produced will not reach the quality that has been targeted by the company [6, 7].

Companies engaged in the pharmaceutical industry produce a variety of drugs ranging from tablets, syrups, ointments, oral powders, external medicine powders, capsules where each of these products has different departments in the manufacturing process. This company is very dependent on the reliability of its production machinery to run the production process. This is because industries that produce drugs must have a high level of precision in producing drugs. If the engine is damaged or its reliability in carrying out the production process decreases, the drugs produced will not match the prepared composition. This can cause a loss because the medicinal products cannot be sold on the market. The company is currently implementing a machine maintenance system in the form of preventive maintenance and corrective maintenance to support the smooth production process. Preventive maintenance is a treatment that is carried out in a scheduled, generally periodic manner where a set of maintenance tasks such as inspection and repair, replacement, cleaning, lubrication, adjustment, and equalization is carried out. Preventive maintenance aims to prevent sudden damage to the engine, increase reliability, and reduce downtime [8]. The implementation of a preventive maintenance system is carried out by cleaning engine parts and lubricating the production machine. While the engine maintenance system in the form of corrective maintenance is carried out by the company by replacing and repairing engine components after damage to the component concerned. If an engine failure occurs during engine working hours, the production process will be disrupted, because time is needed for replacement of engine components. For this reason, planning for maintenance of machine components is carried out in the production process of medicines so that machine damage can be avoided.

Care planning has been carried out in many previous studies. One study was conducted at a company that produces corrugated carton boxes located in Gresik. The study was conducted to design a maintenance schedule with preventive maintenance on corrugating machines and flexo machines. The results of this study indicate that the cost of care before and after scheduling resulted in savings of 22.16% when compared to the company's corrective schedule [5]. Other research was also carried out on shoe sole companies that suffered sudden engine damage so that the planned production schedule was disrupted. The research was carried out by implementing a preventive maintenance system. The results obtained showed that with the implementation of the preventive maintenance system, downtime was reduced by 2.85% and a savings of 38% occurred. [9]. Based on previous studies, this study aims to obtain a machine maintenance schedule by performing critical component replacement of tablet printing machines and increasing the efficiency of engine components.

2. Methodology

The study was conducted at a drug company where the object under study was a tablet printing machine. This type of research is descriptive research that is a type of research that aims to describe systematically, factually and accurately about the facts and characteristics of an object or a particular population [10]. The study was conducted for 10 months with several stages. The research began by conducting observations to observe the state of the drug company directly. Based on the results of observations, the problem formulation will be determined in accordance with the condition of the company, namely how the company carries out maintenance on the production machine. In this stage, various types of machine maintenance can be identified which can be corrective or preventive. With the formulation of this problem, the objectives of the research will be applied as a solution to the problems that occur. Problem solving can be done based on supporting data as input for this research process. The data used to solve the problem of engine damage is the damage to the engine production data and damage data components of the critical engine.

Troubleshooting machine maintenance is done in several stages. The first stage of identifying the components on the critical engine is to find out the types of components that are often damaged so that component replacement must be done. Identification of this critical component is done by using the Pareto diagram with the 80/20 rule. Pareto diagram is an image that sorts the clarification of data from left to right according to the highest to lowest ranking so that important problems are found to be resolved immediately. This diagram helps management to quickly identify the most critical areas that require special and fast attention [11]. The next step is to do a distribution pattern maintenance test to avoid the engine from damage. So that it can be known the appropriate engine maintenance interval. The type of distribution used in order to find out the pattern of data formed is a normal, lognormal, exponential and Weibull distribution. The selection of distribution patterns is using the Least Square Curve Fitting method, which is based on the largest Index of Fit (correlation coefficient) value. Testing the distribution of critical component patterns is done using data time intervals between component damage (MTTF). Mean time to failure (MTTF) is the average time interval of damage that occurs when the engine or component is damaged again [9]. The next stage of this research is to calculate the reliability of critical engine components. The calculation of the reliability of critical engine components on the proposed replacement schedule is used to determine the value of the reliability of engine components when the proposed replacement of the proposed component is made. Calculations are carried out based on the distribution pattern that has been selected for each component. After that, a time interval was calculated so that the machine maintenance schedule was obtained by performing critical component replacement of the tablet printing machine. The maintenance time interval is calculated for each critical component of the tablet printing machine.

3. Result and Discussion

3.1. Identification of Damage to Tablet Printing Components

The selection of critical components that will be the focus of the study was carried out with Pareto 80/20 rules using Minitab software. The results of component identification using the Minitab software can be seen in Figure 1.

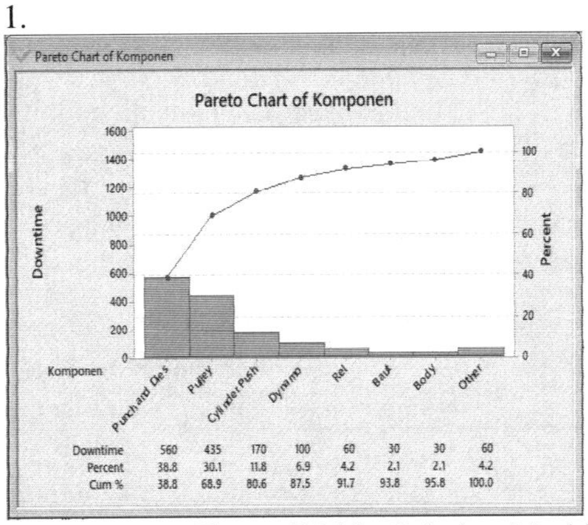

Figure 1. Pareto Chart of Tablet Printing Machine Damage

The diagram above shows 80% loss of production time due to 20% of the damaged components. Thus, there are 2 types of critical components that can cause damage to tablet printing machines namely punch and dies, and pulley.

3.2. Testing the Critical Component Distribution Pattern

Testing the distribution of critical component patterns is done using data time intervals between component damage (MTTF). The distribution used is normal, lognormal, exponential and weibull distribution. The selection of distribution patterns is using the Least Square Curve Fitting method, which is based on the largest Index of Fit (correlation coefficient) value. The recapitulation of the calculation of the Index of Fit for interval distribution patterns of damage to punch and dies and pulley components can be seen in Table 1.

Table 1. Recapitulation of Calculation of Index of Fit Time between Punch and Dies Component Damage

Components	Index of Fit			
	Normal	Lognormal	Exponential	Weibull
Punch and Dies	0,8939	**0,9878**	0,9699	0,9660
Pulley	0,9824	0,9428	0,8751	**0,9842**

From the table, it is obtained that the chosen distribution for punch and dies component is lognormal distribution with index of fit value of 0.9878 while pulley component is weibull distribution with index of fit value of 0.9842.

3.3. Calculation of Mean Time to Failure (MTTF) and Reliability

After the distribution pattern selection is performed for each component based on the largest Index of Fit value then the next stage is calculating the parameters and values Mean Time To Failure (MTTF). The recapitulation of critical component MTTF tablet printing machines can be seen in Table 2 below.

Table 2. Average Time Recapitulation of Component Damage (MTTF)

No.	Component	MTTF (hour)	MTTF (days)
1.	Punch and Dies	791,41	33
2.	Pulley	1694,93	71

Based on the table, it was found that the average time of damage to the punch and dies component was 791.41 hours (33 days) while the pulley component was 1694.93 hours (71 days).

3.4. Determination of Reliability Value

The calculation of the reliability of critical engine components on the proposed replacement schedule is used to determine the value of the reliability of engine components when the proposed replacement of the proposed component is made. Calculations are carried out based on the distribution pattern that has been selected for each component. The recapitulation of reliability value to the use of components can be seen in Table 3.

Table 3. Recapitulation of Reliability Value

No.	Component	MTTF (days)	Valu of Reliability
1.	Punch and Dies	33	0.3429
2.	Pulley	71	0.4183

Based on the table, it obtained that after 791.41 hours (33 days) the use of punch and dies components has a reliability value of 0.3429 and after 1694.93 hours (71 days) the use of pulley components has a reliability value of 0.4183.

3.5. Determination of Schedule for Maintenance of Tablet Printing Machine Components

The maintenance time interval is calculated for each critical component of the tablet printing machine. The maintenance time projection is carried out in the range of the 35th tp hour to 1211 hours along the reliability value (0-1) for the punch and Dies component. While pulley components, the maintenance time projection is carried out in the range of tp 20 hour to 4515 hours along the reliability value (0-1). The maintenance projection results show that the optimum punch and dies component replacement interval is 791 hours per cycle of use. This means that the components of punch and dies must be replaced when the engine operates 791 hours. In pulley components the optimum replacement time interval is 1695 hours per cycle of damage. This means that the pulley component must be replaced when the engine is operating 1695 hours.

4. Conclusion

There are two components that are often damaged IE components Punch and Dies and the Pulley where the value of the components obtained for reliability machine Punch and dies amounted 0.3429, whereas the value of reliability of components engine Pulley is 0.4183. With the application of reliability engineering in determining the treatment interval components engine, obtained that time Interval component replacement punch and dies the optimum is 791 hours per cycle of usage which means that the components of the punch and dies already should be replaced at the time of the engine operating hours, 791. While on the component pulley obtained optimum replacement, intervals are 1695 hours per cycle of damage where the components of the pulley must be replaced at the time of the engine operating hours, 1695.

References

[1] Siregar I, Nasution A A, and Sari R M 2017 Effect of Total Quality Management on the Quality and Productivity of Human Resources. In *IOP Conference Series: Materials Science and Engineering* **180**(1)p. 012116. IOP Publishing.
[2] Sudri N M, Nendissa B C, Herawati Y 2012 *Jurnal Teknik dan Ilmu Komputer* **01** (3).
[3] Ebeling E, Charles 1997 *An Introduction to Reability and Maintanability Engineering* New York: The McGraw-Hill Comapnies Inc.
[4] Mungani D S, Visser J K 2013 *South African Journal of Industrial Engineering* **24** (3).
[5] Pandi S D, Santosa H, Mulyono J 2014 *Jurnal Ilmiah Widya Teknik* **13** (1)
[6] Siregar I, Muchtar M A, Rahmat R F, Andayani U, Nasution T H, and Sari R M 2018, February). Method of calculation overall equipment effectiveness in fertilizer factory. In *IOP Conference Series: Materials Science and Engineering* (Vol. 308, No. 1, p. 012053). IOP Publishing.
[7] Siregar, I., & Nasution, A. A. 2018 Component identification the causes of machinery damage in pharmacy company using Pareto diagram. In *IOP Conference Series: Materials Science and Engineering* **420**(1) 012139 IOP Publishing.
[8] S Assauri 2008 *Manajemen Produksi dan Operasi Edisi Revisi 2008* (Jakarta: Universitas Indonesia)
[9] Y Praharsi, I K Sriwana, D M Sari 2015 *Jurnal Teknik Industri* **14** (1)
[10] S Sinulingga 2014 *Metode Penelitian* (Medan: USU Press)
[11] W D Ariani 2004 *Pengendalian Kualitas Statistik* (Yogyakarta: Penerbit Andi)
[12] Siregar I, Muchtar M A, Rahmat R F, Andayani U, Nasution T H, and Sari R M 2018 Method of calculation overall equipment effectiveness in fertilizer factory. *IOP Conference Series: Materials Science and Engineering* 308(1), p. 012053

Calculation of Physiological Load and Energy Consumption of Warehouse Operators in Manufacturing Companies

Nismah Panjaitan[1], Listiani Nurul Huda[1], Arnita Rahmi[1]
[1]Department of Industrial Engineering, Faculty of Engineering, Universitas Sumatera Utara, Jl. Almamater Kampus USU Medan 20155, Indonesia

E-mail: nismahpjt@gmail.com , [2]Lnurul@gmail.com, [3]arnitarahmi@gmail.com

Abstract. The modest workload provided by operators is the responsibility of every manufacturing company. Operators with excessive workload on an ongoing basis will cause discomfort and even abnormalities in the physiological operator, especially in the muscles and joints. In this study, workload, work frequency and work posture become independent variables while energy consumption becomes the dependent variable that will be reviewed thoroughly to see the workload of operators. Data collected directly about the operator's energy consumption indicates the operator is experiencing a heavy workload. Lifting and moving activities carried out by operators must be immediately carried out repairs. In the near future the improvement of operator activities when lifting must be fixed immediately because it exceeds the energy consumption limit of 8.63 Kcal / minute, 8.063 Kcal/minute dan 8.175 Kcal/minute.
Keywords : Musculoskeletal, Work Posture, Physiological, Workload and Cardiovascular Load (% CVL).

1. Introduction

Manual removal of containers such as wheeled racks, trolleys, hand trucks and socket pallets can cause fatigue and discomfort, such as injuries to the back, shoulders, hands, wrists, or other body parts that can occur if done repeatedly. If this happens in a long time, it will cause musculoskeletal or MSDS (Cal / OSHA, 2007). Lifting activities that provide a great compressive force on the spine and pushing or pulling activities can create a large shear load on the spine and will cause pain in the lower back, consequently it can cause rotating postures that refer to body rotation or torque [1]. Working in an improper position will cause discomfort and greatly endanger workers just to maintain high levels of productivity such as bending and squatting [2].

Manufacturing company engaged in the production of PVC (PolyVinil Chloride) and HDPE (High density polyethylene) pipes. The product transfer activities are carried out manually by the warehouse worker with the help of a four-wheeled manual trolley, the workers carry out work activities such as raising the pipe over the trolley, pulling the trolley containing the pipe from the temporary stacking place, lowering the pipe from the trolley to the finished product warehouse. workers must bend to reach the HDPE (High density polyethylene) pipe roll which is on the floor in a temporary piling up, and then place the HDPE (High density polyethylene) pipe roll over the trolley with a height of 22 cm.

Interesting activities are activities that cause workers to do pull activities with one hand and this will cause the inconvenience of workers' work postures while working. On this basis, a research on the design of a pipe hauling facility in the form of a trolley is needed to create convenience and improvement to the work methods of warehouse workers when carrying pipes and analyze the burden received by workers and improve the design of pipe transportation facilities used by warehouse workers.

2. Literature Review

2.1. Manual Material Handling (MMH)

The United States Department of Labor defines handling as reaching, holding, grasping, rotating, or working with one hand or two hands. The ability of workers to perform work tasks can vary due to differences in age, physical condition, strength, gender, stature, and other factors. There are five things that need to be done to improve work stations so as to create a good working atmosphere, including:

- Reduce or prevent injury
- Reducing workers' efforts by reducing strength in lifting, handling, pushing, and pulling material
- Reducing risk factors for musculoskeletal disorders (for example, awkward postures to reach into containers)
- Increase productivity, quality products and services, and employee morale.
- Reduce costs by reducing or eliminating production delays, error or disability rates, use of medical services due to musculoskeletal disorders, workers compensation claims, excessive employee turnover, absenteeism, and training

The use of muscle power can be optimal, so the regulation of how muscles work must be considered properly. In this case also the muscle activity can be distinguished in 2 ways, namely:

- Dynamic muscle work (rhythmic)
- Static muscle work (work stance / fixed)

Muscles will tighten and contract (loosen) alternately or rhythmically in dynamic work. Whereas static work or attitude here will be in a position to tighten for quite a long time.

Muscles will work alternately in accordance with the tense / compressive and loose rhythm like the work of a "pump" that has an impact on the smooth flow of blood during dynamic work takes place. Muscles will carry a lot of / receive glucose and oxygen when tightening and then discarding metabolism (the result of combustion or metabolism) when loosening because the mechanism of tightening and loosening of muscles occurs alternately. Then the circulation of blood flow + O2 and metabolic will take place smoothly.

3. Method

This type of research is descriptive research that is classified into research work and activity analysis (Job and Activity Analysis). Work and activity analysis research is a descriptive study aimed at investigating in detail the activities and work of a person or group of people in order to obtain recommendations for various purposes. The object of research is the work method and the burden received by warehouse operators who work to transport pipes. The various variables used in the study are as follows:

- Independent Variable :This variable consists of: Load Weight, Transport Frequency, Work Posture
- Dependent Variable :This variable consists of: Energy consumption

4. Discussion

Conduct direct observations of warehouse workers tasked with transferring pipes from temporary stockpiles to the finished product warehouse, obtain information through direct interviews and filling out questionnaires by workers regarding complaints against parts of his body due to carrying out activities as well as personal data from the worker concerned. The activities that take place on the production floor are as follows:

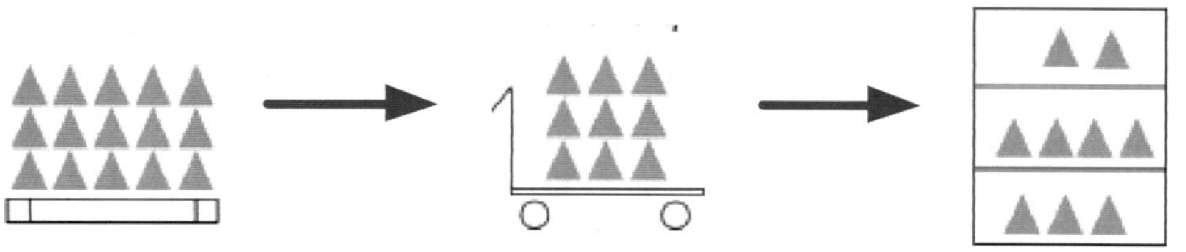

Figure 1: Load Transfer Activity Scheme

Information figure 1:
- Lifting the finished product in the form of a pipe from the temporary stacking place on the trolley.
- Transport the finished pipe from the temporary storage area to the finished product warehouse using a trolley.
- Bring down the pipe from the trolley and arrange it into the finished product warehouse.

4.1. Workload Data Processing Using Physiology
Physiological data processing is carried out to determine the amount of energy consumption needed to carry out HDPE (High density polyethylene) pipe transportation activities. This processing stage is carried out as follows:

4.1.1. Direct Assessment Method
The valuation method is directly used in determining the amount of energy used during work. The amount of energy can be calculated by the follows:

$$E = 1.80411 - 0.0229038\ X + 4.71711\ .\ 10^{-4}\ X^2 \quad (1)$$

Information:
E = Energy (kcal / minute)
X = heart rate (beats / minutes)
Workload categories based on energy consumption are as follows (Minister of Labor Regulation Number 51 of 1999):
Light workload : 100–200 kcal / hour
Medium workload :> 200–350 kcal / hour
Heavy workload :> 350–500 kcal / hour

Table 1. Recapitulation of Calculation Results of Energy Consumption and Workload Categories for Pipeline Transport Activities

No.	Worker	DNK (X) (Pulse/minute)	Energy € (kcal/minute)	Energy (E) (kcal/hour)	Workload Category
1	1	108	4.833	289.952	Medium
2		110	4.992	299.544	Medium
3		138	7.627	457.599	Hard
4		155	9.587	575.213	Hard

5	2	110	4.992	299.544	Medium
6		114	5.323	319.406	Medium
7		140	7.843	470.587	Hard
8		162	10.473	628.397	Hard

Source: Data processing

Table 1 above shows that the activities of lifting pipes ¾ "and 1" are included in the activity with moderate workload categories and the activities of lifting pipes 1 ½ "and 2" are included in the category of heavy workloads.

4.1.2. Indirect Assessment Methods

Calculation of% CVL, Cardiovascular Load (% CVL) can be calculated using the following formula:

$$\%CVL = \frac{100 \times (\text{worker's pulse} - PulseInitial)}{Maximumpulse - PulseInitial} \quad (2)$$

Table 2. Recapitulation of Calculation Results of% CVL and Workload Classification in Pipeline Transport Activities

No.	Worker	DNI	DNK	DN max	%CVL	Information
1	1	83	108	195	22.321	No Fatigue
2			110		24.107	No Fatigue
3			138		49.107	Repair Required
4			155		64.285	Work in Short Time
5	2	88	110	193	24.107	No Fatigue
6			114		27.678	No Fatigue
7			140		50.892	Repair Required
8			162		70.535	Work in Short Time

Source: Data processing

Table 2. shows that HDPE (High density polyethylene) pipe lifting activities ¾ "and 1" are activities that do not cause fatigue to workers who do, HDPE (High density polyethylene) pipe lifting activities 1 ½ "are activities that require repair and HDPE (High density polyethylene) 2 pipe lifting activities" are activities that can be carried out in short time

5. Conclusion

Physiological data processing results show that the activities of lifting pipe 2 "and transporting pipes dan" and 2 "exceed the average energy consumption limit 405.57 Kcal / hour and 429.47 kcal / hour and categorized as work with heavy workloads. The proposed design improvement is the improvement of the pipe haul trolley which data is pulled with two hands and has an electric drive on the wheels so as to reduce energy consumption.

References

[1] Battini D 2017. Ergo-lot-sizing: An approach to integrate ergonomic and economic objectives in manual materials handling. *International Journal of Production Economics.* **185**, issue C, 230-239.
[2] *Cal/OSHA Consultation Service, Research and Education Unit, Division of Occupational Safety and Health, California Department of Industrial Relations. Ergonomics Guidelines for Manual Material Handling*. 2007. *California*: *California Department of Industrial Relations.*

[3] Gyemi D 2016 *3D Peak and Cumulative Low Back and Shoulder Loads and Postures During Greenhouse Pepper Harvesting Using A Video-Based Approach.* Kanada: *Department of Kinesiology, University of Windsor, Windsor* **51**. *Issue 4.*
[4] Harris C 2016 Ergonomic evaluation of standard and alternative pallet jack handless. San Francisco: *International Journal of Industrial Ergonomics* **61**. pp 113-116.
[5] Iridiastadi H 2014 *Ergonomi Suatu Pengantar* Bandung: PT. Remaja Rosda Karya.
[6] Callaghan J 2012 *Possible mechanisms for the reduction of low back pain associated with standing on a sloped surface.* Ontario: University of Waterloo **58** pp 281-286.
[7] Knapik G 2009 *Spine Loading at Different Lumbar Levels During Pushing and Pulling.* Ohio: Biodynamics Laboratory, The Ohio State University, Columbus. 2009. Vol **21** pp 75-89.
[8] Majid A 2016 *Musculoskeletal Model of Awkward Carrying Postures.* Sarawak: Universiti Malaysia Sarawak.. **3**. *pp 1-9.*
[9] Niebel B and Andris F 2007 *Methods Standards and Work Designs.* 2007. New York: McGraw-Hill.
[10] Rahmat R F, Herly E T, Siregar B, Syahputra M F, Sitompul O S, 2018 RFID presence monitoring system as an input to measure the workload of employee *4th International Conference on Computer Applications and Information Processing Technology*, pp. 1-6
[11] Vecchio L 2017 Choosing a Lifting Posture: Squat, Semi-Squat or Stoop. Australia: *Department of Medical and*

Development on Meso Ergonomics in Manufacturing Based on Research Trends in The Last Twenty Years

Nismah Panjaitan[12], Amir Yazid Bin Ali[2], Hasnida Ab Samat[2]

[1]Department of Industrial Engineering, Faculty of Engineering, Universitas Sumatera Utara, Jl. Almamater Kampus USU Medan 20155, Indonesia
[2]School of Mechanical Engineering, Engineering Campus, Universiti Sains Malaysia Seri Ampangan, 14300 Nibong Tebal, Seberang Perai Selatan, Pulau Pinang, Malaysia

E-mail. [12]nismahpjt@gmail.com , [2]meamir@usm.my, [2]hasnida@usm.my

Abstract. Meso ergonomics continues to develop, research journals on meso ergonomics continue to increase in number, meso ergonomic discussions are carried out ranging from the scope of studies to the application of meso ergonomics in various fields. Macro, meso and micro ergonomics do not yet have clear scope limits, this makes the researchers continue to conduct studies on the limits of meso ergonomic studies. The application of ergonomic science continues to be developed by researchers, during the span of 1997 to 2017 a total of 51 studies on ergonomic meso were carried out, the main focus of the research was carried out in the fields of health, services, manufacturing and other fields. Meso ergonomic research in the field of health as many as 20 studies, 6 services and 12 fields of manufacturing and 13 other fields of research. Most research occurred in the health sector with the most research in 2005 as many as 5 studies, research in the field of health, especially in hospitals consisting of, managerial design and improvement, improvement of hospital services and the impact of managerial policies on nurses. In the field of services, the focus of his research is the managerial system and improvement of the customer's perspective on services. In manufacturing the focus of the research is to maximize output by taking into account worker fatigue. All fields of ergonomic meso studies will be evaluated because they still have shortcomings and are not credible
Keywords : Health, Services, Manufacturing, Mesoergonomy.

1. Introduction

This paper aims to summarize studies that have existed in the last ten years. This paper deals with the development of research in the field of Meso Ergonomics. This paper will examine the boundaries of a development perspective or the boundaries between Macro Ergonomics, Meso Ergonomics and Micro Ergonomics. As material for study and comparison which can later be used as a reference for conducting research. This paper was made to see how meso ergonomics tendencies in manufacturing. Meso ergonomics research is not widely known, research has also been conducted in very few in recent years. Ergonomics is a science with very extensive studies ranging from the state of the operator to the policy applied in an organization. Macro ergonomics and micro ergonomics have been known by many circles, the boundaries between macro ergonomics and micro ergonomics are also unclear, so this study seeks to determine the boundaries with the presence of meso ergonomics. The tendency of ergonomics studies will make it easier to determine the criteria of ergonomics as a reference to macro ergonomics, meso ergonomics and micro ergonomics.

The study of ergonomics can be applied in services and manufacturing, because ergonomics always has human aspects, work environment and equipment. In the case of ergonomics services are usually used to assess the service performed by the company to its customers and in the manufacturing sector to measure the workload of operators the company gives workers. Humans must be treated like humans so it is important that measurements are made to see the capabilities of each operator. If there is no measurement, fatigue will occur to the operator which can cause muscle damage. Declining operator performance will affect company productivity because the results for each operator will decrease. Ergonomics is very positive in terms of workers and the company as a whole, meso ergonomics is a study that determines the ease of application of ergonomics in the service sector, especially in the health sector.

2. Literature Review

In various literatures we have obtained various definitions of Ergonomics. Ergonomics definition according to experts can be found in various existing text books. Ergonomics is one scientific discipline that deals with understanding of interactions between humans and other elements of the system, and professions that apply theories, principles, data and methods to design, optimize human well-being and overall system performance. The basic principle of ergonomics is that the demands of work must not exceed work capabilities and limitations to ensure that they will not be exposed to work stresses that can affect the company's safety and health and productivity [1].

Macroergonomics, which is related to sociotechnical system theory (Karsh et al., 2014), and security climate are strong areas in safety and ergonomics research, each of which has unique strengths and constancy when sharing some conceptual overlaps. Macroergonomics has a more holistic-approach to safety through the theory of a good sociotechnical communication system needed but not sufficient in creating a safe and secure work environment. Macroergonomic constraints are the lack of prescribed mechanisms through which the system spreads safety responsibilities across different subsystems. Given the conceptual overlap between macro climate ergonomics and safety, the reluctance that strengths in one area can be exploited while compensating for weaknesses in other areas. Ergonomics are compatible with organizational / machine interface technology and also to research, develop, and apply ergonomic and organizational design principles to macro-level system design [2].

By using construction and many others, macroergonomics seeks to design work systems that fit the characteristics of the organization's sociotechnology system and "then ensure that micro ergonomics is designed to be aligned with the overall structure and process of work systems" [3]. Complex systems, such as an organization, consist of interrelated components, and the nature of those components will change if the system itself changes [4]. The mesoergonomic framework is based on the meso paradigm, the concept of organizational behavior in which two or more levels are examined simultaneously. Karsh et al. (2014) identified safety climate as a good example of mesoergonomic research, which explains that variables are specifically measured at the organizational, unit, and individual levels to determine cross-level relationships.

3. Method

Literature study is often also called a literature survey is a documentation of a comprehensive review of the results on the work contained in the literature either published or not published. The intended forms of literature vary, among others, science books, scientific journals, scientific magazines, conference proceedings, doctoral dissertations, master's theses, government publications, financial reports, marketing and others. Literature surveys can be carried out simultaneously with interviews or before / after the interview.

Literature studies of topics relating to the problem areas to be clarified through initial data collection are very helpful because the data and information obtained together with interviews shorten the time collected in the initial data collection. The reason for the importance of conducting a literature survey is to ensure that no important variables related to research problems will be left behind. That is, if the interview does not reveal the existence of certain variables but found in the literature survey, the researcher can consider it.

The selection of scientific journals is done by adjusting the keywords of the journals relating to ergonomics in manufacturing and services, so that the journals collected have similarities in terms of their scientific studies, making it easier to draw conclusions for each of the journals collected. Journals that have similar studies will be put together to be a reference journal based on previous journals. There are 26 journals related to the journal and the rest only explain in general the study of ergonomics in services and manufacturing, but all journals collected were made as comparison material in the study of ergonomics, judging by the criteria, stages and methods used.

4. Discussion

Table 1. Number of studies per sector in 1997-2017

Year	Health	Service	Manufacture	etc
1997	0	0	1	0
1998	0	0	0	0
1999	0	0	0	0
2000	1	0	0	0
2001	0	0	0	0
2002	0	0	0	0
2003	0	1	1	0
2004	1	0	0	2
2005	0	0	0	1
2006	0	0	1	0
2007	0	0	1	0
2008	0	1	0	2
2009	2	0	0	0
2010	2	1	2	2
2011	1	1	1	0
2012	0	0	1	2
2013	2	0	2	0
2014	0	1	1	1
2015	5	1	0	0
2016	2	0	1	2
2017	4	0	0	1

Table 1. Above shows the research that has been done that shows in which areas the ergonomic meso was conducted or designed with the development of the amount of research in each year. Starting from 1997 until 2017 researchers get the number of journals that research about meso ergonomics with a total of 51 journals that have been obtained by researchers.

With the division in the field of health 20 studies, in the field of services 6 research, in the field of manufacturing 12 fields and in other fields with a total of 13 studies which cumulatively obtained a total of 51 studies.

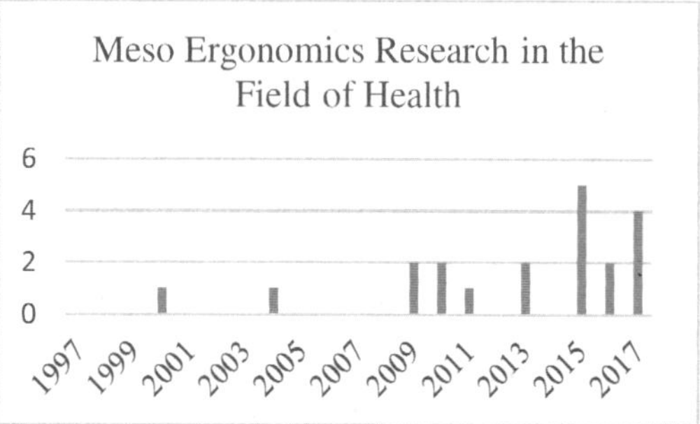

Figure 1. Graph of Development of Meso Ergonomics Research in 1997-2017 in the Health Sector

In Figure 1. It is seen that the development of Meso Ergonomics research in the largest health sector is located in 2015 with a total of 5 studies. Although in previous years there was still a vacuum of ergonomic meso research. For example, in 2014 and others.

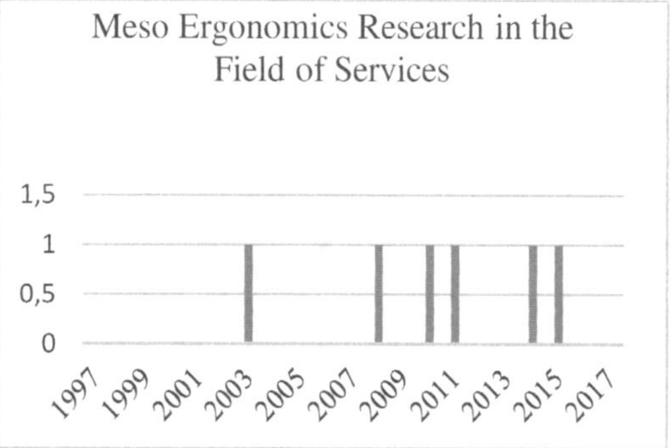

Figure 2. Graph of Development of Meso Ergonomics Research in 1997-2017 in Services

In Figure 2. It is seen that the development of Meso Ergonomics research in the field of service on average only 1 journal is obtained which discusses the Meso Ergonomics in services

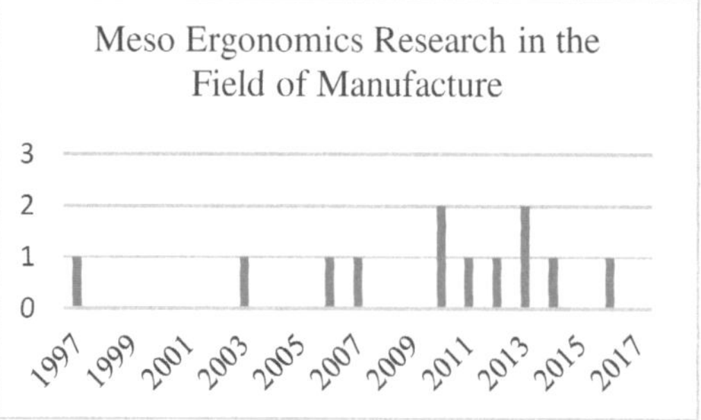

Figure 3. Graph of Development of Meso Ergonomics Research in 1997-2017 in Manufacturing

In Figure 3. It is seen that the development of Meso Ergonomics research in the manufacturing sector was mostly found in 2010 and 2013, namely 2 studies. But it can be seen that research on Meso Ergonomics began in the field of manufacturing because there was 1 journal that discussed Meso Ergonomics in 1997.

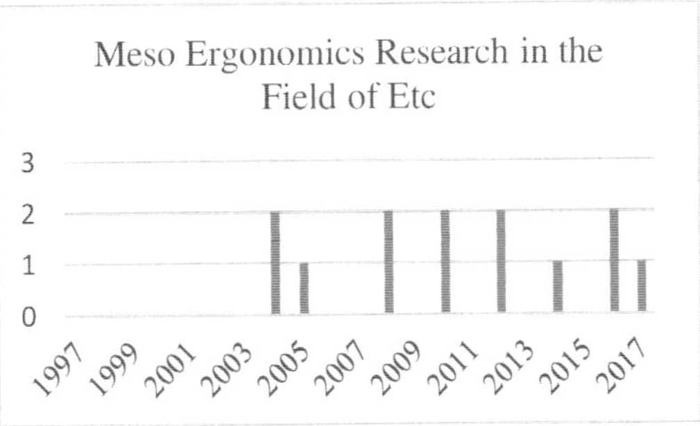

Figure 4. Graph of Development of Meso Ergonomics Research in 1997-2017 in Other Areas

In Figure 4. It is seen that the development of other Meso Ergonomics research which includes examples in the field of telecommunications, consultants and others that are seen also have a stagnant amount each year starting in 2004 to 2017.

5. Conclusion

The objective of this research is to evaluate, through the surveyed literature even though research on Meso Ergonomics in each of these field trends still has shortcomings in its accuracy and reability. After reviewing 51 journals about Meso Ergonomics it can be seen that research on health has the most number, 20 research journals in which have application in the design of improvement or improvement councils from hospitals or from the nursing department itself. And many studies that combine the 3 elements of ergonomics in it to get the best results from these studies. The most research on Meso Ergonomics was found in 2015 in the health sector, which is 5 journals that discuss it.

In the field of research services on meso ergonomics, it is stagnant in 1 journal in several years, whereas in other years there are no journals that discuss the research. The application of this journal much lies in how the managerial system and shareholders in order to maximize the service and customer perception of the services they run.

In the field of manufacturing it can be seen that as a start of research in the field of Meso Ergonomics but in its development the journals obtained were only the 2 highest journals in a few years. With the application of this journal in the field of manufacturing as a way to optimize production and minimize fatigue from workers.

References

[1] Jafffar N, Abdul-Tharim A H, Mohd-Kamar I F, Lop N S, 2011 *A Literature Review of Ergonomics Risk Factors in Construction Industry*, Elsevier 2011

[2] Ayubkhon R, Hai Q, Shupinh X, KyungHyun N,2014 *Ergonomics and Sustainable development in the past two decades (1992-2011) : Research trends and how ergonomics can contribute to sustainable development,* Elsevier

[3] Christina O, Stephan C H, Stefanos M, 2012 *Organizational inscriptions of network pictures : A meSo-level analysis*, Elsevier

[4] Mark A H, Brian E B, 1997 *The Impact of High Performance Work System Implementation Effectiveness, and Alignment with Strategy on Shareholder Wealth*, Rutgers University.

[5] Sue H, 2000 Embedding ergonomics in hospital culture: top-down and bottom-up strategies, Elsevier.

[6] Melroy E. , Joel S. G, 2003 Listening to users in a manufacturing organization: a context-based approach to the development of a computer-supported collaborative work system, Elsevier.

[7] Donald C. C, Richard P. W, 2003 Methodological issues in evaluating workplace interventions to reduce work-related musculoskeletal disorders through mechanical exposure reduction, Sjweh,

[8] Steven A, 2004 *Making Work System Principles Visible and Usable in System Analysis and Design*, AISeL.

[9] Vincent G, Haije W, 2004 *Reliability and validity of Functional Capacity Evaluation methods: a systematic review with reference to Blankenship system, Ergos work simulator, Ergo-Kit and Isernhagen work system*, Springer
[10] Gudela G, 2004 *Uncertainty Management at The Core of System Design*, Elsevier
[11] Performance Work Systems and Organizational Performance: The Mediating Role of Internal Social Structure
[12] Evans W R, Davis W D, High-Performance Work Systems and Organizational Performance: The Mediating Role of Internal Social Structure,
[13] Spyros M, Nicolas M, 2006 *Supporting the Design of Office Layout Meeting Ergonomics Requirements*, Elsevier, 2006
[14] Richard B F, Douglas K, Joseph B, 2007 *The Same Yet Different: Worker Reports on Labour Practices and Outcomes In a Single FirmAcross Countries*, Elsevier.
[15] Metin D, Ihsan Y, Mustafa K, 2008 *A fuzzy analytic network process (ANP) model to identify faulty behavior risk (FBR) in work system*, Elsevier.
[16] Neville M, 2008 *The Good, the Bad, and the Future: On the Archaeology of Ergonomics*, Sage.
[17] Doncho P and Olga P, 2008 *The Work System Model as a Tool for Understanding the Problem in an Introductory IS Project*, ISEDJ.
[18] Waterson P, Eason K, Tzion-Karsh B, 2009 *Crossing the Macro-Micro Divide in Systems Ergonomics*, Loughbrough University.
[19] Samaras G M, Samaras E A, 2009 Feasibility of an e-Health Initiative: Information NWDs of Cancer Survivor Stakeholders, China.
[20] Zjohan B, Sidney W A D 2010 Bridging the Macro and the Micro by Considering the Meso: Reflections on the Fractal Nature of Resilience, Ecology and Society.
[21] Tzu-Shian H, Wan-Chun L, 2010 *Computer-integrated anufacturing and high performance work system: the case of Taiwan*, Routledge.
[22] Steven A, 2010 *Including Work System Co-Existence, Alignment, and Coordination in Systems Analysis and Design*, AMCIS.
[23] Robert P S, Jay M M, and Randolph E S, 2010 Is The Accreditation Council for Graduate Medical Education (ACGME) Resident/Fellow Survey A Valid Tool to Assess General Surgery Residency Programs Compliance With Work Hours Regulations?, Elsevier.
[24] Battini D, Faccio M, Persona A, Sgarbossa F, 2010 New methodological framework to improve productivity and ergonomics in assembly system design, Elsevier, 2010
[25] Samaras E A, Samaras G M, 2010 Nurse Stakeholder Dissonance, AAMI HORIZONS.
[26] Asbjørn F, Hornbæk K, 2010 Work-domain knowledge in usability evaluation: Experiences with Cooperative Usability Testing, Elsevier.
[27] Siregar I, Muchtar M A, Rahmat R F, Andayani U, Nasution T H, and Sari R M 2018 Method of calculation overall equipment effectiveness in fertilizer factory. *IOP Conference Series: Materials Science and Engineering* **308**(1), p. 012053

Corrective Maintenance of Worm Gear a Screw Press Machine with a capacity of 30 tons per hour using Wear Analysis and its response to Conus Pressure

F Ariani[1,*], Tugiman[1], D M Nasution[1], H Sibarani[1]

[1]Department of Mechanical Engineering, Faculty of Engineering, Universitas Sumatera Utara, Padang Bulan, Medan, 20155, Indonesia

[*]Corresponding author: faridaariani@usu.ac.id

Abstract. Worm gear is one of the main components of a screw press machine that is very vulnerable to wear. Based on the survey, it was found that the lifetime of the worm gear of a screw press machine at a palm oil mill owned by PT. Perkebunan Nusantara 4, Adolina estate, is always lower (only 700 to 800 hours) when compared to the recommended lifetime of the manufacturer (1000 hours). The wear on this worm gear needs to be minimized because it affects the performance of the screw press machine and losses on palm oil. Corrective maintenance needs to be applied to worm gear as a repair solution. Temporary corrective maintenance that has been carried out by the operator is to re-weld the surface of the worm gear. Initial analysis has been carried out by calculating the response that occurs in the worm gear due to cone pressure. The analysis results obtained that the highest wear value is located at the end of the worm gear that is in direct contact with the cone pressure. Obtained the value of the rate of wear depth of the worm gear of 0.583 mm per day with the hardness value of the cast steel material that is 215 BHN. After that, at the end of the worm, gear is modified by adding a coating plate that has been given surface hardening treatment using NiKaNa methods. The aims are to make the wear that occurs earlier on the part of the coating plate, so that repair becomes more leisurely. Based on the results of the study, it was found that the value of the hardness of the coating plates increased by 490 BHN, with the rate of wear depth decreasing to 0.256 mm per day.

Keyword: corrective maintenance, worm gear, wear rate, hardness

1. Introduction

A company cannot avoid the process of maintaining a production machine because it is closely related to the smooth production process [1]. The basic concept of maintenance is to maintain or repair equipment or machinery so that it returns to its initial state with a short time and low cost [2]. The palm oil mill in Adolina estate has 4 screw press machines, which consist of 3 screw press machines with a capacity of 10 tons per hour, respectively, and 1 machine with a capacity of 15 tons per hour. A useful worm gear component determines the quality of oil pressed from oil palm fruit. Wear and damage problems are often found on screw press machines. Because, if the wear that occurs on the screw part of a screw press machine is already of high value, then it can cause a wide gap on the outer side of the screw with the press cage. This certainly reduce the performance of oil presses and the impact on the loss of palm oil produced. This condition can be seen through the fibres, which still contain some oil after pressing [3]. Each manufacturer of a worm gear component must provide a lifetime recommendation because it is very vulnerable to wear and damage. The recommended lifetime of a worm gear component is 1000 hours. However, the lifetime provided by the manufacturer is not following the lifetime that occurs in the field, which is only around 700-800 hours. Therefore, intensive maintenance and repairs are needed to avoid undesired severe damage, so that in the future of the lifetime recommended by the manufacturer can be achieved. The focus of this study is (1) to find out the response of the worm gear to cone pressure; (2) analyse the wear that occurs in the worm gear which affects the reduced lifetime, and (3) implement corrective maintenance to minimize the risk of wear on the worm gear.

2. Method

This research was conducted at one of the palm oil mills located in the Adolina estate owned by a national company called PT. Perkebunan Nusantara 4, located at Jalan Perintis Kemerdekaan, Batang Terap village, Perbaungan sub-district, Serdang Bedagai Regency, North Sumatera province. This study was conducted from November to December 2018. The object of study was the worm gear component of a screw press machine located at the press station of the palm oil mill. Photograph of the worm gear is shown in Fig. 1. The specifications of the screw press machine used in this study are shown in TABLE I. The screw press machine includes two screw bars that rotate opposite each other. Palm fruit that has been crushed will be pushed and pressed by the cone on the other side so that the oil palm fruit becomes squeezed. Through the press cage holes, oil is separated from the fibers. The results of the pressing process are fibers and nuts that come out through the gap between sliding or adjusting cone and press cage, which then goes into the Cake Bake Conveyor. Crude palm oil, which still contains impurities such as fibers and water, will then pass the clarification stage through a Sand Trap Tank to separate the impurities from the crude oil. Then enter the Vibrating Screen to separate the fibers from the crude oil and finally transferred to the crude oil storage tank.

2.1. Wear Analysis

Wear of a machine component occurs due to friction between two or more surfaces of a material. This friction occurs between the surface of the worm gear, and the oil palm fruit is squeezed. The surface conditions of materials that are subjected to wear can be observed through surface enlargement using appropriate equipment. Worm gear material is cast steel with a hardness value between 200 to 230 BHN (Brinell hardness number). In this study, the average hardness value of worm gear material is 215 BHN. If converted in pressure units, the value of 1 BHN is equal to 1 kg/mm^2 and is equal to 9.8 MPa. Therefore, the hardness value of 215 BHN is equal to 2107 MPa. Based on Archard wear law, to find the value of the wear rate can be used the following equation [4]:

$$V = KA_rL = KL\frac{W}{H} \tag{1}$$

Where V is the rate of wear volume in m^3, L as sliding distance in m, W is the load in Newton, K is the wear coefficient, then H is the value of material hardness in MPa, and A_r as the contact area in m^2. The value of wear coefficient K, adjusted for abrasive wear on two-component surfaces [5].

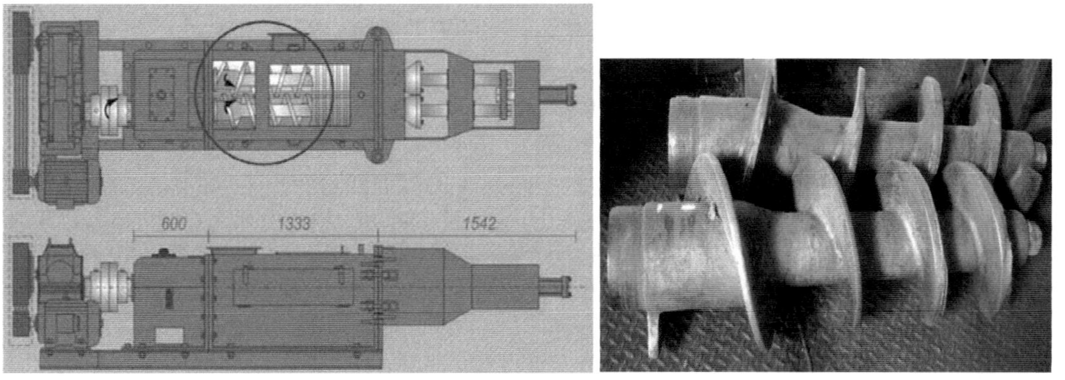

Figure 1.(a) The shape of a screw press machine, unit in mm.(b) The worm gear of a screw press

Table. 1. The Specifications of The Screw Press Machine

Parameter	**Value**
Capacity	10 tons per hour
Type	Continuous double
Cone Pressure	30 to 40 Bars
Clearance	25 mm
Rotation	9 to 11 rpm
Input Cycle	Continuous
Weight of Worm Gear	100 kg is equal to 981 N
Number of Screws	4 or 5

Furthermore, the rate of wear depth that occurs on the surface of the worm gear can be calculated using the following equation [6]:

$$K_a = K\frac{WL}{HA_{ks}} \quad (2)$$

Where K_a is the rate of wear depth that occurs in m, and A_{ks} is the actual contact area in m^2.

2.2. *Response to Conus pressure*

To analyzing worm gear responses, three main factors must be considered, that is (1) dilution water temperature must be maintained at 90^0C; (2) moisture content of oil palm fruit is not more than 20%, and (3) cone pressure must be maintained between 30 to 40 bars. If the cone pressure applied when compressing is too small, then the amount of oil loss will be higher, and if the pressing pressure is too large will cause the number of broken nuts to be higher. The wear always occurs on the surface and the edges of the worm gear. The most critical area, which is often the area of wear, is shown in Fig. 2. In this study, the calculation of the wear rate is only on one screw, which is the outer screw under direct pressure by the conus. It is assumed that the critical point occurs at a maximum distance of 30 mm from the outer side of the screw. The response of the worm gear to the conus pressure can be used by finding the values of torque, nominal shear stress, axial stress, bending stress, and shear stress at the base of the screw. Torque can be calculated using the following equation:

$$T = \frac{F.dm}{2}\left(\frac{p + \pi\mu d_m}{\pi d_m - \mu p}\right) \quad (3)$$

where T is the torque acting on the screw in N mm, F is the axial force acting on the screw in Newton, μ is the coefficient of sliding friction between the material, the screw is 0.49, p is the screw pitch in mm, and d_m is the value average radius area of the screw. Then, the nominal shear stress is calculated using the following equation:

$$\tau_{nom} = \frac{16TD}{\pi(D^4 - d^4)} \tag{4}$$

Where τ_{nom} is the nominal shear stress acting on the screw shaft in N/mm^2 or MPa, D as the outer diameter of the hollow shaft in mm, and d as the inner diameter of the hollow shaft in mm. Furthermore, the axial stress on the screw shaft due to load is calculated using the following equation [7]:

$$\sigma = \frac{4F}{\pi({D_r}^2 - {d_r}^2)} \tag{5}$$

Where σ is the axial stress in N/mm^2, F is the load acting on the worm gear in Newton, D_r, and d_r represent the inner and outer diameters of the hollow shaft in mm, respectively.

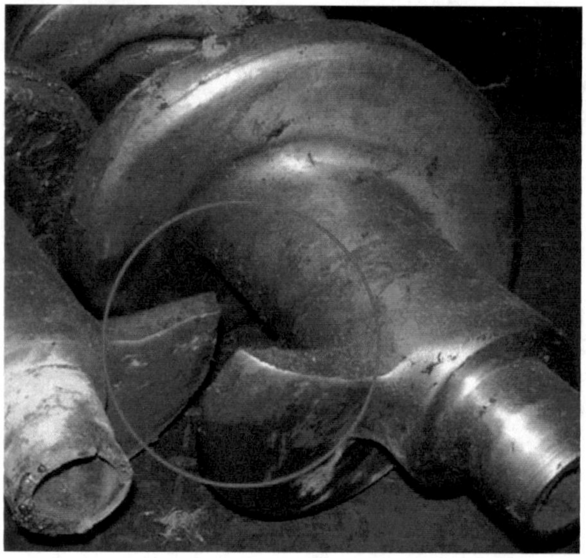

Figure 2. Photograph of the most critical area of the worm gear.

The bending stress can be calculated using the cantilever beam principle, which is projected from the base of the screw beam with the following equation [8]:

$$\sigma_b = \frac{3F.h}{\pi\Delta(D_r - d_r)nb^2} \tag{6}$$

σ_b is the bending stress in N/mm^2 or MPa, h for the beam height in mm, n represents the number of beams, and b as the beam width in mm. Finally, the shear stress acting on the screw base is calculated using the following equation:

$$\tau = \frac{3F}{2(D_r d_r)nb\pi} \tag{7}$$

τ is the bending stress in N/mm^2 or MPa.

3. Results and Discussion

The main cause of wear and tear on worm gear is the result of pressure that occurs on the surface of the screw. The rate of increase in screw pitch occurs because the rotation of the screw causes the oil palm fruit on the side of the screw to be pushed, and on the other side, the hydraulic pressure on the cone presses the crushed oil palm fruit. This certainly makes the oil palm fruit under high pressure on both sides, causing the outer side of the screw under pressure due to cone pressure.

3.1. Worm gear response

By applying equations (3) to (7), the values of torque, nominal shear stress, axial stress, bending stress, and shear stress is 6380707.916 Nmm, 26.95 N/mm^2, -9.58 N/mm^2, 65.515 N/mm^2, and 14.10 1 N/mm^2, respectively.

3.2. Wear rate and depth of wear

Based on equation (1), the value of the rate of wear that occurs for each oil palm fruit to the surface of the worm gear in each cycle along the 0.85094 m is 0.0118786 mm^3. Referring to table 1, the worm gear rotates between 9 to 11 rpm (in this study, it is assumed to be 10 rpm). This means that each rotation of worm gear is equal to one circumference of the worm gear and has the equal value is 0.0118786 mm^3. Under normal conditions, the palm oil mill operates for 22 hours per day or 1320 minutes per day. In 1 minute, the worm gear rotates as much as 10 rpm, so the number of worm gear rotations per day is 13200 rotations per day. So, it can be predicted that the rate of wear volume of worm gear that occurs in 1 day is 156.79752 mm^3 per day.

In calculating the rate of dimensionality reduction due to wear, the first is to determine the actual contact area (A_{ks}) of the friction of the two materials. Based on the survey, the actual contact area is 0.0245862 m^2. The number of oil palm fruits in the critical area of the worm gear is 78. The weight borne by the whole oil palm fruit and worm gear is 269,143 Newton. Thus, the estimated rate of wear depth that occurs on the worm gear surface using equation (2) is obtained by 0.000037685 mm. The estimated value of the rate of wear depth obtained is the estimated rate of wear depth that occurs in the critical area of the worm gear for one screw rotation cycle. In one day operating of the screw press machine, there are 13200 rotation cycles, so the estimated depth of wear that occurs in 1 day is 0.497 mm per day. The value of the rate of wear depth based on operating time for 50 days is shown in Table 2.

Table 2. The Value of The Rate of Wear Depth

Operation (day)	Time (hours)	The rate of wear depth (mm)
1	22	0.497
10	220	4.970
20	440	9.940
30	660	14.92
40	880	19.90
50	1100	24.84

3.3. Corrective maintenance using NiKaNa methods

The NiKaNa method consists of a combination of three metallurgical processes, that is Nitriding, Carbonation, and Quenching using Sodium Chloride (sudden cooling in a salt solution). Three processes that were initially carried out separately, but with this method will get steel with a higher level of hardness. This hardening process occurs because of changes in the phase or atomic structure of the steel. The phase change is done by heating the steel at a specific temperature and cooling at a certain speed by adding new material to the steel [9]. In this study, the handling of worm gear hardening is carried out in three stages, that is (1) the initial treatment stage; (2) NiKaNa treatment stage; and (3) sample testing phase. The initial treatment stage is sanding and heating. A steel sample is heated in a Furnace with variations in temperature and heating time. At the NiKaNa treatment stage, nitrogen gas is sprayed directly on the steel sample that has been heated (nitriding). Then the steel sample is reheated with the same temperature and heating time at the initial treatment stage and dipped quickly in a saturated concentrated oil liquid (carburizing). The steel sample is reheated at the same temperature and heating time at the initial treatment stage and dipped suddenly into a solution of saturated Natrium Chloride (Quenching).

In the process of nitriding, deposition of nitrogen atoms in steel that has been treated with heat, which results in stretching of the material atoms and emptiness, so that the nitrogen atom is filled so that the new atomic bond, that is Fe-N. The infiltrated nitrogen atom occupies the location of interstitials (insertions) and substitutes (displacement). With the entry of nitrogen atoms into the substrate results in changes in the microstructure of the atom, the steel is making atoms more dense and denser. In the carbonation process, the carbon atoms can diffuse into the steel material, and carbon atoms are very easy to infiltrate the substrate because the size of carbon atoms is smaller than that of Fe atoms. With increased carbon content, the hardness increases. The quenching process is very good at hardening metal materials. In the process of sudden cooling, a metal material does not have time to diffuse with its neighboring atoms but will be filled by the quenching media suddenly. This process causes the density of atoms on the surface of the material to be higher and the metal material to be harder.

3.4. Surface Coating process

The surface coating process is an economical method for producing materials, equipment, and machine components that require surface resistance to corrosion, erosion, and wear. The surface coating method used in this study is metal spraying. It is a process wherein a feedstock material is heated and propelled as individual particles or droplets onto a surface. Sprayed particles impinge upon the surface, they cool and build-up, splat by splat, into a laminar structure forming the thermal spray coating. A detonation gun unit mainly consists of a doubled walled barrel, a combustion chamber, a powder feeder, control panels to regulate gas flows and gas operation. The process produces noise levels that can exceed 140 decibels and requires particular sound and explosion-proof rooms. Depending upon the ratio of the combustion gases, the temperature of the hot gas stream can go up to 3890°C, and the velocity of the shock wave can reach 3500 m/s. Depending on the required coating thickness and the type of coating material, the detonation spraying cycle can be repeated at the rate of 1 to 10 shots per second [10].

3.5. Impact of Corrective Maintenance

The results of surface hardening using the NiKaNa method have increased the hardness of the worm gear material to 490 BHN (4802 MPa). Increased hardness of the worm gear, it turns out to have a good impact on reducing the rate of wear. Obtained the rate of wear depth decreased from 0.497 mm per day to 0.218 mm per day. The value of the rate of wear depth after corrective maintenance for 50 days of operation is shown in TABLE III. Decreasing the rate of wear depth will increase the lifetime of the worm gear. Lifetime recommended by the manufacturer will be appropriately achieved. The condition of this wear has been plotted in graphical form, as shown in Fig. 3.

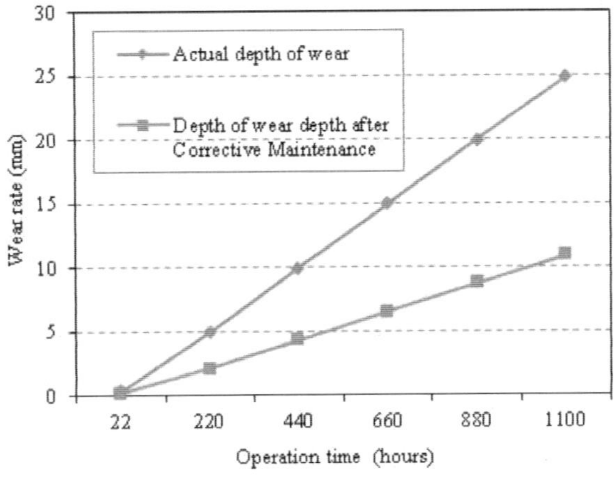

Figure. 3. The comparison of wear rate values.

4. Conclusion

The application of corrective maintenance will provide benefits for a company including (1) emergency maintenance requirements are reduced; (2) heavy downtime losses are reduced; (3) plant availability is increased; (4) results in better

Table 3. The Value of The Rate of Wear Depth After Corrective Maintenence

Operation (day)	Time (hours)	The rate of wear depth (mm)
1	22	0.218
10	220	2.180
20	440	4.260
30	660	6.540
40	880	8.730
50	1100	10.91

Utilization of plant facilities; (5) safety level is improved, and hence there are fewer chances of accidents; and (6) provides sufficient information concerning the maintenance replacement and repair. More intensive studies are needed to find the latest methods in terms of improving the quality of materials, especially screw press machine components. In this study, it can be concluded that to extend the lifetime of the worm gear component from a screw press machine can be done by applying corrective maintenance in the form of increasing the hardness of the worm gear material using the NiKaNa method and coating the surface of the worm gear with metal spraying method.

Acknowledgment

Expression of thanks goes to the Manager of the palm oil mill at the Adolina estate, for the support given to the author in completing this study.

References

[1] Didik W A, 2000 Analisa Penjadwalan dan Biaya Perawatan Mesin Press untuk Pembentukan Kampas Rem *J. Tek. Mesin*, **2**(1) pp. 50–61, 2000.

[2] Tarigan R, Ginting E, and Siregar I, 2013 Perawatan Mesin Secara Preventive Maintenance Dengan Modularity Design Pada Pt. Rxz," *e-Jurnal Tek. Ind.*, **3**(3), pp. 35–39.

[3] Basil O, 2007 Modeling of Metal Wear in Screw Presses in Palm Oil Mills *J. Eng. Appl. Sci.*,

2(3), pp. 481–484

[4] Valentin P 2019 Generalized Archard Law Of Wear Based On Rabinowicz Criterion Of Wear Particle Formation *FACTA Univ.*,**17**(1), pp. 39–45, 2019.

[5] H. Rebai, 2014 Tribology and machine elements *HAMK University of Applied Sciences*.

[6] Zmitrowicz A 2006 Wear Patterns And Laws Of Wear – A Review *J. Theor. Appl. Mech.*, **44**(2), pp. 219–253, 2006.

[7] Ansel C U 2010 *Stresses In Beams, Plates, And Shells*, Third. Boca Raton: CRC Press Taylor & Francis Group.

[8] Budynas R 2014, *Shigley's Mechanical Engineering Design*. McGraw-Hill Higher Education,

[9] Davis J R 2002, *Surface Hardening of Steels Understanding the Basics*. Materials Park: ASM International

[10] Shibe V and Chawla V, 2013 An Overview of Research Work in Surface Coating," *IJRMET*, **3**(2), pp. 85–88.

IOP Conf. Series: Materials Science and Engineering **851** (2020) 012007 doi:10.1088/1757-899X/851/1/012007

Analysis of Production Obstacles in Assembling G Line Process Using FMEA Method (Failure Mode and Effect Analysis)

Khawarita Siregar [1], Farida Ariani[2], Rahmi M Sari[3], Indah Rizkya[4] and Lauren Sopar Simanjuntak[5]

[1,3,4,5] Department of Industrial Engineering, Faculty of Engineering, Universitas Sumatera Utara, Medan, Indonesia
[2] Department of Mechanical Engineering, Faculty of Engineering, Universitas Sumatera Utara, Medan, Indonesia

Abstract. PT. XYZ is a company engaged in the manufacturing industry of manufacturing four-wheeled vehicles. Based on the results of preliminary observations, it was found that the production process in the G Line assembling section in this company had problems that could hamper the smooth production process. Problems that occur during the production process can reduce work time so that the number of products produced by the company becomes reduced. These problems can also result in the number of products produced not according to the number of products planned. To reduce the barriers to production, this study uses the FMEA method to identify the most potential causes of production barriers in the G line assembling process and recommendations for the solutions given. Based on the research conducted, it is known that there are several types of production barriers. Then the calculation of the percentage of disability is done to obtain a pareto diagram. Based on the calculation results, it was found that the types of production barriers analyzed were final line no body with a total resistance of 1400 and a percentage of disability 75.31%. The recommended solution is to provide training to operators so that operators have better skills.

Keywords: *Production Barriers, FMEA, Pareto Diagram, Cause and Effect Diagram*

1. Introduction

Plant II is a new project specifically for four-wheeled vehicles. Here do pressing, welding, painting, and assembling (assembling) four-wheeled vehicles, using a variety of high-tech equipment and is the largest in Southeast Asia for now.

The assembling process is the process of combining body units that have been maintained with the engine and other components, such as wheels, seats, dashboards, interiors, and other components into one unit of the car. This process includes the trimming process, chassis, and final. Chasis is the process of assembling car parts related to the chassis. Triming is the process of assembling the top of the car or the installation of the interior and exterior of the car. Final is an assembling process for completeness after the triming process and chassis such as wheels, fuel, bumpers, and other fittings.

The assembling process at PT. XYZ has problems in the process. Problems that occur during the production process can reduce work time so that the number of products produced by the company becomes reduced. These problems can also make the number of products produced not in accordance with the number of products planned. Based on this, the researcher wants to identify the types and causal factors in the problem of assembling using FMEA so that it can provide recommendations for solutions that must be corrected by the company.

Failure modes and effects analysis (FMEA) first up from studies done by NASA in 1963 and spread to the car manufacturing industry to identify potential defect at the stage of product design [1]. FMEA is a methodical approach to identifying the possibility of failure modes of a manufacturing process or a product to provide key information for improvement and risk assessment [2][3]. Failure modes are the ways, or modes, in which an product can fail/defect. The purpose of FMEA is to identify the possibility of problems in the early design process of a system or product that can influence its safety and performance, and to introduce a treatment to minimize the effects of the identified potential problems (failure modes) [4].

Many previous studies have been carried out to identify type and causal factors using FMEA such as in Steel Industrie Factory [5], in asset maintenance [6], in the supply chain risk environmet [7], for healthcare risk analysis [8], and the others's study. The purpose of this study to identify the types and causal factors in the problem of assembling.

2. Method

The method used in this study is in the form of direct observation in the field of the G Line assembling process, accompanied by interviews with the head of production. The secondary data collected in the form of general data about the company and historical data on production barriers at PT. XYZ.
The steps taken in carrying out this research are as follows:

1. Field Study
   Field studies are conducted to obtain information about the general description and actual conditions of the company
2. Library Studies
   Literature studies are conducted to obtain and better understand theories related to problem solving. The source of the literature comes from books, journals, and studies of previous research on the main topics.
3. Problem Formulation
   Based on literature studies and field studies, existing problems will be known so that the problems being studied can be formulated
4. Determination of Research Objectives
   Determination of research objectives is used to explain what goals you want to achieve with the holding of research.
5. Data Collection
   Data collected in this study include primary data and secondary data, namely: company overview data, organizational structure data, data on current number of workers, job description data for each worker, production barriers data.
6. Data Processing
   This study identifies the most potential causes of production barriers in the G Line assembling process using FMEA. FMEA is a method to identify and analyze potential failures and consequently aim to plan production processes well and can avoid production process failures and unwanted losses. The purpose of implementing FMEA is to prevent problems occurring in processes and products. The result is a better process because corrective actions have been taken and reduced and eliminated failures [9]. The steps using FMEA is:
   - Calculate the percentage of barriers to production and depiction of pareto diagrams
   - Determine the factors that significantly influence production barriers.
   - Calculate the value of Risk Priority Number (RPN).
     Risk Priority Number (RPN) was calculated by multiplyng severity of the error (S), the probability of occurrence (O), and detect ability (D). Errors with RPN >100 were identified as high risk errors. The RPN formula is as follows

$$RPN = S \times O \times D \qquad (1)$$

- Provide a solution recommendation

7. Analysis and Conclusion
   The analysis and conclusions carried out in this study, namely analyze the percentage of production barriers and analyze the factors that significantly influence production barriers. After that, create a conclusions which are the final summaries that are able to answer the research formulations carried out and provide research suggestions

3. Results and discussion

3.1. Pareto Diagram

Pareto diagrams are made to find or find out the problem or cause that is the key to solving problems and comparison to the whole. By knowing the dominant causes, priority for improvement can be determined. This improvement in the dominant causal factors will have a greater effect than the resolution of the causes that are not meaningful. The results of calculating the percentage of obstacles for other types of obstacles can be seen in table 1.

Table 1. Calculation of Barriers Percentage and Cumulative Percentage

Types of Production Barriers	Total Obstacles	Percentage of Obstacles	Cumulative Percentage
Final line no body	1400	75,31	75,31
Lifter suspension error	214	11,51	86,82
PBS error	52	2,8	89,62
Pokayoke error	38	2,04	91,66
DL -1 error	20	1,07	92,73
Shipping problem part	17	0,91	93,64
Repair in line / pull andon	15	0,81	94,45

Based on the data in Table 1. above can be made a pareto diagram for production barriers at PT. XYZ as seen in Figure 1.

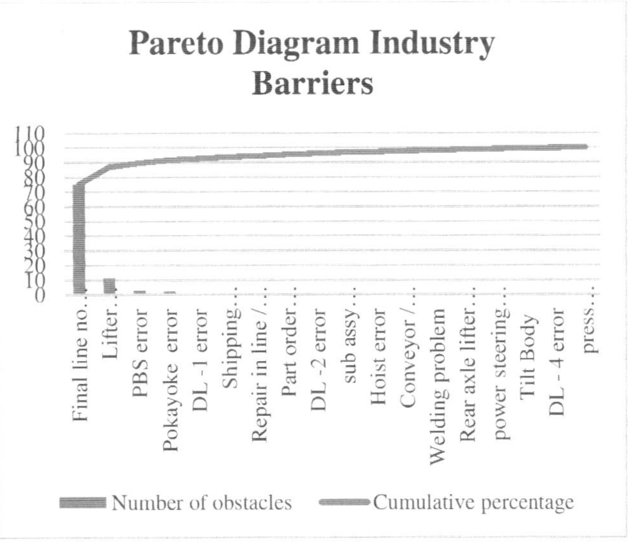

Figure. 1. Pareto chart of obstacles to production of g line assembly parts.

Pareto used in solving this problem is to use Pareto 70% - 30%. Based on Pareto rules 70% - 30%, problems that have a cumulative percentage of more than or equal to 70% must be analyzed further.

Based on the results obtained, the obstacles to be analyzed further are the final line no body because it has a cumulative percentage of 75.31%.

3.2. Cause and Effect Diagram

This diagram is useful for analyzing and finding factors that have a significant effect in determining the characteristics of the quality of work output. In addition, this diagram is useful for finding the real causes of a problem. From the results of observations made at PT. XYZ, then analyzed and determined the factors that significantly influence the search for the real cause of the final line no body problem. The factors that significantly influence the final line no body problem are shown through the Cause and Effect Diagram in Figure 2.

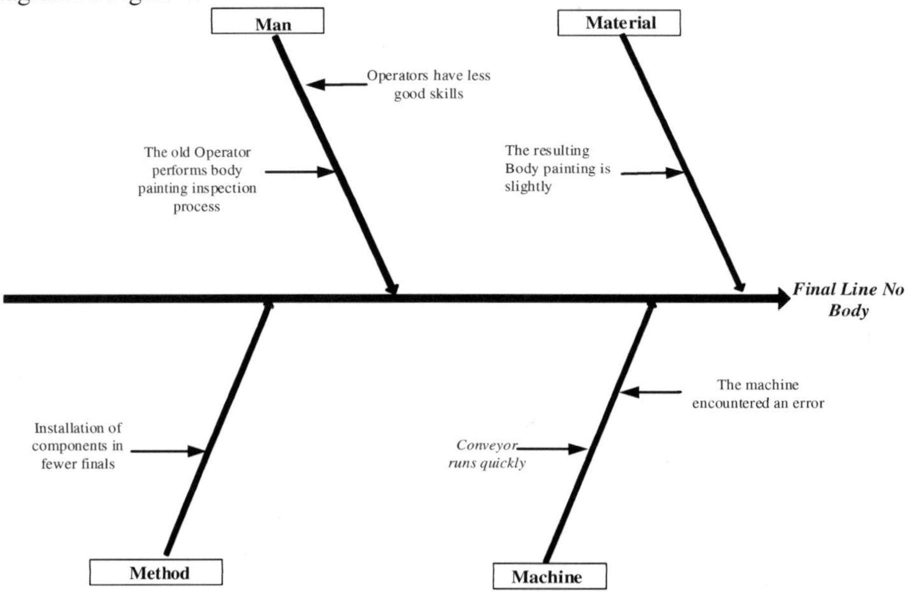

Figure. 2. Cause and effect diagram final line no body.

Based on the image of Cause and Effect, the diagram above shows that the causes of the final line no body are as follows.

1. Humans
   a. Old operators carry out body painting inspection processes because the components that have to be inspected are many, so it takes a long time to complete the inspection process. The long inspection process causes fewer bodies to enter the G line assembly.
   b. Operators have poor skills so that they work on old body painting. The process of making old body painting can cause the body to enter the G line assembly part to be less.
2. Material
   The resulting body painting is a bit so that it will cause the final line no body.
3. Working method
   Installation of components in the final section is less than the other stations, causing the assembly process in the final section to be completed faster and causing a body vacuum.
4. Machine
   a. The conveyor runs quickly so the final line will quickly experience a body vacuum.
   b. An error occurred on the machine which caused body painting to be blocked from entering the line assembly

3.3. FMEA

FMEA is a method to identify and analyze potential failures and consequently aim to plan production processes well and can avoid production process failures and unwanted losses. Making FMEA begins by determining in advance the effects arising from failure in the process, the causes of failure and controls carried out to prevent the effects of the failure of the process. Giving scores on each component is done by giving a prior assessment of the severity, occurance, detection, and the end result in the form of a risk priority number. Calculation of the RPN value and the recommendations for the solutions given to the causes of the occurrence of the final line nobody can be seen in table 2.

Table 2. Calculation of value of risk priority number (RPN) and recommended solution.

Category causes barriers	*Severity*	Factors affecting	*Occurence*	Current Control Process	*Detection*	RPN	Recommended Solution
Human	7	Old operators carry out body painting inspection processes	7	Supervision is done visually	4	196	The inspection process is carried out using more sophisticated equipment
		Operators have poor skills	7	Supervision is done visually	7	343	Training is carried out for operators
Engine	5	Machine error	8	Engine repairs performed	6	240	Preventive maintenance on the engine
		Conveyors run quickly	9	Supervision is done visually	5	225	Regulating the conveyor rate
Material	5	The little body painting produced	9	Supervision of operator performance	7	315	The painting process is carried out using better technology
Working Method	7	Fewer installation of components in the final section	9	Supervision of operator performance	4	252	Workload balancing is done

Based on the calculation of the RPN above, it was obtained that the main priority of repairs at PT. XYZ is paying more attention to the skills possessed by operators, especially new operators in order to produce good body painting. The recommendation given for this problem is training the operators so that operators have better skills.

4. Conclusions

The conclusions obtained from the discussion above are as follows.

1. Types of production barriers that occur in the assembling process at PT. XYZ is the final line no body, lifter suspension error, pbs for errors, pokayoke error, dl -1 error, problem with part delivery, repair in line / pull andon, problem ordering parts, dl -2 error, slow sub assy process, hoist error , conveyor / hanger, welding problems, rear axle lifter error, power steering filling error machine, sloping body, etc -4 error, and broken knuckle press machine.
2. Based on the obstacles in the points above, the most common type of obstacle is the final line no body, which is as much as 1400 with a percentage of 75.31%.
3. Problems Final line no body is caused by operators having poor skills.
4. Recommendation for a solution to the causes of the obstacle is training the operator so that the operator has better skills.

References

[1] Hu A H, Hsu C, Kuo T, and Wu W,2009 Expert Systems with Application 36
[2] Sharma R K, Kumar D, and Kumar P, 2005 International Journal of Quality & Reliability Management **22**, 9
[3] Sayhputri K, Leviza J, Sari R M, Rizkya I, Siregar I, and Tambunan M M, 2018 MATEC Web of Conferences **197**, 14007
[4] Sulaiman S M, Beer A, Felderer M, and Host M, 2019 Software Qual J **27**

IOP Conf. Series: Materials Science and Engineering **851** (2020) 012007 doi:10.1088/1757-899X/851/1/012007

[5] Fattahi R and Khalilzadeh M, 2018 Safety Science **102**
[6] Braaksma A J J, Klingenberg W, and Veldman J, 2013 International Journal of Production Research **51**, 4
[7] Chen P and Wu M, 2013 Computers and Industrial Engineering **66**
[8] Chu K and Hung L, 2014 Journal of Applied Mathematics
[9] McDermot R E, 2009 The Basic of FMEA. Edisi 2, (CRC Press. USA)
[10] Siregar I, Muchtar M A, Rahmat R F, Andayani U, Nasution T H, and Sari R M 2018 Method of calculation overall equipment effectiveness in fertilizer factory. *IOP Conference Series: Materials Science and Engineering* **308**(1), p. 012053

Decision Tree to Predict the Color Quality of Refined Bleached Deodorized Palm Oil (RBPO)

Meilita Tryana Sembiring [1], Steven Chailes[2] and Sawaluddin[3]

[1,2,3] Teknik Industri, Fakultas Teknik, Universitas Sumatera Utara, Medan, Indonesia

Abstract. There are many factors that can affect the quality of Refined Bleached Deodorized Palm Oil produced. Every step in RBDPO processing must be mastered very precisely in terms of process conditions such as temperature, flowrate, dose, pressure, etc. This condition can vary from one facility to another and therefore requires the company's own specifications / rules to implement and integrate all production process facilities until the product is produced. The focus of the study on this particular task is the quality problems faced by companies in the Refinery section where research is carried out using data mining concepts to predict data logsheets that are written every 24 hours by refinery operators. The data can be used as a tool to predict the color quality of Refined Bleached Deodorized Palm Oil (RBPO). The use of Data Mining in this study is to facilitate operators in managing machine settings to achieve the quality required by the company directly. So that the Refinery section does not need to wait for the results of laboratory tests which require approximately 2 hours to find out the RBDPO quality results.

***Keywords**: Data Mining, Decision Tree, Refined Bleached Deodorized Palm Oil*

1. Introduction

Identifying the occurrence of defects as early as possible in the production chain is very valuable, because it saves costs through the physical determination of production process facilities such as temperature, flowrate, dosage, pressure, etc. to get maximum results. According to Arnu's research, RapidMiner software can be used to get information from quality defects faster. The application of RapidMiner is very important in conducting time series analysis of the sensory data of each PT AAJ machine and predicting defects early in the RBDPO production process.

There are many parameters that affect the quality, such as the quality of CPO given, the condition of the engine process such as temperature, pressure on the Deodorizer machine, etc. A slight deviation from the parameters that affect the quality can cause a change in the color of the RBDPO. Identifying the occurrence of quality defects as early as possible in the production chain is very valuable, because it saves costs through remaining production steps for defective end products.

The focus of the study on this particular task is the quality problems faced by companies in the Refinery section where research is conducted using data mining concepts to predict data logsheets that are written every 24 hours by refinery operators. The data can be used as a tool to predict the color quality of Refined Bleached Deodorized Palm Oil (RBPO). The use of Data Mining in this study is to facilitate operators in managing machine settings to achieve the quality required by the company directly. So that the Refinery section does not need to wait for the results of laboratory tests which require approximately 2 hours to determine the RBDPO quality results.

There are many parameters that affect the quality, such as the quality of CPO given, the condition of the engine process such as temperature, pressure on the Deodorizer machine, etc. A slight deviation from the parameters that affect the quality can cause a change in the color of the RBDPO. Identifying the occurrence of quality defects as early as possible in the production chain is very valuable, because it saves costs through remaining production steps for defective end products.

2. Method

2.1. Criteria of production target

In carrying out the production process, there has 4 criteria as determining the success of achieving the production target. The criteria used are as follows:

1. Quality
   Quality is the overall characteristics and characteristics of a product or service whose ability can satisfy needs, whether expressly or disguised (ISO 8402).
2. Capacity
   Capacity is the optimum level of production capability of a facility usually expressed as the amount of output in a given time period
3. Production Costs
   Production costs are all costs associated with the product (goods) obtained, where there are elements of product costs in the form of raw material costs, direct labor costs, and factory overhead costs
4. Losses
   Losses is the reduction in assets or economic resources of the company that is not due to taking the owner and for the reduction of assets or sources there is no benefit obtained by the company

2.2. Data mining

Data mining is the process of discovering patterns in large data sets involving methods at the intersection of machine learning, statistics, and database systems. Data mining is an interdisciplinary subfield of computer science and statistics with an overall goal to extract information (with intelligent methods) from a data set and transform the information into a comprehensible structure for further use. Aside from the raw analysis step, it also involves database and data management aspects, data pre-processing, model and inference considerations, interestingness metrics, complexity considerations, post-processing of discovered structures, visualization, and online updating.

2.2.1. Data Cleaning. Data cleaning is usually called data cleansing or scrubbing. Data cleaning process is carried out to eliminate information errors in the data [1]. So the data cleaning process can be used to determine inaccurate, incomplete or incorrect data and to improve data quality through detecting errors in the data. There are 4 criteria in calculating the quality of data used [2].

1. Missing: A lot of missing data is divided by lots of variable data
2. ID-Ness: Many different values divided by many variable data
3. Stability: Many values are the same (non-missing) divided by many variable data
4. Validity: Divided by values not included by missing, ID, stable.

2.2.2. Data Pre-processing. The preprocessing phase is carried out as an initial and important stage in research [3]. The method used in this research is data transformation using normalization. Normalization is the process of scaling attribute values from data so that they can lie within a certain range.

$$Z=\frac{X-\mu}{\sigma}$$

2.3. Correlation

Correlation is one of the statistical analyzes used to find the relationship between two quantitative variables [4]. Correlation analysis is a study of the degree of relationship or degree of association between two variables, for example the variable X and the variable Y. The more specific understanding of correlation, which implies a relationship that is substantive numeric (numbers / numbers). From this definition, it also shows that the purpose of correlation analysis is to see / determine how closely the relationship between two variables is.

$$\text{Correlation}= r_{xy}= \frac{\Sigma(x_i-\bar{x})(y_i-\bar{y})}{\sqrt{\Sigma(x_i-\bar{x})^2}.\sqrt{\Sigma(y_i-\bar{y})^2}} \quad (1)$$

2.4. Decision Tree

Decision Tree is a prediction model using tree structure or hierarchical structure. The concept of a decision tree is to change data into a Decision Tree and decision rules. The main benefit of using the Decision Tree is its ability to formulate complex decision-making processes to be simpler, so that decision makers will interpret solutions to problems. Decision Tree is also useful for exploring data, finding hidden relationships between a number of potential input variables and a target variable [5].

Decision Tree combines data exploration and modeling, so it is very good as a first step in the modeling process even when used as the final model of several other techniques. In some applications, the accuracy of a classification or prediction is the only thing that is highlighted in this method.

$$\text{Entropi (s)}= \textstyle\sum_{j=1}^{k} - \text{Pj} \log_2 \text{Pj} \quad (2)$$

$$\text{Gain (S,A)} = \text{Entropi (s)}- \textstyle\sum_{i=1}^{n} - \frac{|si|}{|s|} * Entropi\ (Si) \quad (3)$$

2.5. Root mean square error

Root mean square error is an alternative method for evaluating forecasting techniques used to measure the accuracy of the results of a model forecast [6]. RMSE is the average value of the sum of the squares of error, it can also state the size of the error generated by a forecast model. A low RMSE value indicates that the variation in values produced by a forecast model approaches the variation in the value of obesity.

$$RMSE = \sqrt{\frac{\sum_{i=1}^{n}\left(x_{1,i} - x_{2,i}\right)^2}{n}} \quad (4)$$

Absolute Error is the number of errors in your measurement. This is the difference between the predicted value and the actual value. Mean Absolute Error (MAE) represents the average absolute error between forecasting results and actual values [6]. Other variant of error percentage can be obtained using Mean Absolute Percentage Error (MAPE)[8], which can be considered in the future.

$$\text{MAE} = \frac{1}{N}\sum_{i=1}^{N}|y_i - \hat{y}_i| \quad (5)$$

3. Results and discussion

3.1. Variable in data used

The focus of the study on this particular task is the quality problems faced by companies in the Refinery section where research is carried out using data mining concepts to predict data logsheets that are written every 24 hours by refinery operators. The data can be used as a tool to predict the color quality of Refined Bleached Deodorized Palm Oil (RBPO).

Table 1. Variable used

No.	*Variable*
1	Storage Tank CPO No.
2	Flow MT/H CPO
3	Flow MT/H BPO
4	Storage Tank RBDPO No.
5	Storage Tank PFAD No.

6	Level Tank Phosphoric Acid Min 100 L
etc	.....

The variables used in this study are as follows. Problem solving can be done using logsheet data that has been written every hour with 168 data (45 variables) recorded from July 1, 2019 to July 7, 2019. The written sheet will be moved into Excel software which is the source of input for certain algorithms.

3.2. Data cleaning

Data cleaning involves removing the noise and treatment of missing values. The noise is removed by applying smoothing techniques and the problem of missing values is solved by replacing a missing value with the most commonly occurring values for that attribute.

Table 2. Quality data measurement

Variable	*Missing (%)*	*ID-Ness (%)*	*Stability (%)*	*Valid (%)*
Storage Tank CPO No.	1,800	1,800	41,460	59,940
QTY MT Storage Tank CPO	100,000			
QTY MT Storage Tank RBDPO	100,000			
Storage Tank PFAD No.	1,800	1,200	93,900	3,100
QTY MT Storage Tank PFAD	100,000			
Level Tank Phosphoric Acid Min 100	1,800	38,920	5,490	53,790
etc	...	...	...	...

At this stage there were 14 variables that were discarded because the data were completely absent and the data had high stability.

Table 3. Discarded variable

No.	*Variable*	*Parameter*
1	QTY MT Storage Tank CPO	*Missing* 100%
2	QTY MT Storage Tank RBDPO	*Missing* 100%
3	QTY MT Storage Tank PFAD	*Missing* 100%
4	Setting Stroke Phosphoric Acid mm	*Stability* 100%
5	B/E Dosage %	*Stability* 100%
6	Level HE 721	*Stability* 100%
7	Pressure PU 621	*Stability* 100%
8	Pressure PU 701	*Stability* 100%
etc	...	...

So that data with high missing cannot be processed and data with high stability does not have any impact on the data.

3.3. Data preprocessing

The method used in this research is data transformation using normalization. Normalization is the process of scaling attribute values from data so that they can lie within a certain range.

Table 4. Discarded variable

Storage Tank CPO No.	*Flow MT/H CPO*	*Flow MT/H BPO*	*Storage Tank RBDPO No.*
0,8	-0,9	-0,9	-0,5
0,8	-0,9	-0,9	-0,5
-1,5	-0,9	-0,9	-0,5
-1,5	-0,9	-0,9	-0,5
-1,5	-0,9	-0,9	-0,5
-1,5	-0,9	-0,9	-0,5
-1,5	-0,9	-0,9	-0,5
etc	...	...	...

Data pre-processing may affect the way in which the outcomes of the final data processing can be interpreted. This aspect should be carefully considered when interpretation of the results is a key point.

3.4. Correlation

Correlation is a number between -1 and +1 that measures the degree of relationship between two Attributes (say X and Y) [7]. A positive value for correlation implies a positive relationship.
In this case a large X value tends to be associated with a large Y value and a small X value tends to be associated with a small Y value. A negative value for correlation implies a negative or inverse relationship. In this case a large X value tends to be associated with a small Y value and vice versa.

Table 5. Correlation Table

Variables	*RBDPO Red*	*CPO FFA*	*Flow MT/H BPO*	*Flow MT/H CPO*
RBDPO Red	1			
CPO FFA	-0,20311	1		
Flow MT/H BPO	0,15057	-0,150	1	
Flow MT/H CPO	0,15057	-0,150	1	1
etc.	...	...	...	...

Weight by Correlation calculates attribute weights with respect to label attributes using correlations [9]. The higher the weight of the attribute, the more relevant the variable is. Weight by Correlation can only be applied to ExampleSets with numeric or binominal labels.

Table 6. Weight of correlation

Attribute	*Weight*
Phosphoric Acid Dosage %	0,7593
Temp. HE 311 IN	0,7080
Frequency Hz	0,6553
RBDPO Yellow	0,5479
Temp. HE 721 OUT RPO	0,3896
etc.	....

From the calculation of the attribute weights it can be concluded that the Phosphoric Acid Dosage% has the most relevant influence on RBDPO Red. And the lowest attribute relevant to RBDPO Red is Temp. HE 722 OUT.

3.5. Decision tree

The concept of a decision tree is to change data into a Decision Tree and decision rules. The main benefit of using the Decision Tree is its ability to formulate complex decision-making processes to be simpler, so that decision makers will interpret solutions to problems. Here are the results of a comparison between the results of RBDPO Red Decision Tree predictions with the original data.

Table 7. Prediction Results

No.	RBDPO *Red*	*Prediction* (RBDPO *Red*)
1	2,5	2,5
2	2,5	2,4
3	2,5	2,4
4	2,5	2,5
5	2,4	2,2
Etc.	...	...

Performance Decision Tree :

1. root_mean_squaRed_error: 0.081 +/- 0.025 (micro average: 0.084 +/- 0.000)
2. absolute_error: 0.052 +/- 0.017 (micro average: 0.052 +/- 0.066)
3. correlation: 0.936 +/- 0.038 (micro average: 0.934)

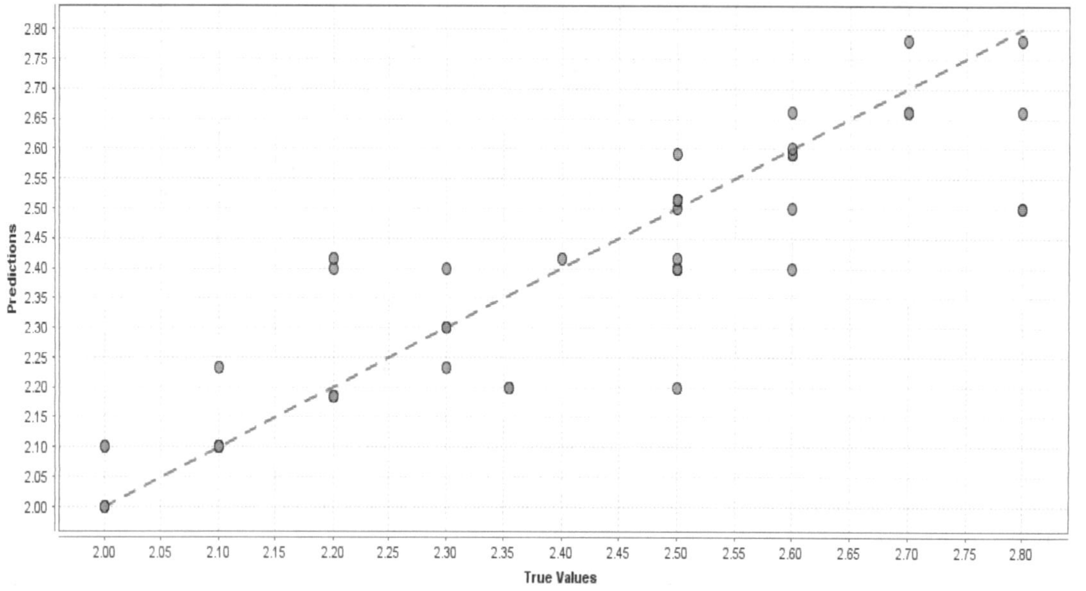

Figure 1. Prediction Chart

4. Conclusions

Data mining is a collection of algorithms that is used by office, governments, and corporations to predict and establish trends with specific purposes in mind. The company has equipped a machine that has sensors measuring pressure, temperature, flow, tank level, stroke settings and so on so that available variations of data recorded periodically every hour in detail by the operator. Continuous processes like these tend to create sensory data which often requires a lot of cleaning and normalization before being able to effectively implement data mining techniques. This is what happens in most heterogeneous industries in their dimensions and physical properties.

In this research, we have presented a summary of data mining predictive analysis and results and focus on the decision tree. In addition, tracking the quality of the RBDPO becomes difficult because the product change is affected by many variables and the resulting color is difficult to examine visually. In this study the researchers also held discussions with the head of the company's laboratory to find out more clearly what factors influence the color quality of the Red RBDPO. Analysis of the problem at PT AAJ was carried out using data mining methods to process the sensory results of each machine that aims to predict the variables that affect the quality of the RBDPO, namely RBDPO Red. So it is likely possible to continuing predicting RBDPO quality result using decision tree.

References

[1] Van Den Broeck J, Cunningham S A, Eeckels R, and Herbst K, 2005 Data cleaning: Detecting, diagnosing, and editing data abnormalities *PLoS Med.*, **2**(10), pp. 0966–0970

[2] O'Donoghue J and Herbert J, 2012 Data management within mHealth environments: Patient sensors, mobile devices, and databases *J. Data Inf. Qual.*, **4**(1)

[3] Chamidah N, and Salamah U, 2016 Pengaruh Normalisasi Data pada Jaringan Syaraf Tiruan Backpropagasi Gradient Descent Adaptive Gain (BPGDAG) untuk Klasifikasi *J. Teknol. Inf. ITSmart* **1**(1), p. 28

[4] Schober P and Schwarte L A, 2018 Correlation coefficients: Appropriate use and interpretation*Anesth. Analg.* **126**(5), pp. 1763–1768

[5] Kakavand S, Mokfi T, and Tarokh M J, 2014 Prediction the Loyal Student Using Decision Tree Algorithms *Int. J. Inf. Commun. Technol. Res* **4**(1), pp. 32–37

[6] Makridakis S *et al.*, 1982 "The Accuracy of Extrapolation (l i m e Series) Methods : Results

of a Forecasting Competition," *J. Forecast.*,**1**, pp. 111–153
[7] Arnu D *et al.* 2017 A Reference Architecture for Quality Improvement in Steel Production," in *Data Science – Analytics and Applications*, Springer Fachmedien Wiesbaden pp. 85–90.
[8] Rahmat R F, Nurmawan, Sembiring S, Syahputra M F and Fadli 2018 Adaptive neuro-fuzzy inference system for forecasting rubber milk production *IOP Conference Series : Materials Science and Engineering* **308**(1), 012014
[9] Waller N G and Jones J A, 2010 Correlation weights in multiple regression *Psychometrika*, **75**(1), pp. 58–69

IOP Conf. Series: Materials Science and Engineering **851** (2020) 012009 doi:10.1088/1757-899X/851/1/012009

Planning of warehouse area for vise manufacturing by considering raw material requirements

Muhammad Haikal Sitepu*[1], Alison McKay[2], Tania Alda[1], Lamsihar Siregar[1] and Ali Akbar Irawan[1]

[1]Industrial Engineering Department, Faculty of Engineering, Universitas Sumatera Utara, Medan, Indonesia, 20155
[2]Institute Design, Robotic and Optimization, School of Mechanical Engineering, University of Leeds, United Kingdom

*Email: mhd_haikalkarana@usu.ac.id

Abstract. Availability of raw materials is important for maintaining the continuity of production. However, in planning inventory for raw materials, several factors such as availability of warehouse space for raw material, handling cost, order cost, transportation time are used in determining amount of raw material need to be stocked. Vise is important tool to support finishing process such as cutting and grinding. Vise consists of several parts such as base, right clamp body and handle. These parts need to be stored in warehouse to support the manufacturing of vise. Hence, the warehouse space must be adequate to store raw materials for manufacturing vise within certain period of time. This paper aims to demonstrate the calculation of warehouse area based on the material requirements that are calculated based on material removal rate for each process.

1. Introduction

Availability of raw materials is important for maintaining the continuity of production. However, in planning inventory for raw materials, several factors such as availability of warehouse space for raw material, handling cost, order cost, transportation time are used in determining amount of raw material need to be stocked [1]. Warehouse space is critical in inventory planning. If the warehouse space too small, it can reduce the quantity of raw materials that can be stored in the warehouse. On contrary, if the warehouse space is too large, although it can store large quantity of raw materials, the handling cost might increase. Hence, the manufacturing needs warehouse with space that can store enough raw material to support the production for certain period of time. For this reason, the calculation of warehouse space needs to consider the material requirements from production [2]. This paper aims to demonstrate the calculation of warehouse space by considering the material requirements from production.

Vise is important tool to support finishing process such as cutting and grinding. Vise consists of several parts such as base, right clamp body and handle. These parts require different production processes and different materials. As the result, to calculate the warehouse space for vise manufacturing, material requirement planning from production need to consider [3]. Planning of raw materials relates to the calculation of the volume of materials, time for material delivery and cost of material [4]. Planning of raw materials is very important and must be aligned with the company's production plans. This paper demonstrates the calculation the volume of raw materials for each process in manufacturing vise. Then this calculation is used in calculating the warehouse space.

Several models are available in the literature such as Economic order quantity and Economic order interval to support the calculation of stock quantity that are influenced by product demand, order cost, and inventory holding cost [5]. However, these models are worked under the assumption, the warehouse has enough space or area to store the raw materials. This paper focuses to calculate warehouse area based on raw material requirements so the warehouse will have enough space to store the raw materials.

2. Methodology

An experiment is used as a research methodology. Experiment is good to use when the objective of research is to investigate the impact different alternatives into system performance. In this research, experiment is conducted to investigate the impact of vise manufacturing process into the warehouse space.

2.1. Research Process

To achieve the aim of the research, this research has been divided into four stages. First stage focuses on observing and collecting material removal rate date for each process in vise manufacturing. Then, the material removal rate data is used to calculate the raw material requirement for each part in second stage. In third stage, the research focuses on converting the raw material requirements from kg to unit of raw material. Then, this is followed by final stage that is focused on calculating the warehouse space based on material requirements.

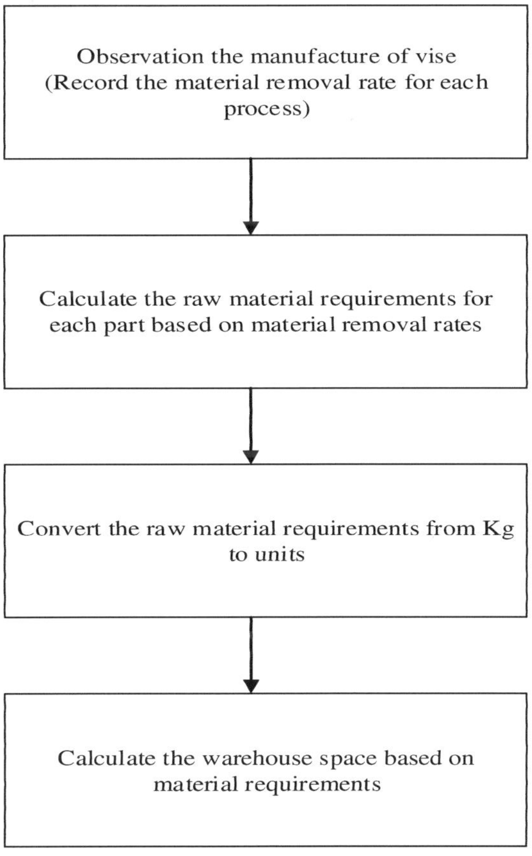

Figure 1. Research Process.

IOP Conf. Series: Materials Science and Engineering **851** (2020) 012009 doi:10.1088/1757-899X/851/1/012009

This research used material removal rate to calculate the raw material required for each process. Material removal rate (MRR) is the amount of material removed per time unit. In roughing operations and the production of large batches, this needs to be maximized. However, in finishing operations, it is a factor to be put on hold, bringing roughness and precision to the forefront. For low roughness, usually low cutting speeds and feed per tooth are both applied, as MRR is usually very low for finishing. In this research the material removal rate data is calculated from the manufacture of vise in laboratory of manufacturing process, Faculty of Engineering, Universitas Sumatera Utara. Material removal rate is calculated by dividing weight of waste material from each process with processing time. Table 1 shows the material removal rate for each process.

Table 1. Material removal rate data

No.	Machines	MRR (gr/s)
1	*Drilling*	0,0625
2	*Milling*	0,2746
3	Grindstone	0,0352
4	*Tap and Dies*	0,0614
5	Lathe	0,033
6	Scrap	0,03

3. Result and Discussion

3.1. Raw Material Requirement Calculation

The calculation of raw material is implemented by multiplying processing/operating time with material removal rate. Backtracking process was implemented to calculate raw material. It means the calculation starts from the last process to beginning process in making part of vise. For example, to make part base, the last process is milling process. Operating time for milling process is 475 seconds. Total material removed during milling process is 0.13 Kg (475 * 0. 2746). Table 2 displays the example calculation of raw material requirements for base part.

Table 2. Calculation of material requirements for Base Part

Processing order	Processing	*Output* (Kg)	*Operation Time* (*Seconds*)	*Input* (Kg)
8	*Milling*	6	475	6.1153
7	*Drilling*	6.1153	300	6.1303
6	*Drilling*	6.1303	300	6.1453
5	Scrap	6.1453	300	6.1525
4	Scrap	6.1525	300	6.1597
3	*Drilling*	6.1597	420	6.1822
2	*Drilling*	6.1822	420	6.2047
1	Grinding	6.2047	660	6.3694

For the next part, the calculation in table 2 can be used in the same way to obtain the amount of input materials required for one part. The following table 3 summarizes the number of inputs required for each part of vise.

IOP Conf. Series: Materials Science and Engineering **851** (2020) 012009 doi:10.1088/1757-899X/851/1/012009

Table 3. Raw Material Requirements for Each Part of Vise

No	Part	Type of materials	Input (kg)
1	Base	Steel ASTM 1	6.3694
2	Clamp Body Left	Steel ASTM 2	1.71932
3	Right Brace Agency	Steel ASTM 2	1.6180
4	The clamp jaw Left	Steel ASTM 4	0.80261
5	Right Clamp jaws	Steel ASTM 4	0.8137
6	Players Board	Steel ASTM 3	2.30832
7	Lahar Agency	Steel ST37	1.4339
8	Handle	Steel ST37	1.0761

3.2. Calculation of Number of Raw Materials

At this stage, the calculation of the number of raw materials is conducted. The calculation is implemented based on the dimension and weight of raw materials available on Indonesian markets. Table 4 displays the specification of raw materials used for producing vise.

Table 4. Specification of vise material

No	Parts	Materials	Dimensions (cm)	Weights (Kg)
1	Base	Steel ASTM 1	24.4x12.2x1.5	50.21
2	Clamp Body Left	Steel ASTM 2	24.4x12.2x2.8	65.43
3	The clamp jaw Left	Steel ASTM 4	24.4x12.2x1.1	65.43
4	Right Clamp jaws	Steel ASTM 4	24.4x12.2x1.1	28.04
5	Right Brace Agency	Steel ASTM 2	24.4x12.2x2.8	28.04
6	Players Board	Steel ASTM 3	24.4x12.2x4.0	93.47
7	Lahar Agency	Steel ST37	d= 3.2. h = 15	2.13
8	Handle	Steel ST37	d= 3.2. h = 15	2.13

The calculation of number of raw material required is done by implementing the following equation 1.

$$\text{Number of raw material} = \text{Total unit produced per week} / (\text{Weight of Material}/ \text{weight of part}) \quad (1)$$

Following is the example of calculation for determining the number of raw material for part base.

$$\text{Number of ASTM 1} = \frac{273 \text{ Unit per week}}{(50.21 \text{ Kg} / 6.369 \text{ Kg})} \quad (2)$$

$$= 39 \text{ Unit}$$

Table 5 displays the calculation of raw material number for each part of vise.

Tabel 5. The calculation of raw material number for each part of vise

Part	Input	Material	Weight	Weeks	Total unit produced Per week	Number of Material
Base	6.3694	Steel ASTM 1	50.21	52	273	39
Clamp Body Left	1.71932	Steel ASTM 2	65.43	52	273	7
Right Brace Agency	1.6180	Steel ASTM 2	65.43	52	273	7
The clamp jaw Left	0.80261	Steel ASTM 4	28.04	52	273	7
Right Clamp jaws	0.8137	Steel ASTM 4	28.04	52	273	7
Players Board	2.30832	Steel ASTM 3	93.47	52	273	6
Lahar Agency	1.4339	Steel ST37	2.13	52	273	273
Handle	1.0761	Steel ST37	2.13	52	273	273

Calculation of warehouse area can be done by multiplying the number of raw materials with the dimension of raw materials. Warehouse area is the total of area required for each part and material handling area. Table 6 shows the calculation of warehouse area for vise manufacturing.

Table 6. Calculation of warehouse area for vise manufacturing

No.	Parts	Type of Material	Number of raw materials	Dimension (cm)	Size (cm^2)
1	Base	ASTM 1	39	24.2x12.2	11,514.36
2	Clamp Body Left	ASTM 2	7	24.2x12.2	2,381.44
3	The clamp jaw Left	ASTM 4	7	24.2x12.2	2,381.44
4	Right Brace Agency	ASTM 2	7	24.2x12.2	2,381.44
5	Right Clamp jaws	ASTM 4	7	24.2x12.2	2,381.44
6	Players Board	ASTM 3	6	24.2x12.2	2,083.76
7	Lahar Agency	ST37	273	3.2 x15	13,104
8	*Handle*	ST37	273	3.2 x15	13,104
9	Iron rack	-	2	200 x 100	40,000
10	Trash can		1	45 x 30	1350
11	Hand Truck Area		1	102x63.5	6477
12	Tool Cart Area		1	138x61	8418
		Total			105,576.88

4. Conclusion

This paper demonstrates the calculation of warehouse space by considering the raw material requirements from production floor. The raw material requirement is calculated for each process required for making the part using material removal rate. However, this research does not consider the level or rack system in storing the raw material. Further research is required to consider uncertainty aspects such as the volatile of demand and the different of delivery times in the calculation of warehouse space. The use of operation research approach might beneficial in optimizing the warehouse space.

References

[1] Lacalle L N L D and Lamikiz A 2011 Modern Machining Technology
[2] Azzam A dkk. Aplikasi Perhitungan Kebutuhan Bahan Baku Produksi Buku Lks dengan Metode Material Requirement Planning (MRP) Studi Kasus pada PT. Harapan Makmur Abadi
[3] Purnomo A 2007 Perencanaan Kebutuhan Bahan Baku dan Penetapan Prioritas Pemasok di PT. Surya Mas Abadi
[4] Supriyanto and Barus T Y 2015 Analisis Perencanaan Kebutuhan Bahan Baku dengan Menggunakan Metode MRP (Material Requirement Planning) dan Metode JIT (Just In Time)
[5] Sitepu M H, Situmorang D P, and Sembiring M T 2019 Inventory Planning at Painting Section in the Indonesian Automotive Industry *IOP Conf. Series: Materials Science and Engineering*

IOP Conf. Series: Materials Science and Engineering **851** (2020) 012010 doi:10.1088/1757-899X/851/1/012010

Material handling selection for vise manufacturing using Hassan algorithm

Muhammad Haikal Sitepu*, Tania Alda, Andri Nasution, Meilita Tryana Sembiring Suci Amalia and Anggi Ridho Habibi

Industrial Engineering Department, Faculty of Engineering, Universitas Sumatera Utara, Medan, Indonesia, 20155

*Email: mhd_haikalkarana@usu.ac.id

Abstract. One type of transportation in industrial companies, for moving raw materials, semi-finished goods or finished goods from their original place to destination can be called as material handling. There are several material handling equipment available for manufacturing industry such as forklift, crane, conveyor, hand truck and trolley. However, not all material handling equipment are appropriate for specific manufacturing such as vise manufacturing. The selection of appropriate and efficient material handling equipment is important for reducing material handling cost. Around 20% of production cost is allocated for material handling activities. Vise consists of several part with different processes and materials. As the result, the flow of materials between processes are high. This paper aims to select appropriate material handling for vise manufacturing. Hassan's algorithm is used as a method for selecting material handling that consider various factors such as price, time and distances. Four types of material handling namely trolley, hand truck, wheeled rack, and hand pallet, are selected in this paper.

1. Introduction

Material handling is a function of moving the right material to the right place, at the right time, in the right amount, sequentially and at the right position or condition to minimize production costs. The aim of material handling is to facilitate transportation and speed up the production process. Based on the formulation by the American Material handling Society (AMHS), the notion of material handling is stated as an art and science that includes handling, moving, packaging or packaging, storing as well as controlling or controlling (controlling) controlling) from materials or materials in all its forms [1]. There is several material handling equipment available for manufacturing such as crane, forklift, conveyer and hand truck. However, not all material handling equipment is appropriate for specific manufacture such as vise manufacturing. The use of inappropriate material handling might increase material handling cost and production time. Hence, the selection of appropriate material handling equipment is necessary [2]. This paper aims to choose the suitable material handling for vise manufacturing.

The Hassan algorithm is a way of selecting material handling by taking into account various factors such as the price of the tool, operating costs, operator costs, tool speed, tool dimensions, tool capacity, hauling distance, and transportation time. Then all types of data are calculated using Hassan algorithm calculations so that the most suitable type of material handling can be selected based on the results of the Hassan algorithm iteration.

At The Plant Layout and Material Transfer Laboratory in the Department of Industrial Engineering, Universitas Sumatera Utara, material handling selection was carried out as one of the practicum activity where the students calculated the appropriate material handling selection based on the design of the production floor [3]. The product unit used is vise, so the design of material handling selection in this study focuses on the selection of material handling for vise manufacturing where there are 9 stations on the production floor and 7 work stations, and 25 displacement in the production process. Four types of material handling are selected for the transportation of material from one station to the next in the design of this vise factory, namely trolley, hand truck, wheeled rack, and hand pallet, where each material handling has a price, cost, dimensions, speed, and capacity differ based on price and speed references obtained from data collection [4].

2. Methodology

An experiment is used as a research methodology. Experiment is good to use when the research focuses to investigate the impact of alternatives into system performance. In this research, the impacts of different material handling equipment on cost of material handling are observed. The inappropriate material handling selection have occurred for several times in manufacturing becomes a reason to conduct this experiment.

2.1. Research Process

To achieve the aim of the research, this research has been divided into three stages. Stages 1 focuses on collecting data and information required to calculate the Hassan Algorithm such as the references for the calculation of the price, cost, dimension, velocity, and capacity of each material handling. In stage 2, the focus of the research is to define the distance of each work station using aisle distance methodology, and then calculate the material handling selection using Hassan algorithm. Then, this is followed by stage 3 that is focused on choosing the selected material handling for each workstation in the production floor. Figure 1 displays the research process used in this research.

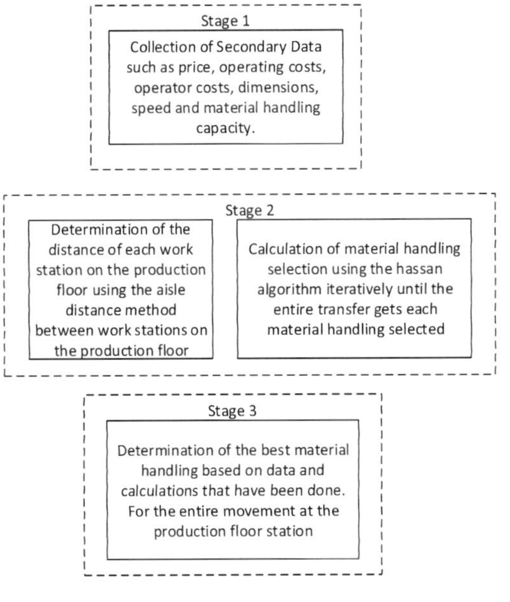

Figure 1. Research process

2.2. *Algoritma Hassan*

This section focuses to introduce the equations for each calculation model in this paper. Each calculation model has different formulation due to different factors that are considered in developing the algorithm. Table 1 shows those equations for each algorithm model [5].

Table 1. Equations for Algorithm Hassan

Algorithm Models	Equations	Number of Equations
For each equipment type, calculate the number of units that would be needed if the equipment performs all the moves	$Y_i = \sum_j \frac{h_{ij}}{H_i}$	(1)
	$\lambda_i = Y_i$	(2)
	$\lambda_i = [Y_i] + 1$	(3)
Calculate the total cost of material handling for each equipment type as	$Z_i = \lambda_i K_i + \sum_{j \in E_i} W_{ij}$	(4)
Calculate the average cost for each equipment type per move as	$\bar{Z}_i = \frac{Z_i}{q_i}$	(5)

First, select the equipment with the smallest Zi and resolve ties by selecting the equipment with the smallest Zi. If ties persist, resolve them by selecting in order of ascending [6]. For the selected equipment type, arrange the moves that can be performed by it in increasing order of operating cost. Assign the moves to the selected equipment starting with the move having the smallest operating cost. After each assignment, check whether the sum of *hij* is equal to *Hi* or within a tolerance **E***i* of it. If the sum of *hij* is equal to *Hi*, go to the next step; otherwise, check either of the following two cases: If the moves are the only remaining moves or cannot be assigned to another piece of equipment, leave the assignment as it is. If the sum of *hij* is greater than *Hi* (or a multiple of *Hi* depending on the number of units required of the equipment so far), check the difference between the least integer multiple of *H* (making it greater than the sum of *hi*) and the sum of *hij*. If the difference, which represents idle time, is less than or equal to *E*2 (a specified acceptable idle time), leave the assignment as it is. If the difference is larger than *E*2, remove moves from the equipment starting with the last assigned move, until the acceptable utilization level is achieved.

3. Result and discussion

This section presents results from the implementation of the hassan algorithm in the vise plant, the results of displacement totaling 25 displacements, 7 iterations and from 4 types of material handling selected and calculated, and 2 types of material handling for the production floor design at the vise factory namely trolley and wheeled shelves.

The following is the selection of material handling using the hassan algorithm:

Salary / annual salary	= IDR. 31.2 million
Monthly salary / wage	= IDR. 31,200,000 / 12 months = IDR. 2,600,000
Salary / wage per day	= IDR. 2,600,000 / 26 days = IDR. 100,000
Hourly salary / wage	= IDR. 100,000 / 7 hours = IDR. 14,286

Table 2 below shows the specification of material handling equipment used in this research.

Table 2. Data Handling Material Used

No	*Material Handling*	Price (IDR)	Dimension (mm)	Speed (m/minute)
1	Trolley	312.000	110 cm x 63 cm	100
2	Hand Truck	305.000	65 cm x 45 cm	98
3	Wheeled Rack	219.000	62 cm x 36 cm	66
4	Hand Pallet	2.860.000	115 cm x 52 cm	58

The following is the result of calculating displacement distance using the aisle distance method from another station in the vise factory based on the production process carried out.

Table 3. Distance between Stations

Displace-ment	Distance (m)	Displace-ment	distance (m)	Displace-ment	distance (m)	Displace-ment	distance (m)
G-D	18	T-G	18	G-S	18	B-D	27
D-S	18	G-D	18	S-D	18	D-G	18
S-D	18	D-S	18	D-G	18	B-G	18
D-M	27	S-G	18	S-D	18	G-A	27
M-A	18	S-D	18	D-M	27		
S-D	18	D-T	27	M-T	18		
D-T	27	T-G	18	T-G	18		

The following is an example of the calculation of the material handling trolley operating costs on the G-D Displacement part.

$$C = \frac{\text{Tool Dimensions}}{\text{Material Dimensions}} = \frac{110 \times 63}{24{,}4 \times 12{,}2} = 23.280 \approx 24 \qquad (1)$$

$$f = \frac{\text{Total Material}}{C} = \frac{4 \text{ unit/hour}}{24} = 0.174 \approx 1 \qquad (2)$$

$$\text{Operationg costs/minute} = \frac{\text{Cost}}{d} = \frac{14286}{66} = \text{IDR } 216{,}455 \qquad (3)$$

Operational costs = r x f x operation cost/m = 18 x 1 x 216.455 = 3896.180 ≈ IDR 3,896 (4)

Where:

C	= Carrying capacity (unit)
Tool dimension	= The size of the conveyance
Material Dimension	= The size of the unit to be moved
f	= transfer frequency
Amount of Material	= Number of units moved
C	= Carrying capacity
Operating costs / m	= Transport costs / meters (IDR / m)
Cost	= Operator Cost
d	= Speed of conveyance
r	= displacement distance
f	= displacement frequency
m	= transport cost / meter

The following is an example of calculating the operating time of a material handling hand truck on the G-D Displacement part.

$$\text{Operation time} = \frac{\text{Distance}}{\text{Speed}} = \frac{18}{100} = 0{,}180 \quad (5)$$

The following calculation of operating costs and operating time of each move can be seen in Table 2.13.

Table4. Calculation of Operating Costs and Operating Times

No	Displacement	Part	Distance (m)	Trolley		Hand Truck		Wheeled Rack		Hand Pallet	
				OperatingCost (Cij)	**Operation Time (Wij)**	**Operating Cost (Cij)**	**Operation Time (Cij)**	**Operating Cost (Wij)**	**Operation Time (Wij)**	**Operating Cost (Cij)**	**Operation Time (Wij)**
1	G-D	Base	18	3896	0.180	2624	0.184	3896	0.273	4434	0.310
2	D-S	Base	18	3896	0.180	2624	0.184	3896	0.273	4434	0.310
3	S-D	Base	18	3896	0.180	2624	0.184	3896	0.273	4434	0.310
4	D-M	Base	27	5844	0.270	3936	0.276	5844	0.409	6650	0.466
5	M-A	Base	18	3896	0.180	2624	0.184	3896	0.273	4434	0.310
6	S-D	Left clamping body	18	3896	0.180	2624	0.184	3896	0.273	4434	0.310
7	D-T	Left clamping body	27	5844	0.270	3936	0.276	5844	0.409	6650	0.466
24	B-G	Handle	18	3896	0.180	2624	0.184	3896	0.273	4434	0.310
25	G-A	Handle	27	5844	0.270	3936	0.276	5844	0.409	6650	0.466
		Total		**109093**	**5.040**	**73471**	**5.143**	**109093**	**7.636**	**124140**	**8.690**

An example of calculating the total cost of a trolley is as follows.

Initial Cost = Rounding Operating Time x material handling costs (6)
= 6 x IDR 312,000
= IDR 1,872,000

Total Cost = Initial Cost + Total Operating Cost (7)
= IDR 1,872,000 + IDR 109,093
= IDR 1,981,093

An example of Bi calculation is as follows.

$$Bi = \frac{\text{Total Cost}}{\text{Total of Displacement}} = \frac{1981093}{25} = 79244 \quad (8)$$

The following table defines Bi Iteration 1.

Table 5. Determination of Bi Iteration 1

Material Handling Type	Total of Displacement	Total Equipment	Rounding	Initial Cost	Total of Operating Costs	Total Cost	Bi
Trolley	25	5.040	6	1872000	109093	1981093	79244
Hand Truck	25	5.143	6	1830000	73471	1903471	76139
Wheeled Rack	25	7.636	8	1752000	109093	1861093	74444

Hand Pallet	25	8.690	9	25740000	124140	25864140	1034566

The following is a table of displacement and parts that experience displacement with the lowest operating costs.

Table 6. Displacement and Parts that Have Displacement with the Smallest Operating Costs

MH Type	Part	Displacement	Operating Costs (Cij)	Operation Time (Wij)	Total of Operation Time (Wij)
Wheeled Rack	Base	G-D	3896	0.273	0.273
Wheeled Rack	Base	D-S	3896	0.273	0.545
Wheeled Rack	Base	S-D	3896	0.273	0.818

So for the above displacement, from stations G to D, D to S and S to D for the base parts, material handling wheel racks are used. Then all of the above calculations are carried out to obtain 7 iterations and each material handling selected. The following is a recapitulation table for each material handling selected for each material handling selected.

Table 7. The Recapitulation of Material Handling Selection

No	Displacement	*Part*	*Material Handling Equipment*
1	G-D	Base	Wheeled Rack
2	D-S	Base	Wheeled Rack
3	S-D	Base	Wheeled Rack
4	D-M	Base	Wheeled Rack
5	M-A	Base	Hand Truck
6	S-D	Left clamping body	Hand Truck
7	D-T	Left clamping body	Wheeled Rack
8	T-G	Left clamping body	Hand Truck
9	G-D	Left jaw clamp	Hand Truck
10	D-S	Left jaw clamp	Hand Truck
11	S-G	Left jaw clamp	Hand Truck
12	S-D	Right clamping body	Hand Truck
13	D-T	Right clamping body	Wheeled Rack
14	T-G	Right clamping body	Hand Truck
15	G-S	Right jaw clamp	Hand Truck
16	S-D	Right jaw clamp	Hand Truck
17	D-G	Right jaw clamp	Hand Truck
18	S-D	Rotation body	Hand Truck
19	D-M	Rotation body	Wheeled Rack
20	M-T	Rotation body	Hand Truck
21	T-G	Rotation body	Hand Truck
22	B-D	Crater body	Wheeled Rack
23	D-G	Crater body	Hand Truck
24	B-G	Handle	Wheeled Rack
25	G-A	Handle	Wheeled Rack

IOP Conf. Series: Materials Science and Engineering **851** (2020) 012010 doi:10.1088/1757-899X/851/1/012010

4. Conclusion

Based on the results, some conclusions are obtained as follows

- Selection of the right material handling can reduce unnecessary waste on the production floor.
- In the calculation of material handling using the hassan data algorithm needed, among others, price, operating costs, operator costs, dimensions, speed and material handling capacity.
- There are 25 displacement distances for the entire vise production process, the distance is obtained based on the distance calculation method using the aisle distance method on the production floor.
- There are 7 iterations to get all material handling selected for each work station on the production floor.
- Of the four types of material handling determined, material handling is chosen, namely the wheel rack and handtruck based on the results of calculations and iterations using the hassan algorithm.

References

[1] Sitecraft 2009 Material Handling Equipment Product Guide

[2] Toyota. Material Handling Hand Pallet Truck BT Pro Litfer L-Series LHT100

[3] Adisuwiryo S, dkk. Perbaikan Tata Letak Dan Material Handling Lantai Produksi PT. Hamson Indonesia Untuk Mencapai Target Produksi Dengan Menggunakan Metode Pairwise Exchange Dan Simulasi *Jurnal Teknik Industri* ISSN: 1411-6340

[4] Sukania I, dkk. 2016 Usulan Perbaikan Tata Letak Pabrik Dan Material Handling Pada PT. XYZ. *Jurnal Ilmiah Teknik Industri* Vol. 4 No. 3, 141 – 148

[5] Hassan M M D, Hogg G L, and Smith D R 1985 A Construction Algorithm for The Selection and Assignment of Materials Handling Equipment *International Journal of Production Research* 23 (2) pp 381-392

[6] Bouh M A and Riopel D 2015 Material Handling Equipment Selection: New Classifications of Equipments and Attributes

Design and analysis of the LDR-controlled device

Suherman Suherman*, Fernando Purba, Riswan Dinzi, Rahmad Fauzi

Electrical Engineering Department, Universitas Sumatera Utara, Medan-Indonesia

*suherman@usu.ac.id

Abstract. Energy efficiency is important for sustainability. However, efficiency cannot rely on human being as human is forgetful. Many energy applications live on only at night. This paper extracted design and analysis of LDR-controlled device based on MXD-122 existing product. Circuits design contains power supply, sensors, comparator and switching device. As analysis performed, it is revealed that system works only for half wave output. The output current is limited by the maximum current of I_{D33}. Many resistors cause dropped voltage, which means system efficiency is very low. In order to enhance system for controlling higher current device, it should be connected to contactor.

1. Introduction

Energy crisis is still an issue as increasing demand is not linear to the energy availability. This also happen in Indonesia even dough this country has unlimited natural resources. According to major energy supplier, there are still more than 3660 villages unreachable by the electricity [1]. The nation energy body confirms the statement by denoting that the economy development rates about 6.2 percent per year with population increment about 1.49 percent per year requires energy development at least 6.2%. Without significant steps taken by the government and public, Indonesia may import the energy in 2027 [2].

Some reasons that triggering energy crisis indeed exist in this country. Customers hold the important role as many energy usages are inefficiently practiced. Even dough the basic electricity price increased from Rp1,392 to Rp1,472.72 per kWh in 2006, and the subsidized price has been eliminated so that users should pay the bill fully, users are still unaware on using electricity efficiently.

On the other hand, energy efficiency cannot count on human awareness as human always forgetful. The efficiency should employ automation such as by using sensors or camera.The previous work examined the use of human presence [3] in order to control electricity usage. Other research also uses passive infrared for detecting human presence and controlling electricity based on the sensor response [4-6]. This paper concentrates on using light depending resistor (LDR) in controlling electrical devices. At this stage, paper observed the light bulb as the controlled device.

Research will explore the design and analysis the existing LDR-based circuit with the market name Muxin MXD-122, then modify the circuit for large current control. However, this paper covers only design and analysis of the observed circuit. This paper will be organized as follows, initial circuit identification is performed at first place, followed by mapping the circuit into blocks and then analyzing each block followed by experimental measurement. Some further steps for controlling larger current will be advised.

IOP Conf. Series: Materials Science and Engineering **851** (2020) 012011 doi:10.1088/1757-899X/851/1/012011

2. Circuit Design and Component Identification

Figure 1 shows the circuit design and component identification within the MXD-122 LDR based device controller driven by 220 Volt. Circuits contains of a power supply circuits, light sensor, comparator and switching circuit. The whole system employs electronic components including LDR, resistor, capacitor, diode, mosfet and operational amplifier.

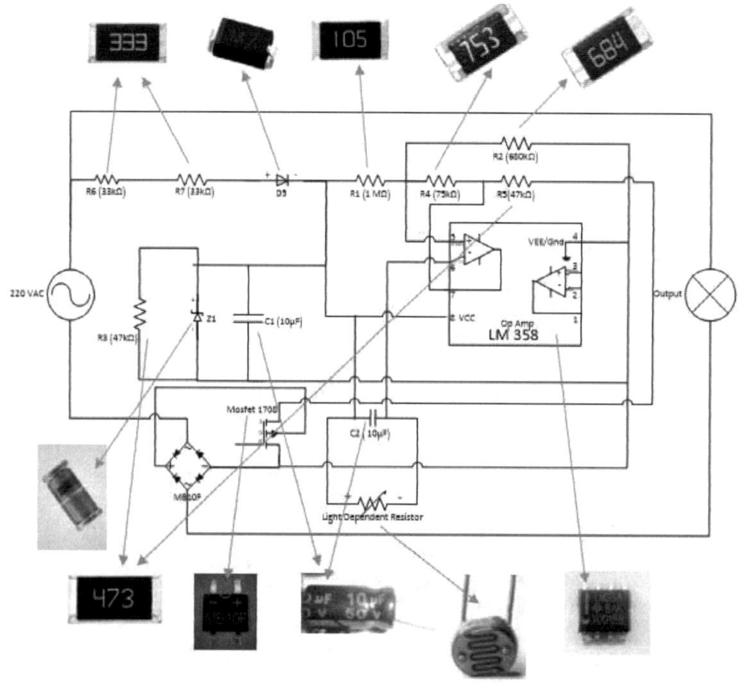

Figure 1. Circuit and component identification

Figure 2 shows the half wave power supply with zener diode as the regulator. Power supply generates sufficient DC voltages to supply operational amplifier and activate the mosfet as switching circuit. Figure 3a shows LDR and comparator circuits. LDR senses the light and adjusts its resistance accordingly. Resistance variation causes input voltage in comparator flipped, as results, operational amplifier produces opposite output voltage. This output triggers switching circuit which is performed by mosfet to either flow or stop current activating the controlled circuit (Figure 3b).

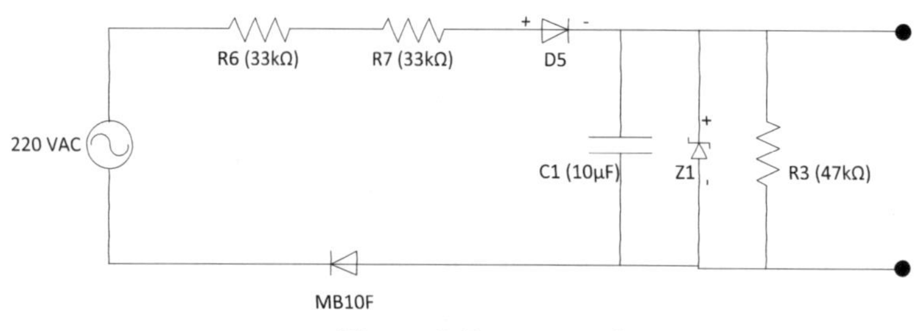

Figure 2. Power supply

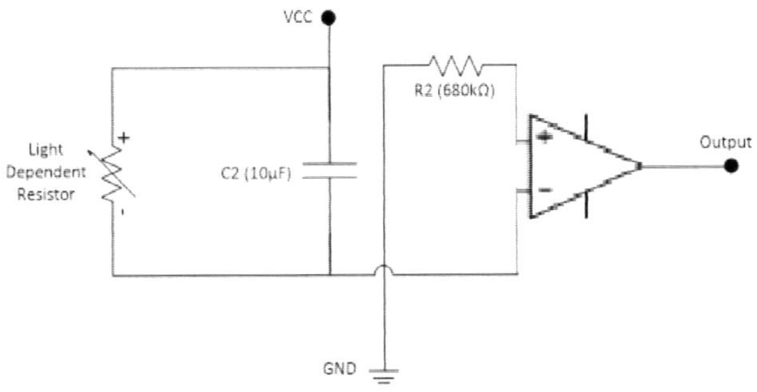

(a) LDR and comparator

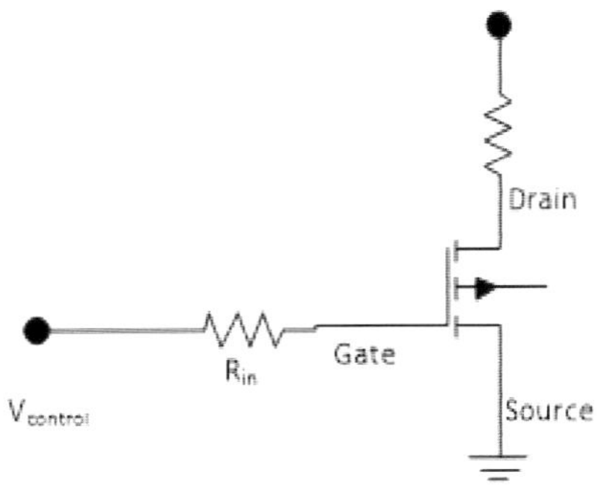

(b) Mosfet switch

Figure 3. LDR, comparator and switch

3. Circuit Evaluation and Analysis

In order to compare circuit analysis and circuit measurement, evaluation circuit is built as in Figure 4. Circuit contains lightbulb as the load, variable resistor to trim the LDR resistance values and resistors to increase load resistances.

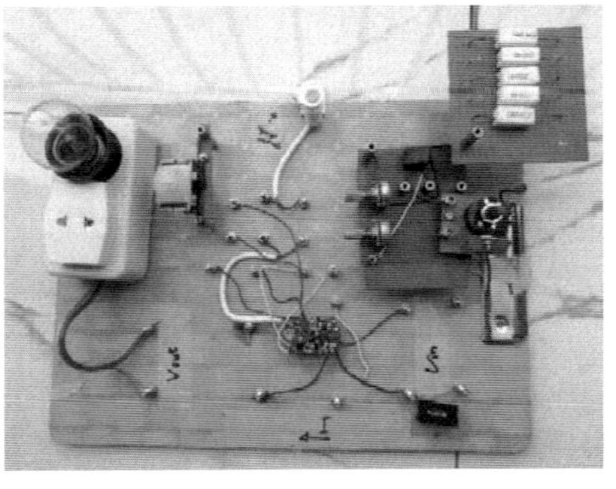

Figure 4. LDR, comparator and switch

For power supply circuit, the voltage regulator, Zener diode has 4.5 volt. If it is assumed that those diodes D_5, D and Z_1 has zero internal resistance, then the current flows through $R_6$and $R_7$is:

$$-E_1 + I(R_6 + R_7) + V_{D5} + V_z + V_D = 0$$

The measured root mean square (rms) voltage source is 216 volts, which means the peak voltage isVp=Vrms/0.707 =216/0.707=305.5 volt. Half wave rectifier generates DC voltage V_{DC} = 0.318 x Vp which gives 97.15 volt. So, the current flow can be determined by:

$$-E1 + I(R6 + R7) + 0.7 + 4.5 + 0.7 = 0$$
$$I = 1.4 \text{ mA}$$

Based on LDR datasheet, the higher light intensity the smaller the resistance. Measurement shows that when resistance higher than 81.80 kΩ, current flows through the controlled device. If resistance is lower than 16.62 kΩ, the current is off. Meanwhile, OpAmp produce lowVout if voltage on pin 6 is higher than voltage in pin 5 and so the opposite. Pin 6 input comes from sensor where:

$$V_{pin\ 6} = V_{CC} - V_{sensor}$$

Measurement shows that $V_{pin\ 6}$=3.43 voltsand V_{pin5} when R_{sensor} = 20kΩ (daylight), so that OpAmp input current is:

$$I_{input6} = \frac{Vcc - Vin6}{Rsensor} = \frac{4.53 - 3.43}{20k\Omega} = 0.0535mA$$

The output voltage is 0.02 volts. Vout is 3.59 volt when sensor resistance 350 kΩ (dark). Output voltage is the control voltage for mosfet switch. When $V_{control}$ is 3.59 volts, current flow through R_5 causes gate voltage drops to $Vg = Vout - I.R_5$. Measurement gets Vg = 0.74 volt, so the current is:

$$I_{RS} = \frac{3.59 - 0.74}{47k\Omega} = 0.06\ mA$$

This current causes bigger current from drain to source activate the controlled device. In Figure 5, current turns to A (not B, ground) as A is more negative than B, but cannot flow in the opposite direction.

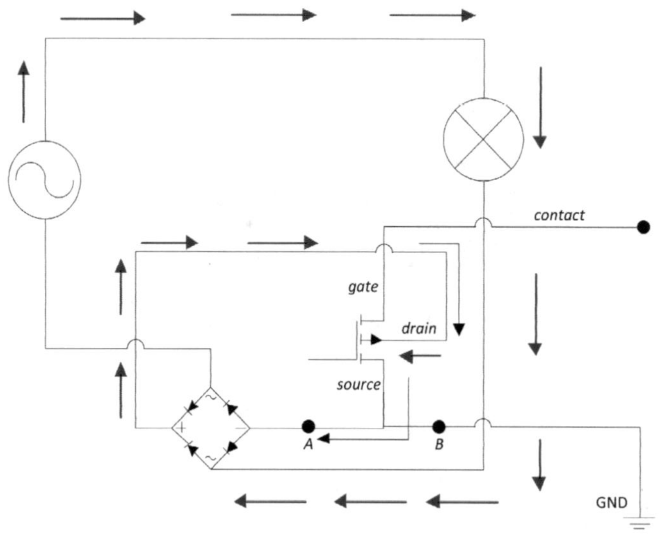

Figure 5. Control current flow

4. Conclusions

This analysis shows that, the control current is half wave, current flows in one direction as high as the mosfet I_{DS}. The maximum controlled device is $0.5\ V_{AC}*I_{DSmax.}$ In order to control larger current, this circuit requires a contactor. This circuit has high loss. For instance, voltage drops at R_6 and R_7.

Acknowledgement

This research has been supported by DRPM DIKTI under the Penelitian Terapan Unggulan Perguruan Tinggi research grant 2019.

References

[1] Nefril. 2019. Puluhan Desa Belum Teraliri Listrik di Kawasan Lumbung Energi". 2019. Liputan 6
[2] Setiyo B. 2017. "BP Global: Konsumsi Indonesia Meningkat 5,9 Persen di 2016". Technology Indonesia, September 14.

[3] Ahmed N. U., Taha Z. B., Khairuddin I. B. M., Rabbi M. F., Rahaman S. A. M. M., and Sundaraj K., 2016 Fuzzy logic controller design for intelligent air-conditioning system *2016 2nd International Conference on Control Science and Systems Engineering (ICCSSE)*, no. October. pp. 232–236,.
[4] Zeebaree H. M. Y., S. R., 2014 Arduino Based Remote Controlling for Home: Power Saving, Security and Protection, *Int. J. Sci. Eng. Res*., **5**(8), pp. 266–272
[5] Harikrishnan P., Sivagami R. 2017. Intelligent power saving system using PIR sensors, in International conference of Electronics, *Communication and Aerospace Technology*, **2**, pp. 573–577.
[6] Bakhtiar B., Suherman S, 2015. Realisasi Sistem Switch Lampu Penerangan Ruangan Otomatis Untuk Meningkatkan Efisiensi Energi Listrik, *J. Arus Elektro Indonesi*a., **1**(2), 2015.

The Development of Budgeting Module on Android based on Backpacker BackInd Applications: Sorting and Filtering Features

Yoel F. Tampubolon, Rachmadita Andreswari and *Muharman Lubis

Telkom University, Jalan Telekomunikasi No. 1, Bandung, 40257, Indonesia
*Email: muharmanlubis@telkomuniversity.ac.id

Abstract. Decision making problems often occur in tourists to determine the activity within the tourist destinations or location. For tourists, information about attractions also is needed to allow their planning to be precise and meaningful. The BackInd (Backpacker Management System) have been developed as an application to help and support tourists by providing comprehensive information about tourist attractions through filtering and sorting mechanism. In addition to provide those type of information, tourists also allowed to buy tickets directly without much more burden to wait physically in that attractions counter. The development of this application is expected to help tourists get recommendations based on their preferences that are in accordance with the budget of the travellers. Thus, this study uses the Rapid Application Development (RAD) method, which the results focused on the featured added to the application in term of sorting, filtering and budgeting features so the homestay that tourist want will match accordingly with their plan.

1. Introduction

In the current era, traveling activities have become one of the most prominent and sought leisure activities by the public eyes from children to even the elderly. Actually, doubled with the ever increasing development and especially in very rapid technological developments, traveling activities become intense and meaningful with more stylish and diverse for example in finding information about desired tourist destinations that related to their spiritual enhancement or look for soul-mate. According to the current developments, there are many emergence of start-ups that are engaged in travelling activities, which one of them is BackInd as Startup Company that still in the incubation phase. It has objective to to facilitate tourists in choosing tourist attractions precisely and accordingly to reduce headache and of course the burden expense that at certain occasion exceed the plan. The application are called Backpacker System Management (BackInd), where this system have two kind of platform; web-based application used by business managers and android-based applications used by tourists. The website service have given its user the full right to do several primary activities in the menus such as create, read, update and delete (CRUD) processes for the products sold. Meanwhile, android services allow the tourists to search attractions and lodging as desired, as well as do comparison and survey regarding those are being visited. Then, the tourist can determine their choice with they find more appropriate and suitable by purchasing the tickets directly within the system.

Based on the questionnaire delivered to potential customer in 2017, the researcher obtained interesting result regarding the main problem faced based on 45 correspondents, which can be concluded that the problems experienced by tourists are a matter of budgeting or the price is in line with capital pocket. In short, tourist find difficulty regarding the proper prices according to their budget owned. The need for this feature is also supported by the adjustment of the current user characteristics, especially in e-commerce and the traveling field, for example Traveloka, Tiket.com, booking.com, etc., which all of have budgeting features in it as most of the site realize the importance of this features to satisfy the user. Meanwhile, the tourist also have another problem regarding the difficulty of finding study or packet tours that are suitable for their budget especially to compare each of them in the detail. Thus, to overcome this problem, the development of the budgeting module with filtering and sorting features is a solution especially by adding mobile-based features to increase mobility. The system was built using the Java programming language as the basis for making Android applications. Furthermore, Android Studio as a tool for creating applications wherein the development expected to advance the prediction and forecasting of the tourists so they can anticipate and prepare reliable and confidence budget before going to travel. This study uses the Rapid Application Development (RAD) method because application development using this method is fairly short and very helpful in improving customer needs for application development.

2. Theoretical Basis

Mobile application development has three development approaches, namely native applications, web-base applications and hybrid applications. The application development approach chosen in this study is native applications. Native Applications are developed in programming languages for certain OS, using certain IDEs based on the selected OS. With this approach, applications can access and take advantage of all the capabilities of mobile devices that can be accessed for development. The developed application provides the best use, the best features, and the best overall usage experience. This native applications development approach uses the Java programming language and supports mobile application platforms [3][4]. Meanwhile, android is a Linux-based mobile device operating system that includes operating systems, middleware and applications. Android is an operating system application for cellphones based on Linux (Nazruddin, 2012). Android provides an open platform for developers to create their own applications for use by various mobile devices [5]. Here's the Android version from the first to the present, which are Alpha (1.0), Beta (1.1), Cupcake (1.5), Donut (1.6), Eclair (2.0 - 2.1), Froyo (2.2 - 2.2.3), Gingerbread (2.3 - 2.3.7), Honeycomb (3.0 - 3.2.6), Ice Cream Sandwich (4.0 - 4.0.4), Jelly Bean (4.1 - 4.3.1), KitKat (4.4 - 4.4.4), Lollipop (5.0), Marshmallow (6.0), Nougat (7.0) and Oreos (8.0 - 8.1).

Analysis of capital budgeting calculations is useful for analysing a plan and decisions in using funds. Regarding capital budgeting, it is emphasized strongly towards a decision-making process used in the acquisition of fixed assets. Basically, capital budgeting is the entire investment planning process for fixed assets that will provide the level of income expected by the company for a period of more than one year. Capital budgeting analysis describes the benefits of a fixed asset investment plan to be carried out by the company, so that it can be used as a basis for companies in making decisions to invest in fixed assets [2]. Rapid Application Development (RAD) is a method of developing information systems with a short time. RAD uses an iterative method (repetitive) in developing a system where the working model (model works) system is constructed at the beginning of the development stage with the aim of determining the requirements (requirements) of users and subsequently removed [6]. There are three (3) stages in this method, namely requirement planning, where at this stage the user and analyst meet to identify the objectives of the system and information needs to achieve the objectives, and this stage is the most important thing because of the involvement of both parties. Then, system design process, where at this stage the activity of the user determines the achievement of goals because this process is a design process and can make improvements if there is a design mismatch between the user and analyst. After this stage was conducted comprehensively, the output can be in the formed of the software specification. At last, the implementation where the programmer who develops the design of a program that has been approved by the user and analyst. Before being applied to an organization, the testing process of the program is first carried out whether there is an error or not. At this stage the user usually gives a response to the system that has been made and gets approval about the system.

IOP Conf. Series: Materials Science and Engineering **851** (2020) 012012 doi:10.1088/1757-899X/851/1/012012

Functional testing is a test carried out on the proposed system seen as a "black box", because this test is based on the specifications of the requirements and does not need to update the code system. This test uses applications that have been completed [5]. Therefore the experiment in this study is more aimed at each functional system that is built that can be done correctly or not. Unit testing is a code-based test carried out by the developer, the purpose of this test is to test individually and each unit separately. This unit testing can be done for small code units or generally not larger than the class [7]. Therefore testing in this study is more directed at whether each code in the function features developed can run well or not. Integration testing is validating that two and more units or other integrated department to work together correctly and tend to focus on interfaces specified in low-level designs [8]. The test in this study was used to validate the output data from each feature function developed between the android application and the use of an API (application program interface).

3. Results and Discussions

Therefore, one type of the trips that are common carried out by tourists is backpacking, which is to travel independently and cheaper than usual with matters more regarding the cost or expense that they will afford from beginning until the end in the specific location. This character type of user become the main persona of this application. Based on the system development process used is the rapid application development method, as follows, an explanation of the results of all phase based iterations:

A. Plan Requirements

The result of this stage is to get a user requirement, namely the need for software development obtained from the results of observations from existing applications. Based on observations, there are no features that accommodate relationships with customers. Therefore, the primary attribute that developer should consider regarding the preparation of plan requirements that it should be accessible, clear, specific, precise and accurate. The requirements management plan is used to document the information necessary to effectively manage project requirements from identification, monitoring to delivery. The responsible authorities have discretion to decide what information should be included within their local list, of course the collaboration with other tourism attraction should be done accordingly and contractual before the list of budget have been calculated. The exact and compact strategy is extremely important to support the implementation of features related to the requrement analysis, which it must ensure that the information requirements are clear, justified and proportionate. On the other hand, every local validation requirements should help understand on how the proposed development is implemented. Undermine the policy of the state or of the adopted development plan where possible, local validation requirements are identify non-informational size thresholds, if necessary, take a phased approach required information such as scale or proposed sensitivity. This process have been conducted in the system internally without customer realization to avoid burden and confusion. Therefore, clarify what information is needed to meet the requirements and foçus on promoting a proportional approach and simplicity. Meanwhile, you need to provide assistance for each element of the requirement. Of course, if you get more information or answers to queries, you need to be clear. Therefore the application requires some additional features. Additional feature specifications can be seen in table 1.

Table 1 User Requirement

No.	Details of System Requirements	Description	Actor
Budgeting Management			
1.	Budgeting	The process of doing budgeting by inputting data on a mobile application	Customer
2.	See budgeting results	See the results of budgeting	Customer
3.	Update budgeting	The process of updating by entering data back on the mobile application.	Customer
Filtering Management			
4.	Filtering	The process of doing budgeting by inputting data on a mobile application	Customer
5.	See filtering results	See filtering results	Customer
6.	Update filtering	Update filtering process by inputting data back.	Customer

Sorting Management			
7.	Sorting	The process of sorting by selecting the sorting criteria on the mobile application.	Customer
8.	See the results of sorting	The process of displaying a list of data that has been sorted	Customer
9.	Update sorting	Update sorting process by re-selecting the sorting criteria on the mobile application.	Customer

B. Design Process

To do a user-design analysis, it is necessary to make a diagram that describes the actors involved and what access rights can be performed by the actor based on the results of the functional requirements analysis that has been done before. The diagram created is use case diagram, which is created to have one actor involved, namely the customer. Customers have four activity rights, namely, doing budgeting on tourist attractions (tourism), filtering tourist attractions (tourism) and lodging (homestay), doing sorting on tourist attractions (tourism) and lodging (homestay), and the last is see the list of products whether it is a list of tourist attractions (tourism) or a list of homestays. Industry 4.0 refers to increasing competitive pressures, reduced margins, the availability of new technologies and marketing development techniques to shows more sophisticated decisions to be successful and sustained. Data management is absolutely essential for an organization because the management processes considering that relevant and important data is acquired, validated, stored, protected, presented and processed through accessibility, reliability, and timing, which meet the customer needs. It is also the process of processing the data and developing data structures, practices and procedures that regularly implement these aspects. All this process facilitates and facilitates the workflow, as a result, many of the data management faces to maintain huge amounts of data, take an interactive approach, take no manipulation and data processing, have ownership of segmented data and lead the data culture Issues [1][10].

C. Implementation

At this stage, the developer develops the design into a program. Information system design is carried out based on needs analysis and user design analysis. In the usecase diagram, it has one actor involved, the customer to do sorting, filtering, and budgeting for tourism (tourist attractions) and homestays (lodging places). The design made in the form of a user interface display developed in a mobile application. The results of designing this information system can be seen in figure below. The ongoing check process with the target user actually leads to a progressive enhancement of the product as a form of hill climbing and a well-known mathematical procedure for finding local optimizations [11][12][13]. Gradual product innovation is a small change in the product that results in improved performance, reduced costs, increased desirability, or simply a new model release. By implementing search, filtering, and sorting as a budgeting form, you can expand your alternative navigation to your customers and get a travel experience.

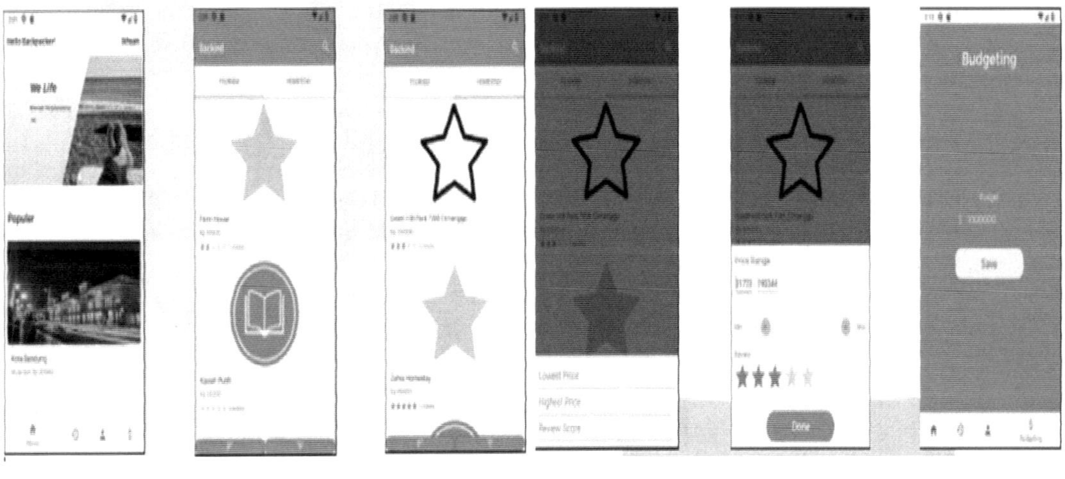

Figure 1. Interfaces Sample of Sorting, Searching and Filtering in Backind Apps

4. Conclusion

The development of the budgeting module on the mobile-based Backind application at Backind startup is designed using the Rapid Application Development method which includes 3 phases, namely the requirement planning phase, design process and implementation to facilitate users in finding tourism (homestay) according to the wishes and budget of the user. The features developed in this application are features of sorting, filtering, and budgeting. This application development uses the Java programming language and Android Studio code editor. In actual, most successful products have undergone continuous incremental innovation aimed at reducing costs and increasing effectiveness.

References

[1] Lubis M and Azizah AH. 2018. Towards Achieving the Efficiency in Zakat Management System: Interaction Design for Optimization in Indonesia. *User Science and Engineering*, pp. 289–301.
[2] Covic, Z. & Babic, D. (2014). Development and Implementation of Location Based Native Mobile Application. *International Journal of Electrical and Computer Engineering Systems*, 27-31.
[3] Doloksaribu, E., AR., M. D., & Rahayu, S. M. (2016). Analisis capital budgeting sebagai dasar pengambilan keputusan investasi aktiva tetap (Studi Kasus Pada PT. Blue Sky Travel Surabaya). *Jurnal Administrasi Bisnis (JAB),* 96-102.
[4] Rahmat R F, Dennis, Sitompul O S, Purnamawati S, Budiarto R 2019 Advertisement billboard detection and geotagging system with inductive transfer learning in deep convolutional neural network *Telkomnika* **17**(5) pp 2659-2666
[5] Ningrum, M., Fajar, S., & Ambarsari, N. (2017). Membangun aplikasi backpacker management system berbasis android mobile application development pada startup backind. *Jurnal Telkom University*, 1-7.
[6] Sulihati, & Andriyani. (2016). Aplikasi akademik online berbasis mobile android pada universitas tama jagakarsa. *Jurnal sains dan teknologi*, 15-26.
[7] Wahyuningrum, T., & Januarita, D. (2014). Perancangan WEB e-Commerce dengan Metode Rapid Application Development (RAD) untuk Produk Unggulan Desa. *Seminar nasional teknologi informasi & komunikasi terapan* 2014, 81-88.
[8] Khan, M. E. (2012). A Comparative Study of White Box, Black Box and Grey Box Testing Techniques, 3(6), 12–15.
[9] Lexmond J and Leadbeater C. 2013. *A Gift for Innovation*. Nesta, Innovation Foundation, November.
[10] Lubis M, Lubis AR, Lubis B and Lubis A. 2018. Incremental Innovation towards Business Performance: Data Management Challenges in Healthcare Industry in Indonesia. *MATEC Web of Conferences* 218, ICIEE.
[11] Norman DA and Verganti R. 2012. Incremental and Radical Innovation: Design Research versus Technology and Meaning Change. Design Issues: *Designing Pleasurable Products and Interfaces Conference*. March.
[12] Lyons G. 2016. *Radical Innovation is Collaborative, Disruptive and Sustainable*. Chapter 7. The Global Innovation Index 2016.
[13] Tellis GJ, Prabhu JC and Chandy RK. 2009. Radical Innovation Across Nations: The preeminence of Corporate Culture. *Journal of Marketing*, 73(January), pp. 3-23.
[14] Ardevol LM. 2013. *User Experience Methodology for the Design and Evaluation of Interactive Systems*. PhD Thesis, Universitat de Lleida.
[15] Lubis AR, Fachrizal F and Lubis M. 2017. The Effect of Social Media to Cultural Homecoming Tradition of Computer Students in Medan. *Procedia Computer Science* 124, 423-428.

IOP Conf. Series: Materials Science and Engineering **851** (2020) 012013 doi:10.1088/1757-899X/851/1/012013

Effect of Small Amount SiC on Mechanical Properties Composite Fabricated by Casting and Subsequent Rolling Process

Tugiman[1,*], F Ariani[1], M Z Arbi[1], R Hermawan[1], K Y Iswara[1]

[1]Department of Mechanical Engineering, Faculty of Engineering, Universitas Sumatera Utara, Padang Bulan, Medan, 20155, Indonesia

[*]Corresponding author: tugiman.karmani@gmail.com

Abstract. Aluminum matrix composites (AMCs) are promising material to develope for engineering application in last a few decade. Combination between aluminum matrix and reinforced particles was succesfull increase the mechanical properties. Casting method have been used to fabricate the AMCs due to their simple process and low cost, however the nonuniform distribution elements and some defects were often observed in the microstructure by this traditional casting process. Furthermore, these problems have influenced the mechanical properties of composites. In other hand, the thermomechanical treatment and dispertion strengthening by particles such SiC are attractive technique to improved mechanical properties of composite such AMCs. Thermomechanical such rolling process was successfully modified the properties of composite, high energy from rolling machine induced plastic deformation and grain refinement of microstructure. The modification of this microstructure is the important role to enhance the mechanical properties. The aim of this research to investigate the effect of rolling process on microstructure and mechanical properties of aluminum cast reinforced with SiC particles. In the current work, Al-Si alloy ingot used as a matrix and reinforced with SiC particles for different composition (1.5, 2.0, 2.5 wt.%). Remelted ingot and SiC have been done in crucible grafite to prevent from contaminant. Composites material were fabricated by using pressure casting at 35 MPa, 657°C pouring temperature and the permanent mold made from mild steel with holding temperature 300°C. Further, a hot rolling process was performed at 550°C with 20% thickness reduction through multi pass process to obtain the thickess of sheets. The microstructure of composites were investigated by optical microscope (OM) with different magnification. Brinell hardness test was carried out to obtained the hardness on the surface of composites rolled. The impact test was conducted in room temperature by using charpy model. Furthermore, the ultimate tensile strength of composites were perfomed by universal tensile machine.The results shows the increase of SiC particles upto 2.5 wt.% promoted the finer microstructure was observed after rolled. Further, the sligthly increase both of hardness and tensile strength have been achieved after hot rolling with different silicone particle. The maximum hardness 71 HB of composite rolled was obtained for 2.5 wt.% SiC. The impact strength 19 joule was observed in 2.0 wt.% SiC and sligthly decrease with the increasing of silicone carbide. In other hand, the ultimate tensile strength 130 MPa was obtained for 2.5 wt.% SiC, the addition of a small amount SiC promote a slightly change the mechanical properties.

Keyword: Aluminum, rolling process, hardness, tensile strength

1. Introduction

Aluminum A356 alloys are a common nonferrous metal widely used in engineering field. Owing to they have a good corrosion resistance, thermal conductivity and mechanical properties. Nevertheless, the extensive growth in engineering application encourage higher mechanical strength requirements are needed in aluminum alloy. Some researcher were developed mechanical properties of aluminum alloys by using stir casting [1][2], colling slope [3] and cold rolling method [4][5]. Rolling process have been used to modify microstructure and mechanicall properties for many alloy system such as Nickel-Aluminium-Bronzes (NAB) and Al alloy [6][7]. A. Shahsavari et al. was reported that the cryogenic rolling improved the mechanical properties of aluminum alloy [8]. It is believed that the enhanced mechanical properties correlated with microstructure change of aluminum during rolling process [9]. Further, a higher thickness reduction after rolling reduce porosity and grain size reduction [10][8]. In other hand, the thermomechanical treatment such rolling technique has been used to synthesized aluminum composite [11][12][13]. Composites reinforced with different particles such as SiC, TiB2 and TiC [11][14][15] were fabricated with different route and subsequent rolling process. J. Nie et al. proposed that the hot rolling process promote distribution of TiB2/TiC particles dispersed more homogeneous between aluminum matrix composite [11] and the microstructure refinement of aluminum lead to increases of microhardness [16]. The homogenity aluminum composite was obtained with the increase in pass number of rolling process [17]. Meanwhile, impact, tensile, hardness and corrosion resistance of aluminum alloys are influenced by degree of rolling and temperature heat treatment [6][8]. The thermomechanical treatment have been used in varying such as Al-Mg-Si [18], 2024 aluminum [8], AZ31 [12], Al-TiB2/TiC composites [11] and other hand SiC particles were used to improved mechanical properties aluminum matrix composite [19]. Furthermore, B.M. Viswanatha et al [20] was reported that the addition of SiC particles into A356 by liquid metallurgy process was succesfully improve the hardness of composites. The presence of SiC particles in A356 matrix composite fabricated by semi-solid stir casting followed rolling process could influence the corrosion behaviour [21]. Fabrication aluminum composite reinforced with 15 vol.% SiC by using hot and cold rolled was conducted, the result shows that cold rolled sheet has lower ultimated tensile strength due to the presence of micro-cavities [22]. Thermomechanical treatments are attracted method to modify the mechanical properties of aluminum matrix composite. However, only a few paper studies on the properties of aluminum composite reinforced with SiC particles produced by casting, further, the combination pressure casting and subsequent rolling process are needed to developed for aluminum matrix composite.

In the present work, aluminum A356 reinforced with SiC particles was fabricated by casting route and subsequent single stage rolling. The aim of this research to investigate the effect of SiC contents on microstructure and mechanical properties of composite in room temperature. Optical microscope (OM), Brinell hardness, uniaxial tensile and Charpy impact test were conducted to analysis the properties of composite material.

2. Materials and method

In the present work, aluminum A356 ingot are used as matrix with the chemical composition as presented in table 1. The composites reinforced with (1.5, 2 and 2.5wt.%) SiC particles were produced by pressure (35 MPa) casting at mold temperature 300°C and pouring temperature 675°C, as starting material 12.5 mm thickness of composites have been used for rolling process, see Figure. 1a. Furthermore, as cast composite materials were hot rolled at 550°C, 18 rpm in speed with 20% thickness reduction (± 10 mm) in final stage rolling. The microstructure of rolled composite was investigated using optical microscope (OM) at different magnification, SiC paper with 100 ~ 1500 grid used to prepare sampels for microstructure followed water cleaned. Brinell hardness with 500 kg load and Charpy impact test were conducted in room temperature. Tensile test carried out by Universal tensile machine with plate shape sample as shown in Figure. 1b.

Table 1. Chemical composition of aluminum matrix composites (AMCs)

Elements	Si	Fe	Cu	Mn	Mg	Ni	Ti	Al
%wt.	7.67	0.127	0.0032	0.0014	0.0687	0.0025	0.0889	Bal.

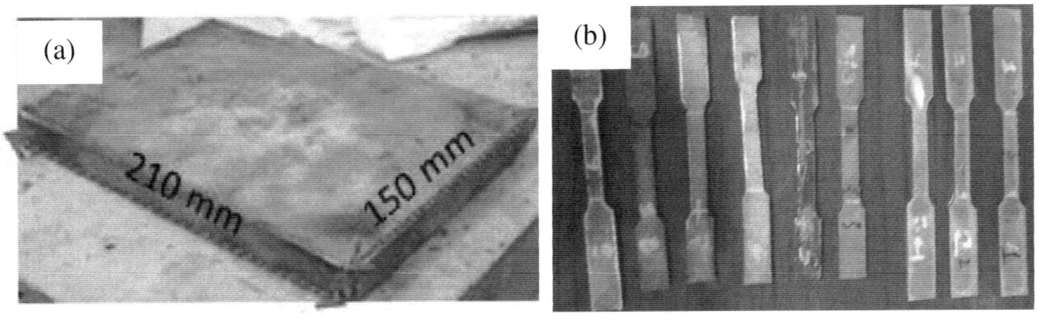

Figure. 1 (a) Composite material before rolling process and (b) samples for tensile test

3. Results and discussion

Figure. 2 shows the microstructure of A356 matrix reinforced with SiC particles fabricated by casting and subsequent with rolling process (20% reduction). It is clearly was observed that the microstructure composed of α-aluminum and Si-rich, both of appear as bright and dark area respectively. Low concentration of SiC (1.5wt.%) particles demonstrated non uniform distribution of the dark particles dispersed between aluminum matrix, see Figure. 2a. Further increasing of SiC particles (2 and 2.5 wt.%) shows the the microstructure change was obtained as seen in Figure. 2b-c. A homogeneous distribution with finer particles (Si/SiC-rich) less then 10 μm in size was observed in the microstructure. It is believed that the presences of the fine Si/SiC particles between aluminum matrix play important role to enhanced the mechanical properties. The aluminum reinforced with small amount of SiC fabricated by pressure casting and subsequent rolling process could reduced the defects in microstructure such as pores, inhomogeneous coarse grain and gas trapped. However, in the present work a small pore (< 3μm) still observed between microstructure, see Figure. 2b, it is believed that the increase of thickness reduction during rolling could reduce the number of pore defects. Nevertheless, more attention are needed due to heavy reduction their generate severe deformation, further the crack formation occured in the microstructure.

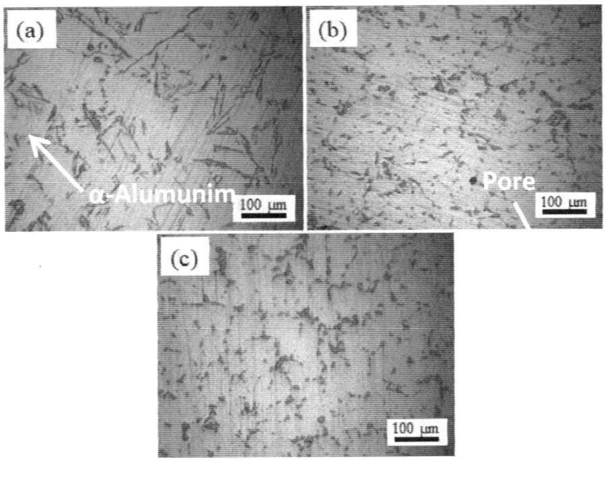

Figure. 2. Microstructure of A356 with 20% reduction of rolling for different silicon carbide contents; (a) 1.5 wt.%, (b) 2 wt.% and (c) 2.5 wt.%

Figure. 3 shows the hardness (HB) of aluminum alloy for varying SiC particle contents, the hardness increases with the increasing of silicone carbide. The maximum hardness 71 HB was achieved at 2.5% SiC, the improved hardness more than 30% of the A356 raw material (as cast), about ±55 HB. In the current research a small amount of SiC (1.5 ~ 2.5wt.%) introduced to the aluminum A356 matrix, low hardness at 69 HB obtained for 1.5 wt.%SiC. It is believed that the addition of a small amount SiC have not change the hardness significantly, see graph of hardness in Figure.3. Furthermore, the standard deviation of 2 wt.% SiC higher than others model aluminum composites, it is suggesting that the distribution and presence of pore s in microstructure are influence the deviation of the hardness.

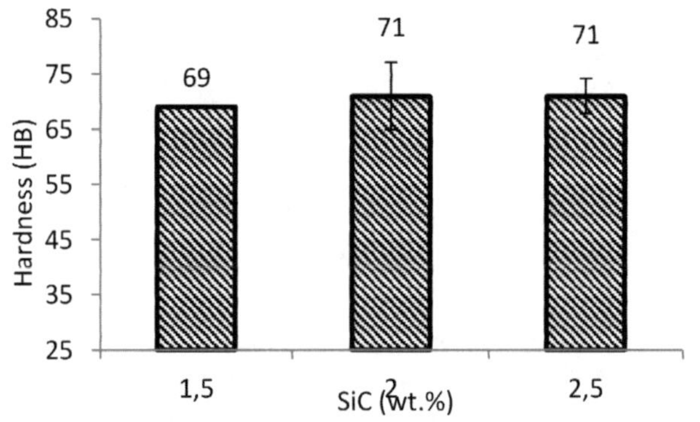

Figure. 3 Hardness of A356 with 20% reduction of rolling for different silicon carbide contents; (a) 1.5 wt.%, (b) 2 wt.% and (c) 2.5 wt.%

The fracture properties of Al composite reinforced with SiC particles evaluated by Charpy V-notch test method in room temperature condition. The effect of SiC on energy adsorbed for Al-composite rolled at 20% reduction as given in Figure. 4. It is clearly obtained that the impact energy of Al composite slightly increases with the increase of SiC contents. The maximum impact strength 19 joule at 2 wt.% SiC, which is corresponding with the fine microstructure as shown in Figure. 2a. However, the result is similar to the hardness, higher deviation also obtained in this composition. Furthermore, the increase of particles upto 2.5% leads to slightly reduce the impact strength (16 joule). Figure. 2b shows the fracture surface of composite materials with difference SiC content, the surface tends to flat with bright color were clearly observed. Low impact strength could be identified by this fracture surface, which is corresponds to energy adsorbed in Figure. 2a.

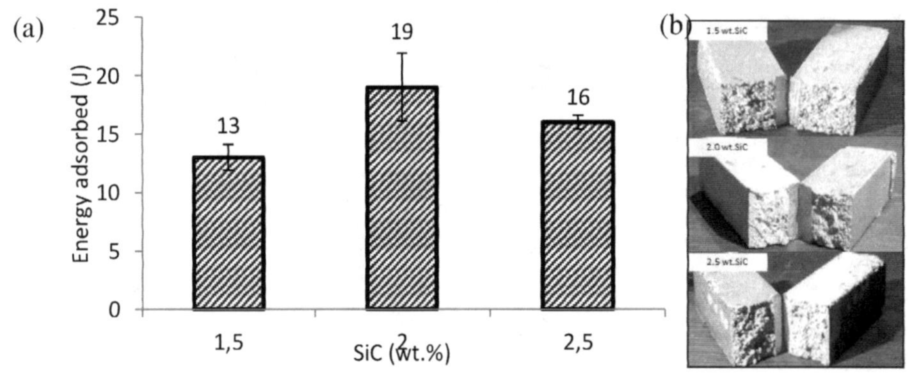

Figure. 4 (a) Impact strength and (b) fracture surface of A356 with 20% thickness reduction for different silicon carbide contents

Figure. 5 shows the strength of aluminum composite reinforced by different SiC particles content. The maximum ultimate tensile strength (UTS) 130 MPa was achieved for 2.5 wt.% SiC at room temperature. A lower standard deviation obtained at higher SiC content, It believed that the deviation correlated with fine and homogenous distribution particles in the microstructure, as seen in Figure.2. The addition of a small amount particles have not change tensile strength significantly, this results corresponds with the hardnes and impact strength in Figure. 3-4.

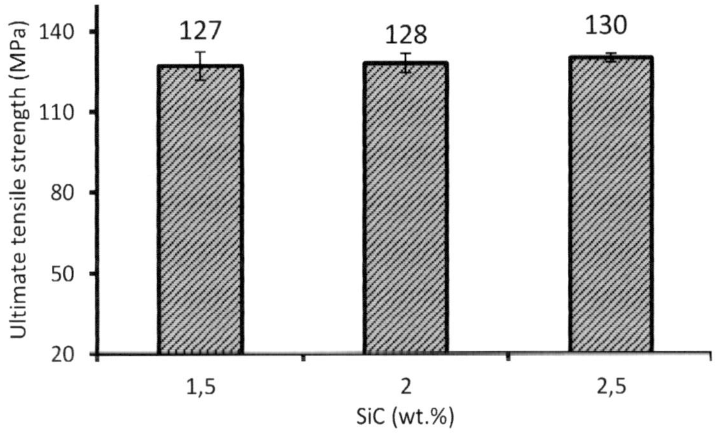

Figure. 5 Tensile strength of A356 with 20% reduction of rolling for different silicon carbide contents

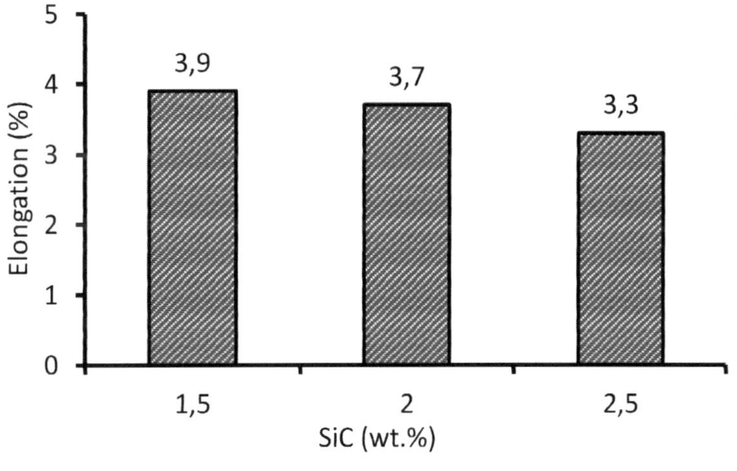

Figure. 6 Elongation of the A356 with 20% thickness reduction of rolling for different silicon carbide contents

The elongation of samples was obtained by using uniaxial tensile test, see Figure. 6, a slightly decreases with the increasing of SiC content was observed. The minimum elongation (3.3 %) obtained at higher silicon carbide particles, this result confirmed that the addition a small amount (1.5 ~ 2.5wt.%) of SiC has not change the mechanical properties significantly. In the future, further increase of SiC needed to see more detail the effect of particles on mechanical properties AMC fabricated by rolling process.

4. Conclusion

The microstructure of aluminum matrix composite reinforced with SiC particles composed of α-aluminum and Si/SiC particles dispersed between matrix. Microstructure refinement and more homogeneous distribution of particles were obtained for 2.5 wt.% SiC after 20% thickness reduction at 550°C rolling process. The increase of SiC content exhibit slightly increase of hardness and tensile strength. However, the impact strength and elongation were decreasing. It can be concluded from all of data, the addition a small amount of SiC particles was successful change the microstructure and mechanical properties in the small scale.

Acknowledgment
This current project was financialy supported by TALENTA program of the Universitas Sumatera Utara, Ministry of Education and Culture of the Republic of Indonesia under the grant number 101/UN5.2.3.1/PPM/KP-TALENTA USU/2019

References

[1] Alaneme K K, Bodunrin M O, Casting S, 6063 Metal Matrix Composites Developed By Two Step – Stir Casting Process, (n.d.) 105–110.

[2] Tugiman, Ariani F, Taher F, Hasibuan M S, Suprianto, 2017 The analysis of composite properties reinforced with particles from palm oil industry waste produced by casting methods, IOP Conf. Ser. Mater. Sci. Eng. 277 (2017). doi:10.1088/1757-899X/277/1/012028.

[3] Haga T, Nakamura R, Tago R, Watari H, Effects of casting factors of cooling slope on semisolid condition, Trans. Nonferrous Met. Soc. China (English Ed. 20 (2010) s968–s972. doi:10.1016/S1003-6326(10)60615-2.

[4] Pan Y, Shen Y, Zhao P, 2018 The Microstructure and Deformation Behavior of Al-Fe-Mn Alloys with Different Fe Contents during Cold Rolling, 1–9. doi:10.3390/met8100753.

[5] A. Alloy, Microstructure, Mechanical Properties, and Corrosion Resistance of Thermomechanically Processed AlZn6Mg0.8Zr Alloy, (n.d.). doi:10.3390/ma11040570.

[6] Gundu D T, Tile J M, Folorunso D O, 2016 Open Access Effect of Cold Rolling and Annealing Treatments on Microstructure , Impact Toughness and Corrosion Resistance of Cu-12Al-2Ni-5Fe NAB Alloy, 307–312.

[7] Mhedhbi M, Khlif M, Bradai C, 2017 Investigations of microstructural and mechanical properties evolution of AA1050 alloy sheets deformed by cold-rolling process and heat treatment annealing, 8, 2967–2974.

[8] Shahsavari A, Karimzadeh F, Rezaeian A, Heydari H, 2015 Significant Increase in Tensile Strength and Hardness in 2024 Aluminum Alloy by Cryogenic Rolling, Procedia Mater. Sci. **11,** 84–88. doi:10.1016/j.mspro.2015.11.111.

[9] Yu H, Wang H, Lu C, Tieu A K, Li H, Godbole A, Liu X, Kong C, 2016 Microstructure evolution of accumulative roll bonding processed pure aluminum during cryorolling, 797–805. doi:10.1557/jmr.2016.70.

[10] Mittal R, Singh D, 2014 Microstructure , Porosity and Hardness of Spray Deposited Cold Rolled Al-6Si Alloy, 1–5.

[11] Nie J, Wang F, Chen Y, Mao Q, Yang H, Song Z, 2019 Results in Physics Microstructure and corrosion behavior of Al-TiB 2 / TiC composites processed by hot rolling, Results Phys. 14 102471. doi:10.1016/j.rinp.2019.102471.

[12] Zhao Z, Gao Q, Hou J, Sun Z, Chen F, 2016 Determining the microstructure and properties of magnesium aluminum composite panels by hot rolling and annealing, J. Magnes. Alloy. 4 , 242–248. doi:10.1016/j.jma.2016.07.003.

[13] Indexed S, Saketh P K, Students U G, Effect Of Rolling On Microstructure Of Aluminium Metal Matrix Composites With Magnesium Particulat Particulates-An An Experimental Study, 8 (2017) 583–587.

[14] Lu Y, Zhou D, Li X, Wang J, Li W, Bai Z, 2018 Aluminum Matrix Composites Reinforced With Nano Tic Particles : Microstructure And Wear Behavior, 52, 673–678. doi:10.17222/mit.2018.001.
[15] Alaneme K K, Bamike B J, 2018 Characterization of mechanical and wear properties of aluminium based composites reinforced with quarry dust and silicon carbide, Ain Shams Eng. J. 9 , 2815–2821. doi:10.1016/j.asej.2017.10.009.
[16] Cieslar M, Republic C,2014 Annealing Effects in Twin-Roll Cast AA8006 Aluminium Sheets Processed by Accumulative Roll-Bonding, 8058–8069. doi:10.3390/ma7128058.
[17] Gharechomaghlu M, Mirzadeh H, 2018 Computational evaluation of the homogeneity of composites processed by accumulative roll bonding (ARB), J. Comput. Appl. Res. Mech. Eng. 0 351–358. doi:10.22061/jcarme.2018.3039.1324.
[18] Martinova Z, Zlateva G, 2002 Microstructure Development During Thermomechanical Treatment Of Al-Mg-Si Alloy, 38, 153–162.
[19] P.O. Babalola, C.A. Bolu, A.O. Inegbenebor, K.M. Odunfa, Development of Aluminium Matrix Composites: A review, 2 (2014) 1–11. http://www.onlineresearchjournals.org/OIJETR.
[20] Viswanatha B M, Kumar M P, Basavarajappa S, Kiran T S, 2013 A356 / SiCp / Gr METAL MATRIX COMPOSITES, J. Eng. Sci. Technol. 8, 754–763.
[21] Sheykholeslami S O R, Mousavian R T, Brabazon D, 2016 Corrosion behaviour of rolled A356 matrix composite reinforced with ceramic particles, Int. J. Mater. Res. 107 , 1100–1111. doi:10.3139/146.111443.
[22] Chen L Q, Lü Y X, Lee C S, Bi J, Li R K Y, 1998 Mechanical Properties and Textures of Particulatereinforced Aluminum Alloy Matrix Composite Under Hot- and Cold-Rolling Conditions, Textures Microstruct. 31, 43–52. doi:10.1155/tsm.31.43.

IOP Conf. Series: Materials Science and Engineering **851** (2020) 012014 doi:10.1088/1757-899X/851/1/012014

Determination of Inventory Policy based on ABC Classification

I Rizkya*, R M Sari, Erwin, R F Sari
Departement of Industrial Enginnering, Faculty of Engineering, Universitas Sumatera Utara

*e-mail: indahrizkya@usu.ac.id

Abstract. Inventory control is very necessary to prevent supply shortages. The right policy in managing inventory will reduce inventory costs. The type of inventory classification is needed to carry out appropriate handling in managing inventory in both the manufacturing and service industries. Drug inventory classification is done by ABC analysis to manage inventory so that no stock out occurs. ABC analysis tends to profit oriented product because it is based on the funds needed from each type of supply. From the results of data processing analysis, in category A (Always) were 46 types (10,8%) with the annual investment of drugs 70,39%, category B (Better) as many as 74 types (17,3%) with the annual investment is 19,99%, and category C (Control) as many as 307 types (71,9%) with of the total annual investment is 9,61%.

1. Introduction

Logistics management is very important in managing inventory in both the manufacturing and service industries including health services (1). One factor that contributes to the quality of health services is drug management. A drug stock that is too large or too small will increase hospital operating costs. Management of drugs supply needs to be done to prevent shortages. If the patient does not receive proper treatment because of the drugs availability are always absent, it makes the patient feel dissatisfied and has a negative impact on the hospital image.

The planning of drugs supply has been carried out is seen based on diseases often occur in hospitals, the quantity of Drug stocks, usage in the previous year, and seasonal outbreaks. However, data on diseases are difficult to obtain with certainty as well as recording and reporting are not good cause the planning of drugs supply to be not good enough. This condition lead to stock out of Drugs and Drugs are not available must be purchased accidentally to other distributors and the purchasing cost Drugs increase. For this reason, it is necessary to manage the drugs supply.

Inventory management can be done using various methods, one of them Always, Better, and Control (ABC) method. Analysis of ABC, also known as Pareto Analysis is one of the methods used in logistics management to classify a category of Goods into three categorys (category A, B and C) [2]. The ABC analysis is the "Inventory categorization method" which requires items to be divided into three categories, A, B and C. A is the most valuable item and C is a less valuable item, whereas B contains items that range between A and C. It aims to focus on a few critical (A-item) and not on many trivial (C-item) [3]. In the ABC analysis, items in category A contributed to the majority (70-80%) of the total inventory value of the item. Category B contributes (10-15%) And category C consists of 5% of the total inventory value of the item. However, it is not appropriate for one criterion now one day. The inventory

IOP Conf. Series: Materials Science and Engineering **851** (2020) 012014 doi:10.1088/1757-899X/851/1/012014

of companies depends on various criteria, such as unit price, annual demand rate, critical nature, scarcity, endurance, and etc [4].

ABC analysis or classification methods are commonly used in supply control for companies have various types of materials in inventories have different usage values [5]. For example, ABC analysis is also used for the analysis of inventory management of sponge iron plant [6]. ABC analysis can also be implemented in managing spare parts inventory in one of the printing industries in Pakistan. The result is ABC analysis showed that 15% items were from each category A and category B, remaining 70% items were from category C accounting for around 66%, 15% and 19% of total annual expenditure of technical store respectively [7].

ABC analysis is also applied in inventory control of consumables in the service industry. This is done to find out the inventory of consumables become categorys A, B, and C based on ABC analysis of usage, investment, and critical index. The results obtained indicates that the the inventory control of consumables in one of the study programs is sufficient.

Based on previous research, it is known that the application of ABC analysis can be done on any product, including Drugs. Control of drug supplies to a clinic or hospital is required to fulfill orders in the right amount and time so that the total inventory cost can be reduced by the optimal order quantity and period [8]. This research was conducted to manage drugs supplies in hospitals to avoid stock outs. ABC Analysis is used to identify the drugs types that require the greatest cost because of the high level of usage and/or high prices by grouping.

2. Methodology

This study was conducted at one of the hospitals in the Medan city, especially in pharmaceutical installations. The object under study is in the form of data on medication needs in the hospital. Research begins with observations directly to the pharmaceutical installation department. Activities carried out at this stage are observing the conditions occur in pharmaceutical installations. After observation, the topic and purpose of the study are determined according to the conditions of the installation. After that, data collection is needed to analyze the supply of drugs in the pharmaceutical installation. Data collection in the data form on drug needs and the cost of ordering drugs.

Based on these data, an analysis of the management of drug supplies was carried out. The analysis was carried out using ABC Analysis. ABC analysis is a technique for prioritizing the management of inventory [9]. The ABC Inventory Control technique is based on the principle that a fraction of the items may typically represent most of the use of the total material of the total inventory in the process, while a relatively large number of items may be from a fraction of the value of money [10]. Drug grouping is based on 3 categories, namely:

a. Category A is a drug group to absorb 70% of the budget with the quantity of drugs not more than 20%. Drugs belonging to the category A are very critical drugs category need to be tightly controlled, and continuously monitored. Category A orders can be done with a small quantity but the order frequency is more often and because the investment value is quite large has the potential to provide a large profit for the hospital, this category requires strict drug monitoring and monitoring, accurate and complete record.
b. Category B absorbs the budget of 20% with the quantity of drug around 10-80%. Drugs belonging to category B, inventory control are not too strict as category A, but reports on their usage and the remaining drugs must be reported and inventory control always be controlled.
c. Category C absorbs the 10% budget with the quantity of drug around 10-15%. Category C has more medicinal items but has no impact on warehouse and financial activities because the cost is cheap and usage less. Monitoring of this category can be more relaxed, for example, done six months or once a year.

Drug supply classification with ABC analysis begins by determining the total annual investment value of each type of drug based on the total demand in a year and the purchase value of each type of drug. Then the percentage of annual investment absorption is calculated for each type of drug. The percentage of investment absorption is determined by the annual investment absorption of each type of drug to the total annual investment for procurement of all drugs in the hospital.

The next step is to sort the drugs based on the percentage of annual investment absorption, starting from the largest to the smallest percentage. Then calculate the cumulative percentage of annual investment absorption of each type of drug. The final step classifies each drug types in categories A, B, or C based on the cumulative percentages. Drugs type with cumulative percentage of 1-70% included in category A, 71-90% included in category B drugs and 91-100% included in category C.

3. Result and Discussion

This research was conducted by collecting the required data in the form of the drugs quantity, the drugs purchase cost, and the total usage of generic Drugs during the months of January-March 2019 as many as 427 types. The Drug price is taken based on the purchase transaction of drugs supply to the distributor and the usage quantity based on the drug usage for 3 months in one of the hospital in Medan City.

Most companies use the annual inventory of goods usage rankings in the application of the ABC principle, namely how much annual investment for inventory procurement is on the ABC classification [11]. The results recapitulation of the ABC Drugs classification in one of the government hospitals in Medan is as follows:

Table 1. ABC Classification Results Based on the Quantity of Drug Usage

Category of Drugs Supply	Investment Value		Quantity of drugs type
	IDR	Percentage %	
A	1.048.207.800	70,39	46
B	297.740.400	19,99	74
C	143.150.500	9,61	307
Total	1.489.098.700	100,00	

Based on the table above, it can be seen that drugs including category A only accounted for 44 of drug types, but this drug absorbed the largest hospital budget, which was 70,39% compared to drugs type in category B and category C. Drugs type in category B accounted for 74 types and category C accounted for 307 types. Drugs type in category A must have stricter inventory control, more accurate records verification, and stricter physical supervision carried out every month.

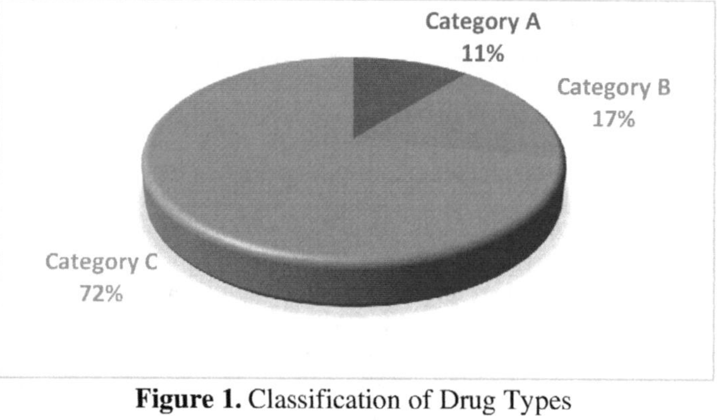

Figure 1. Classification of Drug Types

Based on Figure 1, it can be seen that category A only consists of 11% (10.8%) of drugs types, category B is 17% (17.3%) of drugs types and category C is 72% (71.9%) of drugs types. The drugs type included in category A with the highest investment value must be a concern for the hospital.

Category A is a drugs type that need to be controlled carefully, because the quantity cost and the Drug usage in category A is the highest compared to the other categorys of Drug, which is 35,1%, and it needs strict control. Category B and C drugs type have a small impact on warehouse and financial activities but must still get good monitoring and control. Pharmacy installations must be able to overcome not to run out of stock for this category A, because this will affects patient satisfaction. This category requires close attention and monitoring in inventory control, by carrying out careful calculations in needs determination, requires a complete and accurate recording system, as well as evaluations carried out every month.

Drug supply policy in category B is not as strict as Category A. However, drug inventory levels must still be controlled and recorded. The amount of drug used and drug stock must be reported so that inventory control can always be controlled. There are more items in category C drugs but they do not have an impact on warehouse and financial activities because they are cheap and use less. The inventory policy for this category can be done every six months or once a year. Orders for drugs in this category can be done once or twice a year. In the drugs supply management in hospitals not only is the money issue as priority but the urgency of drugs usage. For this reason, this article can be deepened by adding the urgency of the drugs usage in the future.

4. Conclusion

Based on ABC analysis, Drugs included category A (Always) were 46 types (10,8%) with an investment of 70,4% of total Drug usage, category B (Better) as many as 74 types of Drugs (17,3%) with total investment of 20% of total Drug usage, and category C (Control) as many as 307 types of Drugs (71,9%) with an investment of 9,6% of total Drug usage. The results of this study indicates that the application of ABC analysis can help in Drug management supply effectively. This will help to understand the problems occurring due to purchase, inventory and stock safety.

Acknowledgement

Authors would to thank to Universitas Sumatera Utara for its financial support in the 2019 TALENTA program so that this research can be published. The author also thanks all the participants who helped carry out this research.

References

[1] E L Magad and J M Amos 1995 *Total Material Management* (New York: Chapman & Hall)
[2] H Puspasari, S S Danu, and E Sulistyani 2017 *Jurnal Farmasi dan Ilmu Kefarmasian Indonesia* **4** (2), pp 60-67
[3] S S Mahagaonkar and A A Kelkar 2017 *International Research Journal of Engineering and Technology (IRJET)* **4**, pp 614-620
[4] K Balaji and V S S Kumar 2014 *The 47th CIRP Conference on Manufacturing Systems* **17** (2014), pp 463-468
[5] S Assauri 1999 *Manajemen Produksi dan Operasi* (Indonesia : Universitas Indonesia)
[6] Y Kumar, A Lilhare, A Sahu, B Lal, and Y Khaperde 2016 *International Journal for Research in Applied Science & Engineering Technology (IJRASET)* **4**, pp 32-36
[7] M Ayat 2017 *Merit Research Journal of Business and Management* **5** (1), pp 001-007
[8] P Djunaidi 2005 Manajemen Logistik dan Farmasi Rumah Sakit (Jakarta: Universitas Indonesia)
[9] H Ravinder and R B Misra 2014 *American Journal of Business Education* **7** (3), pp 257-264
[10] Y S Indrasan, V Rajput, and K Chaware 2018 *IAETSD Journal for Advanced Research in Applied Sciences* **5**, pp 134-137
[11] G W Zimmerman 1975 The ABC's of Vilfredo Pareto, Production and Inventory Management **16** (3), pp 1-9

IOP Conf. Series: Materials Science and Engineering **851** (2020) 012014 doi:10.1088/1757-899X/851/1/012014

[12] Siregar I, Muchtar M A, Rahmat R F, Andayani U, Nasution T H, and Sari R M 2018 Method of calculation overall equipment effectiveness in fertilizer factory. *IOP Conference Series: Materials Science and Engineering* **308**(1), p. 012053

IOP Conf. Series: Materials Science and Engineering **851** (2020) 012015 doi:10.1088/1757-899X/851/1/012015

Development Strategy Analysis of Fertilizer Company Using BCG Matriks Method

I Yahya[1], Anizar[2], G Khatami[1], T Al Khansa[2]
[1]Department of Accounting, Faculty of Economic and Business, Universitas Sumatera Utara, Medan 20155, Indonesia
[2]Department of Industrial Engineering, Faculty of Engineering, Universitas Sumatera Utara, Medan 20155, Indonesia

Email: idhar@usu.ac.id

Abstract. This study aims to analyse the strategy of a fertilizer company that produces urea-based fertilizer and organic fertilizer. A company should always have a certain development strategy by observing its growth rate and comparing it with other competitors. Boston Consulting Group (BCG) Matrix can recommend regarding the strategy needed for resource allocation and company development. Two competitors are chosen to decide this company position: the first one is a market leader with the best company growth rate and highest relative market share, and the second one is an equivalent to the company being studied in terms of its production and sales. The growth of 5.63% in fertilizer sales indicates a high market share growth rate. The relative market share of this fertilizer company compared to the first competitor and the second competitor is 0.117 and 0.476 respectively, meaning that it has a low market share as the number is smaller than 1. In BCG Matrix, this company is in the Question Marks position (quadrant II).

1. Introduction

In a company, a development strategy is needed to maximize its growth potential. Many skills, technology, and intellectual resource can be utilized accordingly to improve the company's capacity, so that business development can be focused on the implementation of a business plan, using equity funding, acquisition, technology divestment and product [1]. To find the right development strategy, Boston Consulting Group (BCG) Matrix is used as a method to analyze the company position among its competitors, so that the company does not implement the wrong strategy later [2]. The BCG matrix can be used to analyse both short term and long-term work performance. BCG matrix incorporates the use of static analysis and time adaptation [3]. The BCG matrix will give one out of four possible positions; each possesses distinct meaning and symbolled differently. The symbols are Question Marks, Stars, Cash Cows, and Dogs [4].

Question Marks, the symbol of quadrant 1, means the company is still in a growth phase, as can be seen from how market growth rate and low market share. The company needs to spend a large amount of capital to increase market share, to move to Stars position and finally to Cash Cows position[5]. Usually, this is the position of the newly-established company [6]. Stars, the symbol of quadrant II, is the position of a company that already has proper market growth rate and market share, but still not profitable enough. In this case, market share grows rapidly due to the company already has a good market growth rate. The next step is the company must increase its position carefully to reach the Cash Cow position and does not fall to the Question Mark position [7]. Cash Cows, the symbol of quadrant III, is the position of a company with a relatively high market share and low market growth rate. The strategy that should be taken in this position is a harvesting strategy because in this position the company is very profitable. However, the company must take extra precautions to maintain its position and not fall to the Dog's position. This position is the most advantageous, yet also the most vulnerable. Usually, the company becomes carefree and not careful in this position [8]. Dogs, the symbol of quadrant IV, is the position of a company that has a low relative market share and low market growth rate. This is a position where a company is usually liquidated and divested. Therefore, the company must choose a wise decision because the choice is hard. The company must choose either to run the company properly or be divested. Usually, this position does not give any advantage and many companies choose to quit and abandon the company [9].

This study aims to decide the position of the corresponding company in the BCG matrix, to decide the best marketing strategy needed to increase the company's sales volume higher than its competitors. In this study, two competitors are chosen, the first competitor is the market leader having the best market growth rate and highest relative market share, and the second one is an equivalent to the company being studied in terms of its production and sales volume.

2. Method

This study is conducted in a company producing urea-based fertilizer and organic fertilizer. This study aims to calculate the market growth rate and the relative market share of this company using the BCG matrix method. The Market growth rate is the percentage increase of sales volume in the last two years,

$$MGR = \frac{SVN - SVN_1}{SVN_1} \times 100\,\% \qquad (1)$$

where MGR is the market growth rate, SVN is last year's sales volume, and SVN_1 is the last two-year sales volume [10].

The relative market share is calculated with formula,

$$RMS = \frac{SVN}{SVCN} \qquad (2)$$

where RMS is relative market share, SVN is last year sales volume, SVCN is last year sales volume of a competitor. If the relative market share is lower than one, it means the relative market share is lower than its competitor. Based on market growth rate and relative market share, the company position in the BCG matrix is then decided [11].

3. Results and Discussions

3.1. Market growth rate

The market growth rate calculated using the BCG matrix method is 5.63%, a quite high number for market growth rate. This number is obtained from data of fertilizer sales volume in 2017 and 2018 as shown in figure 1.

	2017	2018
Competitors 2	900.038 ton/year	838.650 ton/year
Competitors 1	2.826.328 ton/year	3.393.970 ton/year
Fertilizer Co.	378.645 ton/year	399.975 ton/year

Figure 1. Fertilizer sales volume of the studied company, competitor 1, and competitor 2.

Relative market share indexes a firm's or a brand's market share against that of its leading competitor. Analysis of relative market share is conducted by comparing its sales volume of fertilizer compared to competitor 1 and competitor 2. The data used are the sales volume of the fertilizer company being studied and its first and second competitor in 2017 and 2018, as shown in figure 2.

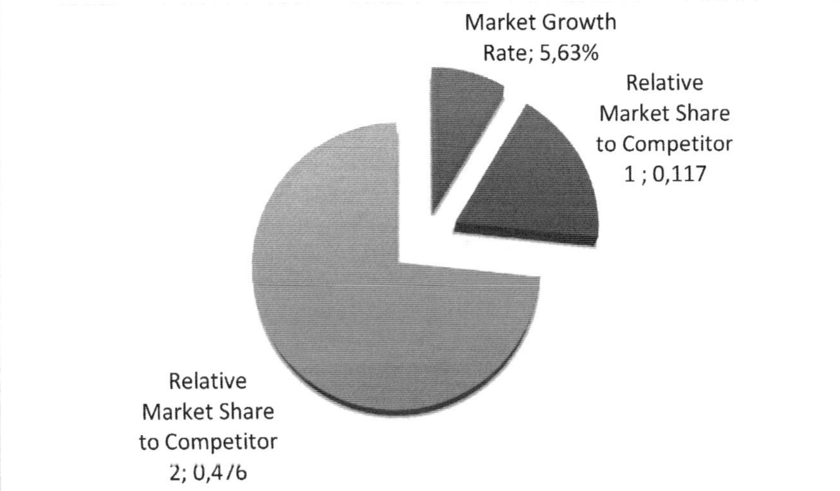

Figure 2. Relative market share of fertilizer company compared to competitor 1 and competitor 2.

Based on the BCG matrix calculation, the relative market share of the corresponding company is 0.117 when compared to competitor 1 and 0.476 when compared to competitor 2. This means the company has a small market share since the relative market share is smaller than 1.

3.2. The position of the fertilizer company compared to competitor 1 based on BCG matrix

The relative market share of the fertilizer company compared to competitor 1 (market leader) is 0.117. Therefore, the company's position in the quadrant II, i.e. in Question Marks position, as shown in figure 3.

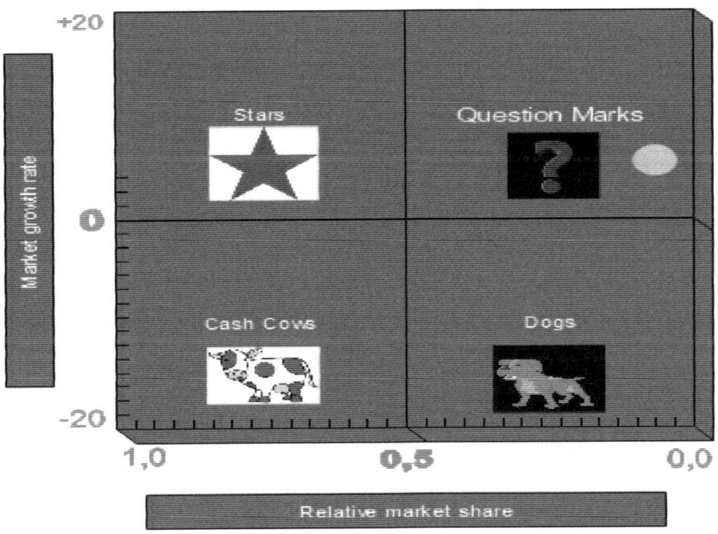

Figure 3. Position of fertilizer company compared to competitor 1 based on the BCG matrix.

This Question Marks position is the condition where the company has a high market growth rate, yet low relative market share. In other words, the company's problem is high cash needs and low cash earnings.

3.3. The Position of the fertilizer company regarding competitor 2 in BCG matrix
The relative market share of the company compared to Competitor 2 is 0.477. In this case, the position of the company is still in the quadrant II, i.e. Question Marks position, although the relative market share is slightly higher than when compared to Competitor 1. The result is shown in figure 4.

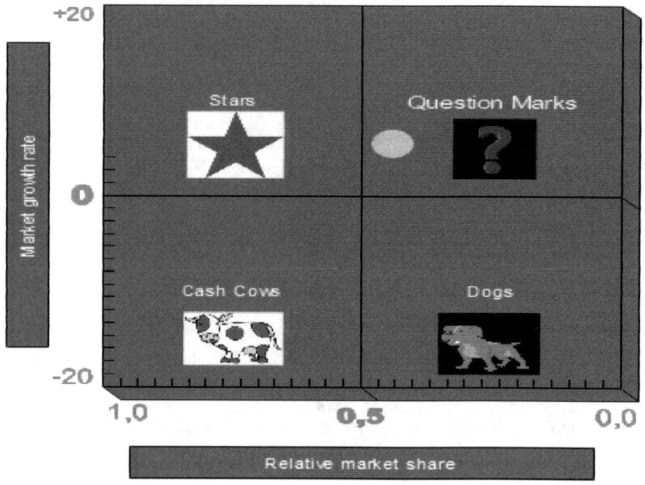

Figure 4. Position of fertilizer company compared to competitor 2 based on the BCG matrix.

3.4. *Analysis of strategy based on the BCG matrix*
The fertilizer company is in the quadrant II, i.e. in the Question Marks position where the market growth rate is quite high, but the relative market share is still low. Therefore, the company should make a precise marketing strategy to solve this problem. If taken correctly, the marketing strategy can help the company to gain maximum profit. A strategy that fits this problem is to build a strategy. Build strategy helps to build relative market share during the growth phase of a business unit in an industry. A business unit is the main unit that needs a large amount of cash. The cash is used to develop a product, induce market growth, and expand activities. The goals of this expenditure are to be the market leader in the short term and to increase market share in the long term.
For this company, some examples of build strategy may include
1) Develop new types of fertilizer, such as NPK fertilizer.
2) Increase the quality of fertilizer by improving its ingredients quality and its making process.
3) Maintain good customer service.
4) Expand marketing networks, both locally and internationally.
5) Promotes the product better by giving proper information on its website and its social media.
6) Improve investation in either fertilizer or business unit to help increase market share.

4. Conclusions
Based on the BCG Matrix, the corresponding fertilizer company is in the Question Marks position. This position means that the company has a high market growth rate and low market share. The growth rate of 5.63% in fertilizer sales is quite high. Meanwhile, the relative market share in comparison with the first and second competitors is 0.117 and 0.476 respectively, which means that it has a low market share as the number is smaller than one.

IOP Conf. Series: Materials Science and Engineering **851** (2020) 012015 doi:10.1088/1757-899X/851/1/012015

Due to the company is in the Question Marks position, the company has to make a new strategy that will either move its position higher to the star position or lower to the dog position. Strategies than being chosen are selling the company or strengthen the company divisions by performing intensive strategies, such as market penetration, market development, and product development. The strategy chosen may potentially waste liquid resources if the target can not be met[12]. Based on the BCG Matrix, the right strategy that should be taken by the company is the build strategy. This building strategy helps to build relative market share during the growth phase of a business unit in an industry.

References

[1] Konečný Z and Zinecker M 2015 *Acta Univ. Agric. Silvic. Mendelianae Brun.* 63 1 p 235–243
[2] Debrecht D and Levas M 2014 *J. High. Educ. Theory Pract* 14 3 p 65–70
[3] Hersen A, da Silva JCGL and Timofeiczyk Junior R 2018 *Rev. Árvore* 42 3
[4] Madsen DØ 2017 *Probl. Perspect. Manag* 15 1 p 19–34
[5] Wahyuandari W 2013 *J. Bonorowo* 1 1 p 88–104
[6] Mohajan H 2017 *Noble Int. J. Bus. Manag. Res.* 2 84237 p 1–6
[7] Puteri FS Nuringwahyu S and Krisdianto D 2019 *JIAGABI* 8 3 p130–135
[8] Sarjono H and Kuncoro EA 2013 *Binus Bus. Rev.* 4 1 p 414
[9] Sari H and Sultan MA 2019 *J. Ilmu Manaj. Bisnis* 10 2 p 213–222
[10] Mulyani, Kusnandar, Antriyandarti E 2019 Analisis Pasar Ekspor Karet Indonesia dengan Matriks BCG 3 1 p 67–74
[11] Guntara SR, Yogaswara B and Aurachman R 2017*eProceedings Eng* 4 3 p 4436–4444
[12] Barksdale HC and Harris CE 1982 *Long Range Plann* 15 6 p 74–83

IOP Conf. Series: Materials Science and Engineering **851** (2020) 012016 doi:10.1088/1757-899X/851/1/012016

Management Support in Enhancing Participatory Ergonomic Interventions in Palm Oil Industry

Anizar[1], AR Matondang[1], R Ismail[2], Nazaruddin[1]
[1]Department of Industrial Engineering, Faculty of Engineering, Universitas Sumatera Utara, Medan 20155, Indonesia
[2]Department of Sociology, Faculty of Social Political Science, Universitas Sumatera Utara, Medan 20155, Indonesia

Email: anizar_usu@usu.ac.id

Abstract. This study is conducted in a palm oil industry in North Sumatera. The palm oil industry, which has great potential to end poverty, is usually performed by manual workers in such a way that it may cause musculoskeletal disorders. This is a challenge for future palm oil industry's continuity. Participatory ergonomic intervention is one way to reduce this risk and needs management support for its success in implementation. A questionnaire is then spread to find the most influential type of management support that may promote active participation from workers. The recapitulation of the questionnaire shows that "being involved" as the highest score, followed by work authority and management commitment. Workers' participation in ergonomic program implementation can only be realized if the management is committed.

1. Introduction

Indonesia, a country known as the world's largest producer of palm oil, has roughly 19.5 million palm oil workers. Palm oil is a mainstay of Indonesia commodity and contributes toward 3.5% total of Gross Domestic Product (GDP) with income from palm export amounts to approximately USD 20.54 billion. Palm is the most productive oil type of crop, in which the value chain ranges from upstream plantation to downstream industry of processing. The increase of vegetable oil demand domestically and the big potential of palm oil export value (CPO) trigger the rapid growth of palm plantation nationally. In 1980, the total area of palm oil plantation is only about 295 thousand hectares and it soars to approximately 14.68 million hectares, an almost fifty-fold increase. The potential of the palm industry to end poverty but has an impact on the musculoskeletal risk that may happen to its workers. World Bank states that employers' health and safety is the main concern of future palm oil industry continuity [1].

There are two types of activities in the palm industry, the first is in the palm plantation and the second is in the palm processing industry, which for the most part is performed manually [2,3]. The process of harvesting oil palms is done manually starting from cutting the fronds and fresh fruit bunches (FFB) with egret, inserting FFB into the sprouts with corners, pushing the sprouts containing FFB to the yield bin and loading the FFB into the transport trucks with the corner.

A study conducted in 3 palm industries in Malaysia shows that musculoskeletal disorders are more prevalent in seasoned employees (71.7%) than employees that have been working less than one (50.8%) [4]. When harvesting, the workers usually experience musculoskeletal pain around the neck area, lower back, right shoulder, right forearm, left forearm, and bottom. Factors related to musculoskeletal disorders are work posture, age, and overall work duration, thus corrective action is needed to prevent the increase of increase the spreading of musculoskeletal disorders. Musculoskeletal disorders generally happen due to excessive muscle contraction; the muscle experiences. excessive loading or contract for a long duration [5]. Epidemiological study shows that there exists a correlation between musculoskeletal disorders with lifting heavyweight, weird work posture, repetitive work, and whole-body vibration [6].

Musculoskeletal disorders can seriously affect workers and contribute to workers' absence or workhour loss, which negatively affects economic activity [7]. Problems caused by work-related activity; especially musculoskeletal disorders greatly affect productivity. A study, *Global Burden of Disease, Injuries, dan Risk Factors 2016* (GBD 2016), states that musculoskeletal disorders contribute to the second-highest reason for physical disability. Meanwhile, musculoskeletal condition differs greatly across different generation and a study says that 20-30% of the world population lives with the painful musculoskeletal disorder [8]. To prevent the increase of musculoskeletal prevalence in the future, a corrective activity that can be taken care of the participative approach for prevention and searching the correct solution.

The effective participatory ergonomic program needs cooperative involvement from both management and workers [9]. How to involve both parties to actively participate in the ergonomic program remains a big question [10]. Prevention of musculoskeletal complaints should be taken from a management system approach, therefore all parties are involved and weighed in achieving maximal results [11]. Participatory ergonomics interventions must be directed at risk factors at the organizational level so that program success requires ongoing management commitment at all levels of the organization as well as the participation of workers and internal staff. Thus, the adoption of participatory ergonomics into the organization becomes an integrated organizational management system. It is predicted that management support will affect the effectiveness of participatory ergonomics intervention in reducing musculoskeletal disorders experienced by palm industry workers. Management support is management that is committed, actively participates, and has the authority to change the structure, so that planned progress can be carried out [12, 13].

2. Method

Observation is conducted in some palm industries in 9 districts in North Sumatera. Respondents in this study are 280 male workers that perform production activities at different stations, including loading ramp station, sterilizing station, threshing station, pressing station, clarification station, and kernel plant station. The study is conducted as follows, doing direct observation, giving interviews, spreading questionnaires regarding management support, which may affect work climate, which may improve participatory ergonomic intervention.

Management support is management's willingness to facilitate and to allocate the time needed to review plans and get results. Indicators used to measure management support variables are management commitment, worker participation, and workers' authority [14-16]. Management commitment consists of 4 statements, workers' participation consists of 4 statements, and workers' authority consists of 2 statements. The questionnaire is designed as structured questioned to be answered by each respondent. Validity and reliability test are conducted before the questionnaire is spread.

Respondents fill the questionnaire by giving a check sign (√) on the listed measured scale. The questionnaire consists of ten statements and uses a Likert scale (interval scale), where 1 means strongly disagree, 2 means disagree, 3 means neutral (neither disagree nor agree), 4 means agree, and 5 means strongly agree. Data analysis is performed using the method of Structural Equation Modeling (SEM) Partial Least Squares (SEM-PLS) with WarpPLS software. The contribution of statement items is calculated using a convergence validity test, consistency of statement items is calculated using the Composite Reliability (CR) test, and how good statement items are used to measure management support is evaluated using Cronbach's Alpha (CA) test.

3. Results and Discussions

3.1. Respondents judgment regarding management support

Respondents judgment regarding management support show the respondents answer about the statements given. From these answers, average score and scoring criteria are then calculated. In determining the scoring interpretation from respondent answers, the Likert scale is used. The scale states that score of 1.00-1.70 means a very bad, score of 1.80-2.59 means bad, score 2.60-3.39 means neutral, 3.40-4.19 means good, and 4.20-5.00 means very good.

Respondent's judgment regarding variables of management support in 10 statements mainly obtains good criteria (90%). The only indicator of participation1 that obtains an average score of 4.25 – a very good criterion. This means workers feel respected by management concerning active involvement in problem-solving, especially in the ergonomic intervention program. The next indicator, authority1 obtains an average score of 4.16 – a good criterion. Workers do feel management support existence – they obtain authority to perform work using their method. The lowest score of 3.99 is obtained by the commitment1 indicator concerning management support about ideas given by workers to improve ergonomic intervention effectivity. Respondent answer recapitulations concerning management support can be seen in table 1.

Table 1. Respondent answer recapitulation concerning management support.

Indicator	Strongly Agree		Agree		Neutral		Disagree		Strongly Disagree		Total		Average Score	Criteria
	F	%	F	%	F	%	F	%	F	%	F	%		
Commitment1	74	26	159	57	25	9	13	5	9	3	280	100	3.99	Good
Commitment2	77	28	157	56	26	9	13	5	7	3	280	100	4.01	Good
Commitment3	89	32	145	52	30	11	12	4	4	1	280	100	4.08	Good
Commitment4	89	32	159	57	12	4	14	5	6	2	280	100	4.11	Good
Participation1	103	37	153	55	16	6	6	2	2	1	280	100	4.25	Very Good
Participation2	84	30	157	56	25	9	12	4	2	1	280	100	4.10	Good
Participation3	87	31	160	57	22	8	7	3	4	1	280	100	4.14	Good
Participation4	77	28	162	58	19	7	16	6	6	2	280	100	4.03	Good
Authority1	95	34	150	54	23	8	9	3	3	1	280	100	4.16	Good
Authority2	90	32	151	54	23	8	14	5	2	1	280	100	4.12	Good

Head management support is needed to prevent the severed management role in the implementation process. Management involvement and full commitment to support the project of participatory ergonomic intervention will reduce musculoskeletal disorders experienced by workers in the palm industry. However, previous studies conducted reveal that not all studies concerning management support will positively affect participatory ergonomic intervention effectivity. Many other factors need to consider, as only management support does not have a significant effect on participatory ergonomic intervention.

3.2. Convergency validity test

Convergency validity test is performed to calculate the contribution of ten indicators of management support, which consist of 4 management commitment indicators, 4 workers participation indicators, and 3 workers authority indicators with the criteria of loading factor above 0.7 and significant P-Values [17]. An indicator with a high loading factor has a high contribution and vice versa. All loading factors of management support is valid. The lowest score of 0.773 is obtained for participation4 indicator, a statement stating that workers give input to solve the problem about participatory ergonomic intervention and the highest score of 0.872 is obtained for commitment3 indicator, a statement stating that management push workers to participate in developing the ergonomic program. From table 2, it can be seen that the highest contribution indicator toward management support is commitment3 indicator, followed by commitment1, authority1, participation2, etc.

Table 2. Factor loading value.

Indicator	Loading Factor	P-Value	Criteria
Commitment1	0.835	< 0.001	Valid
Commitment2	0.811	< 0.001	Valid
Commitment3	0.872	< 0.001	Valid
Commitment4	0.804	< 0.001	Valid
Participation1	0.801	< 0.001	Valid
Participation2	0.834	< 0.001	Valid
Participation3	0.819	< 0.001	Valid
Participation4	0.773	< 0.001	Valid
Authority1	0.835	< 0.001	Valid
Authority2	0.782	< 0.001	Valid

3.3. Reliability test

The reliability of the questionnaire is tested using Composite Reliability (CR) and Cronbach's Alpha (CA) test with a critical value higher than 0.7 [18]. Composite Reliability (CR) depicts the statement consistency in every indicator in measuring management support and Cronbach's Alpha (CA) is a value to evaluate how good the indicator used to measure management support. CR of 0.952 implies that the statement of every indicator is consistent and CA of 0.994 implies that all statements are reliable, as can be seen in table 3.

Table 3. Result of CR and CA toward management support.

	Value	Critical Value	Criteria
Composite Reliability (CR)	0.952	0.700	Reliable
Cronbach's Alpha (CA)	0.944	0.700	Valid

4. Conclusions

The success of participatory ergonomic intervention needs continuity in management commitment at all levels, and also workers' participation and authority involvement. Management commitment and support are important because the management party is a policymaker, making policy regarding every aspect of production and development activity. Workers' participation in a participatory ergonomics program can only be realized if the management party is fully committed.

If a company wants to develop an effective participatory ergonomic intervention model, the factor of management support must be weighed first. This can be done by planning the necessary things. Analysis of problems faced by workers regarding ergonomic intervention programs is crucial, so that rejection due to changes does not happen. As widely known, ergonomy is a process of designing, modifying, arranging materials, tools, workroom, assignments, works, products, systems, and environments to meet human psychology, society, anatomy, biomechanics, and physiological ability, needs and limits. Some examples of support management parties can give are work controlling, work redesigning, autonomy in working, so that laborers can work within their capacity and health risk can be minimized. Giving what workers needs will improve their participation and in turn an increase in satisfaction and acceptance of ergonomic intervention program.

Acknowledgments
Thank you to Universitas Sumatera Utara for funding the publication of this paper. The author would like to thank also the various parties who have helped carry out this research.

References
[1] Myzabella N, Fritschi L, Merdith N, El-zaemey S, Chih H and Reid A 2019 *Int. J. Occup. Environ. Med* 10 p 159–173
[2] Anizar, Matondang AR and Sibarani JA 2018 *IOP Conf. Ser. Mater. Sci. Eng* **309** 1
[3] Andriani M and Erfani E 2017 *J. Integr. Sist. Ind* **4** 2 p 119–128
[4] Razak S, Karuppiah K and Tamrin SBM 2014 *Adv. Environ. Biol* **8** 15 p 277–284
[5] Hossain MD *et al. 2018 PLoS One* **13** 7 p 1–18
[6] Charles LE, Ma CC, Burchfiel CM and Dong RG 2018 *Saf. Health Work* **9** 2 p 125–132
[7] Dehghan N, Aghilinejad M, Nassiri-Kashani MH, Amiri Z and Talebi A 2016 *Med. J. Islam. Repub. Iran* **30** 1
[8] Vos T *et al. 2017 Lancet* **390** 10100 p 1211–1259
[9] Dale A, Welch L and Evanoff B 2018 *Occup Env. Med* **75** p A83–A84
[10] Wang L and Lau HYK 2012 *Work* **41** Suppl 1 p 2866–71
[11] Yazdani A and Wells R 2018 *Appl. Ergon* **73** June p 122–140
[12] Motamedzade M 2013 *Int. J. Occup. Saf. Ergon* **19** 3 p 475–484
[13] Tappin DC, Vitalis A and Bentley TA 2016 *Appl. Ergon* **52** p 151–159
[14] Khan MNAA and Lee HS 2014 *J. Sci. Res* **21** 8 p 1357–1364
[15] Knies E, Leisink P and van de Schoot, R 2017 *Int. J. Hum. Resour. Manag* **5192** p 1–33
[16] Lepold A, Tanzer N, Bregenzer A and Jiménez P 2018 *Int. J. Environ. Res. Public Health***15** 7
[17] Hair JF, Black WC, Babin BJ and Anderson RE 2014 *Multivariate Data Analysis***16** 1
[18] Ghozali I and Latan H 2014 *Partial Least Squares Konsep, Metode dan Aplikasi Menggunakan Program WARPPLS 4.0* 2nd ed. Semarang: Badan Penerbit Universitas Diponegoro
[19] Siregar I, Muchtar M A, Rahmat R F, Andayani U, Nasution T H, and Sari R M 2018 Method of calculation overall equipment effectiveness in fertilizer factory. *IOP Conference Series: Materials Science and Engineering* **308**(1), p. 012053

IOP Conf. Series: Materials Science and Engineering **851** (2020) 012017 doi:10.1088/1757-899X/851/1/012017

Environmental Sensor Assessment for a Light Quadcopter Drone

Suherman Suherman*, Rizky Ananda, Maksum Pinem

Electrical Engineering Department, Universitas Sumatera Utara, Medan-Indonesia

*suherman@usu.ac.id

Abstract. This paper presented drone design and environmental sensor assessment. The assessment was conducted for temperature and pressure. The measurements were performed offline and online. Offline measurement was by locating drone inside a controlled chamber, while online measurement was when drone is flying. Assessment shows that there were significant error differences in temperature, where offline drone exerted only 0.75% error, but online drone produced 6.23% error. Error increases as temperature increases. Pressure assessment exerted a slight error different between offline and online drone, 0.16% and 0.17%.

1. Introduction

Light drones have been popular being used for many applications such as toys, image retrieval, or even for farming applications [1]. Light drones are available from several dollars to thousands dollar depending upon the services provided. For specific or environmental monitoring, drone is usually equipped by some sensors [2]. Monitoring drones are mainly equipped by camera to take images from the air [3]. Temperature, pressure, humidity are some sensors to obtain environmental conditions at the surrounding of flying drones.

Drone is a smaller type of the unmanned aerial vehicle (UAV). UAV has been considered as a flying robot, as it requires control system as in robot. Drone design is dominated by propeller driven [4]. Military drones can fly several hundred kilometers, while small drones are usually restricted by local government, flying only up to several feet [5].

This paper examines environmental sensors attached on the designed drone that has been equipped by an obstacle avoidance system. The designed obstacle avoidance system makes use the ultrasonic's sensors. Although camera based [6] and wind based [7] obstacle avoidance systems are exist, ultrasonic sensor was chosen for its simplicity. Fuzzy logic was employed to decide drone maneuver, instead of neural network [8], again for drone simplification. Those simplifications must be done as the intended designed drone is the light one, so that environmental sensors attached to it still allow drone to fly normally.

Environmental sensors as well as ultrasonic sensors are connected to arduino. Measurement data was transmitted through telemetry radio to ground station. Radio of 433 MHz was chosen rather than WiFi as the 2.4GHz spectrum is more crowded at the assessment location [9]. In order to measure environmental properties precisely, sensors were operated and validated by using some standard laboratory measurement equipment. Transmitted data that were received by the ground station were then compared to the standard value. Sensor deviation was measured to ensure valid monitoring data.

2. Drone design and measurement method

Drone design is shown in Figure 1. Ultrasonic sensors were attached to drone body, and continuously ping arduino. Arduino forwards control signal from remote receiver to flight controller based on sensor ping to apply collision avoidance. Meanwhile, environmental sensors were connected to arduino, ardunio forwards measurement data to ground station by using 433 MHz telemetry radio.

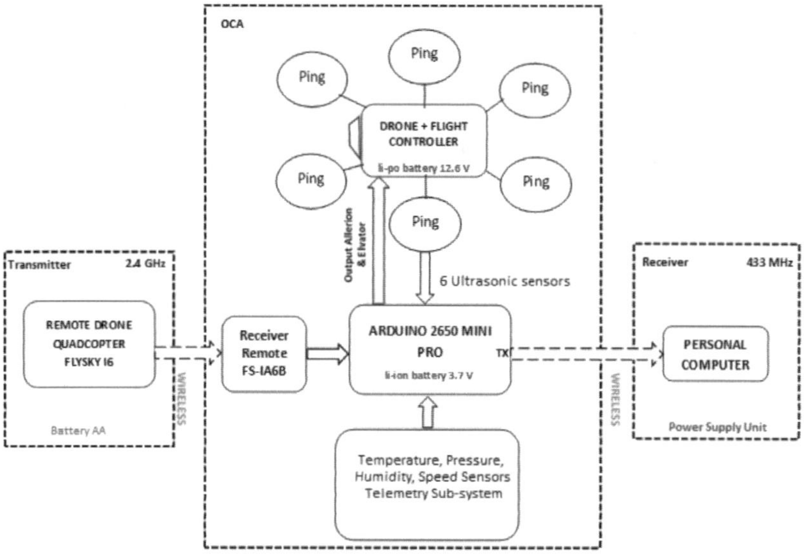

Figure 1. Designed Drone System

Figure 2 shows the realization of the drone designed using four propellers. Drone uses frame 250 made from fiber, four brushless motor MT2204 and ES20, four units three blade propellers, CC3D circuit as flight controller, FS-I6 remote control module, HC-12 433 SI4463 for telemetry and Li-Po battery1500 mAh for drone power source. Temperature and pressure sensor uses BME 280, air quality sensor uses MQ135. The fiber frame 250 was chosen to produce total weight of only 800 grams, including additional battery for arduino and 433 MHz radio system.

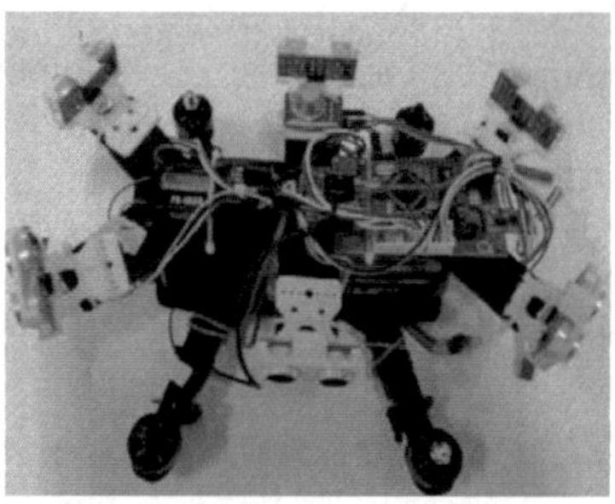

Figure 2. Drone realization

Next step is to assess environmental sensor, in this case, temperature and pressure. Offline temperature sensor assessment uses temperature chamber and a standard temperature sensor fluke type 5615with temperature variations of 20 , 30 and 40 °C. Pressure validation uses pressure chamber and standard pressure sensor Vaisala type PTB 330 as shown in Figure 3.

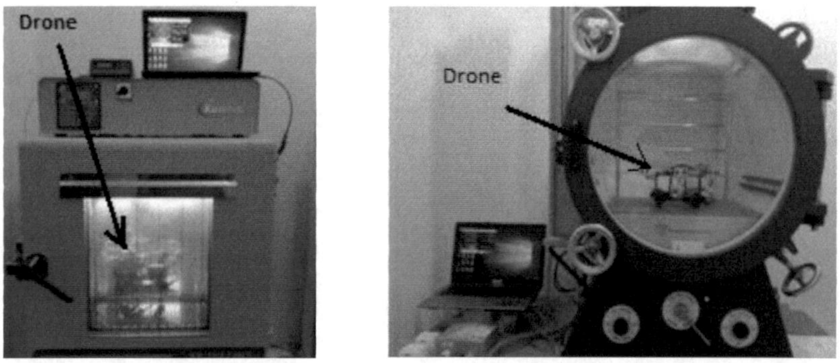

(a) Temperature sensor test (b) Pressure sensor test

Figure 3. Environmental sensor assessments

IOP Conf. Series: Materials Science and Engineering **851** (2020) 012017 doi:10.1088/1757-899X/851/1/012017

3. Assessment Results

Offline assessments by placing drone inside the chamber, setting some set points, and then comparing the telemetry data and the standard temperature gives values as shown in Table 1. There are slight differences between standard and the transmitted temperatures. Average temperature deviation is 0.23 °C, with the highest deviation is 0.38 °C. Deviation values increase to measured temperatures. This deviation produces measurement error 0.75% in average.

Table 1. Offline temperature assessment

Set Point	**Standard Temperature (°C)**	**Measured Temperature (°C)**
20 °C	20.095	20.21
	20.095	20.21
	20.095	20.21
	20.093	20.22
	20.093	20.22
30 °C	29.861	30.09
	29.859	30.09
	29.859	30.09
	29.861	30.08
40 °C	39.939	40.31
	39.939	40.31
	39.938	40.30
	39.938	40.32

Online assessments by flying the drone in a reachable height, measuring surrounding temperatures, and then comparing the received data and the measured temperature produces values as depicted in Table 2. Online measurements exert higher deviation than offline measurement. Average temperature deviation when drone fly is 1.88 °C, 7.19% higher than offline value. The highest deviation is 1.91 °C and measurement error is 6.23% in average.

Table 2. Online temperature assessment

Standard Temperature (°C)	**Online Assessment (°C)**
30.2	32.05
30.2	32.07
30.21	32.08
30.21	32.1
30.22	32.12
30.22	32.1
30.24	32.13
30.24	32.14
30.26	32.14
30.26	32.17

The same experiments were applied for pressure which gave values depicted in Table 3 and Table 4. Average offline pressure deviation is 1.62 mBar with average error is 0.16%. Deviation and error increased when drone flies, producing deviation of 1.73 mBar and error of 0.17%.

Table 3. Offline pressure assessment

Set Point	**Standard Pressure (mBar)**	**Measured Pressure (mBar)**
950 mBar	950.06	951.62
	950.06	951.62
	950.06	951.63
	950.06	951.64
1000 mBar	999.77	1001.32
	999.77	1001.34
	999.77	1001.35
	999.78	1001.35
1050 mBar	1050.05	1051.76
	1050.04	1051.75
	1050.04	1051.75
	1050.05	1051.76

Table 4. Online pressure assessment

Standard Temperature (°C)	**Online Assessment (°C)**
1011.51	1013.47
1011.51	1013.3
1011.51	1013.18
1011.5	1013.16
1011.5	1013.15
1011.49	1013.19
1011.49	1013.21
1011.49	1013.18
1011.49	1013.22
1011.48	1013.2

4. Conclusions

This paper presented environmental monitoring drone with focus on temperature and pressure sensor measurement. The offline measurement in a controlled chamber gave errors 0.75% and 0.16% for temperature and pressure subsequently. Meanwhile, when drone flies, temperature error increased significantly to 6.23%, while pressure error is only 0.17%. The increment may cause by drone vibration and surroundings. These error increments should be consideration for those measuring these parameters using drone.

Acknowledgement

This research has been supported by TALENTA USU research grant 2020.

References

[1] Giones F. Brem A 2017 From toys to tools: The co-evolution of technological and entrepreneurial developments in the drone industry. *Business Horizons*. **60**(6). 875-884.

[2] Chand G S L K. Lee M. Shin S Y 2018 Drone based wireless mesh network for disaster/military environment. *Journal of Computer and Communications*. **6**(04). 44.

[3] Pobkrut T. Eamsa-Ard. T. Kerdcharoen T (2016. June). Sensor drone for aerial odor mapping for agriculture and security services. In 2016 13th *International Conference on Electrical Engineering/Electronics. Computer. Telecommunications and Information Technology*. pp. 1-5

[4] Bhardwaj S 2015. Sensor Siystem Implementation for Unmanned Aerial Vehicles. Indian Journal of Science and Technology vol **8** p 7-11.

[5] Syaifudin A. 2016. Rancang Bangun Quadcopter untuk pencarian Rute Optimum pada Kebakaran Lahar Gambut menggunakan metode particle swarm optimization. Universitas islam negeri Maulana malik Ibrahim. Malang.

[6] McGuire K. De Croon G. De Wagter C. Tuyls K. and Kappen H 2017. Efficient optical flow and stereo vision for velocity estimation and obstacle avoidance on an autonomous pocket drone. IEEE Robotics and Automation Letters. **2**(2). 1070-1076.

[7] Gonzalez R D L G. Pohl D. Gutierrez D G. Hippelein A. Macias L C. Lopez R A. ... and Vervoorst J W 2019. U.S. Patent Application No. 16/020.338.

[8] Yang X. Chen J. Dang Y. Luo H. Tang Y. Liao C. and Cheng K T. 2019. Fast Depth Prediction and Obstacle Avoidance on a Monocular Drone Using Probabilistic Convolutional Neural Network. IEEE Transactions on Intelligent Transportation Systems.

[9] Al-Akaidi M, 2017, September. A review on transport layer protocol performance for delivering video on an adhoc network. In IOP Conference Series: Materials Science and Engineering (Vol. **237**, No. 1, p. 012018). IOP Publishing.

Electronic Snellen Chart with Bluetooth Connection and Smartphone App

H R Fajrin[1*], R N Adnan[1*], M Irfan[2], I P Sari[3]

[1]Department of Medical Electronics Technology, Vocational Program, Universitas Muhammadiyah Yogyakarta, Bantul, Yogyakarta, Indonesia
[2]PKU Muhammadiyah Hospital, Gamping, Yogyakarta, Indonesia
[3]Department of Informatics and Computer Engineering, Jakarta State Polytechnic, Depok, Indonesia

E-mail: hanifah.fajrin@vokasi.umy.ac.id*

Abstract. Snellen chart is a tool to measure a person's visual acuity consisting of a series of black capitalized letters on a whiteboard arranged in rows. Snellen charts commonly used in many hospitals today are still in the form of posters, which are inefficient because to show the letters on the Snellen board, we still have to input it manually. Therefore, an electronic Snellen chart tool, which is considered to be more efficient and practical is made. The control device is an application on a cellphone connected via Bluetooth by utilizing LED as lighting while showing letters on the Snellen chart. It will facilitate ophthalmologists to perform examinations, and also show the letters by pressing a button on the cellphone so that they can install a trial lens without having to return to the Snellen board to indicate the letters. The testing method used in this study tested the Bluetooth connection and the tool's performance with seven questions on a questionnaire testing method to determine the eligibility of the instrument. The feasibility test results by using a questionnaire show that it has 82.8% feasibility percentage with 0.9 validity rate and 0.87 reliability; moreover, the results of 20 trials of Bluetooth connection testing have a 100% success rate. The questionnaire results showed that the tool is eligible and proper to use because the installation of the application is easy to do.

Keywords—Bluetooth; Snellen chart; visual acuity;

1. Introduction

Visual acuity (VA) test [1] is a simple procedure performed in an eye examination based on optical principles. A visual acuity test aims to determine the smallest letters that can be read by patients on a standard chart (Snellen chart) in 20ft. The Snellen chart consists of a series of black capitalized letters arranged in rows on a whiteboard [2]. This chart is a helping tool to measure visual acuity. It is a poster containing letters and numbers of various sizes.

In the present time, the Snellen chart that commonly used in hospitals is in the form of a poster. Practically, the ophthalmologist must point the numbers or letters on the poster directly by using a pointer. It can reduce the efficiency of the examination because they cannot put the trial lens sets at the same time. They must designate the letters or numbers listed on the poster then return to the patient to put or replace the trial lens. It took more time and considered less practical [3,4].

Several Snellen charts in Hospital use selector control and displaying letters on the Snellen board is still set manually by pressing the selection button on the control board connecting via a cable connection with the Snellen board. The appointment of letters uses lights that will move using a motor, which will then illuminate the letters on the Snellen board. Despite that, the selector control is less efficient and practical as well as the risk of damaging the driving motor due to the intensity of the tool usage. In the study conducted by [5], the Snellen letters are displayed on a PC screen and have been equipped with a visual acuity inspection results displayed on the LCD control device. The control tool in this study is connected to the monitor/PC screen using a cable, so this is considered to be less practical in the installation. Because of this, the Snellen chart is able to point to letters or numbers more easily and efficiently. It is intended that the doctor does not need to stand in front of the patient to be able to point the letters or numbers, the doctor can stand next to the patient and put on a trial lens to measure the patient's visual acuity. Therefore, an electronic Snellen chart tool is the answer to these problems [2].

Electronic Snellen chart uses LED lights that will light up on the letters or numbers intended by the doctor. An application on a mobile phone controls it. In the control application, there is a button menu that can be selected to show letters or numbers on the Snellen chart board. The usage of Bluetooth will make it easier to maintain the device because there is no need to maintain or replace cable [6,7]. The use of electronic charts can increase efficiency in performing the visual acuity test. However, to the best of our knowledge, we design an easier and more practical control of letter on the Snellen chart. The author creates an electronic Snellen chart with lighting using LED and controls the appointment of letters using Bluetooth remote [8]. Letter appointment is made by adjusting the light in the designated letters only. Using a Bluetooth connection and application in a smartphone make it easier for the ophthalmologist to carry out checks, they can do the examination as well as testing using a trial lens.

2. Research Method

2.1. Software Designing

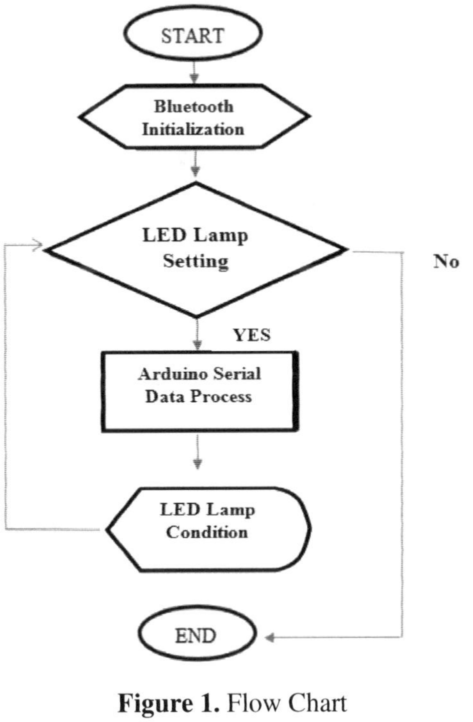

Figure 1. Flow Chart

Based on Figure 1, the process starts with Bluetooth initialization by connecting Bluetooth on the smartphone with Bluetooth on the Snellen chart board, then setting the LED lights on the appliance control application, which will then send the serial data that will be processed by Arduino. If the serial data transmitted by the app is in the form of High, so the LED will light and the Low value to turn off the LED.

2.2 Tool designing

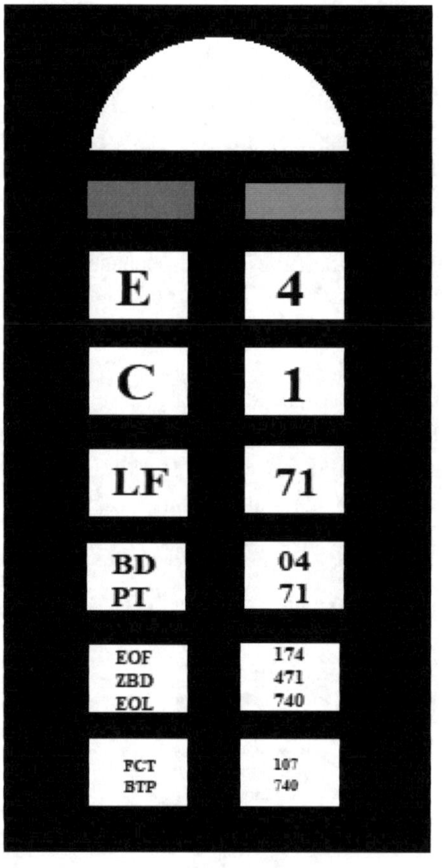

Figure. 2 Snellen Chart Board

Figure 2 is a Snellen chart board design. In the front, there is an acrylic board with a variety of letters and numbers in various sizes in black with white background. On the Snellen board, there is a power supply, an Arduino minimum system, a Bluetooth module, and an LED light. On the board, there are ten types of font sizes arranged from the top down from the largest to the smallest letter size. Ranging from sizes 20/200, 20/100, 20/70, 20/50, 20/40, 20/30, 20/25, 20/20, 20/15 and 20/13 [9,10].

Figure 3 is a tool control application with a text box, Bluetooth connection settings, and a letter or number selection button that will be directed or shown during the inspection, and a Turn Off button to turn off the LED lights. The function of the application is to adjust the lights to show letters on the Snellen board.

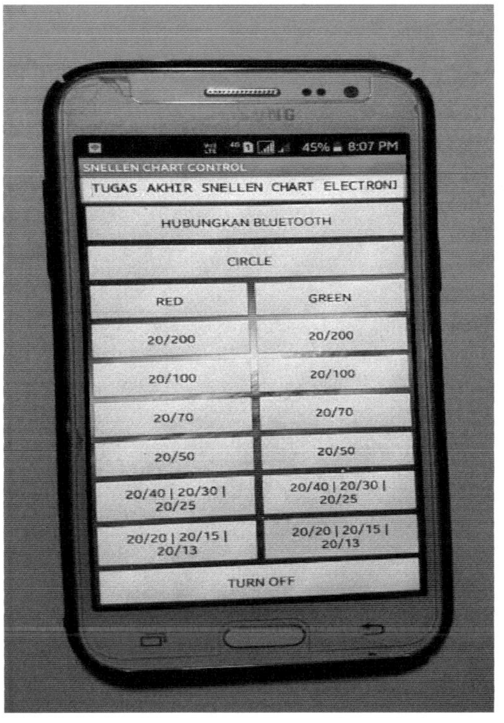

Figure 3 Control Tool Application Display

3. Results And Discussion

3.1. Bluetooth Connection Test

This test aims to measure the success rate of Bluetooth connection with the use of a control application. Testing is performed by turn on/off the Bluetooth connection, turn on the LED, which is done 20 times with the distance between the Snellen board and the control device as far as 7 meters. In the usage, the need to use the Bluetooth range of the Snellen chart is as far as 6 meters. It is based on the visibility between the Snellen board and the patient. Bluetooth connection test uses a distance of 7 meters to measure the maximum ability of a Bluetooth connection. By conducting this test, it will prove that the device is able to make a good connection, even though it is less than 7 meters. Hence, this ensures that the range of the Bluetooth connection will work well within a range of 6 meters. The following are the Bluetooth test results in Table 1.

Table 1. Bluetooth Connection Test

Trial	Connecting	Stop	Accuracy
1.	Passed	Passed	
2.	Passed	Passed	
3.	Passed	Passed	
4.	Passed	Passed	
5.	Passed	Passed	
6.	Passed	Passed	
7.	Passed	Passed	

IOP Conf. Series: Materials Science and Engineering **851** (2020) 012018 doi:10.1088/1757-899X/851/1/012018

8.	Passed	Passed	
9.	Passed	Passed	
10.	Passed	Passed	
11.	Passed	Passed	100%
12.	Passed	Passed	
13.	Passed	Passed	
14.	Passed	Passed	
15.	Passed	Passed	
16.	Passed	Passed	
17.	Passed	Passed	
18.	Passed	Passed	
19.	Passed	Passed	
20.	Passed	Passed	

According to the test results, as can be seen in Table 1, by testing to turn on and off the Bluetooth connection, it indicates that the Bluetooth connection got a 100% success rate. Based on these results it can be ensured that the performance of Bluetooth on the device is very good.

3.2 Tool Feasibility Test

The test is done by conducting a performance test tool that is assessed by the ophthalmologist in the form of a questionnaire. The test was conducted to 5 respondents (ophthalmologist), with the following test results in Table 2. The following are the values/points from each answer to the questions listed in the questionnaire tool:

a. Very Suitable (SS) = 5
b. Suitable (S) = 4
c. Quite Suitable (CS) = 3
d. Unsuitable (TS) = 2
e. Very unsuitable (STS) =1

Table 2. Tools Feasibility Testing

No.	**Questions**	**Score**				
		SS	S	CS	TS	STS
1.	Is the lighting on the device good?	1	2	1	1	
2.	Is the font on the Snellen board appropriate?	1	3		1	
3.	Is the tool appropriate and can be used in the checkup procedure?	2	2		1	
4.	Is the procedure for turning on the device easy to do?	2	3			
5.	Is the control application easy and	1	4			

IOP Conf. Series: Materials Science and Engineering **851** (2020) 012018 doi:10.1088/1757-899X/851/1/012018

	practical to use?					
6.	Is it easy to install the control application?	3	2			
7.	Is it easy to connect the Bluetooth?	2	3			
Total			76	3	6	0
Total Score				145		
Feasibility Percentage				82.8%		

Based on Table 2, in the first question, respondents answered the most questions with the "appropriate" range, with the number of respondents is two people, while three respondents each chose very appropriate, quite appropriate and inappropriate. In the second question, most respondents answered in the "appropriate" range, with the number of respondents who chose three people, and one person chose "very appropriate," and one person chose "inappropriate." In the third question, two respondents chose "very appropriate," 2 respondents chose "appropriate," and one person chose "inappropriate". Whereas in the fourth question, respondents chose more in the range "appropriate" with the number of voters, three people and two respondents chose "very appropriate." In the fifth question, respondents who chose "appropriate" as many as four people and one person chose "very appropriate." Then in the sixth question, respondents who chose the range "very appropriate" amounted to 2 people, and three people chose "very appropriate." In the last question, 3 respondents chose "appropriate" range, and 2 people chose "very appropriate".

According to the number of observation scores in Table 2, the sum of the scores of each observation result question is multiplied by the weight of the score using the Likert scale [11]. The maximum score is the maximum score on the Likert scale multiplied by the number of questions, so 5x7=35. The expected total score is the maximum score multiplied by the number of respondents, so 35x5=175. The calculation of the feasibility percentage from the data above has a percentage value of 82.8%. The total observational score of the tools' feasibility data was 145 (82.8%) of the expected score of 175 (100%). Based on the criteria in the feasibility table [12], the total score percentage is included in the Very Feasible category. The following is the feasibility scale in Figure 4.

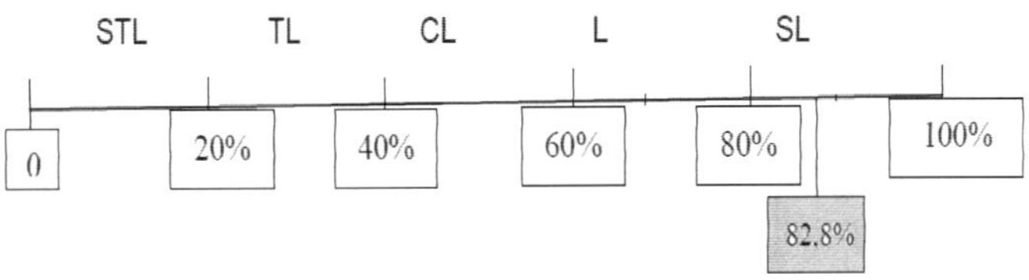

Figure 4. Feasibility Scale [13]

Explanation:
STL = Very Inappropriate
TL = Inappropriate
CL = Less Feasible
L = Feasible
SL = Very Feasible

Table 3. Score of Questionnaire Testing

Respondent	**Questionnaire Question (x)**							**Total (y)**
	1	2	3	4	5	6	7	
1	4	5	4	4	4	5	4	30

2	5	4	5	5	5	5	5	34
3	4	4	5	4	4	4	4	29
4	2	2	2	4	4	4	4	22
5	3	4	4	5	4	5	5	30
Total	18	19	20	22	21	23	22	145

Based on Table 3, it can be seen that the test results show that questions number 3 and 6 get high points. The results of this test prove that the electronic Snellen chart tool is appropriate and can be used in inspection and control tools are easy to use.

According to the validity and reliability test, it obtains that the validity coefficient is 0.90. Thus, the data is considered as very high validity criteria. High or low coefficient correlation [14] can be determined by the number of correlation coefficients ranging between several levels, such as (1). 0.00-0.20: very low; (2). 0.20-0.40: low; (3). 0.40-0.70: medium; (4). 0.70-0.90: high; (5). 0.90-1.00: very high. High or low correlation coefficient value determines the accuracy of a measuring instrument in making measurements. For reliability testing, the reliability value is 0.87. In line with it, it can be said that the test results have high reliability because the value meets the level of reliability value >0.60 [15,16]. Measuring the reliability value aims to measure the accuracy of the measurement results, the relationship between the variables tested, and the consistency of a measuring instrument. According to two things above (reliability and validity), it can be ensured that the tool testing is appropriate, the tool also has compatibility with the existing standards, and it is suitable to use.

4. Conclusion

It can be concluded that the electronic Snellen chart is able to be created by the control through the application on a smartphone connected to Bluetooth. The electronic Snellen chart can function adequately, it is appropriate and suitable for a checkup. It is according to the scores obtained on the tool feasibility test with a questionnaire method containing seven questions and five ophthalmologists as respondents. The questionnaire of the electronic Snellen chart feasibility test results from 82.8%. It means that it meets the standards and can be used in checkups. The questionnaire data used in tools testing are validated, and reliable based on the validity and reliability test obtains from 0.90 on validity and 0.87 on reliability. The tool uses a smartphone application as the control tool for the user to conduct easier checkups.

References

[1] Leitman M W 2007 *Manual for eye examination and diagnosis* (USA: Blackwell)

[2] Radner W 2017 Reading charts in ophthalmology *Graefe's Arch. Clin. Exp. Ophthalmol.* **255** 1465–82

[3] Martin H G 1941 Illumination Standardization of the Snellen Chart* *Am. J. Ophthalmol.* **24** 287–91

[4] Regan D and Neima D 1983 Low-Contrast Letter Charts as a Test of Visual Function *Ophthalmology* **90** 1192–200

[5] Johnson A T, Dooly C R and Simpson C R 1998 Generating the Snellen chart by computer *Comput. Methods Programs Biomed.* **57** 161–6

[6] Agarwal A, Abhishek K, Kumar V, Kumar V, Prasad N and Singh M P 2015 Dr. Eye: An Android Application to Calculate the Vision Acuity *Procedia Comput. Sci.* **54** 697–702

[7] Pathipati A S, Wood E H, Lam C K, Sáles C S and Moshfeghi D M 2016 Visual acuity measured with a smartphone app is more accurate than Snellen testing by emergency department providers *Graefe's Arch. Clin. Exp. Ophthalmol.* **254** 1175–80

[8] Fajrin H R, Ilahi M R, Handoko B S and Sari I P 2019 Body temperature monitoring based on telemedicine *J. Phys. Conf. Ser.* **1381**

[9] Keating D and Parks S 2008 Sight Measurement *Assist. Technol. Vis. Impair. Blind People* 103–33

[10] Oduntan A O 1996 Arabic near test chart for partially sighted patients *Ophthalmic Physiol. Opt.* **16** 450–2

[11] Wadgave U and Khairnar M R 2016 Parametric tests for Likert scale: For and against *Asian J. Psychiatr.* **24** 67–8

[12] Kimbrel N A, Flynn E J, Carpenter G S J, Cammarata C M, Leto F, Ostiguy W J, Kamholz B W, Zimering R T and Gulliver S B 2015 Internal consistency, test-retest reliability, and predictive validity for a Likert-based version of the Sources of occupational stress-14 (SOOS-14) scale *Psychiatry Res.* **228** 961–2

[13] Louis Cohen L M and K M 2007 *Research Methods in Education* (New York: Taylor & Francis Group)

[14] Ma D J, Yang H K and Hwang J M 2013 Reliability and validity of an automated computerized visual acuity and stereoacuity test in children using an interactive video game *Am. J. Ophthalmol.* **156** 195–201.e1

[15] Douma S G, Bombois X and Van Den Hof P M J 2005 Validity of the standard cross-correlation test for model structure validation *IFAC Proceedings Volumes (IFAC-PapersOnline)* vol 16 (IFAC) pp 898–903

[16] Vanslette K, Tohme T and Youcef-Toumi K 2020 A general model validation and testing tool *Reliab. Eng. Syst. Saf.* **195** 106684

Human Reliability Assessment Analysis with Human Error Assessment and Reduction Technique (HEART) Method on Sterilizer Station at XYZ Company

Chalis Fajri Hasibuan, Yudi Daeng P, Reza Rinaldi Hasibuan

Industrial Engineering Departmenent, Faculty Of Engineering, Universitas Medan Area, Indonesia

chalisfajri@yahoo.co.id

Abstract. The matter of human reliability is a subject that has great influence on work-related accident levels or productivity levels of a product. Work-related accidents at XYZ company has 3 types of accidents that occur at boiling station and most of it are caused by human error from work operators, such as boiling overtime, lorry that went off track, inputting inaccurate boiling pressure standards, low discipline levels, not using protection equipment. XYZ company is one of many companies that processes palm oil. Human Reliability Assessment is done using HEART (Human Error Assessment and Reduction Technique) method. At first, an analysis is done on work operator's tasks by organizing HTA (Hierarchical Task Analysis). The resultant description of tasks at boiling station reveal that there is 11 tasks being handled by operator. Based on the calculation, the value of HEP (Human Error Probability) is highest on setting boiling time task with the value of 0,5324. The second highest is 0,5251 on setting steam pressure task. Researcher has analysed which task with the highest HEP value using Fishbone diagram, and continues to compile prevention methods for human error based from EPC, HEP, and Fishbone diagram.

1. Introduction

The HRA method that is being used in this research is HEART (Human Error Assessment and Reduction Technique) which is a part of quantifying human reliability method developed by Williams on 1985. HEART is proven to be more advantageous due to its flexible implementation in various situation or industry such as chemical, aviation, oil, medical field and so [1]. HEART method describes the task that should be done by an operator and its difficulty well and detailed [2]. This method has also been tested for its validity [3], where his research results claims that HEART method has an acceptable level of accuracy. Human Reliability Assessment done with HEART method is a technique aiming to evaluate the possibility of human error that occurs during completing a certain task, with the basis that each time a task is completed, there is a chance of failure that is being influenced by one or more EPC (Error Producing Condition) such as fatigue, unstable work place, etc. These factors that has significant influence to performance is graded with the highest HEP value. These conditions then can be implemented to "best-case scenario" then calculating the probability of failure under the ideal condition to receive the final chances of errors.

The calculated result will help in communicating the chance of error in analysing higher risks or safety cases. HEART method also has an indirect effect on how to increase reliability from ergonomic stand point.

IOP Conf. Series: Materials Science and Engineering **851** (2020) 012019 doi:10.1088/1757-899X/851/1/012019

HEART method is a technique aiming to evaluate the possibility of human error that occurs during completing a certain task, with the basis that each time a task is completed, there is a chance of failure that is being influenced by one or more EPC (Error Producing Condition) such as fatigue, unstable work place, etc. These factors that has significant influence to performance is graded with the highest HEP value. HEART method is built upon several subjects, which are:

1. Human reliability depends on generic nature of the tasks that should be completed.
2. In ideal conditions, reliability levels that is reached tends to be consistent with nominal probability in probabilistic limit.
3. Considering that ideal conditions cannot be reached in every situation, human reliability can be predicted to decrease as the identification of EPC increase.[4]

2. Methods of Research:

The type of research that is being used in this journal is descriptive research. This research is done at. XYZ company that moves in palm oil production. This research is done at Boiling Station. First, researcher identifies tasks done by operator at boiling station using HTA (Hierarchical Task Analysis). And then researcher will identify failure or human error at boiling station. Afterwards, researcher will grade Human Error Probability (HEP) with HEART method. This method is done in several stage by classifying tasks into generic task. Subsequently identifying conditions that cause errors with EPC (Error Producing Conditions). In the end, predicting the effect of every EPC with value of Assessed Proportion, Assessed Proportion value can be obtained from interview experts to list tasks at boiling station. The final stage is calculating Assessed effect and HEP (Human Error Probability) value.

In HEART method, there a several stages to be completed am the first stage is HTA (Hierarchy Task Analysis). According to Mandal, et al. The First Stage is the identification work process which is crucial to do. If the process does not divide into sub-processes, it would result in unidentified cause when drafting alternative solutions and would leave errors unchecked in those processes.

The second stage of this method is to classify the tasks that has been analyzed into nominal levels according to tasks unreliability. Classifying the tasks into its general categorization and inputting the nominal value of its level of human unreliability corresponding to table 1

Table 1. Generic Task

Code	**Task Catagory**	*Nominal Human Unreliability*
A	Totally unfamiliar, performed at speed with no real idea of likely consequences	0,55
B	Shift or restore system to a new or original state on a single attempt without supervision or procedures	0,26
C	Complex task requiring high level of comprehension and skill	0,16
D	Fairly simple task performed rapidly or given scant attention	0,09
E	Routine, highly practised, rapid task involving relatively low level of skill	0,02
F	Restore or shift a system to original or new state following procedures, with some checking	0,003

Code	**Task Catagory**	*Nominal Human Unreliability*
G	Completely familiar, well-designed, highly practised, routine task occurring several times per hour, performed to highest possible standards by highly motivated, highlytrained and experienced personnel, with time to correct potential error, but without the benefit of significant job aids	0,0004
H	Respond correctly to system command even when there is an augmented or automated supervisory system providing accurate interpretation of system state	0,00002

The third stage of HEART method is identifying conditions that cause errors (EPC's) which will be applied into scenarios or activities of the tasks that is being analyzed, so that adjustments could be made corresponding to EPC table. This EPC value identification could be done by observing directly with emphasis on details how operators on the field work. Table 2 is EPC value table.

The fourth stage of HEART method is to determine Assessed Proportion value with the range from 0 – 1. Determining this value is done according to interview results from experts and the operator concerned. The continued with calculating the HEP (Human Error Probability) value.

defines human error as an action or behaviour of human that is inappropriate, causing effectivity, work safety, and system performance to decrease [5]. Human error can occur due to manifestation of characterization that is categorised as unwanted, such as tardiness, clumsiness, forgetfulness, that could cause miscommunication at work or inaccurate form of risk-taking action.[6].

Claim that human error is an action that fails to achieve expected results, producing consequences differing from expectations [7]. Explains that there are a few similarities that seems to be driving factor that define human error. One of these driving factors is deviations such as trip/stumble, slip/lapses, mistake

Based on these opinion, human error definition can be simplified as human failure in performing tasks or producing result that unsuited with the main goal that is trying to be achieved.[8]

Fishbone diagram is a method to analyse cause of problem or condition. Often times this diagram is also called cause effect diagram. The inventor, Professor Kaoru Ishikawa, is a Japanese researcher who is alumni of chemical engineering, University of Tokyo on 1943.
So, this diagram could be called Ishikawa diagram as well. Fishbone diagram could be used for:

1. Identifying root of cause of a problem.
2. Achieving ideas for solutions for solving a problem.
3. Helping in investigating and researching facts further.

3. Result and Discussion

Form calculation human eror probality in XYZ company can see in table 2

Table 2. HEP calculation

Task	*Generic Task*	*Sterilizer Station Calculation*						HEP
Check Capstand and Mobile Centiliver	D	EPC	31	36	38			0,1047
		Proportion	0,7	0,2	0,4			
		Assessed Effect	1,140	1,012	1,008			
Clean The Rail Track	D	EPC	25	31				0,1272
		Proportion	0,4	0,7				
		Assessed Effect	1,240	1,140				
Link a link to a truck/Lori	E	EPC	31	33	34	38		0,0256
		Proportion	0,7	0,5	0,3	0,7		
		Assessed Effect	1,140	1,075	1,030	1,014		
Attract a Lori with a capstand	G	EPC	31	34	38			0,0005
		Proportion	0,7	0,2	0,2			
		Assessed Effect	1,140	1,020	1,004			
Replace lori with mobile centiliver	G	EPC	31	34	38			0,0005
		Proportion	0,7	0,2	0,2			
		Assessed Effect	1,140	1,020	1,004			
Check and clean condensate filter plate	D	EPC	31	36	38			0,1047
		Proportion	0,7	0,2	0,4			
		Assessed Effect	1,140	1,012	1,008			
Look at pressure and temperature gauges	E	EPC	11	23	31	38		0,1110
		Proportion	0,6	0,7	0,7	0,4		
		Assessed Effect	3,4	1,420	1,140	1,008		
open and close the door stew/Sterilizer	D	EPC	27	31	33	34		0,1480
		Proportion	0,7	0,7	0,7	0,2		
		Assessed Effect	1,280	1,140	1,105	1,020		
Dispose of condensate water	E	EPC	23	31	34	38		0,0344
		Proportion	0,7	0,7	0,5	0,6		
		Assessed Effect	1,420	1,140	1,050	1,012		
steam in and out (regulate steam pressure)	D	EPC	11	23	31	34	38	0,5251
		Proportion	0,6	0,7	0,7	0,6	0,6	
		Assessed Effect	3,4	1,420	1,140	1,060	1,012	
Set the stew/sterilize time	D	EPC	11	23	31	34	38	0,5324
		Proportion	0,6	0,7	0,7	0,6	0,7	
		Assessed Effect	3,4	1,420	1,140	1,060	1,014	

Based on figure 1, it could be seen the highest rated HEP type of task to the lower one.

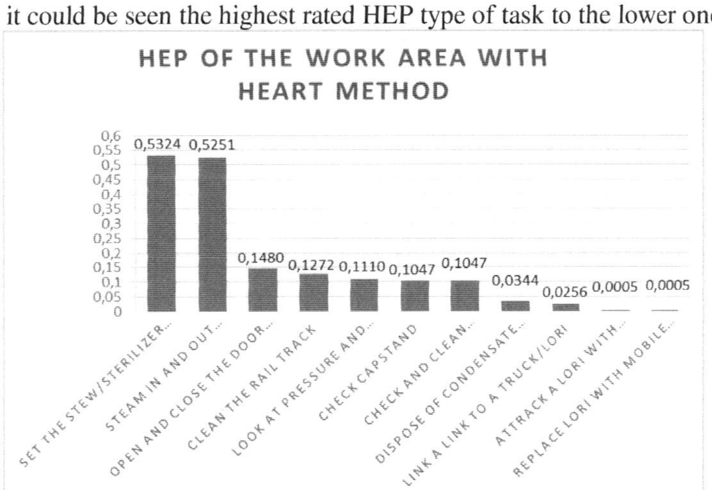

Figure 1. Order of HEP values from high to low

Based from measurement from the HEART method, the biggest HEP (Human Error Probability was on the task which set the boiling time, which at the value of 0,5324, where the failure deemed from a very small or too much amount of time spending on boiling. The second highest HEP value was on the task which putting in and out steam (setting steam pressure), at the number 0,5251. Then the total HEP would be analysed using the *fishbone* diagram to determine the cause and factor which contributes to the operator failure in performing the task.

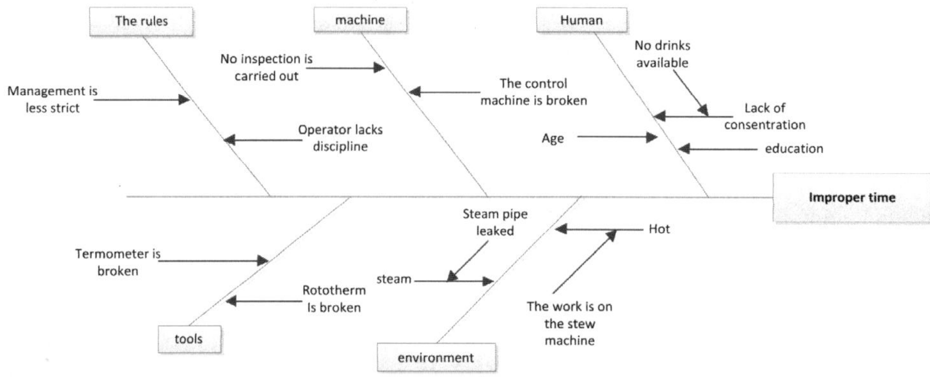

Figure 2. Fishbone Diagram for Stew/Sterilizer Time Settings

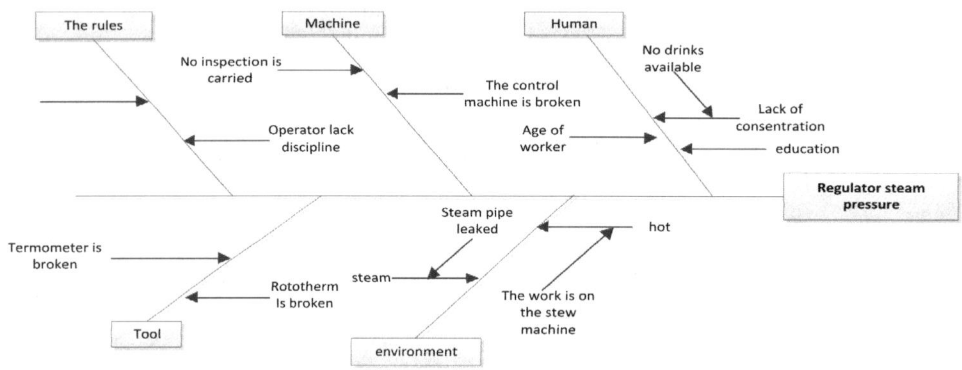

Figure 3. Fishbone Diagram for Regulator Steam Pressure

Based on HEP data, researchers would analyze the factor which contributes into failure of the task using fishbone diagram.

a. Human

Related causes from the human side were education background, where in performing the task, knowledge and skill is needed. Age of the operator would contribute a critical aspect on the endurance of the operator, the older they get, the lower the productivity rate, and at a risk of faster rate in getting tired and lower the concentration. Low concentration would increase the failure rate. Hot environment causing dehydration and unavailability of drinks play a critical role also in the rate of failure.

b. Machine

Rarity of checking up the boiling control machine causing failure of the machine. When there is malfunction of the machine and can't be used, operator forced to do the task manually, leading into higher human error rate when performing the task, such as setting the steam pressure which didn't meet the standard, or wrongly setting the boiling time, etc.

c. Environment

Environment in the boiling station isn't suitable as a working environment, caused by steam leaking out of pipes from steam turbine to boiler, causing distraction to operator and making the work environment hotter, thus lowering concentration. Temperature also affect operator at various degree (Wignjosoebroto, 2008);

1. ± 49° C: Operator could endure it at least for 1 hour, even though it exceeds the physical and mental endurance.
2. ± 30 ° C: mental and concentration lowers, and tend to make mistake in performing the task, and causing physical fatigue.
3. ± 24 ° C: optimum condition.
4. ± 10 ° C: extreme weird physical behavior would be apparent.

Temperature at the boiling station would range between 30 ° C - 34 ° C, causing fatigue on physical mental state and increasing the tendency of making mistake. In the hot environment it also increases injury or work accident related to dehydration.

d. Equipment

The broken boiling control machine forcing the operators to manually setting the boiling time cycle which had 3 peak boiling time. When there's unavailability of the boiler setting machine, operator would assume and count the boiling time with the working clock at the working space, leading to a very short amount or exceed the boiling time, inaccuracy in putting in steam, etc. The mistake committed by operators would contribute into lowering the productivity in the boiling station or lowering the desired oil result from boiling.

When tasked of setting the steam pressure, rarity of checking up the equipment and machine causing malfunction and failure of the equipment. Broken Thermometer and Rototherm would impact the performance of operator who can't determine the exact temperature or pressure fit for boiling thus using manual means by feeling and assuming it, thus lowering the productivity of TBS (Tandan Buah Segar) from boiling.

e. Rules

The faulty company management in organizing the operators contribute into less disciplined operator. A lot of reports not using the protective equipment, thus causing work accident. Operator working on the boiling station reports of getting injured by steam caused by leaking pipe, opening hose or the steam pressure regulating pipe without protective equipment such as gloves, which lead into injury and work accident.

4. Conclusion

Factors which contributes into the rate of operator failure in performing the task in boiling station are as follows, Age of operator plays an important role into stamina and endurance of the operator, the older they get the lower the productivity and the faster getting tired thus causing low concentration which increasing the high rate of human error., Education plays an important role because in performing the task knowledge, discipline and skill in performing the task. Operators in the boiling station still haven't the required amount of discipline, caused by unused Protection Equipment , Equipment and Machine, plays a critical role in the rate of human error, due to broken machine and equipment in the boiling station forcing the operators using manual means and feeling without the help of the equipment/machine, leading to mistake in the boiling time, too much steam pressure, etc., The type of task also become a contributing factor an operator commits a mistake or not. Harder tasks need more people to finish it, and it counts on the age of operator, equipment/machine, and operator's education. And Based on the result of HEART (Human Error Assessment and Reduction Technique), the task with the highest HEP is setting the boiling time, at the number 0,5323, then followed by putting in and out the steam, at the number 0,5251.

5. References

[1]. Bell J, and Holroyd J. 2009. Review of human Reliability Assessment Method. Health and Safety laboratory.
[2]. Safitri D M 2015.*Human Reliability Assessment dengan Metode Human Error Assesment and Reduction Technique pada Operator Stasiun Shourd PT.X.* 4(1):1-7
[3]. Kirwan B 1994. A Guide To Practical Human Reliability Assesment. Taylor& Francis, London
[4]. Dhillon B, 2007 Human Reliability and Error in Transportation System. London: Springer-Verlag.
[5]. Sanders M S and McCormick E J, 1993, Human Factors In Engineering and Desain, 7thedition, McGraw-Hill Book Company, New York.
[6] Iridiastadi,H dan Yassierli 2016. *Ergonomi Suatu Pengantar.* Bandung: Rosda
[7]. Hollnagel E, 1998. Cognitive Reliability and Error Analysis Method. Edisi I.Norway: Elsevier.
[8]. Nurhayati R, Ma,ruf I, Hartanti R 2017. Penilaian *Human Error Probability* dengan Metode *Human Error Assessment and Reduction Technique* di PT. Eratex Djaja,Tbk. 5(3):1-11

IOP Conf. Series: Materials Science and Engineering **851** (2020) 012020 doi:10.1088/1757-899X/851/1/012020

Effect of pressure to rice plant fibre reinforced composite on coefficient of friction of brake lining

Agung Setyo Darmawan[1,*], Pramuko Ilmu Purboputro[1], Bambang Waluyo Febriantoko[1] and Abdul Hamid[2]

[1]Department of Mechanical Engineering, Faculty of Engineering, Universitas Muhammadiyah Surakarta, Jl. Ahmad Yani, Tromol Pos 1 Pabelan, Surakarta 57162, Indonesia
[2]Department of Mechanical Engineering, Faculty of Engineering, Universitas Batam, 29464 Batam, Indonesia

*Corresponding author: Agung.Darmawan@ums.ac.id

Abstract. Brake lining of motor cycle made from asbestos can produce powder when friction occurs. The resulting powder can enter the human body and cause disease. For this reason it is necessary to find alternative materials that can replace asbestos. One material that has the potential to replace asbestos as brake lining is a rice plant fibre reinforced composite. This study aims to determine the effect of pressure on the coefficient of friction of a rice plant fibre reinforced composite. The rice plant fibre reinforced composite are made from fibre rice plant, aluminium powder and fibreglass as a dispersed phase and polyester resin as a matrix. Pressure varies from 10 kg/cm^2 to 30 kg/cm^2, with a speed of 2000 rpm. Micro photographs were also taken to investigate the phase and bond failure on the composite. The results showed an insignificant increase in the coefficient of friction by 1.74% in dry conditions and 6.23% in wet conditions. In addition, the results also show the coefficient of friction in dry conditions is higher than in wet conditions.

1. Introduction

Both motorcycle and car brakes are usually equipped with brake lining components, and are placed near the wheels. This brake lining has very vital functions, including helping to control braking, making it easier to cross the downhill and uphill roads, and providing easy braking facilities. Nowadays, brake lining of motor cycle made from natural resources and waste can replace classic materials that are difficult to process and help protect the environment [1, 2]. Brake lining material is designed to meet various performance requirements, including high and stable coefficient of friction, wear, strength, high recovery, low fade, noise and vibration in various work environments [3]. One of them is natural fiber ingredients [4]

Natural materials combined with metals, polymers and ceramics form a composite material. Composite materials can be defined as a combination or a mixture of two phases in a macroscopic and microscopic manner in order to obtain the desired combination of properties. To achieve the desired properties, various combinations of metals, ceramics and polymers are made and produce various types of composite materials. In general, composite materials are made to improve the predominance of mechanical properties such as strength, toughness, stiffness at room temperature and at high temperatures [5-7].

Composite materials are widely used in industry. Its uses include the fields of aircraft and military [8, 9], aerospace applications [10, 11], the automotive industry [12, 13], packaging materials [14] and sports equipment [15].

Composites in the current context are multi-phase materials that are made and not formed naturally. The forming phases are chemically different. This is different from metal alloys which, although also a multi-phase material (e.g. carbon steel with ferrite and cementite phases), are formed by natural phenomena. Most composite materials consist of only two phases: Matrix and Dispersed Phase. The matrix is the phase that binds to the dispersed phase. The properties of a composite material are a function of the properties of its constituent materials, their relative quantities, and the geometrical shape of the dispersed phase (including its distribution and orientation) [16, 17].

The use of composites as a motor cycle brake lining material will cause the composite to experience friction. This friction is affected by pressure as well as speed and operating conditions. Therefore, this study aims to investigate the effect of pressure on the coefficient of friction of a fiber reinforced composite rice plant. Rice plant fiber reinforced composite are selected because it has hardness and wear resistance suitable for use as brake lining of the motor cycle.

2. Materials and Methods

The materials used in this study are rice plant fiber reinforced composite containing 40% fiber rice plant, 40% polyester resin, 10% fiberglass and 10% Aluminum powder. The mesh size of aluminum is 60.

The first step of the study was to weigh the material using a digital scale according to the composition and then manually mixed until the material was evenly mixed. Brake lining molds are installed and brake lining plates are installed on the mold.

The next step is the brake lining plate is given polyester resin which is useful for attaching the material to the surface of the canvas. Then the composite material that has been mixed was put into the mold.

After that the mold is pressed using a press machine. Pressing is done by giving heat to a temperature of 90 ^{o}C. Then the composite is sintered at temperature of 180 ^{o}C with duration of 15 minutes. After it cools the brake lining is released from the mold.

After the specimen is obtained into a brake pad, then a micro photograph is taken. Micro photographs are carried out using optical microscope. The next test is testing the coefficient of friction. The friction tests were conducted in dry condition and wet condition. The coefficient of friction testing is carried out at 2000 rpm. In this test, the effect of pressure variations is examined by applying pressure variations of 10 kg/cm^2, 15 kg/cm^2, 20 kg/cm^2, 25 kg/cm^2 and 30 kg/cm^2. Failures that occur with the bond in the composite are also examined using optical microscopy.

3. Results and Discussions

Brake linings made of the composites are shown in figure 1. The composite is composed by 40% fiber rice plant, 40% polyester resin, 10% fiberglass and 10% Aluminum powder. Figure 2 shows a photo micro of a rice plant fiber reinforced composite. Rice plant fiber, aluminum powder and fiberglass play the role of transferring the load inside the polyester resin matrix.

IOP Conf. Series: Materials Science and Engineering **851** (2020) 012020 doi:10.1088/1757-899X/851/1/012020

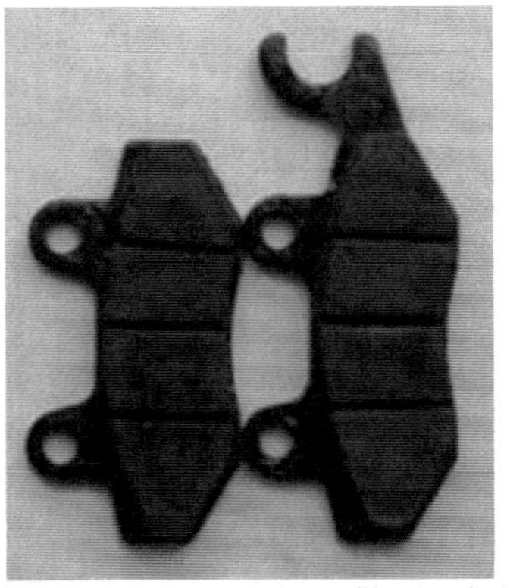

Figure 1. Brake lining form rice plant fiber reinforced composite

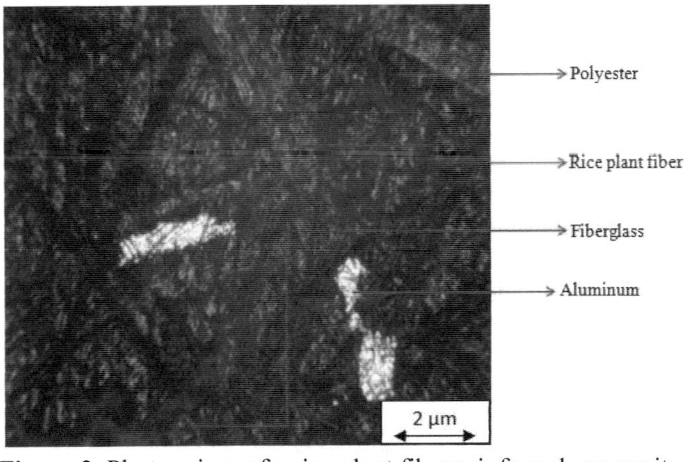

Figure 2. Photo micro of a rice plant fiber reinforced composite.

The relationship between the coefficient of friction and pressure is shown in figure 3. The higher pressures in the brake lining the greater the coefficient of friction will be exist. However, the increasing coefficient of friction was not significant. This trend is shown both in dry conditions and wet conditions.

The wet conditions show a smaller coefficient of friction than dry conditions. Dry conditions produce a coefficient of friction of 0.632, 0.635, 0.634, 0.64 and 0.643 for pressures of 10 kg/cm^2, 15 kg/cm^2, 20 kg/cm^2, 25 kg/cm^2 and 30 kg/cm^2, respectively. While wet conditions produce a coefficient of friction of 0.594, 0.597, 0.598, 0.628 and 0.631 for pressures of 10 kg/cm^2, 15 kg/cm^2, 20 kg/cm^2, 25 kg/cm^2 and 30 kg/cm^2, respectively.

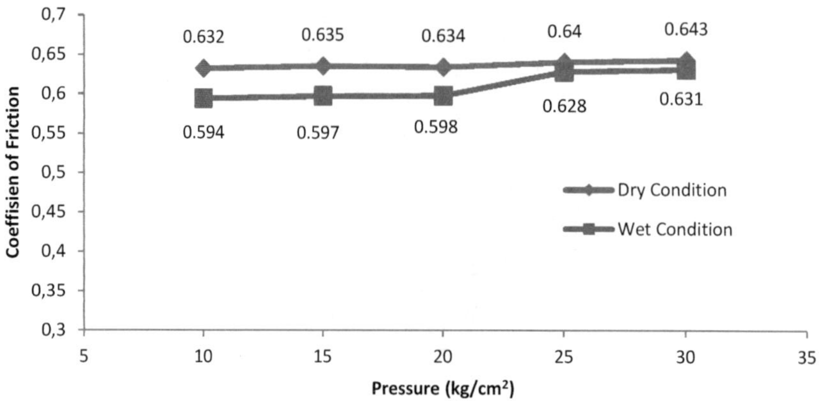

Figure 3. Relationship between coefficient of friction and pressure.

Figure 4 shows the bond failures that occur in composites when experiencing friction. Failures that occur in the form of adhesive bond failure. In the rice plant fiber reinforced composite, adhesive failure was an interfacial bond failure between the adhesive and the adherent.

Figure 4. Failure of adhesive bonding in rice plant fiber reinforced composite

4. Conclusions

An increase in pressure will result in an increase in the coefficient of friction, although not significantly. Operating conditions affect brake lining. Wet operating conditions will result in a lower coefficient of friction than dry conditions. In dry conditions, an increase in pressure from 10 kg/cm^2 to 30 kg/cm^2 will result in an increase in the coefficient of friction from 0.632 to 0.643. While in a wet condition an increase in pressure from 10 kg/cm^2 to 30 kg/cm^2 will result in an increase in the coefficient of friction from 0.594 to 0.631.

Acknowledgment

The authors would like to acknowledge The Directorate of Research and Community Service, The Ministry of Research, Technology and Higher Education of the Republic of Indonesia for research funding [Project Number = 199.24/A.3-III/LPPM/V/2019]. The author also would like to thank the Mechanical Engineering Laboratory of Universitas Muhammadiyah Surakarta and Universitas Gajah Mada for mechanical testing carried out in this research.

References

[1] Yun R, Filip P and Lu Y 2010 Performance and evaluation of eco-friendly brake friction materials, *Tribology International* **43(11)** 2010–2019

[2] Darmawan A S, Purboputro P I and Febriantoko B W 2020 The aluminum powder size' effect on rice plant fiber reinforced composite to hardness, wear and coefficient of friction of brake lining, *IOP Conference Series: Materials Science and Engineering* **722** 012002.

[3] Singh T, Kumar N J A R, Grewal J S, Patnaik A and Fekete G 2019 Natural fiber reinforced non-asbestos brake friction composites: Influence of ramie fiber on physico-mechanical and tribological properties, *Materials Research Express* **6** 115701.

[4] Cavalcanti D K K, Banea M D, Neto J S S, Lima R A A and Carbas R J C 2019 Mechanical characterization of intralaminar natural fibre-reinforced hybrid composites, *Composites Part B: Engineering* **175** 107149.

[5] Sarikaya E, Çallioğlu H and Demirel H 2019 Production of epoxy composites reinforced by different natural fibers and their mechanical properties, *Composites Part B: Engineering* **167** 461-466.

[6] Salernitano E and Migliaresi C 2003 Composite Materials for Biomedical Applications: A Review, *Journal of applied biomaterials & biomechanics* **1(1)** 3-18.

[7] Verma D, Gope P C, Shandilya A, Gupta A and Maheshwari M K 2013 Coir Fibre Reinforcement and Application in Polymer Composites: A Review, *J. Mater. Environ. Sci.* **4(2)** 263-276.

[8] Ren Y, Qiu L, Yuan S and Fang F 2020 Gaussian mixture model and delay-and-sum based 4D imaging of damage in aircraft composite structures under time-varying conditions, *Mechanical Systems and Signal Processing* **135** 106390.

[9] Turczyn R, Krukiewicz K, Katunin A, Sroka J and Sul P 2020 Fabrication and application of electrically conducting composites for electromagnetic interference shielding of remotely piloted aircraft systems, *Composite Structures* **232** 111498.

[10] Zhou J, Li Y, Cheng L and Zhang L 2019 Indirect Microwave Curing Process Design for Manufacturing Thick Multidirectional Carbon Fiber Reinforced Thermoset Composite Materials, *Applied Composite Materials* **26(2)** 533–552.

[11] Kappel E. 2019 Distortions of composite aerospace frames due to processing, thermal loads and trimming operations and an assessment from an assembly perspective, *Composite Structures* **220** 338-346.

[12] Pietroluongo M, Padovano E, Frache A and Badini C 2020 Mechanical recycling of an end-of-life automotive composite component, *Sustainable Materials and Technologies* **23** e00143.

[13] Khatkar V, Behera B K and Manjunath R N 2019 Textile structural composites for automotive leaf spring application, *Composites Part B: Engineering* **182** 107662.

[14] Mohamed S A A, El-Sakhawy M, Nashy E H A and Othman A M 2019 Novel natural composite films as packaging materials with enhanced properties, *International Journal of Biological Macromolecules* **136** 774-784.

[15] Wang J L 2012 Application of Composite Materials on Sports Equipments, *Applied Mechanics and Materials* **155-156** 903-906.

[16] Ding Z, Li Y, Lu C and Liu J 2018 An Investigation of Fiber Reinforced Chemically Bonded Phosphate Ceramic Composites at Room Temperature, *Materials* **11** 858.

[17] Williams G I and Wool R P 2000 Composites from Natural Fibers and Soy Oil Resins, *Applied Composite Materials* **7** 421-432.

IOP Conf. Series: Materials Science and Engineering **851** (2020) 012021 doi:10.1088/1757-899X/851/1/012021

Analysis of slope stability by the planning of cantilever retaining wall reinforcement using the application of Plaxis (case study: substation Panyabungan sta 0+060)

Ika Puji Hastuty[1] and Rizky Prambudi[1]

[1]Department of Civil Engineering, Universitas Sumatera Utara, Medan, Indonesia

ika.hastuty@usu.ac.id

Abstract. Slope is a natural surface appearance that has height differences. Slope stability is very closely related to landslide or ground movement which is a natural process of soil mass displacement from higher ground to lower ground. Unstable slopes are very dangerous to the surroundings; therefore, an analysis of slope stability is highly needed. In this research, Plaxis are used to planning cantilever retaining wall reinforcement with the aim to determine the safety factor value of the slope condition before and after was given reinforcement. Slope stability was analyzed in three conditions: the original condition before reinforcement, the condition with the planning of cantilever retaining wall reinforcement. The safety factor value in the slope condition with cantilever retaining wall reinforcement was 1,567 > 1,3 (Safe). Deformation value was 453,280 x 10-3 m. Cantilever retaining wall safety factor against gulling was 2,528 > 2 (Safe). Safety factor against shear was 2,683 >1,5 (Safe). Safety factor against bearing capacity was 3,704 > 3 (Safe).

1. Introduction

Retaining wall is a structure that is designed and built to withstand lateral (horizontal) soil pressure when there is a change in ground elevation which exceeds the at-rest angle in the ground. A crucial factor in designing and constructing a retaining wall is to ensure that the retaining wall does not move or collapse due to gravity [1][2][3].

The ground lateral force behind the retaining wall depends on the internal friction angle of soil (φ) and cohesion (c). The lateral force increases from top to the very bottom of the retaining wall. If not planned properly, the ground pressure will push the retaining wall, causing construction failure and landslide. Failure is also caused by the groundwater behind the retaining wall that is not dissipated by the drainage. For this reason, it is very crucial for a retaining wall to have a good drainage system to reduce hydrostatic pressure and increase soil stability [4][5].

The retaining wall design using Lateral soil pressure analysis, it is a force caused by the force of soil behind the retaining wall. The value of lateral pressure is highly affected by the displacement of retaining wall and the properties of soil. Safety Factor (SF) of slope can be calculated using various methods. Safety Factor is a comparative value between restraining force and driving force. The data needed in the calculation of the safety factor of slope are [1][6]:

a. Data of slope (especially needed for the making of slope cross section) namely the angle of slope and the height of slope or the length of slope from the foot to its peak.
b. Data of soil mechanics such as internal friction angle (Ø), bulk density of soil (ɣ), cohesion (c) and water content (w).

2. Methodology

The Soil Investigation was carried out at the substation Panyabungan, covering the stages:

1. Cone Penetration Test (CPT) and hand boring data.
2. Laboratory data like weight dry density, weight wet density, cohesion.
3. Retaining wall data as retaining wall type, elevation of bottom, base width and retaining wall height.

This analysis was conducted to determine the slope safety factor of the original condition and of the condition with cantilever retaining wall reinforcement using Plaxis. Plaxis (Finite Element Code for Soil and Rock Analysis) is the result of a finite element program for analyzing deformation and geotechnical stabilization. In this study, plain strain modeling is commonly used for longitudinal structure modeling. The steps in calculating the safety factor on Plaxis include the following [7][8]:

1. Plaxis input, like plain strain, basic element with 15 nodes to analyze by producing accurate stresses and collapsed loads.
2. Geometry modelling
   Modelling used 2D, where the geometry obtained is soil layers and retaining walls.
   The parameters of cantilever retaining wall are as follows:
   1. γ (specific gravity) = 24 kN/m^3
   2. E (modulus of elasticity) = 3,028 x 107 kN/m^2
   3. v (Poisson ratio) = 0,15
3. Input parameter of material

Table 1. Soil parameter

layer	depth (m)	type of soil	layer thickness (m)	MAT (m)	γ_{dry} (kN/m^3)	γ_{sat} (kN/m^3)	Kx (m/day)	Ky (m/day)	Es (kN/m^2)	v	c (kN/m^2)	ϕ	Ψ
1	0,00-0,80	*Organic clays & mixed soils*	0,80		17	17	7,77x10^{-3}	1,56x10^{-3}	5377,91	0,3	20	31	1
2	1,00-2,00	*Clayey-sands and silts*	1,00		18	18	8,64x10^{-3}	1,73x10^{-3}	14341,1	0,3	17	33	3
3	2,20-4,00	*Moderate sands*	1,80		16,42	16,42	8,64	1,72	22065	0,3	1	38,7	8,7
4	4,20-6,00	*Dense or cemented sands*	1,80	5,00	5,89	15,89	12,96	2,59	19613,3	0,3	1	38,1	8,1
5	6,20-10,00	*Moderate sands*	3,80		6,89	16,89	8,64	1,72	22555,3	0,3	1	39	9
6	10,20-11,40	*Clayey-sands and silts*	1,20		7,33	17,33	8,64x10^{-3}	1,73x10^{-3}	18160,7	0,3	18	32	2
7	11,60-13,00	*Moderate sands*	1,40		6,84	16,84	8,64	1,72	24026,3	0,3	1	40,6	10,6
8	13,20-14,20	*Dense or cemented sands*	1,00		7,37	17,37	12,96	2,59	26478	0,3	1	41,8	11,8
9	14,40-56	*Very shell sands, lime-rocks*	41,60		12	22	100	20	51484,9	0,2	27	46	16

4. General meshing

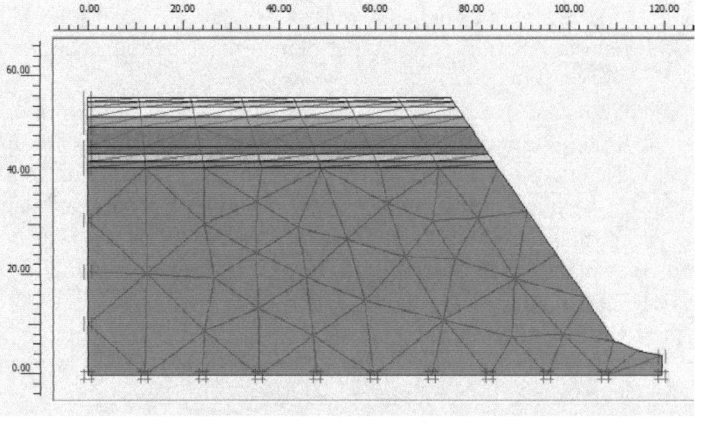

Figure 1. General meshing

5. Initial condition.
   Initial condition requires a calculation of water pressure and initial water pressure.

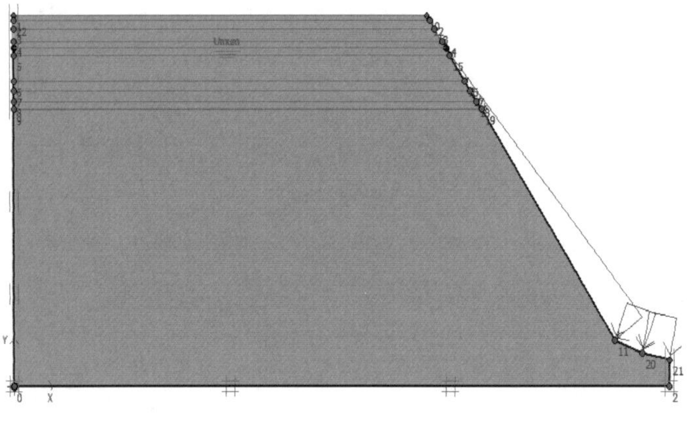

Figure 2. Initial Condition

6. Plaxis calculation
   The calculation phase is created automatically by the program, where the safety factor value will be obtained.
7. Plaxis output
   The output plaxis is a deformation value

3. Research results

3.1. Initial Condition of Slope

The Plaxis calculation process in the initial condition consists of two phases; those are the slope initial condition phase and the calculation of safety factor. In this initial condition, the slope safety factor was 1.16 < 1.3 (Not Safe). The output of Plaxis was a deformation value of 216.88 x 10^{-3} m. Geometry modelling of slope is in accordance with field investigation data where evenly distributed loads equal to 10 KN / m^2 in accordance with SNI 8460: 2017.

IOP Conf. Series: Materials Science and Engineering **851** (2020) 012021 doi:10.1088/1757-899X/851/1/012021

3.2. Slope Condition with Cantilever Retaining Wall Reinforcement

In the planning of this cantilever retaining wall reinforcement, using SNI 8460:2017 geotechnical design requirements regulation.

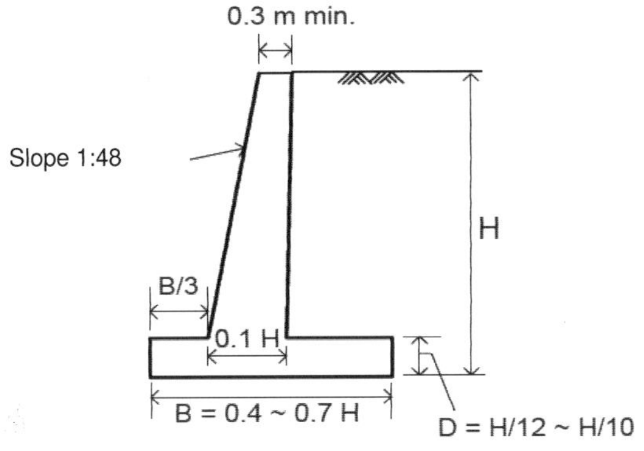

Figure 3. Dimension Requirements for The Planning of Cantilever Retaining Wall
(*Source: SNI 8460:2017)*

On this slope, cantilever retaining wall with the height of 10 meters was planned.

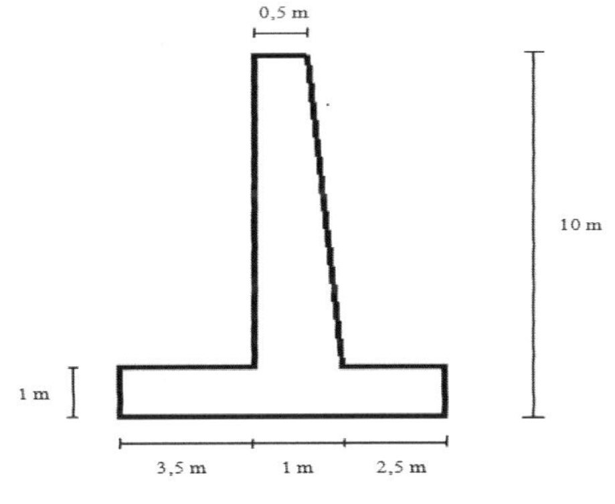

Figure 4. Dimension of Cantilever Retaining Wall

This standard reinforcement used cantilever retaining wall with the height of 10 m and was constructed at the points where quite high deformation values were found in the original slope analysis. In this condition the cantilevered soil retaining wall is placed at points x = 104 m and y = 15 m. The foundation of the front at the point x = 100.5 m and y = 5 m. While the foundation of the back at the point x = 107.5 and y = 5 m. The thickness of the cantilevered retaining wall foundation is 1 m.

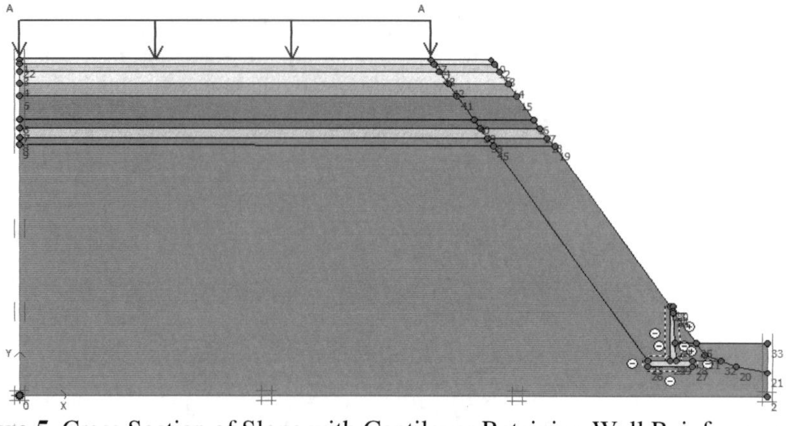

Figure 5. Cross Section of Slope with Cantilever Retaining Wall Reinforcement

The result of the running of Plaxis 2D can be seen in **Figure 8** below.

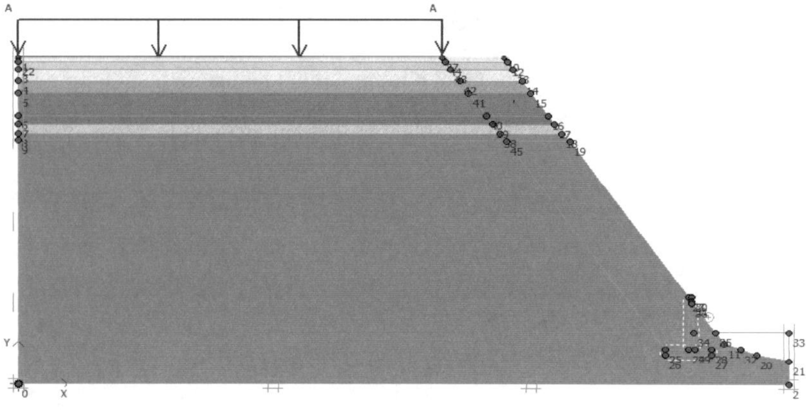

Figure 6. Modelling of Loading Phase

For the safety factor with the application of cantilever retaining wall, with Plaxis Calculation a value of 1,56 > 1,3 (safe) was obtained. The Plaxis result also showed the slope total deformation with the planning of cantilever retaining wall reinforcement value of 453,28 x 10-3 m.

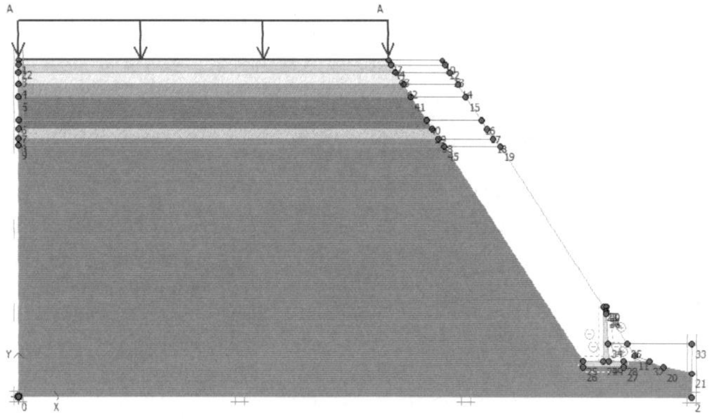

Figure 7. Modelling of Cantilever Retaining Wall Construction Phase

The safety factor value in the condition of slope with cantilever retaining wall reinforcement using Plaxis was 1.567. The safety factor value obtained was bigger than 1.3. This value indicated that the slope condition became safe with the use of cantilever retaining wall reinforcement. Plaxis output also produce deformation value where the total deformation value of the slope with cantilevered retaining wall retaining plan is 453.28 x 10^{-3} m.

4. Conclusion

1. The safety factor value in the initial condition of slope by Plaxis analysis was 1,19. The value showed that the slope condition was still not very stable.
2. The safety factor value in the condition of slope with cantilever retaining wall reinforcement by Plaxis was 1.57, the value showed that the slope condition became safe with the application of cantilever retaining wall reinforcement. Deformation value of 453.28 x 10^{-3} m was obtained on Plaxis.

Acknowledgments
The author would like to thank the related parties especially Soil Mechanics Laboratory of Universitas Sumatera Utara.

5. References

[1] Bowles J, 1979 *Physical and Geotechnical Properties of Soils*. McGraw-Hill, Tokyo.
[2] Das B M and Sobhan K, 2014 Principles of Geotechnical Engineering, SI Version, *Animal Genetics*.
[3] Hardiyatmo H C,1992 *Mekanika Tanah 1*.
[4] Amin M, 2017 Reviews in Civil Engineering Analisis Stabilitas Lereng Di Kaki Gunung Sumbing, no. 1937, pp. 8–12.
[5] Adriyati M, Dananjaya R H, and Silmi N, 2017 Analisis Stabilitas Lereng Menggunakan Sofware Geo Studio 2007 Dengan Variasi Kemiringan (Studi Kasus : Bukit Ganoman Kab Karanganyar), no. 2011, pp. 97–102.
[6] Wulandari P S and Tjandra D, 2015 Analysis of geotextile reinforced road embankment using PLAXIS 2D, *Procedia Eng.*, vol. 125, pp. 358–362.
[7] Kasus S and Jember D I, 2008 Analisa Balik Kelongsoran (Studi Kasus di Jember), *Media Komun. Tek. Sipil*, vol. 16, no. 2, pp. 130–147.
[8] Vijayalakshmi R, Student G, and Engineering E, 2011 Numerical Modeling of Rock Slopes in Siwalik Hills Near Manali Region : a Case Study, pp. 763–783.

IOP Conf. Series: Materials Science and Engineering **851** (2020) 012022 doi:10.1088/1757-899X/851/1/012022

Slope stability analysis using sheet pile reinforcement with the Bishop method

Ika Puji Hastuty[1], Roesyanto[1], and Ayuditha Maulidza Suwandi[1]

[1]Department of Civil Engineering, Universitas Sumatera Utara, Medan, Indonesia

ika.hastuty@usu.ac.id

Abstract. Slope is a sloping ground surface, and forms a certain angle to a horizontal plane and is unprotected. With non-horizontal soil conditions, a component of gravity will tend to move the ground down, so that landslides cannot longer be avoided. The location of this research is located at 150 KV substation, Payabungan. The selection of sheet piles as an alternative treatment is expected to minimize the occurrence of landslides at the substation, where if there is landslide at that location will have an impact on the power outages that will disrupt the activities of residents around. This research was conducted with Bishop's analytical method, aiming to determine the safety factor of the slope before and after reinforcement. In this analysis, sheet pile reinforcement is done with three conditions. From the calculations obtained the value of the safety factor using the Bishop method obtained by 1.11. The value of the safety factor under conditions 1, 2 and 3 was 1.45; 1.50; 1.58. From this study, it was concluded that adding sheet pile as reinforcement would increase the value of the safety factor in order for the slope to be stable compared to the slope without any reinforcement.

1. Introduction

Landslide problems are often encountered in the field of geotechnical science. Landslide events or also known as land mass movements, rocks or their combinations often occur on natural or artificial slopes. Slides on slopes are generally caused by a decrease in the shear strength of the soil, hence it is not capable to carry the burden acting on it [1][2].

Instability can result from rainfall, groundwater levels and changes in geological activities such as faults, fractures and lineages. Likewise, natural slopes that have been stable for years may suddenly fail due to changes in geometry, external forces and loss of shear strength. In addition, long-term stability is also related to weathering and chemical effects that can reduce shear strength and cause shear strength tension cracks. In these circumstances, evaluation of slope stability conditions is quite important. In choosing the type of landslide countermeasure, there will be several alternatives. For this reason, this study tries to provide an alternative to slope reinforcement that can be used as a solution to overcome slides on slopes. The type of slope reinforcement used as an alternative in this study is sheet pile.

Sheet pile is designed to withstand lateral (horizontal) soil pressure when there is a change in land elevation. that exceeds the at-rest angle in the soil. Lateral soil pressure behind the retaining wall depends on the friction angle in the ground (φ) and cohesion (c). Lateral pressure increases from the top to the very bottom of the retaining wall. If not well planned, soil pressure will push the retaining wall causing failure of construction and landslides [3][4].

The slope analysis in this study was conducted to minimize the occurrence of landslides in the substation, where if there is a landslide at that location will have an impact on the power outages that will disrupt the activities of residents around. Primary data needed for the analysis of slope stability are

CPT data, unit weight (γ), friction angle(ϕ), cohesion (c), permeability coefficient (k). The secondary data used is the project location map and slope layout. This analysis uses initial conditions and slope conditions after being given sheet pile reinforcement.

The Bishop Method is a method that was introduced by A.W. Bishop that uses the cutting method in which the forces acting on each piece Bishop Method is used to analyse circular slip surfaces. In this method, it is assumed that the total normal forces are at the center of the piece and can be determined by describing the forces on the piece vertically or normally. Balance requirements are applied to the pieces that make up the slope. The Bishop Method assumes that the forces acting on the slide have zero resultant in the vertical direction [5][6][7].

In general, there are three kinds of assumptions that can be made namely, assumptions about the normal stress distribution along the surface of the slip, assumptions regarding the inclination of the forces between the pieces, and assumptions about the position of the resultant lines of the forces between the pieces. In most analytical methods, the normal force is assumed to work in the center of the base of each piece, due to the thin section. This applied to a number of assumptions. This Bishop Method uses as many assumptions as (2n – 1) [7]. The Bishop method is a calculation method based on the Limit Equilibrium Method using slices in a circular landslide field. Before the calculation is performed, an attempt is made beforehand to calculate the center point of the landslide. After that, the slope safety factor is performed for the initial conditions and conditions with Geogrid reinforcement by trial and error until the appropriate safety factor is obtained.

The equation used to find the safety factor in the Bishop method is as follows:

$$SF = \frac{\Sigma\,[c.b+(W-u.b)\tan\emptyset]\,(\frac{1}{Mi})}{\Sigma\,W\,sin\,a}$$

Where :

c : cohesion

ϕ : friction angle

F : FK (safety factor)

w : unit weight of each element

b : width of each element

Mi : coefficient value

: $(\frac{1}{\cos\,a\,(\,1+\tan\emptyset\,\tan\frac{a}{F}})$

2. Methodology

This research was conducted at Panyabungan Substation 150 KV. The data needed is data from the results of soil investigations namely, the Cone Penetrometer and hand boring, where laboratory testing is carried out to obtain the soil property index value of the analytical method that is used to analyze slope stability, the Simplified Bishop Method. The Simplified Bishop Method assumes where the landslide field is considered to be in the form of a circular arc and then the mass above the landslide field is divided into a number of vertical slices. This method also assumes friction forces acting on the slice can be ignored so that only horizontal forces are calculated.

Step of Methodology on Bishop Method

1. Determine the center point of the landslide and the value of R
2. Determine the point of landslide with a circle
3. Divide the landslide area into 10 parts
4. Calculate the area and weight of each segment/slices
5. Find the calculated value of the variable for safety factor
6. Determine the value of 1 / Mi (α)
7. Draw a line from point R geometry
8. Find to safety factor value (SF)

IOP Conf. Series: Materials Science and Engineering **851** (2020) 012022 doi:10.1088/1757-899X/851/1/012022

3. Research results

3.1 Slope condition with sheet pile reinforcement (original slope condition)
In this condition the sheet piles are mounted at the edge of the slope plane with the coordinate of X=68,43m and Y=43,92m along 30m as shown in Figure 1.

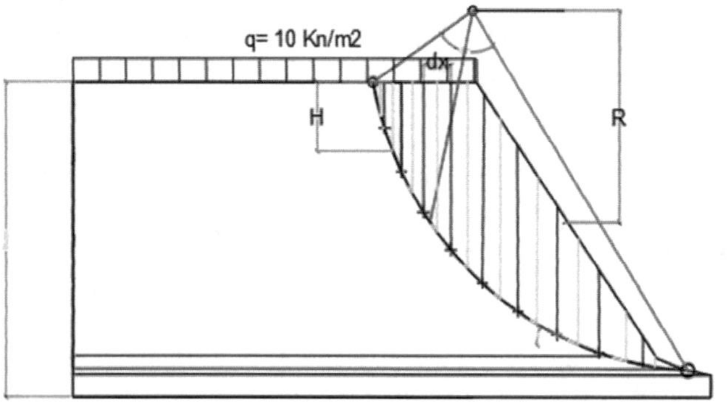

Figure 1. Divisions of Landslide Planes in the Form of Slices

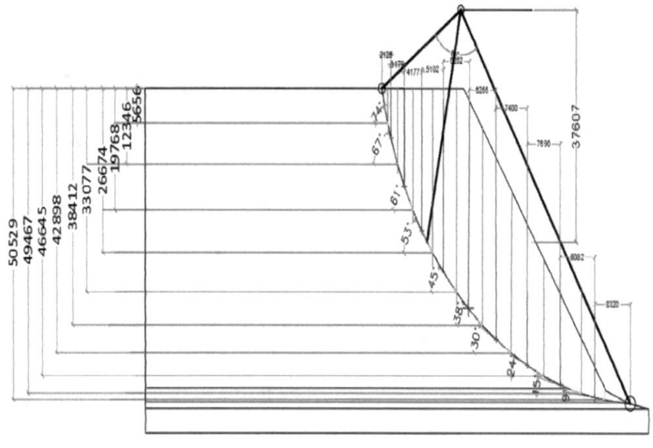

Figure 2. Dimensions of Slope Slices in Original Condition

Table 1. Tabulation of Safety Factor Calculations Using *Bishop* Method for original condition

W tan Φ	cΔx+W tan Φ	Mi	(cΔx+W tan Φ) / Mi	rm	dm
113,54	151,79	42,72	6485,96	670,20	753,47
585,84	589,02	15,02	39,19	1360,57	1602,85
1310,21	1314,37	7,97	164,75	2267,35	2412,25
2151,11	2156,21	4,33	497,86	2864,07	2663,68
3151,04	3157,30	2,81	1120,33	3483,30	2044,30
3225,77	3232,03	2,14	1507,61	3183,33	1590,08
3026,58	3033,98	1,69	1784,88	2753,25	1210,96
2341,59	2349,28	1,48	1582,02	2066,96	820,46
1275,94	1284,03	1,26	1018,19	1130,09	424,87
220,85	229,17	1,14	199,69	208,12	88,84

Total	14400,53	19987,29	13611,82
		F	1,11

3.2 Calculation Analysis with sheet pile reinforcement.

In this analysis with sheet pile reinforcement, 3 (three) conditions of sheet pile mounting were attempted at the points with the greatest deformation. The planning used sheet piles with predetermined length of 30 m.

3.2.1 Condition 1. In this condition, sheet piles are mounted at the edge of slope plane with the coordinate of X=68,43m and Y=43,92m along 30m as shown in Figure 4.

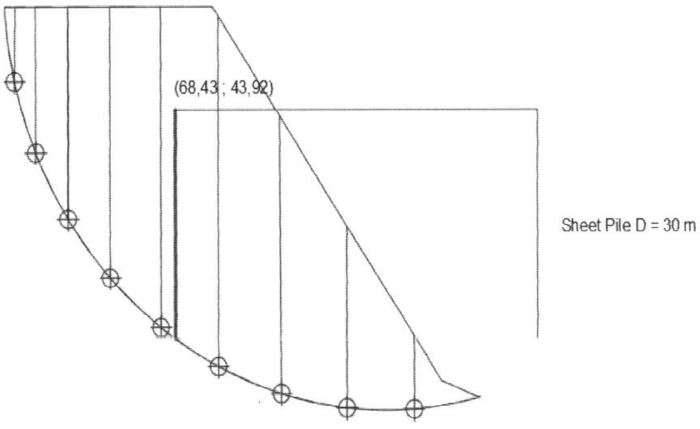

Figure 3. Dimensions of Slope Slices with Sheet Piles in Condition 1

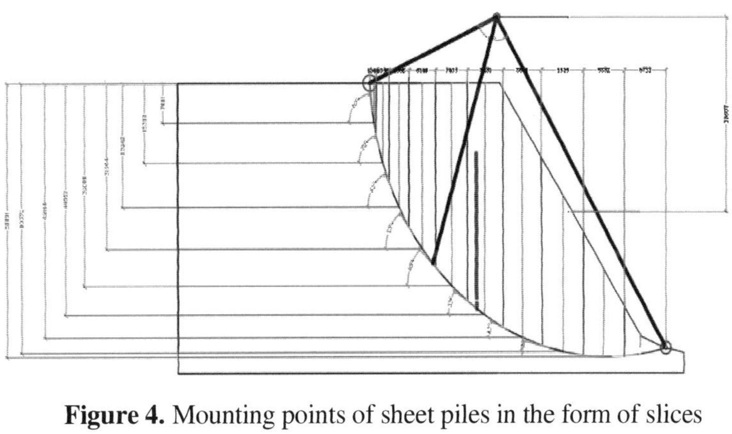

Figure 4. Mounting points of sheet piles in the form of slices

3.2.2 Condition 2. In this condition, sheet piles are mounted at the edge of slope planes with the coordinate of X=71,85m and Y=41m along 30m as shown in Figure 6.

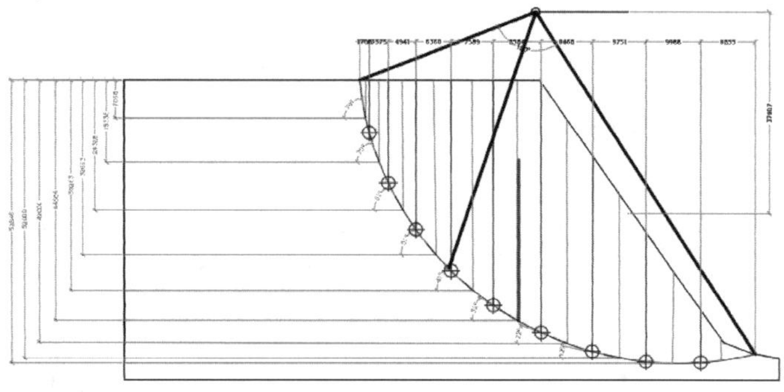

Figure 5. Dimensions of Slope Slices with Sheet Piles in Condition 2

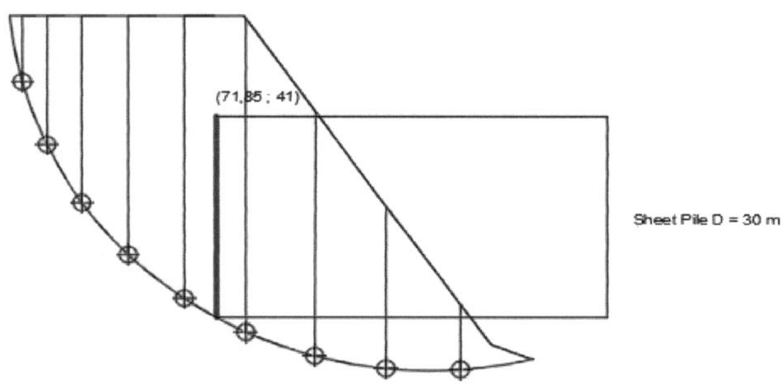

Figure 6. Mounting points of Sheet Pile in the form of slices

Tabulation of calculations for the analysis of slopes with sheet piles in condition 2 can be seen in the Table 2. The initial value of safety factor that was attempted (trial and error) was 1,51.

Table 2. Tabulation of Safety Factor Calculations in Condition 2

W tan Φ	cΔx+W tan Φ	Mi	(cΔx+W tanΦ) / Mi	rf	df
297,17	327,92	95,20	31218,75	2865,58	1014,31
1002,88	1006,18	16,65	60,40	3346,99	2066,78
2110,67	2115,61	6,11	345, 59	4337,16	3094,13
3450,88	3457,24	3,07	1122,94	5028,23	2849,91
4934,41	4942,00	2,017	2449,63	5715,15	3301,55
6191,72	6200,30	1,57	3926,13	6281,61	1898,09
6293,33	6302,80	1,31	4799,14	5908,68	900,14
4981,13	4990,88	1,14	4347,05	4629,78	67,93
2922,13	2932,11	1,00	2932,11	2932,11	0
841,17	851,02	1,00	851,02	851,02	0
Jumlah			52053,08	41896,35	15192,89
				F	1,51

3.2.3 Condition 3. In this condition, sheet piles are mounted at the edge of the plane with the coordinate of x=73,68m and y=39m along 30m as shown in ***Figure 8.***

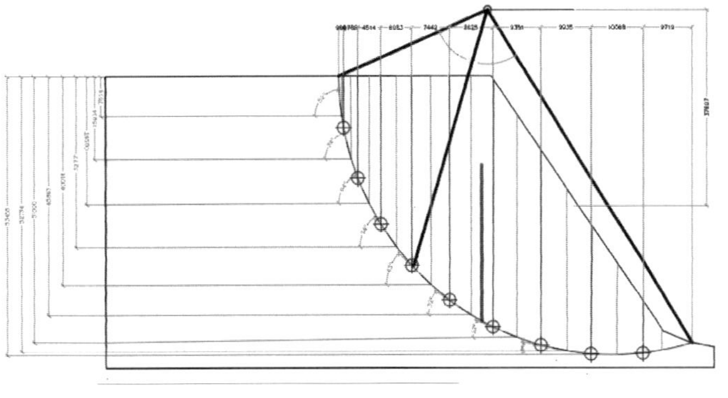

Figure 7. Dimensions of Slope Slices with Sheet Pile in Condition 3

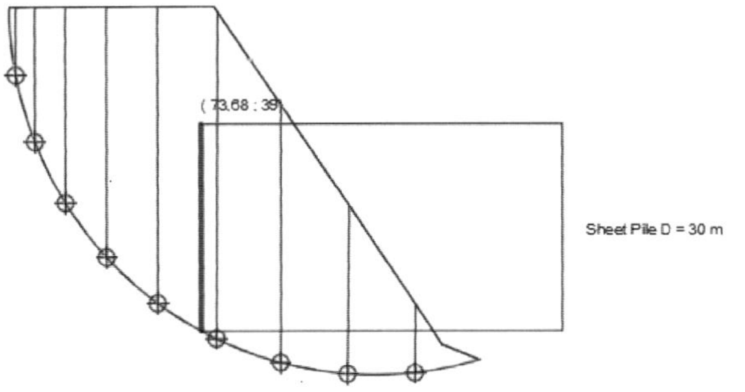

Figure 8. Mounting points of sheet piles in the form of slices

Tabulation of the calculations for the analysis of slopes with sheet piles in condition 3 can be seen in the Table 3. The initial value of safety factor that was attempted (trial and error) was 1,58.

Table 3. Tabulation of Safety Factor Calculations in Condition 3

W tan Φ	cΔx+W tan Φ	Mi	(cΔx+W tan Φ) / Mi	rf	df
308,37	482,25	234,98	113321,48	6497,42	1014,31
885,32	888,11	30,23	29,37	4025,45	2066,78
1996,27	2000,79	7,79	256,65	4803,22	3094,13
3388,44	3394,52	3,53	959,58	5507,90	2849,91
4925,15	4932,59	2,11	2337,21	6023,21	3301,55
6475,85	6484,47	1,55	4174,34	6649,96	1898,09
6413,62	6422,97	1,29	4943,22	6073,90	900,14
5151,33	5161,26	1,12	4573,61	4822,44	67,93
2960,34	2970,41	1,00	2970,41	2970,41	0
862,34	872,06	1,00	872,06	872,06	0
Total			134437,97	48246	15192,89
				F	1,58

Of the three conditions above where the condition of one sheet pile mounted on the slope with X = 68.43 m and Y = 43.92 m along 30 meters obtained safety factor value is 1.1. Condition two the value X = 71.85 m and Y = 41 m along 30 m obtained by safety factor value is 1.50, and condition three with value X = 73.68 m and Y = 39 m along 30 m obtained safety factor value is 1.58.

4. Conclusion

The value of Safety Factor in the initial condition of the slope using the Bishop method is 1,11 and the condition of the slope with sheet pile reinforcement in Condition 1, 2, and 3 are 1.45; 1.50; 1.58, respectively where the value of the biggest safety factor in condition 3 with a value of 1.58.

5. References

[1] Wulandari P S and Tjandra D, 2015 Analysis of geotextile reinforced road embankment using PLAXIS 2D, *Procedia Eng.*, vol. 125, pp. 358–362, 2015.

[2] Hardiyatmo H C, 2006 *Penanganan Tanah Longsor & Erosi*. Yogyakarta: Gadjah Mada University Press, 2006.

[3] Hastuty I P, Roesyanto, and Sihite A B, 2018The decline of soil due to the pile of highway project Medan-Kualanamu (STA 35 + 901) with the finite element method, *IOP Conf. Ser. Mater. Sci. Eng.*, vol. 308, no. 1.

[4] Roesyanto, Iskandar R, Silalahi S A, and Fadliansyah, 2018 Soil settlement analysis in soft soil by using preloading system and prefabricated vertical draining runway of Kualanamu Airport, *IOP Conf. Ser. Mater. Sci. Eng.*, vol. 309, no. 1.

[5] Das B M and Sobhan K, 2014 Principles of Geotechnical Engineering, SI Version, *Animal Genetics*.

[6] Hardiyatmo H C,1992 *Mekanika Tanah 1*.

[7] Bowles J, 1979 *Physical and Geotechnical Properties of Soils*. McGraw-Hill, Tokyo

IOP Conf. Series: Materials Science and Engineering **851** (2020) 012023 doi:10.1088/1757-899X/851/1/012023

User experience study on crane operator erection simulator using senso glove in a virtual reality environment

F Fahmi[1], F Nainggolan[2], B Siregar[2], Soeharwinto[1], M Zarlis[2]

[1]Department of Electrical Engineering, Faculty of Engineering, Universitas Sumatera Utara, Medan Indonesia
[2]Department of Information and Technology, Faculty of Computer Science and Information Technology, Universitas Sumatera Utara, Medan Indonesia

Corresponding author: fahmimn@usu.ac.id

Abstract – Failure or mistakes in operating the crane can cause an accident while working on the project site, mainly when it deals with heavy and extensive material, for example, in the steel erection process. Proper crane operation for steel erection training using Virtual Reality (VR) technology can overcome this problem effectively and efficiently. The level of interactivity in a VR system can be increased by using a controller that can observe the user's hand and finger movements. Senso Glove is a wireless glove that is able to read hand and finger movements and is able to provide vibrations. In the system developed in this study, Senso Glove is used as a system control tool to read hand movements and become the system input. In addition to input to the system, Senso Glove can also provide vibration to the user when the virtual hand is holding the lever on a crane in a virtual world or when the steel hit some part of construction during the process. Senso Glove's ability as a crane operator simulator control tool can be tested using the Usability Test. Therefore, USE Questionnaire was used with a Likert Scale as a media survey for system users. From the survey results, it was found that the user was easy to remember how to operate a crane with a Senso Glove with an average value of 4.25 per 5 Likert scales. However, users also find Senso Glove challenging to learn by everyone with an average value of 2.95 per 5 Likert scales. This is due to the precision that is still difficult given by Senso Glove with the average value of virtual hand movements produced by Senso Glove is 3.15 per 5 Likert scale.

1. Introduction

As one of the heavy equipment, Cranes requires specialized training in order to reduce the potential accidents that can occur while working on the project site. Accidents that occur in the project field can cause massive losses and even loss of life [1], mainly when it deals with massive and extensive material, for example, in the steel erection process. The problems that occur in the construction field can be overcome effectively and efficiently by using Virtual Reality (VR) [2-4]. The risk of accidents that can occur when using heavy equipment such as cranes can also be reduced by using VR as a training medium [4].

In addition to its immersive appearance, VR can also increase the level of user interactivity with the system. Many applications have been developed in other areas, including in education, such as anatomy learning systems [5,6] in the medical field [7-9] or in culture preservation [17]. Some technology of VR controllers such as the VIVE controller, leap motion controller, and many others

IOP Conf. Series: Materials Science and Engineering **851** (2020) 012023 doi:10.1088/1757-899X/851/1/012023

now become widely used [10,11]. One way to increase the level of user interactivity with a VR system is by stimulating the touch senses [12,13]. The sense of touch can be triggered by vibrations that can be felt by the hands of VR system users. Users can increase the feel the VR system that is getting closer to reality with the combination of the appearance and vibration felt.

Senso Glove is a tool that is able to observe hand and finger movements with an Inertial Measurement Unit (IMU) sensor. With Senso Glove's wireless capability with Bluetooth networks, users can move their hands freely. In addition to freehand reading, Senso Glove is also equipped with a motor that is capable of being vibrated by the system [14]. Vibration on Senso Glove, when worn by the user, is expected to increase the level of interactivity between the user and the VR system.

Increasing the level of user interactivity with the VR system is expected to be obtained by using the Senso Glove as a control tool. In addition to input to the system, Senso Glove can also provide vibration to the user when the virtual hand is holding the lever on a crane in a virtual world or when the steel hit some part of construction during the process. However, to obtain the level of acceptability and level of satisfaction of Senso Glove as a control device on the VR system requires a test to the user. Tests proposed in this study are Usability Tests using the USE Questionnaire [15]. In this paper, a user experience study of the level of usability, level of ease of use, level of ease to learn, and level of satisfaction was obtained and analyzed.

2. Material and Method

The system developed was carried out by using VR headsets for display media and haptic glove for system control devices. The VR headset used on the system is the HTC VIVE VR headset, and the haptic glove used on the system is the Senso Glove. The use of a VR headset gives users the sensation of being able to see around the virtual world on a 360-degree system. The Senso Glove (Figure 1) is used as a control device by observing the user's hand and finger movements, which are used as input to grasp and move the control lever on the virtual crane. The movement of the virtual control lever functions in accordance with the features contained in Table 1.

Figure 1. A pair of Senso Glove and Senso Glove when used on Unity3D

Table 1. Features and Function

Features	Function
The right joystick	To move the pole to go up or down when moving left or right
The right joystick	To move the boom to go up or down when moving back or front
The left joystick	To move the upper body of the crane to rotate counter-clockwise when left or clockwise when right

As for the movement of the boom (an arm of the crane) and the anchor, which is moved by the right lever on the system functions with equation (1) and equation (2) follows.

$$\theta_1 = |\sin\alpha| \times \alpha \times \delta \tag{1}$$

IOP Conf. Series: Materials Science and Engineering **851** (2020) 012023 doi:10.1088/1757-899X/851/1/012023

$$\theta_2 = |\sin\beta| \times \beta \times \delta \tag{2}$$

Where in Equation (1) is the position of the boom on the z-axis coordinate degrees, and α is the right joystick position on the x-axis. While δ is a constant to set the comfort level of the boom to the right lever. The right lever is not driven by a button but by holding the user's right hand wearing Senso Glove. When holding the right lever, the sensation of vibration will be felt by the user's right hand.

Figure 2. Crane parts which can be controlled by a joystick

In Eq (2), θ_2 is the position of the anchor on the y-axis coordinates that can go up and down. While β is the degree of coordinates of the z-axis by the right lever and δ is a constant to adjust the comfort level of the movement of the anchor to the movement of the right lever. When it is moved, vibrations can also be felt by the user, which is triggered by grasping the right lever by the user's hand.

The movement of the top of the crane, which is controlled by the left lever on the system functions with the following Equation (3).

$$\theta_3 = |\sin\gamma| \times \gamma \times \delta \tag{3}$$

In Equation (3), the degree of rotation of the top of the crane is θ_3. While γ is the degree of rotation of the left lever on the z-axis coordinates and δ is a constant to set the comfort level of the rotation of the top of the crane to the rotation of the left lever. Just as when the right lever is held, the user will also feel the feedback in the form of vibrations when grasping the left lever with a virtual hand.

Figure 3. Right Joystick and Left Joystick as virtual controls

In the system that has been developed, users can operate cranes to make steel erections. Steel erection is a process for raising, adjusting, and tightening, which can involve cranes [16]. This process can be done by the user by moving the virtual lever and operating the crane so that the anchor reaches a certain distance from the steel. Then the steel will be connected to the anchor automatically by the system. Steel connected to the anchor can be raised by the user by moving the anchor and boom on the crane. When steel is placed on the foundation, the steel will be installed and tightened automatically by the system.

To obtain a quantitative value of user experience, a survey of 20 respondents was conducted to obtain a level of acceptability and user satisfaction. This survey was conducted by giving a number of questionnaires to respondents after respondents were asked to use the VR simulation system. The questionnaire used was the USE Questioner using the 4-Point Likert Scale [15]. The statements used in the survey in this study are in Table 2, and the Likert Scale used is found in Table 3.

Table 2. List of Crane Simulator Survey Statements

Nr.	Statement
1.	In my opinion, the simulation of the Crane with Senso Glove is handy to be a Crane operation training simulation.
2.	In my opinion, a Crane simulation with Senso Glove can facilitate the use of cranes in the real world.
3.	I think the Crane simulation with Senso Glove is very interactive.
4.	Crane Simulation with Senso Glove functions according to my expectations.
5.	The thrill that Senso Glove gave me was exciting and interactive.
6.	Crane Simulation with Senso Glove is easy to use.
7.	I think the crane simulation with Senso Glove is easy for anyone to use.
8.	I need several times to fail to use Crane simulation with Senso Glove successfully.
9.	I quickly learn gestures to control a Crane simulation system with Senso Glove.
10.	I can easily remember how to use the Crane simulation system with Senso Glove.
11.	I like the look of the virtual hand used.
12.	The virtual hand movements in the system move according to my hand movements.
13.	I am satisfied with the whole Crane simulation system with Senso Glove.
14.	I am satisfied with the vibration feedback from Senso Glove.
15.	I am satisfied with the look of the virtual hand that matches my hand movements.

Table 3. A 4-Point Likert Scale Used in Surveys

Scale of Agreement	Satisfaction Scale	Score
Strongly Disagree (STS)	Very Dissatisfied (STP)	1
Disagree (TS)	Dissatisfied (TP)	2
Agree (S)	Satisfied (P)	4
Strongly Agree (SS)	Very Satisfied (SP)	5

3. Results and Discussion

The output of this study was the values of a survey of User Experience with the statements presented in Table 2. Before conducting a survey, respondents are given a brief explanation of how the system works and how to control it. Then the respondent was asked to grab the lever and hold it. Respondents were also asked to try to erect steel in the system. There are 20 respondents in this survey with the age range of respondents between 20 to 30 years and various professions. The results of the survey conducted on these respondents are in Table 4 and Figure 4.

Table 4. List of Crane Simulator Survey Statements

Nr.	Statement	Score Average
1.	In my opinion, the simulation of the Crane with Senso Glove is handy to be a Crane operation training simulation.	3.8
2.	In my opinion, a Crane simulation with Senso Glove can facilitate the use of cranes in the real world.	3.65
3.	I think the Crane simulation with Senso Glove is very interactive.	3.6
4.	Crane Simulation with Senso Glove functions according to my expectations.	3.3
5.	The thrill that Senso Glove gave me was exciting and interactive.	3.95
6.	Crane Simulation with Senso Glove is easy to use.	3.35
7.	I think the crane simulation with Senso Glove is easy for anyone to use.	2.95
8.	I need several times to fail to use Crane simulation with Senso Glove successfully.	3.85
9.	I quickly learn gestures to control a Crane simulation system with Senso Glove.	3.5
10.	I can easily remember how to use the Crane simulation system with Senso Glove.	4.25
11.	I like the look of the virtual hand used.	3.85
12.	The virtual hand movements in the system move according to my hand movements.	3.15
13.	I am satisfied with the whole Crane simulation system with Senso Glove.	3.65
14.	I am satisfied with the vibration feedback from Senso Glove.	3.9
15.	I am satisfied with the look of the virtual hand that matches my hand movements.	3.55

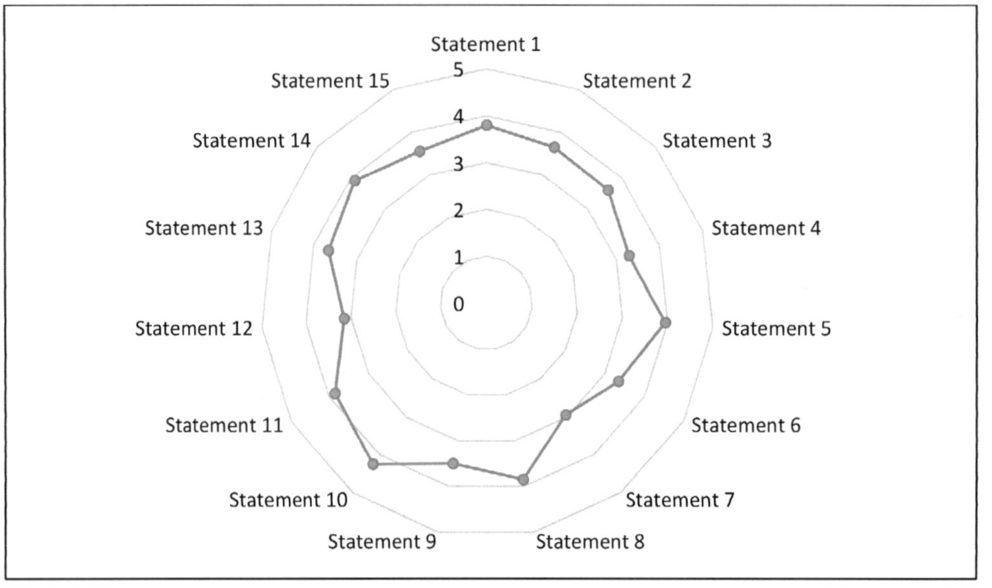

Figure 4. Graph of user experience survey results

4. Conclusions

From the results of a survey conducted, we obtained the highest average value of 4.25 for ease of remembering how to use the system. This statement is the only question that reaches value 4, the value of agreement on the statement, while no other statements reach scale 4. The lowest average value in the system is in statement 7, which reads, 'I think the crane simulation with Senso Glove is easy to use by anyone' with an average value of 2.95. This is related to the precision of the Senso Glove, which is challenging to control. Users find it difficult to move Senso Glove to their preferences. This low level of precision from Senso Glove makes the system difficult to learn and satisfies the user.

Even so, the level of acceptability to vibration and satisfaction of vibration feedback in the system can reach an average value of 3.95 and 3.9, respectively. This means that the vibrations in the crane simulation system almost meet the level of user satisfaction. Vibration sensation in the crane simulation system increases the level of user interactivity so that users feel they are operating a real crane.

Acknowledgment

This research is fully funded by Universitas Sumatera Utara through TALENTA USU Research Program financial year 2019 based on contract Nr. 4167/UN5.1.R/PPM/2019 dated 01 April 2019

Reference

[1] Moore and Gheisari 2019 A Review of Virtual and Mixed Reality Applications in Construction Safety Literature *Safety* vol **5** no 3 p 51 doi: 10.3390/safety5030051.

[2] Ahmed S 2018 A Review on Using Opportunities of Augmented Reality and Virtual Reality in Construction Project Management *Organization, technology & management in construction : an international journal* vol **10** no 1 pp1839–1852 doi: 10.2478/otmcj-2018-0012.

[3] Fahmi F, Nainggolan F, Andayani U and Siregar B 2018 Development of excavator training simulator using leap motion controller *Journal of Physics: Conference Series* Vol. 978, No. 1, p. 012034

[4] Nainggolan F, Siregar B, and Fahmi F 2019 User Experience in Excavator Simulator using Leap Motion Controller in Virtual Reality Environment *The 4th International Conference on Computing and Applied Informatics (ICCAI)* under publication

[5] Nainggolan F, Siregar B, and Fahmi F 2016 Anatomy learning system on human skeleton using Leap Motion Controller *3rd International Conference on Computer and Information Sciences (ICCOINS)* (pp. 465-470) IEEE
[6] Nainggolan F, Siregar B and Fahmi F 2019 User experience in using vive controller as a controller in the anatomy learning system in the virtual reality environment The 4th International Conference on Computing and Applied Informatics (ICCAI) under publication
[7] Andayani U, Syahputra MF, Muchtar MA, Sattar M, Prayudani S and Fahmi F 2019 3D Modelling Intestine Anatomy with Augmented Reality for Interactive Medical Learning *IOP Conference Series: Materials Science and Engineering* Vol 648 No 1 p 012035
[8] Fahmi F, Nainggolan F, and Siregar B 2019 3D anatomy learning system using Virtual Reality and VR Controller *Journal of Physics: Conference Series* Vol. 1235 No. 1 p012043
[9] Nainggolan F, Siregar B, and Fahmi F 2016 Anatomy learning system on human skeleton using Leap Motion Controller *3rd International Conference on Computer and Information Sciences (ICCOINS)* (pp. 465-470) IEEE
[10] Tanjung K, Nainggolan F, Siregar B, Panjaitan S and Fahmi F 2019 The use of virtual reality controllers and comparison between vive, leap motion and senso gloves applied in the anatomy learning system *TALENTA-International Conference on Science and Technology* under publication
[11] Fahmi F, Tanjung K, Nainggolan F, Siregar B, Mubarakah N and Zarlis M 2020 Comparison study of user experience between virtual reality controllers, leap motion controllers, and senso glove for anatomy learning systems in a virtual reality environment *The 2nd International Conference on Information Technology and Engineering Management (ITEM)* under publication
[12] Benko H, Holz C, Sinclair M and Ofek E 2016 NormalTouch and TextureTouch: High-fidelity 3D Haptic Shape Rendering on Handheld Virtual Reality Controllers *in Proceedings of the 29th Annual Symposium on User Interface Software and Technology* pp717–728 doi: 10.1145/2984511.2984526
[13] Nainggolan F, Siregar B, and Fahmi F 2019 Design of interactive virtual reality for erection steel construction simulator system using senso gloves *TALENTA-International Conference on Science and Technology* under publication
[14] Perret J and Vander P 2018 Touching Virtual Reality: a Review of Haptic Gloves *in Proceedings of 16th International Conference on New Actuators* pp. 270–274
[15] Lund AM 2001 Measuring Usability with the USE Questionnaire pp 5
[16] Irizarry J, Simonsen KL, and Abraham DM 2005 Effect of Safety and Environmental Variables on Task Durations in Steel Erection *Journal of Construction Engineering and Management* vol. 131 no 12 pp 1310–1319 DOI: 10.1061/(ASCE)0733-9364(2005)131:12(1310).
[17] Syahputra M F, Annisa T, Rahmat R F, and Muchtar M A 2017 Virtual Application of Darul Arif Palace from Serdang Sultanate using Virtual Reality *Journal of Physics: Conference Series* **801**(1) 012009.

IOP Conf. Series: Materials Science and Engineering **851** (2020) 012024 doi:10.1088/1757-899X/851/1/012024

Comparison study of user experience between virtual reality controllers, leap motion controllers, and senso glove for anatomy learning systems in a virtual reality environment

F Fahmi[1], K Tanjung[1], F Nainggolan[2], B Siregar[2], N Mubarakah[1], and M Zarlis[2]

[1]Department of Electrical Engineering, Faculty of Engineering, Universitas Sumatera Utara, Medan Indonesia

[2] Department of Information and Technology, Faculty of Computer Science and Information Technology, Universitas Sumatera Utara, Medan Indonesia

Corresponding author: fahmimn@usu.ac.id

Abstract. Virtual Reality (VR) can be used as an assistive system to learn with immersive visualization and interactive control. Anatomy is a branch of biology that is difficult to learn using textbooks because of its limitations in displaying anatomical objects in two-dimensional media. VR can display anatomical objects in the form of a three-dimensional virtual world. Besides being assisted by its immersive appearance, the interactivity learning system can also be improved by using the appropriate VR control device. In this study, three controls were used with different features and functions, namely VIVE Controller, Leap Motion Controller, and Senso Glove, and a user experience study was conducted. The user was asked to use the anatomy learning system with three different control devices, and then the user was asked to fill out a survey form, which uses the Liker scale. The survey focuses on the level of acceptability and user satisfaction in the usability factor, ease of learning, the suitability of movement, suitability of display, and haptic feedback provided by each VR controllers. From this study, it was found that the VIVE Controller is superior compared to the Leap Motion Controller and Senso Glove in terms of usability, ease of learning, movement suitability, and display suitability. The senso glove is superior in terms of haptic feedback satisfaction.

1. Introduction

Virtual Reality (VR) can be used as a learning medium on multimedia systems [1]. Learning by using VR can support the level of user interaction with the system. In addition to the use of immersive displays, control tools in virtual reality can further increase the level of user interaction with the system [2,3].

Anatomy is a branch of biology that is difficult to learn from textbooks because of the limitations of the textbook media to display anatomical objects [4]. Anatomy is a lesson that can be learned with a multimedia system [5]. One of the multimedia systems that can be used is a VR system for learning anatomy [6-8]. The use of control devices in the VR system can affect the level of interactivity of the anatomy learning system with the users. With the availability of various VR control devices, we can produce different levels of interactivity according to the features and functions of each tool [9,10].

VR control devices have different features and functions. By using the different VR system control devices, it is expected that the results will be in the form of acceptability and satisfaction level of users. This result can be obtained by conducting a survey for selected users [8,11-14]. In this study, three VR control device was used with different features and functions. The three tools are VIVE Controller, Leap Motion Controller, and Senso Glove. These three devices were tested by a user experience study and then will be rated on a Likert scale to obtain a level of acceptability and satisfaction.

2. Material and Method

2.1 VIVE Controller

VIVE Controller is a pair of wireless joysticks whose movements are observed by a pair of base stations or also called lighthouses [15]. From the base station, light is transmitted periodically to the VIVE Controller. Gap or time difference between the light will be input as a result of observing the movement of the VIVE Controller. The VIVE Controller uses buttons to trigger instructions on the system. In addition, the VIVE Controller can provide feedback in the form of vibration to the user.

2.2 Leap Motion Controller (LMC)

The Leap Motion Controller is a hand motion sensor observer that is connected by a USB cable to the computer [18]. This sensor uses an infrared monochromatic camera that takes pictures periodically. When the user's hand is in the observation area, this tool will track the user's hand gestures and fingers [16]. Because no device is connected from the Leap Motion Controller to the user's hand, no feedback can be felt by the user's hand.

2.3 Senso Gloves

Senso Glove is a device in the form of a wireless glove that can observe the movements of the user's hands and fingers wearing it [17]. Senso Glove users' hand and finger movements are observed with the Inertia Measurement Unit (IMU) Sensor found on the Senso Glove. This glove also does not use a cable to send data to a computer because it has been equipped with a Bluetooth network. In addition, the haptic feedback feature in the form of vibration is also found in the Senso Glove. The vibrator motor on the Senso Glove is on each of the Senso Glove fingertips, backs, and wrists.

2.4 The use of controllers and comparison

In this system, three anatomical learning systems were developed with three different controls. Each tool uses a different virtual hand model display (Figure 1). The primary function of this system is to display the name of the selected anatomical object. The selection of anatomical objects is in accordance with the features of each tool, such as pressing a button on the VIVE Controller and holding the object on the Leap Motion Controller and Senso Glove. In addition, the VIVE Controller and Senso Glove users can feel the feedback in the form of vibrations, while the Leap Motion Controller does not.

2.5 The User Experience Study

In the anatomical learning system control system data collection, a survey was conducted to obtain the level of acceptability and the level of user satisfaction. This survey was conducted after each respondent was given a little introduction to the program. The respondents in this survey were 20 students aged between 18-23 years. All respondents tried the learning system with each control device then filled out a survey using the 4-Point Likert Scale. As stated in Table 1 is a translation of the scale for the level of acceptability and the level of satisfaction in the USE Questionnaire [11]. The statements for the level of acceptability and the level of user satisfaction on each tool are in Table 2, Table 3, and Table 4.

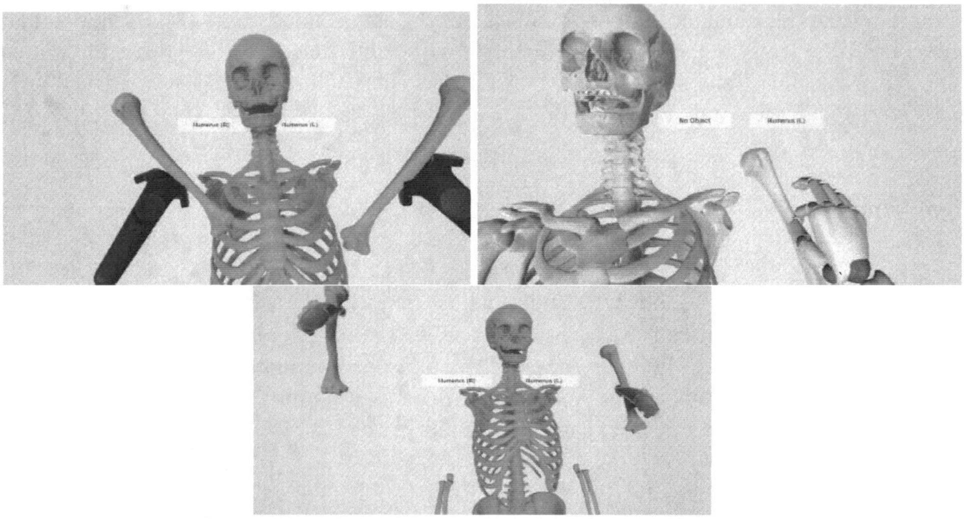

Figure 1. Display of virtual hands on the system using (a) VIVE Controller,(b) Leap Motion Controller, and (c) Senso Glove

Table 1. A 4-Points Likert Scale Used in Survey

Scale of Agreement	Satisfaction Scale	Value
Strongly Disagree (STS)	Very Dissatisfied (STP)	1
Disagree (TS)	Dissatisfied (TP)	2
Agree (S)	Satisfied (P)	4
Strongly Agree (SS)	Very Satisfied (SP)	5

Table 2. List of Statements for User Experience Level on the VIVE Controller

Nr.	Statements
1.	The use of the VIVE Controller in learning Anatomy in a VR environment is very interactive.
2.	I quickly learn how to use VIVE Controller.
3.	The virtual hand movements correspond to the movements of my hands.
4.	Vibration on the VIVE Controller when retrieving bone objects is very interactive.
5.	The combination of the VIVE Controller on the VR headset is very suitable.
6.	The use of the VIVE Controller is appropriate to be the control of the Anatomy system in a VR environment.
7.	Pressing the trigger button on the VIVE Controller to display the name of the object is easy to use.
8.	I managed to get the bone object that I wanted with the VIVE Controller.
9.	I am satisfied using the VIVE Controller for learning Anatomy.
10.	I am satisfied with the combination of the VIVE Controller and VR Headset.
11.	I am satisfied with the virtual hand movements of the VIVE Controller.
12.	I am satisfied with pressing the button to retrieve the anatomical object.
13.	I am satisfied with the vibration from the VIVE Controller when retrieving objects.
14.	I am satisfied with taking the object by pressing the button on the VIVE Controller.
15.	I am satisfied with the virtual hand display with the VIVE Controller.

Table 3. List of Statements for User Experience Level on the Leap Motion Controller

Nr.	Statements
1.	The use of Leap Motion in learning Anatomy in a VR environment is very interactive.
2.	I quickly learned about using Leap Motion.
3.	The virtual hand movements correspond to the movements of my hands.
4.	The absence of vibration when using Leap Motion does not reduce the system's interactive level.
5.	The combination of Leap Motion on a VR headset is very suitable.
6.	The use of Leap Motion is feasible to be the control of the Anatomy system in a VR environment.
7.	Hold a virtual hand with Leap Motion to display the name of the object, easy to use.
8.	I managed to take the bone object that I wanted with Leap Motion.
9.	I am satisfied using Leap Motion for learning Anatomy.
10.	I am satisfied with the combination of Leap Motion with VR Headset.
11.	I am satisfied with the virtual hand movements of Leap Motion.
12.	I am satisfied with grasping an object to pick up an anatomical object.
13.	I am satisfied WITHOUT vibration from Leap Motion when taking an object.
14.	I am satisfied with taking the object done by holding a virtual hand from Leap Motion.
15.	I am satisfied with the appearance of a virtual hand with Leap Motion.

Table 4. List of Statements for User Experience Level on the Senso Glove

Nr.	Statements
1.	The use of Senso Glove in learning Anatomy in a VR environment is very interactive.
2.	I can quickly learn how to use the Senso Glove.
3.	The virtual hand movements correspond to the movements of my hands.
4.	Vibration on Senso Glove when taking bone objects is very interactive.
5.	The Senso Glove combination on the VR headset is very suitable.
6.	The use of Senso Glove is worthy of being a control in the Anatomy system in a VR environment.
7.	Hold a virtual hand with Senso Glove to display the name of the object, easy to use.
8.	I managed to take the bone object that I wanted with Senso Glove.
9.	I am satisfied to use Senso Glove for learning Anatomy.
10.	I am satisfied with the combination of Senso Glove with VR Headset.
11.	I am satisfied with the virtual hand movements of Senso Glove.
12.	I am satisfied with grasping an object to pick up an anatomical object.
13.	I am satisfied with the vibration from the Senso Glove when taking an object.
14.	I am satisfied with taking the object done by holding a virtual hand from Senso Glove.
15.	I am satisfied with the look of the virtual hand with Senso Glove.

IOP Conf. Series: Materials Science and Engineering **851** (2020) 012024 doi:10.1088/1757-899X/851/1/012024

3. Results and Discussion

The output of this study was the average values obtained from surveys with a Likert scale on each tested device. In each statement that has been surveyed to 20 respondents, we then obtained and concluded the results. The results of the user experience survey for anatomy learning systems in a VR environment with the VIVE Controller, Leap Motion Controller, and Senso Glove as control devices are shown in Table 5 and Figure 2.

Table 5. Comparison of user experience study

Nr.	Statement	VIVE	Leap	Senso
1.	The use of controls on the system is very interactive.	4.50	4.30	4.10
2.	Controls are easy to learn.	4.45	3.75	3.55
3.	The suitability of virtual hand movements with the original hand.	4.40	3.20	2.60
4.	Approval of the interactive level of vibration given	3.60	3.80	3.85
5.	The combination of control devices on the VR headset is very suitable.	4.10	3.75	3.50
6.	The appropriateness of the tool to be in control of the Anatomy system in a VR environment.	4.30	3.95	3.80
7.	Ease of gesture on the device to display the name of the object.	4.35	3.55	3.15
8.	Success in retrieving anatomical objects with VR control devices.	4.60	3.80	2.95
9.	The satisfaction of using control devices for learning Anatomy.	4.40	4.20	3.85
10.	The satisfaction blend of tools with VR Headset.	4.30	4.05	3.60
11.	The satisfaction of virtual hand movements.	4.15	3.45	2.80
12.	The satisfaction of gestures to pick up anatomical objects.	4.55	3.75	3.35
13.	Haptic feedback satisfaction when taking objects.	3.50	3.65	3.65
14.	The satisfaction of gestures on the device to retrieve anatomical objects.	4.20	3.90	3.35
15.	The satisfaction of the display of virtual hand models.	3.80	3.75	3.45

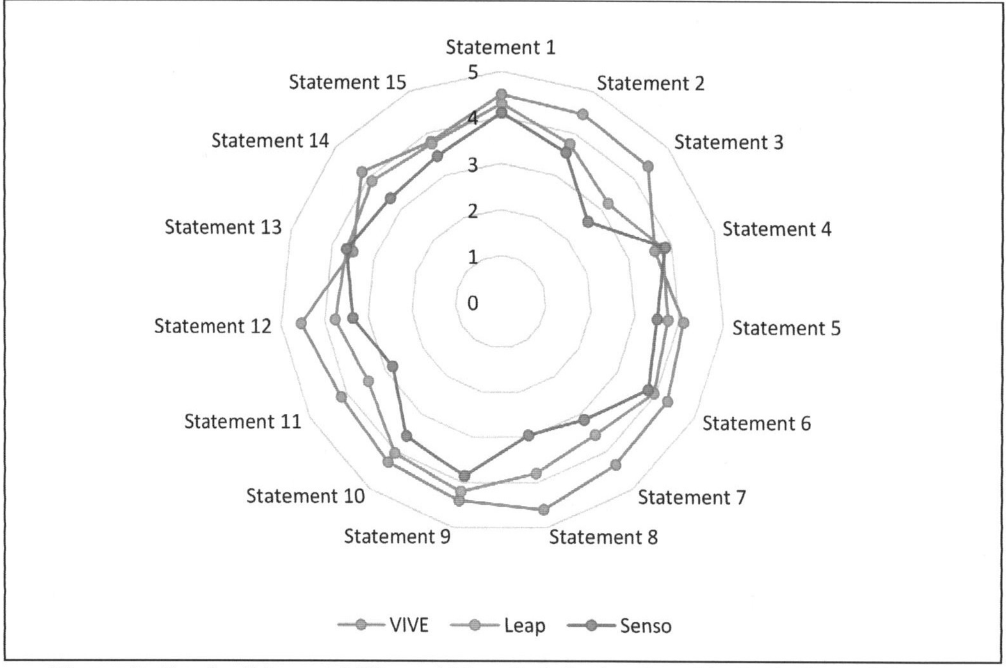

Figure 2. Graph of Acceptability and User Satisfaction Level in Each Control Device

4. Conclusions

In terms of its usefulness to be an interactive control tool, the three tools used, namely VIVE Controller, Leap Motion Controller, and Senso Glove, reached values of 4.50, 4.30, and 4.10 respectively, meaning that all three were accepted by respondents as useful tools as control devices interactive. But in terms of user satisfaction, VIVE Controller is superior to Leap Motion Controller and Senso Glove with values of 4.40, 4.20, and 3.60, respectively. Then in terms of ease of use VIVE Controller is also superior compared to Leap Motion Controller and Senso Glove with values of 4.35, 3.55, and 3.15, respectively.

In the results shown in Table 4 and Figure 2, Senso Glove excels at the two statements of satisfaction with the haptic feedback provided. Compared to the VIVE Controller and the Leap Motion Controller, Senso Glove excels at the vibrations provided. Even so, the value of Senso Glove's excellence in the statement about vibration is still low because it has not yet reached a value of 4 where the user agrees or is satisfied with the statement given. It is expected in the future, the availability of tools that are easy to learn, precise, and able to provide haptic feedback that satisfies its users.

Acknowledgment
This research is fully funded by Universitas Sumatera Utara through TALENTA USU Research Program financial year 2019 based on contract Nr. 4167/UN5.1.R/PPM/2019 dated 01 April 2019.

References

[1] Sutcliffe A 2003 Multimedia and Virtual Reality : Designing Multisensory User Interfaces *Psychology Press*
[2] Fahmi F, Nainggolan F, Andayani U and Siregar B 2018 Development of excavator training simulator using leap motion controller *Journal of Physics: Conference Series* Vol. 978, No. 1, p. 012034
[3] Andayani U, Syahputra MF, Muchtar MA, Sattar M, Prayudani S and Fahmi F 2019 3D Modelling Intestine Anatomy with Augmented Reality for Interactive Medical Learning *IOP Conference Series: Materials Science and Engineering* Vol 648 No 1 p 012035
[4] Notebaert AJ 2009 Student perceptions about learning anatomy *Theses and Dissertations* doi: 10.17077/etd.q0k5zpz3.
[5] Thiriet P 2011 3D Human Anatomy Learning - Demonstration of 3D Tools used in Teaching: 3D Videos, Podcasts, PDF *CSEDU* doi: 10.5220/0003339804080411.
[6] Nainggolan F, Siregar B, and Fahmi F 2016 Anatomy learning system on human skeleton using Leap Motion Controller *3rd International Conference on Computer and Information Sciences (ICCOINS)* (pp. 465-470) IEEE
[7] Fahmi F, Nainggolan F, and Siregar B 2019 3D anatomy learning system using Virtual Reality and VR Controller *Journal of Physics: Conference Series* Vol. 1235 No. 1 p012043
[8] Nainggolan F, Siregar B and Fahmi F 2019 User experience in using vive controller as a controller in the anatomy learning system in the virtual reality environment The 4th International Conference on Computing and Applied Informatics (ICCAI) under publication
[9] Nainggolan F, Siregar B, and Fahmi F 2019 Design of interactive virtual reality for erection steel construction simulator system using senso gloves *TALENTA-International Conference on Science and Technology* under publication
[10] Tanjung K, Nainggolan F, Siregar B, Panjaitan S and Fahmi F 2019 The use of virtual reality controllers and comparison between vive, leap motion and senso gloves applied in the anatomy learning system *TALENTA-International Conference on Science and Technology* under publication
[11] Lund AM 2001 Measuring Usability with the USE Questionnaire p. 5

[12] Nainggolan F, Siregar B and Fahmi F 2019 User experience in using vive controller as a controller in the anatomy learning system in the virtual reality environment The 4th International Conference on Computing and Applied Informatics (ICCAI) under publication
[13] Nainggolan F, Siregar B, and Fahmi F 2019 User Experience in Excavator Simulator using Leap Motion Controller in Virtual Reality Environment *The 4th International Conference on Computing and Applied Informatics (ICCAI)* under publication
[14] Fahmi F, Nainggolan F, Siregar B, Soeharwinto and Zarlis M 2020 User experience study on crane operator erection simulator using senso glove in a virtual reality environment *The 2nd International Conference on Information Technology and Engineering Management (ITEM)* under publication
[15] Niehorster DC, Li L, and Lappe M 2017 The Accuracy and Precision of Position and Orientation Tracking in the HTC Vive Virtual Reality System for Scientific Research *i-Perception*, vol. 8, no. 3, p. 2041669517708205, DOI: 10.1177/2041669517708205.
[16] Nowicki MR, Pilarczyk O, Wasikowski J and Zjawin K 2014 Gesture Recognition Library For Leap Motion Controller
[17] Perret J and Vander PE 2018 Touching Virtual Reality: a Review of Haptic Gloves *Proceedings of 16th International Conference on New Actuators*, pp. 270–274.
[18] Rahmat R F, Hasibuan RH, Siregar B, Syahputra MF 2018 A traditional bekel game using leap motion controller *Journal of Physics: Conference Series*, **1116**(2), 022036.

Effect of operation conditions to rice plant fiber reinforced composite on coefficient of friction and wear rate of brake lining

Pramuko Ilmu Purboputro[1], Agung Setyo Darmawan[1,*], and Bambang Waluyo Febriantoko[1]

[1]Department of Mechanical Engineering, Faculty of Engineering, Universitas Muhammadiyah Surakarta, Jl. Ahmad Yani, Tromol Pos 1 Pabelan, Surakarta 57162, Indonesia

*Corresponding author: Agung.Darmawan@ums.ac.id

Abstract. Composites made from a combination of rice plant fiber, fiberglass, aluminium powder and polyester resin are potential materials to be used as brake lining. In addition to the properties of the composite material itself, operating conditions also affect the performance of the brake lining. Therefore, this study is aimed to investigate the effect of operating conditions on the coefficient of friction and the wear rate of the composite when used as brake lining. The coefficient of friction and the wear rate tests were carried out on composites made of 40% fiber rice plant, 40% polyester resin, 10% fiberglass and 10% aluminium. The tests were carried out at a pressure of 30 kg/cm^2 and a speed of 2000 rpm. The operating conditions applied in this study were dry, water, salt water and brake oil. The test results show that the highest coefficient of friction was the dry operating conditions of 0.643 and the lowest coefficient of friction was the operating condition using brake oil of 0.586. While the wear test results showed the highest wear rate in dry operating conditions of 81.25 mm^3/hour and the lowest wear rate in operating conditions using brake oil of 68.34 mm^3/hour.

1. Introduction

Composite materials defined as a combination of two or more phases in a macroscopic and microscopic manner in order to obtain the desired combination of properties. To achieve the required properties, various combinations of metals, ceramics and polymers are mixed. This will produce various types of composite materials. In general, composite materials are made to improve the predominance of mechanical properties such as strength, toughness, stiffness at room temperature and at high temperatures [1-3].

Composite materials play an important role in industry. For example, composites are widely used in the fields of aerospace applications [4, 5], aircraft and military [6, 7], the automotive industry [8, 9], sports equipment [10] and packaging materials [11].

Nowadays, natural fiber reinforced composites developed by numerous industries are utilized in various applications [12-15]. In order to find new possibilities, the natural fibers have been recently studied as a reinforcing material in automotive brake friction composite materials and showed significant potential. The researchers highlighted that apart from tribological properties, the economic and ecological benefits of natural fiber reinforced friction composites helps them to be utilized in automotive braking systems.

IOP Conf. Series: Materials Science and Engineering **851** (2020) 012025 doi:10.1088/1757-899X/851/1/012025

Brake lining is a friction material made from a mixture of materials that serves as a barrier to friction between two plates. The main function of the brake lining is to reduce or even stop the rotational speed of an object. One of potential material to replace common material is natural composite that help protect the environment and human health [16, 17].

Matějka et al. [18] investigate effects of silicon carbide particle sizes on friction and wear properties of composites. The composite is applied for brake lining of car. The coefficient of friction is significantly increased with the presence of SiC. Increasing of particle size caused decreasing of specific wear rate.

The coefficient of friction and the rate of wear become important properties to be investigated when composites are used as components that experience friction with other components. One of its uses is as a material for brake lining. Therefore, this research is intended to investigate the effect of brake lining operating conditions on the coefficient of friction and corrosion rate of the composite

2. Materials and Methods

In this study, composite materials are used as alternative materials for brake lining. This composite is a combination of 40% polyester resin, 40% of rice plant fiber, 10% fiberglass and 10% Aluminum powder.

The first step of this research is to weigh material for composites. Furthermore, rice plant fiber, fiberglass and aluminum powder are mixed until blended. Then the steel brake lining mold is mounted on the plate. In the second step, polyester resin is placed in the mold. Then the other material that has been mixed is put into the mold.

After all components of the composite were mixed in the mold, the mold is then pressed using a pressing machine. During the compression process, the composite mixture was given heat at a temperature of 90 C. Furthermore, the composite was sintered at a temperature of 180 C for 15 minutes. After the composite in the mold has cooled, the brake lining is then removed from the mold.

Metallographic testing is conducted by taking micro photos of brake lining, and then proceeds with friction testing to get the coefficient of friction. The coefficient of friction testing is carried out at a pressure of 30 kg/cm^2 and a speed of 2000 rpm. The operating condition for the coefficient of friction test and wear rate test were dry, water, salt water and brake oil.

3. Results and Discussions

Figure 1 shows a photo of a brake lining made of composites as a product of this research. This composite is a combination of metals, ceramics, polymers and natural materials. The composition of this composite is 40% polyester resin, 40% of rice plant fiber, 10% fiberglass and 10% Aluminum powder.

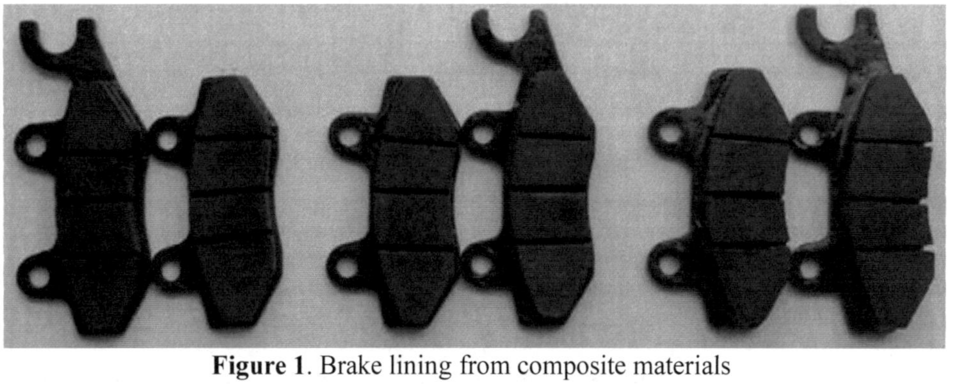

Figure 1. Brake lining from composite materials

Micro photo of brake linings made of composites are shown in figure 2. Aluminum powder with a size of 3 μm, fiberglass and rice plant fiber is surrounded by polyester resin matrix. Aluminum powder, fiberglass and rice fiber plant have a function as composite reinforcement and transferring the load received by the component.

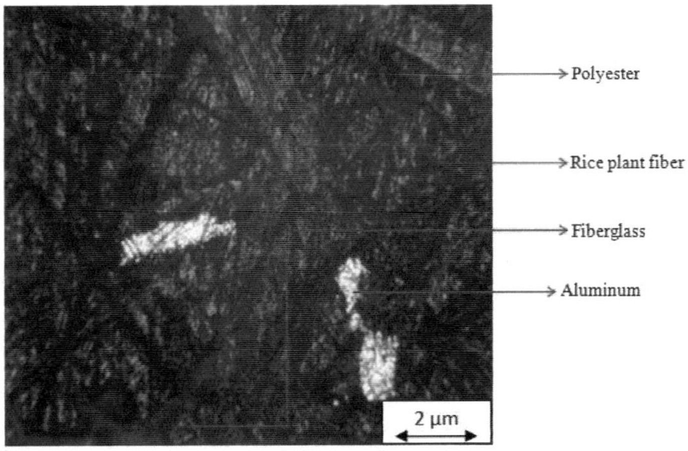

Figure 2. Photo micro of brake lining from composite materials.

The coefficient of friction of composite at various operating condition is shown in figure 3. The coefficient of friction of the composites under dry, water, salt water, and brake oil operating conditions is 0.643, 0.631, 0.603 and 0.586, respectively.

The coefficient of friction of the composites under dry, water, salt water, and brake oil operating conditions is 0.643, 0.631, 0.603 and 0.586, respectively. Compared to other operating conditions, the coefficient of friction in the brake oil operating conditions has decreased the greatest coefficient of friction by 9.7%. This is due to the function of the lubricant from brake oil which reduces friction.

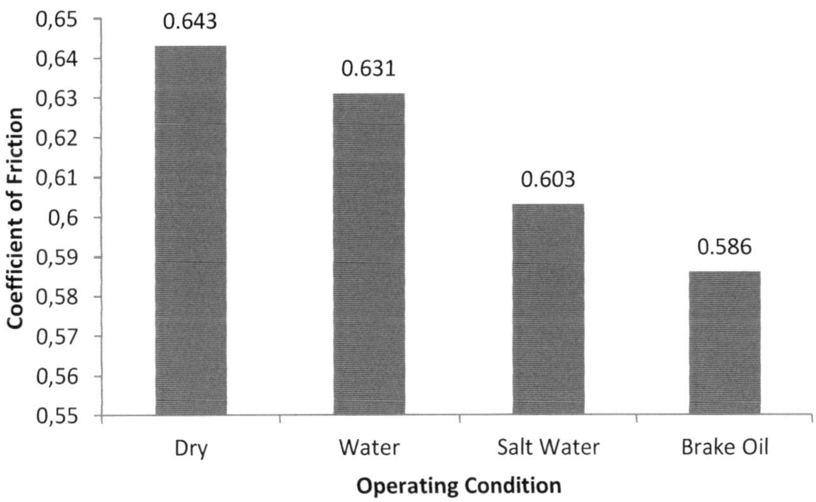

Figure 3. Coefficient of friction of composite at various operating condition.

The wear rate test shows a similar tendency to the results of the friction coefficient test (figure 4). The wear rate of the composites under dry, water, salt water, and brake oil operating conditions is 81.25 mm^3/hour, 78.66 mm^3/hour, 72.38 mm^3/hour and 68.34 mm^3/hour, respectively. The highest decrease in wear rate also occurs in brake oil operating conditions by 16%. Figure 5 shows the surface of the former wear test on composite materials. It can be seen in the picture that the wear on the composite material is caused by a cohesive failure on the composite

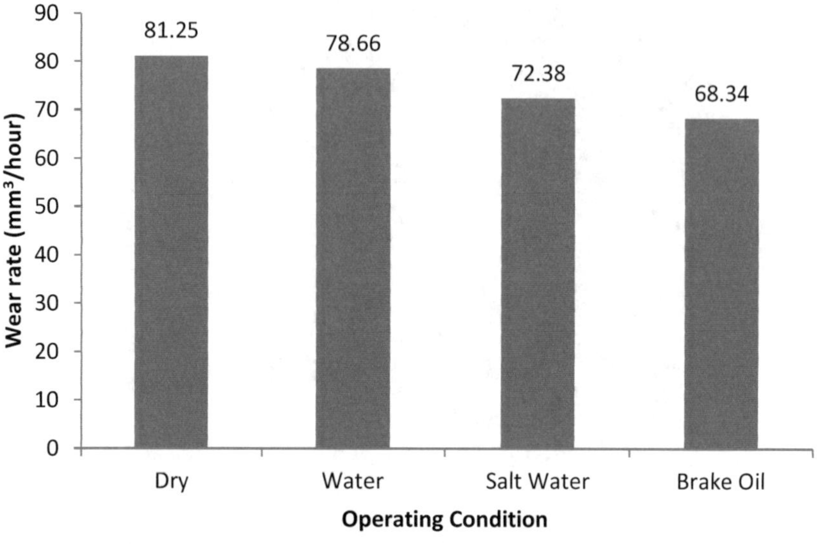

Figure 4. Wear of composite at various operating condition

Figure 5. Cohesive failure

4. Conclusions

From the results of the study it can be concluded that operating conditions affect the coefficient of friction and the rate of wear. The highest coefficient of friction of 0.643 occurs in dry operating conditions and the lowest coefficient of friction occurs in operating conditions using brake oil of 0.586. In the wear test it was found that the highest wear rate occurred in dry operating conditions of 81.25 mm^3/hour, while the lowest wear rate occurred in conditions using brake oil of 68.34 mm^3/hour.

Acknowledgment

The authors thank The Directorate of Research and Community Service, The Ministry of Research, Technology and Higher Education of the Republic of Indonesia for research funding [Project Number

= 199.24/A.3-III/LPPM/V/2019]. The authors also thank the Metallurgy Laboratory of Universitas Muhammadiyah Surakarta and Metallurgy Laboratory Universitas Gajah Mada for specimen preparations and mechanical testing carried out in this research.

References

[1] Sarikaya E, Çallioğlu H and Demirel H 2019 Production of epoxy composites reinforced by different natural fibers and their mechanical properties, *Composites Part B: Engineering* **167** 461-466.

[2] Salernitano E and Migliaresi C 2003 Composite Materials for Biomedical Applications: A Review, *Journal of applied biomaterials & biomechanics* **1(1)** 3-18.

[3] Verma D, Gope P C, Shandilya A, Gupta A and Maheshwari M K 2013 Coir Fibre Reinforcement and Application in Polymer Composites: A Review, *J. Mater. Environ. Sci.* **4(2)** 263-276.

[4] Zhou J, Li Y, Cheng L and Zhang L 2019 Indirect Microwave Curing Process Design for Manufacturing Thick Multidirectional Carbon Fiber Reinforced Thermoset Composite Materials, *Applied Composite Materials* **26(2)** 533–552.

[5] Kappel E. 2019 Distortions of composite aerospace frames due to processing, thermal loads and trimming operations and an assessment from an assembly perspective, *Composite Structures* **220** 338-346.

[6] Ren Y, Qiu L, Yuan S and Fang F 2020 Gaussian mixture model and delay-and-sum based 4D imaging of damage in aircraft composite structures under time-varying conditions, *Mechanical Systems and Signal Processing* **135** 106390.

[7] Turczyn R, Krukiewicz K, Katunin A, Sroka J and Sul P 2020 Fabrication and application of electrically conducting composites for electromagnetic interference shielding of remotely piloted aircraft systems, *Composite Structures* **232** 111498.

[8] Pietroluongo M, Padovano E, Frache A and Badini C 2020 Mechanical recycling of an end-of-life automotive composite component, *Sustainable Materials and Technologies* **23** e00143.

[9] Khatkar V, Behera B K and Manjunath R N 2019 Textile structural composites for automotive leaf spring application, *Composites Part B: Engineering* **182** 107662.

[10] Wang J L 2012 Application of Composite Materials on Sports Equipments, *Applied Mechanics and Materials* **155-156** 903-906.

[11] Mohamed S A A, El-Sakhawy M, Nashy E H A and Othman A M 2019 Novel natural composite films as packaging materials with enhanced properties, *International Journal of Biological Macromolecules* **136** 774-784.

[12] Duigou A L, Requile S, Beaugrand J, Scarpa F and CastroM2017 Natural fibres actuators for smart bio-inspired hygromorph biocomposites, *Smart Mater. Struct.* **26** 125009

[13] Zheng Y, Cai C, Zhang F, Monty J, Linhardt R J and Simmons T J 2016 Can natural fibers be a silver bullet? Antibacterial cellulose fibers through the covalent bonding of silver nanoparticles to electrospun fibers, *Nanotechnology* **27** 055102

[14] Graupner N, Labonte D, Humburg H, Buzkan T, Dörgens A, KeltererWand Müssig J 2017 Functional gradients in the pericarp of the green coconut inspire asymmetric fibre-composites with improved impact strength, and preserved flexural and tensile properties, *Bioinsp. Biomim.* **12** 026009

[15] Singh T, Tejyan S, Patnaik A, Singh V, Zsoldos I and Fekete G 2019 Fabrication of waste bagasse fibre reinforced epoxy composites: study of physical, mechanical and erosion properties, *Polym. Compos.* **40** 3777–86

[16] Yun R, Filip P and Lu Y 2010 Performance and evaluation of eco-friendly brake friction materials, *Tribology International* **43(11)** 2010–2019

[17] Darmawan A S, Purboputro P I and Febriantoko B W 2020 The aluminum powder size' effect on rice plant fiber reinforced composite to hardness, wear and coefficient of friction of brake

lining, *IOP Conference Series: Materials Science and Engineering* **722** 012002.
[18] Matějka V, Lu Y, Jiao L, Huang L, Simha Martynková G and Tomášek V 2010 Effects of silicon carbide particle sizes on friction-wear properties of friction composites designed for car brake lining applications, *Tribology International* **43(1-2)** 144–151.

IOP Conf. Series: Materials Science and Engineering **851** (2020) 012026 doi:10.1088/1757-899X/851/1/012026

Standard Distance Detector Tool for TV users based on Fuzzy Decision Tree

Iswanto, Nia Maharani Raharja

Departement of Eletrical Engineering, Universitas Muhammadiyah Yogyakarta, Yogyakarta, Indonesia
Departement of Eletrical Engineering, Universitas Islam Negeri Sunan Kalijaga, Yogyakarta, Indonesia

iswanto_te@umy.ac.id

Abstract. Eye health is a very important thing, so it is fitting that we take good care of it. With the frequent viewing of television programs, we should be able to maintain eye health from the effects of light released by a TV tube that can make our eyes tired quickly which can interfere with daily activities. This standard distance detector device based on microcontroller TV users is designed and uses ultrasonic sensors and PIR sensors as detectors, so that it can anticipate fatigue in the eye due to television viewing distance that is too close. The design of this tool uses the literature study method, by conducting library research to obtain a theoretical foundation from supporting books, internet, articles, journals and exchanging ideas with people who are experienced in the field of electronics. The results of the design of a Standard Distance Detector Microcontroller-based TV Users are devices that can measure and detect safe distances for the eyes in watching television activities, based on the size of television resolution. With the security of using electronic equipment that has been designed, so that in its implementation can minimize cases of damage to eye health in the general public as a whole.

1. Introduction

TV is often used as a scapegoat to damage the sense of sight in children who like watching TV. When sitting and watching close to the TV can cause eye strain and fatigue, watching TV in the wrong lighting can also cause eye strain. Sitting too close to the TV does not cause permanent damage to the eye but it causes eye strain and fatigue according to the Canadian Optometrist Association. At least there are some recommendations from CAO that you can do to minimize the effects of any type of TV lighting. Watching television has rules that must be obeyed if you don't want bad effects to approach. One of them is the distance of the television monitor screen to the eye must follow the standard calculation that applies internationally The distance between the television screen and the audience is 5 times the screen diagonal (for tube television). For this reason, the distance to watch television must follow the calculation of internationally recognized standards.

Distance sensors from previous researchers are used as the reference of standard distance detector TV users. Several researchers have used proximity sensors to measure distance. Miah examined unique smart glasses for people with the visual disorder. The glasses detect obstacles and measure distance perfectly using ultrasonic sensor and microcontroller. SIM900A GSM/GPRS module was used to collect information from the internet [1]. Hatthasin examined measuring wheel speech for the blind. The work aimed to design and make a prototype of a low-cost speech measuring wheel for individuals with visual impairments. By providing a Braille appearance and stating the distance traveled, the proposed device helps educate the blind to understand the concepts of distance and measurement [2].

Razvan-Daniel designed non-contact scouts using flexible and inexpensive ultrasonic. These scouts can be used in a variety of applications, such as toxic environments, hard-to-reach spaces, electromagnetic-polluted environments, etc. The core of the scout was the Atmega 328 P-PU microcontroller, which has an RISC architecture, a 32 8-bit register, namely sufficient resources to implement a highly accurate cost-effective product [3]. Tianyu researched a control method to prevent falls from the treadmill based on STM32 microcontroller and ultrasonic transducer. Treadmills have been widely used in homes or health entertainment centers; safety is very important. If the body's gait is not appropriate or the level of body tendency is too large, it causes movement instability. The intelligent electronic control method was based on the STM32 microcontroller and ultrasonic transducer, which automatically control the speed of the treadmill in real-time through detection of upper body position or the tilt angle of the human body [4].

The real-time embedded system for accident prevention was investigated by John. It was autonomous accident prevention with techniques that allow safety, speed control and accident detection system. The proposed system consisted of two separate design units: the transmitter unit, and the receiver unit. The security system included alcohol sensors, eye sensors and smoke sensors. The piezoelectric sensor will detect the signal, and send it to the ATmega328P microcontroller [5]. An analysis of the performance of laser distance finder and blue LEDs for Autonomous Underwater Vehicles (AUV) was carried out by Aras. The new technique for avoiding obstacles or depth control used a laser range finder and blue LED and was then applied to the Autonomous Underwater Vehicle (AUV). The finder was to measure and detect obstacles in a certain range. The components used were Arduino Nano as a microcontroller, 5 mm blue LED (receiver) and LRF (transmitter) able to focus on depth control [6].

The design and implementation of traffic signs and barriers detection on self-driving cars using SURF detectors and brutal force matching were examined by Renjith. One of the most amazing innovations of smart city traffic management and control is the increasing technological advancements in the field of automatic vehicles. This is made possible through an image matching algorithm using a combination of SURF detectors, brute force matching and RANSAC. This feature matching algorithm was preferred because it depended at least on the lighting conditions and rotation or tilt [7]. Papa investigated the detection of obstacles and start sensor integration for small unmanned aircraft systems. Barrier detection and attitude estimation for small quadrotors using low-cost sensors, namely the Sonic Ranging Sensor (SRS) and InfraRed Sensor (IRS), widely used in mobile applications for short distance measurements, were controlled and managed by a microcontroller (Arduino Mega 2560) and synchronized on 2-Hz sampling [8].

The development of autonomous robots for inspection systems was investigated by Saat. The system was equipped with an Inertial Measurement Unit (IMU) sensor that measured the orientation and direction of the robot, and a rotary encoder that measured the distance traveled by the robot. Proportional Derivative (PD) controller was applied for the autonomous navigation system. The feedback signals from the IMU sensor represented the orientation of the robot, and the accumulation of the distances was measured from the rotary encoder [9]. Collision detection and vehicle avoidance in the VANET environment were investigated by Anadu. Physical experiments involved making VANET implemented with microcontrollers, sensor systems, and radio-controlled cars. Integrating all of them resulted in a collision detection system that helped improve traffic efficiency on a small scale. This physical model was executed using Arduino UNO, RGB LCD Shield, GPS Logger Shield, SR04 Ping Distance Sensor, RF 433 MHz Transmitter Module, RF 433 MHz Receiver Module and DC motor [10].

Based on previous research, the use of proximity sensors in this paper is different from previous studies. Previous researchers to determine a distance using ultrasonic sensors SRF04, IMU, GPS, and Infrared. Ultrasonic sensors with PIR sensors are used to detect the distance of people watching TV. In addition, this paper presents a proximity sensor combined with a fuzzy decision tree algorithm for the detection of a standard distance for TV use.

IOP Conf. Series: Materials Science and Engineering **851** (2020) 012026 doi:10.1088/1757-899X/851/1/012026

2. Method

The system designed includes hardware and software that provide standard TV viewing information and give a warning if the distance is less than the internationally defined standard. The system diagram block is shown in figure 1.

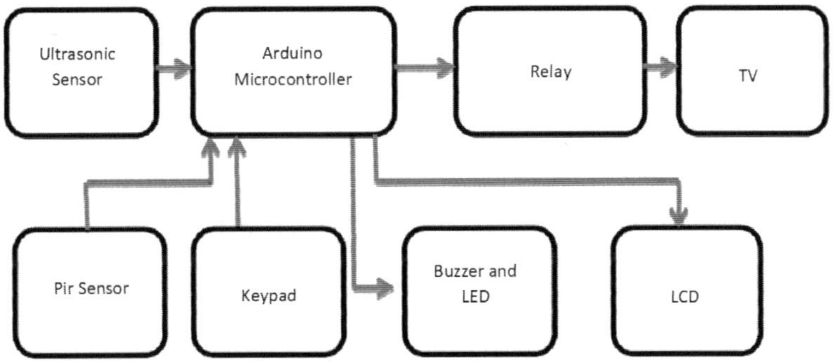

Figure 1. System block diagram

The working principle of the block diagram shown in Figure 1. A keypad is used to enter the type of television namely the screen width in inches used as the input from the microcontroller [11], [12], [21], [13]–[20]. When there is an object that does not emit infrared rays exposed to the ultrasonic sensor[22]–[28], the LCD shows the distance of the object to the TV. When there is an object that emits infrared light, as long as the distance is met, the LED will turn off and the buzzer will not ring. When the standard distance is not met, the LED will blink and the buzzer will sound, and when the warning given is more than 10 seconds, the relay will work and turn off the power from the television automatically.

2.1. Hardware design

The device in a wiring system along with the devices and electrical components used shown in Figure 2. It can be seen in the figure that the system consists of an Arduino Uno [11], [29], [38]–[43], [30]–[37] development board used as the controller for the entire workflow of this device, a 4x8 keypad module used as an input for programming variables, a 16x2 LCD used as programming variable data viewer, a buzzer circuit as an indicator of control conditions, and a relay circuit to control the on/off of the television.

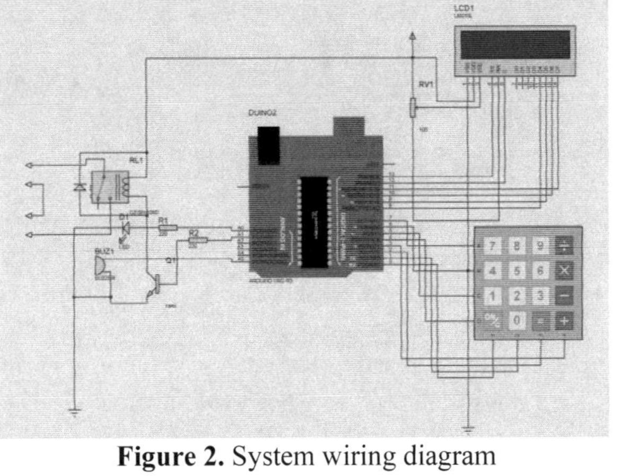

Figure 2. System wiring diagram

2.2. Software design

The programming design is explained in the workflow diagram shown in Figure 3. The diagram shows that the first pin initialization namely Arduino pins used for LCD, keypad, PIR, buzzer, and ultrasonic sensors has been done. After that library declarations and variables are used in the software program. The second stage is the program opening screen to set TV size options. Then the third is the main program to read the condition of the proximity sensor logic and the process of displaying distance on the LCD. After the distance processing and displaying it on the LCD, there are sub-programs as safe distance detectors that are distinguished by the size of the television used and infrared sensor detectors to detect the presence of living things. If the object has exceeded the safe limit of watching television, the buzzer and LED give a warning based on the distance of the object, if the warning process is ignored for more than 10 seconds, the relay will be activated to turn off the television. But if it meets the safe distance again, the relay will be deactivated so that the television turns on again.

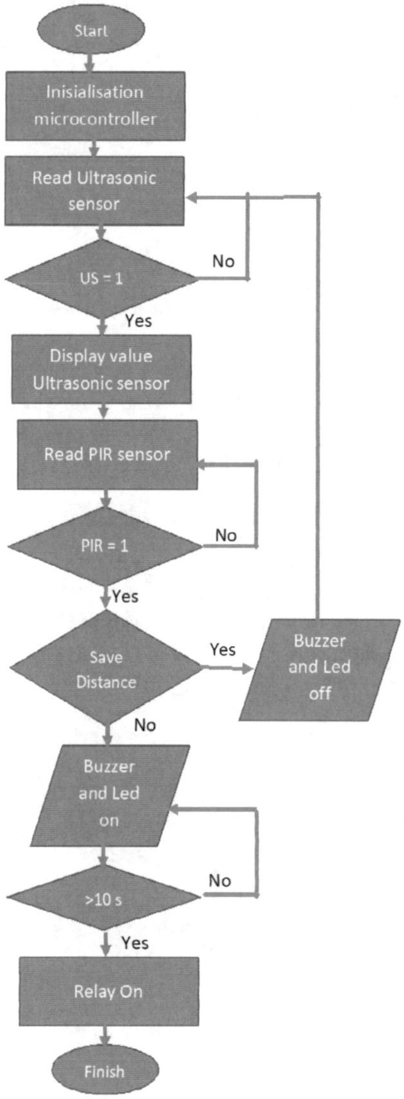

Figure 3. System workflow diagram.

3. Implementation

At this stage 2 test implementations were carried out namely testing the sensor and the whole device. Sensor testing is performed to determine the accuracy of detection and measurement, while the device testing is performed to determine the actual performance of the tool in the field.

3.1. Sensor testing

This test was conducted on two sensors namely PIR and ultrasonic sensors. The PIR sensor is performed to adjust the sensitivity in the minimal position with variations in height and angle of measurement. In testing with a height of 75 cm from the floor, the state of the PIR sensor is always appropriate and accurate to the distance from the TV. As for testing the distance with ultrasonic sensors was carried out by varying the distance of the sensor from the TV compared to the measurement of standard measuring instruments. This test shows a maximum error of 0.15% and an accuracy rate of 98%.

3.2. Testing tool

Device testing is performed to determine the success of the overall performance of the hardware and software designed. This test is carried out with 3 television size variations, namely; 14 inches, 17 inches and 20 inches.

Table 1. Table for 14 Inch television testing

No.	Watching Distance (cm)	Device Condition	TV Condition
1	160	Normal	On
2	140	LED blink (medium) & Beep (low frequency)	On
3	120	LED blink (faster) & Beep (medium frequency)	On
4	40	LED blink (very fast) & Beep (high frequensinya)	Off

In this variation, the device performance is in accordance with the design and workflow

Table 2. Table for 17 Inch television testing

No.	Watching Distance (cm)	Device Condition	TV Condition
1	200	Normal	On
2	180	LED blink (medium) & Beep (low frequency)	On
3	140	LED blink (faster) & Beep (medium frequency)	On
4	50	LED blink (very fast) & Beep (high frequensinya)	Off

In this variation, the device performance is in accordance with the design and workflow

Table 3. Table for 20 Inch television testing

No.	Watching Distance (cm)	Device Condition	TV Condition
1	255	Normal	On
2	200	LED blink (medium) & Beep (low frequency)	On
3	170	LED blink (faster) & Beep (medium frequency)	On
4	75	LED blink (very fast) & Beep (high frequensinya)	Off

In this variation, the device performance is in accordance with the design and workflow

4. Conclusion

Having designed, implemented and tested the device, it can be concluded that the device has been running normally in accordance with the purpose of the study namely measuring the distance of television viewers to objects. The results of the 14inch television testing at the distance of 160 cm, the television is still on and at the distance of 45 cm and 40 cm the television is off. The 17inch television testing at the distance of 200 cm, the television is still on and at the distance of 48 cm the television is off. The test results of a 20inch television testing at the distance of 255 cm, the television is still on and at the distance of 75 cm, the television is off. The distance that make the television off is 50 cm and closer for either 14inch, 17inch, or 20inch televisions.

References

[1] M. R. Miah and M. S. Hussain, "A Unique Smart Eye Glass for Visually Impaired People," in 2018 International Conference on Advancement in Electrical and Electronic Engineering (ICAEEE), 2018, pp. 1–4.

[2] U. Hatthasin, N. Setamung, P. Piyawongwisal, and S. Tisom, "A Talking Distance Measuring Wheel for the Visually Impaired," in 2018 15th International Conference on Electrical Engineering/Electronics, Computer, Telecommunications and Information Technology (ECTI-

CON), 2018, pp. 517–520.
[3] A. Razvan-Daniel and D. Nicolae, "Designing a flexible and low cost ultrasonic non-contact rangefinder," in 2017 14th International Conference on Engineering of Modern Electric Systems (EMES), 2017, no. 2, pp. 244–247.
[4] B. Tianyu, "A control method to prevent falling from a treadmill based on STM32 microcontroller and ultrasonic transducer," in 2017 29th Chinese Control And Decision Conference (CCDC), 5092, pp. 5025–5028.
[5] A. John and P. R. Nishanth, "Real time embedded system for accident prevention," in 2017 International conference of Electronics, Communication and Aerospace Technology (ICECA), 2017, pp. 645–648.
[6] M. S. M. Aras et al., "Analysis performances of Laser Range Finder and blue LED for Autonomous Underwater Vehicle (AUV)," in 2016 IEEE International Conference on Underwater System Technology: Theory and Applications (USYS), 2016, pp. 171–176.
[7] R. Renjith, R. Reshma, and K. V. Arun, "Design and implementation of traffic sign and obstacle detection in a self-driving car using SURF detector and Brute force matcher," in 2017 IEEE International Conference on Power, Control, Signals and Instrumentation Engineering (ICPCSI), 2017, pp. 1985–1989.
[8] U. Papa, G. Del Core, G. Giordano, and S. Ponte, "Obstacle detection and ranging sensor integration for a small unmanned aircraft system," in 2017 IEEE International Workshop on Metrology for AeroSpace (MetroAeroSpace), 2017, pp. 571–577.
[9] S. Saat, A. R. M. Kamil, M. Z. M. Tumari, and A. S. R. A. Subki, "Development of an autonomous robot for inspection system," in 2018 IEEE 14th International Colloquium on Signal Processing & Its Applications (CSPA), 2018, no. March, pp. 272–276.
[10] D. Anadu, C. Mushagalusa, N. Alsbou, and A. S. A. Abuabed, "Internet of Things: Vehicle collision detection and avoidance in a VANET environment," in 2018 IEEE International Instrumentation and Measurement Technology Conference (I2MTC), 2018, pp. 1–6.
[11] H. U. Zaman, M. S. Hossain, M. Wahiduzzaman, and S. Asif, "A novel design of a robotic vehicle for rescue operation," in 2015 18th International Conference on Computer and Information Technology (ICCIT), 2015, pp. 507–510.
[12] B. Ahmad and I. Hussain, "Design and hardware implementation of ball & beam setup," in 2017 Fifth International Conference on Aerospace Science & Engineering (ICASE), 2017, pp. 1–6.
[13] M. D. Khairunnas, E. Ariyanto, and S. Prabowo, "Design and Implementation of Smart Bath Water Heater Using Arduino," in 2018 6th International Conference on Information and Communication Technology (ICoICT), 2018, vol. 0, no. c, pp. 184–188.
[14] S. Widadi, M. K. Huda, I. Ahmad, and O. Tanane, "Atmega328P-based X-ray Machine Exposure Time Measurement Device with an Android Interface," J. Robot. Control, vol. 1, no. 3, pp. 81–85, 2020.
[15] S. Patranabis, J. Breier, D. Mukhopadhyay, and S. Bhasin, "One Plus One is More than Two: A Practical Combination of Power and Fault Analysis Attacks on PRESENT and PRESENT-Like Block Ciphers," in 2017 Workshop on Fault Diagnosis and Tolerance in Cryptography (FDTC), 2017, vol. 2017-Janua, pp. 25–32.
[16] P. Prakash, R. Sharma, S. Sindhu, and T. Shankar, "Visible Light Communication using Solar Panel," in 2017 2nd International Conference On Emerging Computation and Information Technologies (ICECIT), 2017, pp. 1–5.
[17] K. Kunal, A. Z. Arfianto, J. E. Poetro, F. Waseel, and R. A. Atmoko, "Accelerometer Implementation as Feedback on 5 Degree of Freedom Arm Robot," J. Robot. Control, vol. 1, no. 1, pp. 31–34, 2020.
[18] M. Penna, Shivashankar, B. Arjun, K. R. Goutham, L. N. Madhaw, and K. G. Sanjay, "Smart fleet monitoring system using Internet of Things(IoT)," in 2017 2nd IEEE International Conference on Recent Trends in Electronics, Information & Communication Technology

(RTEICT), 2017, vol. 2018-Janua, pp. 1232–1236.
[19] A. S. Tikhonenkov, A. V. Peresypkin, A. S. Toporskaya, and E. S. Suloeva, "Modeling of measuring systems based on programmable debugging circuits Arduino," in 2017 XX IEEE International Conference on Soft Computing and Measurements (SCM), 2017, pp. 519–521.
[20] A. Latif, H. A. Widodo, R. Rahim, and K. Kunal, "Implementation of Line Follower Robot based Microcontroller ATMega32A," J. Robot. Control, vol. 1, no. 3, pp. 70–74, 2020.
[21] S. A. Anil and R. K. Megalingam, "A comparative analysis of a navigation platform designed in Arduino and FPGA," in 2016 International Conference on Communication and Signal Processing (ICCSP), 2016, pp. 1306–1310.
[22] T. P. Tunggal, A. W. Apriandi, J. E. Poetro, E. T.Helmy, and F. Waseel, "Prototype of Hand Dryer with Ultraviolet Light Using ATMega8," J. Robot. Control, vol. 1, no. 1, pp. 7–10, 2020.
[23] S. Ghosh, P. K. Panigrahi, and D. R. Parhi, "Analysis of FPA and BA meta-heuristic controllers for optimal path planning of mobile robot in cluttered environment," IET Sci. Meas. Technol., vol. 11, no. 7, pp. 817–828, Oct. 2017.
[24] H. P. Tripathy, D. Bej, P. Pattanaik, D. K. Mishra, S. K. Kamilla, and R. K. Tripathy, "Measurement of Zone Temperature Profile of a Resistive Heating Furnace Through RVM Model," IEEE Sens. J., vol. 18, no. 11, pp. 4429–4435, Jun. 2018.
[25] A. Latif, K. Shankar, and P. T. Nguyen, "Legged Fire Fighter Robot Movement Using PID," J. Robot. Control, vol. 1, no. 1, pp. 15–19, 2020.
[26] S. Hwang and B. P. Kintigh, "Implementation of an intelligent roving robot using multiple sensors," in Proceedings of 1994 IEEE International Conference on MFI '94. Multisensor Fusion and Integration for Intelligent Systems, 1994, pp. 763–770.
[27] R. Manoj and A. Fernandez, "Rapid Prototyping IoT End Applications Using Software Development Kits and Add on Plugins," in 2017 IEEE International Symposium on Nanoelectronic and Information Systems (iNIS), 2017, vol. 2018-Febru, pp. 263–267.
[28] I. Prasojo, P. T. Nguyen, O. Tanane, and N. Shahu, "Design of Ultrasonic Sensor and Ultraviolet Sensor Implemented on a Fire Fighter Robot Using AT89S52," J. Robot. Control, vol. 1, no. 2, pp. 59–63, 2020.
[29] W. Ahmed, R. T. Naser Khan, F. Riaz, and S. Y. Arafat, "Model Intelligent Vehicle System for Life Safety at T, Y -Junctions," in 2018 IEEE 21st International Multi-Topic Conference (INMIC), 2018, pp. 1–6.
[30] R. Dey, U. Thakur, L. Sunny, L. D'Almeida, and K. Chakravarty, "Digital chest expansion measurement & its biomedical application," in 2015 E-Health and Bioengineering Conference (EHB), 2015, pp. 1–4.
[31] S. Sharma, K. Chand, D. Sharma, and P. Guha, "Development of an early detection system for fire using Wireless Sensor Networks and Arduino," in 2018 International Conference on Sustainable Energy, Electronics, and Computing Systems (SEEMS), 2018, pp. 1–5.
[32] S. Murti, P. Megantoro, G. D. B. Silva, and A. Maseleno, "The Design and Analysis of DC Electrical Voltage-Current Datalogger Device Implemented on Wind Turbine Control System," J. Robot. Control, vol. 1, no. 3, pp. 75–80, 2020.
[33] M. Borghetti, M. Serpelloni, E. Sardini, and O. Casas, "Multisensor System for Analyzing the Thigh Movement During Walking," IEEE Sens. J., vol. 17, no. 15, pp. 4953–4961, Aug. 2017.
[34] A. Juliano, A. H. Hendrawan, and R. Ritzkal, "Information System Prototyping of Strawberry Maturity Stages using Arduino Uno and TCS3200," J. Robot. Control, vol. 1, no. 3, pp. 86–91, 2020.
[35] R. Asnawi and M. Said, "Testing of other languages usage in addition to the default languages for the easy voice recognition module," in 2018 International Conference on Electronics Technology (ICET), 2018, pp. 321–324.
[36] I. Prasojo, A. Maseleno, O. Tanane, and N. Shahu, "Design of Automatic Watering System Based on Arduino," J. Robot. Control, vol. 1, no. 2, pp. 55–58, 2020.
[37] S. Sawidin, D. S. Pongoh, and A. A. S. Ramschie, "Design of Smart Home Control System Based

on Android," in 2018 International Conference on Applied Science and Technology (iCAST), 2018, pp. 165–170.

[38] S. Widodo, E. A. Pratama, S. Pramono, and S. B. Basuki, "Outdoor propagation modeling for wireless sensor networks 2.4 GHz," in 2017 IEEE International Conference on Communication, Networks and Satellite (Comnetsat), 2017, vol. 2018-Janua, pp. 158–162.

[39] M. W. Hariyanto, A. H. Hendrawan, and R. Ritzkal, "Monitoring the Environmental Temperature of the Arduino Assistance Engineering Faculty Using Telegram," J. Robot. Control, vol. 1, no. 3, pp. 96–101, 2020.

[40] Rahmat R F, Yusuf A, Lini T Z 2019 Real-Time Sensor Application for the Measurement of the Vertical Profiles of Atmosphere *International Conference on Electrical, Telecommunication and Computer Engineering, ELTICOM* 8943853, pp. 182-188.

[41] K. Ummah and K. Mutijarsa, "Design and development prototype of electronic payment system for angkot case study: City of Bandung, Indonesia," in 2015 International Conference on Information Technology Systems and Innovation (ICITSI), 2015, pp. 1–6.

[42] S. Chugh and J. Kaur, "Non-invasive hemoglobin monitoring device," in 2015 International Conference on Control Communication & Computing India (ICCC), 2015, no. November, pp. 380–383.

[43] J. Juhariansyah, R. Ritzkal, and A. H. Hendrawan, "Design Of An Automatic Bell Warning System For Prayer Times In A Net Centric Computing Lab," J. Robot. Control, vol. 1, no. 3, pp. 92–95, 2020.

Engineering Economics Evaluation For Manufacturing Competitiveness : A case study.

Ivo Andika Hasugian [1], Evita Dewi [1], Vandrick [1]

[1] Department of Industrial Engineering, Faculty of Engineering, Universitas Sumatera Utara, Indonesia

*Email: ivo.andika@usu.ac.id

Abstract. Cassava is one of the valuable commodities in Indonesia that can be used as various types of food. One of the foods made from cassava is Tapai which is the result of boiling and fermentation of cassava and has a high carbohydrate content. Tapai's business Mr. Legimin is a business that produces Tapai and sells it in plastic packaging. This business has been producing tapai since 20 years ago and has never conducted an economic analysis to prepare itself to enter the development of a rapidly growing industry. The research method used in this study consists of depreciation calculations, production costs consisting of fixed costs and variable costs, production costs, breakeven points and investment feasibility analysis using the Net Present Value (NPV) method, the benefit cost ratio (Net B) / Ratio C), Internal Rate of Return (IRR) and Payback Period (PBP). Based on an economic analysis with a project life of five years, it was found that the total production cost was Rp38,533,566.7 / month with production costs reaching Rp3,952.16 / kg, breakeven point 2,101 kg / month, NPV Rp2,355,523,695, B / C Net 2.41, IRR> MARR and the payback period is two months.

1. Introduction

Indonesia is an agrarian country, the livelihoods of its people are in the field of agricultural products. The development process in Indonesia drives industrial growth based on agricultural products [1]. Raw materials produced by the agricultural industry include cassava tubers. Cassava is included in food that is often produced by Indonesians and has become a common or staple food that is favored by many Indonesians [2]. Cassava is liked by many people because of its diverse nutritional content and is beneficial for human health. Cassava is rich in calories, carbohydrates, nutrients and more [3]. Because the nutritional content of the many benefits of cassava for health is maintaining the health of the digestive tract, controlling the content of pure sugar in the body and many other benefits of cassava for the body [4]. Because of the many benefits of cassava, many Indonesian people change cassava into various types of food one of which is tapai [5].

Cassava is fermented with the help of yeast sprinkled on the surface of yams that have been steamed into sweet ribbons. Tapai is a traditional Indonesian food which is a modification of cassava, tapai is produced from cassava fermented with the help of yeast as a substrate [6]. Tapai production is used by small and medium businesses as an opportunity to open a business, as it is known that tapai is a food that is preferred by many Indonesians even outside Indonesia [6]. One of the small and medium enterprises reviewed is Tapai Pak Legimin, which produces tapai in one area in Medan. Tapai production is sold to the surrounding community using 13 units of pedicabs. During 20 years in running this business, Tapai Legimin has never conducted an economic analysis of its business activities. The development of industry in Indonesia, especially the city of Medan, which is getting faster and the demand to reach is increasing, making Mr. Legimin want to expand production capacity and add business branches [8]. This study aims to conduct an economic analysis of production prices, breakeven points and the feasibility of investing in businesses owned by Bpk. Legimin. The results of the study will be given to the Tapai Legimin business with the hope that the results of the study can help evaluate the Tapai Legimin business so that improvements and developments in the business can be done.

2. Methodology

2.1. Time and Place

The research was conducted in December 2019 at Mr. Legimin's Tapai business located on Setiabudi Street Pasar 1, Amal Hallway No. 25, Medan, North Sumatra.

2.2. Method of Collecting Data

The data used are primary and secondary data. Primary data were obtained through interviews with tapai business managers, while secondary data were collected through scientific publications that discussed cassava. The research procedures carried out are assumptions and approaches as a basis for doing calculations and analysis. The assumptions and approaches used consist of: (1) the economic life of production equipment is 5 years, (2) the final price of the equipment is 10% of the initial price, (3) the interest rate is 12% effective per year, (4) the equipment maintenance costs 2 % / year, (5) working time 12 hours per day and 26 days per month, (6) the price of cassava raw materials is IDR 1,200 / kg and yeast IDR 200 / item, (7) the price of packaging IDR 5,000 / packaging, (8) electricity consumption is 12.33 kW / day.

2.3. Analysis Method

2.3.1. Depreciation Analysis

The amount of the cost of depreciating production equipment is calculated using the straight-line method. The straight-line method is a method in which the amount of depreciation in each period is the same that can be formulated as follows:

$$\text{Depreciation} = \frac{\text{Acquisition Cost-Residual Value}}{\text{Economic Age}} \tag{1}$$

2.3.2. Production Cost Analysis

Production costs are the sum of fixed costs with variable costs. The amount of production costs is calculated using equation 3 as follows:

$$TC = FC + VC \tag{2}$$

Symbol :
TC = Total Production Cost (IDR/year)
FC = Fixed Cost (IDR/year)
VC = Variable Cost (IDR/year)

2.3.3. Cost of Goods Sold Analysis

The cost of production is all costs that have been incurred from the production process until it is finished becoming goods ready for sale. The base price can be calculated using the following equation:

$$\text{Cost of Goods Sold} = \frac{\text{Total Production Cost}}{\text{Total Production}} \quad (3)$$

2.3.4. Break Even Production Analysis

The breakeven analysis is a way for the company to determine sales volume and product mix in order to achieve the desired profit level. The formula for calculating the break-even point for production is as follows:

$$\text{Break Even Production} = \frac{\text{Total of Fixed Cost}}{\text{Selling price per unit-Variable Cost Per Unit}} \quad (4)$$

2.3.5. Investment Feasibility Analysis

An investment is assessed for its feasibility from four factors namely NPV, BCR, IRR and PBP. An effort is said to be feasible if NPV> 0, BCR> 1, IRR> MARR interest rate.

a) Net Present Value (NPV)

Net Present Value is the present value derived from all cash flows from production to completion. The Net Present Value can be formulated as follows:

$$\text{NPV} = \sum_{t=0}^{n} \frac{(C)t}{(1+i)^{I}} - \sum_{t=0}^{n} \frac{(Co)t}{(1+i)^{I}} \quad (5)$$

where :
(C)t = t-year cash inflows
(Co)t = cash flow out of the t-year
n = investment age
I = transition current
t = year

b) Benefit Cost Ratio

Benefit cost ratio is done using two approaches for calculating Net BCR. Net BCR is the ratio between all value benefits to the cost value. The following is a mathematical formula from BCR.

$$\text{Net BCR} = \frac{\sum_{t=1}^{n} \frac{B_t - C_t}{(1+i)^t}}{\sum_{t=1}^{n} \frac{C_t - B_t}{(1+i)^t}} \quad (6)$$

where:
B_t = cash inflows in the t-year
C_t = cash outflows in the t-year
I = the actual interest rate

c) Internal Rate of Return (IRR)

The IRR method is used to solve interest rate problems by projecting to the present value of all cash flows from both receipts and expenses. The eligibility requirements are if IRR> MARR interest rates. The interest rate used as a reference in this study is the interest rate for PT. Bank Central Asia, Tbk, which is 12% effective annually. To calculate IRR you can use the following formula:

$$\sum_{k=0}^{n} R_K \left(\frac{P}{F}, i\%, k\right) - \sum_{k=0}^{n} D_k \left(\frac{P}{F}, i\%, k\right) = 0 \quad (7)$$

where :
R_k = K-year reception

D_k = K-year expenditure
n = Project Age

d) Payback Period

Payback period (PBP) is the period of the initial refund fee. The faster the return, the alternative is more attractive compared to other alternatives. The formula of the Payback period (PBP) is as follows:

$$\text{Payback Period} = \frac{\text{Investation}}{\text{Cashflow}} \text{ x 1 year} \quad (8)$$

3. Result and Discussion

3.1. Overview of Tapai Processing Business in Mr Legimin Business Tapai

Tapai is a superior product in the UKM Legimin Tapai. Tapai is the result of boiling and fermenting from cassava. Cassava (M. esculenta) is a plant that has a high carbohydrate content and high energy (Wardany, 2012). According to the Ministry of Agriculture (2012), the water and carbohydrate content of cassava is 60 grams and 37.9 grams per 100 grams of cassava.

Mr. Legimin's tapai UKM runs its business by selling tapai in the form of 500 gr plastic packaging. Tapai sold is in a ready-made state so that the tapai can be eaten without further processing.

The tapai production process at UKM Legimin Tapai is carried out continuously and is not dependent on orders (carried out continuously). Workers needed in the tapai production process at the business are 6 workers in the stripping section, 8 workers in the cutting section and 2 workers in the cassava boiling section. All tapai production activities are carried out by manual with the help of simple tools such as knives and cassava slicer. The production process of tapai is difficult and is not optimally done by machine due to several factors, namely: 1). The size of cassava is different so it is not suitable for slicing and cutting using a machine. 2). Lack of knowledge and experience of workers to operate machinery.

3.2. Economy Evaluation

Economic evaluation is carried out on tapai production which is still done manually without the aid of a machine. This is done manually due to differences in the size of cassava so that the cutting process cannot be done using a cutting machine.

a. Tapai Business Investment

Investment is an investment in the form of money or goods in a company with the aim of obtaining profits. Equipment investment in tapai production can be seen in Table 1.

Table 1. Tapai Business Investment

No	Equipment Name	Amount	Unit	Unit Cost (IDR)	Total Cost(IDR)
1	Knife	7	Piece	10.000	70.000
2	Wrapping tapai	3	Kg	37.000	111.000
3	Large basin	8	Piece	20.000	160.000
4	Refrigerator	3	Unit	3.300.000	9.900.000
5	Balance	3	Unit	63.000	189.000
6	Stewpan	6	Unit	200.000	1.200.000
7	Cassava peeler	7	Unit	12.000	84.000
			Total (IDR)		**11.714.000**

Calculation of the cost of tapai production is based on data and several assumptions that can be seen in Table 2.

Table 2. Data Variables and Assumptions of Tapai Production Business

No	Equipment Name	Value	Unit	Information
1	Age of the project	5	Year	Based on the age of the engine
2	Bank interest rates	12	%	People's business credit (kur)
3	Value of the beat of the engine	10	% of investment costs	Assumption
4	Machine maintenance costs	2	% of investment costs	Assumption
5	Business hours (hours)	12	Hour	Real condition
6	Business Day (days)	26	Day / month	Real condition
7	Cassava raw material	500	Kg / day	Real condition
9	Cassava Prices	1.200	IDR / kg	Real condition
10	Cost of cassava raw materials	600.000	IDR / day	Cassava raw material x price of cassava
11	Yeast raw material	2	Grain / kg of cassava	Real condition
12	Yeast Prices	200	IDR	Real condition
13	Cost of raw yeast	200.000	IDR / day	Yeast raw material x yeast price
14	Manual stripping capacity	41,67	Kg / hour	The calculation results
15	Stripping results	75	% of cassava raw materials	Measurement results
16	Tapai produced	375	Kg / day	Raw material x
17	Price of packaging (500 gr)	5.000	IDR. / Packaging	Real condition
18	The price of electricity is 900 VA	1.352	IDR / kwh	PLN
19	Electricity energy costs	16.666,67	IDR / day	Calculation
20	Consumption of charcoal fuel	5	Kg / day	Assumption
21	Price of charcoal	8.500	IDR / kg	Tokopedia
22	Fuel costs	42.500	IDR / day	Calculation
23	Wages	1.000.000	IDR / month	Real condition
24	Total manpower	16	Person	Real condition
25	Labor costs	16.000.000	IDR / month	Employment wage x number of workers

b. Total Tapai Production Costs

The total cost of producing tapai can be calculated by adding up the total fixed costs with the total variable costs. Fixed costs in the tapai production process are the costs of employee salaries. While the variable costs are determined from the total use of electricity, charcoal, cassava and yeast. Depreciation costs can be calculated using equation [1]. Data on the calculation of fixed costs can be seen in Table 2. The results of the calculation of fixed costs can be seen in Table 3.

Table 3. Fixed Costs of Tapai Production Business

No	Cost Components	Fixed Cost (IDR/Year)	Fixed Cost (IDR/Month)
1	Labor costs	192.000.000	16.000.000
2	Depreciation costs of equipment	2.108.520	175.710

3	Equipment maintenance costs	234.280	19.523,33
	Total (IDR)	**194.342.800**	**6.195.233,33**

Based on Table 3, it is known that the fixed costs of the tapai business are IDR.16,195,233.33 /month or IDR. 194,342,800/year. While the details of the variable costs of the tapai business can be seen in Table 4.

Table 4. Variable Costs of Tapai Production Business

No	Cost Components	Variable Costs (IDR/Day)
1	Cassava raw material	600.000
2	Yeast raw material	200.000
3	Electrical energy	16.666,67
4	Charcoal fuel	42.500
	Total (IDR)	**859.166,67**

Based on Table 4, it is known that the variable cost of the tapai business is IDR859,166.67 / day or IDR22,338,333.4 / month. Tapai production costs can be calculated by adding up fixed costs with variable costs using equation [2], namely: TC = FC + VC
Based on calculations it is known that:
FC = IDR 16.195.233,33/month
VC = IDR 22.338.333,4/month
So the amount of production costs (TC) is IDR 38.533.566,7/month

c. Cost of Goods Sold Tapai Business

The cost of Tapai business production is influenced by the number of products that can be produced in a certain period of time. The cost of production can be calculated using equation [5]. Based on calculations it is known that:
TC = IDR 38.533.566,7/month
TP = 9.750 kg/month
So that the cost of producing the Tapai business is : IDR 3.952,16 /kg

d. Break Even Point (BEP) Tapai Business

Break Even Point is the point where a business return on investment. The break-even point of a business is influenced by the total cost of production, the total production and the selling price of a product. Based on the results of the interview, the tapai is sold at IDR 10,000 / kg. By using equation [6], the break-even point for Tapai business can be known that is 25,211 kg / year or 2,101 kg / month.

e. Estimated Tapai Business Revenues

The revenue of the tapai business was obtained from the tapai sales which were assumed in the first month to the third month only 50% of the tapai was sold, the 4th month to the 6th month as many as 75% of the tapai was sold and in the 7th to the month 60% of 100% of the production tapai was sold.

Details of the Tapai business income and expenditure over the life of the project which is 5 years can be seen in Table 5.

Table 5. Business Benefits of Tapai Production

Year	Incomes (IDR/Year)	Expenseses (IDR/Year)	Profit (IDR/Year)

0	0	11.714.000	11.714.000
1	950.625.000	462.402.800	476.508.200
2	1.170.000.000	462.402.800	707.597.200
3	1.170.000.000	462.402.800	707.597.200
4	1.170.000.000	462.402.800	707.597.200
5	1.170.000.000	462.402.800	707.597.200

Based on Table 5, it can be seen that in the 0 th year, the profit gained is not yet available due to the UKM having to pay investment costs for the equipment to be used in the amount of IDR. 11,714,000, but in the 1st year, this Tapai business has been get a profit of IDR. 476,508,200 where in the first year tapai sales were assumed to be still unstable. Whereas for the 2nd to 5th year the tapai sales are assumed to be stable so that the profit of the Tapai business is IDR 707,597,200 / year.

f. Investment Feasibility Analysis

Before making an investment in a business, we must first analyze its feasibility. This analysis is carried out so that we can obtain information about the level of success (prospects) of the business. The analysis can also provide certainty of income from the business so that we avoid investments that do not provide benefits. In this study 4 methods were used to assess the feasibility of the Tapai SME investment, namely: 1) Net Present Value (NPV), 2) Internal Rate of Return (IRR), 3) Net Benefit Cost Ratio (Net B / C), 4) Payback Period (PBP). The business is declared feasible if NPV> 0, Net B / C> 1, IRR> MARR and PBP interest rates are needed, the shorter the better.

1) Net Present Value (NPV)

Expenditures incurred during the observation period are equipment investment costs at the beginning of the business, namely IDR. 11,714,000 and production costs as much as IDR. 462,402,800 / year, and income derived from tapai sales is IDR. 950,625,000 in the first year and IDR. 1,170,000,000 in the second year to the fifth year with the final value of the equipment which is assumed to be 10% of the initial investment cost, which is IDR 1,171,400. A cash flow diagram that contains details of income and expenses per period can be seen in Figure 1, where income is indicated by an up arrow while expenses are indicated by a down arrow.

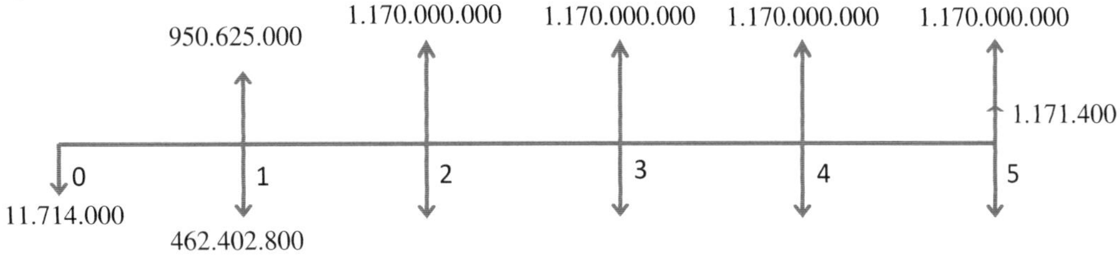

Figure 1. Cash flow diagram of Mr Legimin Tapai Business

Based on Figure 1, the income and expenditure at the present value is in accordance with KUR Bank BCA of 12% so that it is found that the income of the SME Tapai is at the present value of IDR4,022,382,304 while the expenditure of the SME is at the present value of IDR1,666,858,609. The NPV value can be calculated using equation [5] and the results show that NPV = IDR2,355,523,695 where NPV> 0 then the business is declared feasible.

2) Net Benefit Cost Ratio (Net B/C)

Net Benefit Cost Ratio is a comparison between total income at present value with total expenditure at present value within a certain time period in which this study is 5 years. The amount of Net Benefit Cost Ratio (Net B / C) can be calculated using equation [6] of 2.41 where Net B / C> 1, then the business is declared feasible.

3) Internal Rate of Return (IRR).

Internal Rate of Return abbreviated as IRR is an indicator of the efficiency level of an investment. An effort can be made if the rate of return is better than the rate of return if an investment is made elsewhere. Other places referred to are interest rates on bank deposits, mutual funds and others or commonly denoted as market interest rates or Minimum Attractive Rate of Return (MARR). In this study the MARR used is the Bank BCA interest rate of 12%. IRR calculation can be done using equation [7] so that the calculation results obtained are the amount of IRR> 12% then the effort is declared feasible.

4) Payback Period (PBP)

Payback Period (PBP) can be used to find out at what period the initial investment capital of a business can cover the initial capital of money already issued. PBP indicates in what period the business has generated greater revenue compared to business expenses. Details of SME Tapai income and expenditure in the initial period to the period in which the total cost of income has been greater than the total cost of expenditure can be seen in Table 6.

Table 6. Tapai Business Income and Expenses

Month	Income (IDR/month)	Expenses (IDR/month)	Balance (IDR/month)
0	0	11.714.000	-11.714.000
1	48.750.000	38.533.566,7	-1.497.566,7
2	48.750.000	38.533.566,7	8.718.866,6

Based on Table 6, it can be seen that the balance is positive in the 2nd month. This indicates that the business has already received initial investment costs in the 2nd month. So in the second month of the business has earned more revenue than expenses.

4. Conclusion

1) The amount of the cost of producing tapai is IDR 38,533,566.7 / month and the cost of production is IDR 3,952.16 / kg.
2) The break-even point for tapai production is 25,211 kg / year or 2,101 kg / month where the break-even point is affected by fixed costs, variable costs and total production.
3) Based on investment feasibility analysis, NPV> 0, Net B / C ratio> 1, IRR> 12%, PBP in the 2nd month.

Acknowledgments

The author would like to thank all participants who have helped and supported the completion of this paper, especially the Universitas Sumatera Utara, which has supported the funding of this paper.

References

[1] Needy K L, Lavelle J P, Nachtmann H, and Eschenbach T G, 2000 An Empirical Analysis of Engineering Economy Pedagogy," *The Engineering Economist*, **45**(1) pp. 74 – 92.
[2] Evans E, Nachtmann H, and Needy K 2010 A Look into Engineering Economy Education Literature *American Society for Engineering Education Annual Conference Proceedings*
[3] Rantung AH, Pakasi CBD, and Talumingan C 2015 Analisis Struktur Biaya Dan Efisiensi Pada Agroindustri Bawang Goreng 'Ud. Sri Rejeki' Di Kota Palu Provinsi Sulawesi Tengah, P. 3

[4] Anita U 2014 Analisis Perhitungan Harga Pokok Produksi Sebagai Dasar Penetapan Harga Jual Produk Furniture (Studi Kasus Pada Pt. Hanin Designs Indonesia - Indonesian Legal Wood) *J. Ekon. Akunt* 2014.

[5] Tangeren R, Sondakh J J, And Pontoh W, 2018 Analisis Titik Impas Dan Batas Aman Sebagai Dasar Perencanaan Laba Jangka Pendek Pada Pt. Soho Industri Pharmasi Cabang Manado," *J. Ris. Akunt. Going Concern*, **13**(4), P. 376, 2018, Doi: 10.32400/Gc.13.03.20274.2018.

[6] Sari C F K, Sawaki M E, And Sabarofek M S, 2018 Pengaruh Analisis Investasi Terhadap Kelayakan Penambangan Batu Mangan Di Pt. Berkat Esa Mining *J. Sci. Tech*, **4**(11), P. 13

[7] Wahyudin Y And Lesmana D 2016 Analisis Kelayakan Ekonomi Pengembangan Bisnis Pemanfaatan Kima Secara Berkelanjutan *J. Mina Sains*, **2**(2) P. 56

[8] Ginting E, Ariani F, And Rahmi M S 2019 *Ekonomi Teknik*. Medan: Graha Ilmu

[9] Purnatiyo D 2016 Analisis Kelayakan Investasi Alat Dna Real Time Thermal Cycler (Rt-Pcr) Untuk Pengujian Gelatin *J. Pasti*, **8**(2), P. 215, 2016.

[10] Hasugian I, Fernando F, and Supriadi S, 2020 Analisis Studi Kelayakan Usaha Pendirian Industri Rumahan (Studi Kasus Di Ukm Mie Pangsit XYZ Medan) *Jurnal Simetri Rekayasa*, **1**(2), pp. 46-50

[11] Hasugian I, Ingrid F, and Wardana K, 2020 Analisis Kelayakan Dan Sensitivitas : Studi Kasus UKM Mochi Kecamatan Medan Selayang) *Buletin Utama Teknik*,**15**(2), pp. 159-164

[12] Siregar I, Muchtar M A, Rahmat R F, Andayani U, Nasution T H, and Sari R M 2018 Method of calculation overall equipment effectiveness in fertilizer factory. *IOP Conference Series: Materials Science and Engineering* **308**(1), p. 012053)

IOP Conf. Series: Materials Science and Engineering **851** (2020) 012028 doi:10.1088/1757-899X/851/1/012028

Analysis Of Queuing Models Of Fast Food Restaurant with Simulation Approach

Ivo Andika Hasugian [1], Vandrick [1] , Evita Dewi [1]

[1] Department of Industrial Engineering, Faculty of Engineering, Universitas Sumatera Utara, Indonesia.

*Email: ivo.andika@usu.ac.id

Abstract. Queues in the cashier area in food purchases is one of the problems that often occurs in business processes in the community. Queues occur because of an imbalance between those served and their services. In fast food restaurants like XYZ, queues generally occur during lunch breaks or at dinner time. The purpose of this study is to determine the queuing model that occurs in the XYZ cashier area and analyze whether the queuing model is appropriate and efficient by comparing the previous model with the simulation results that have been carried out with the help of Anylogic software. The restaurant queue system is 3 queue lines and 3 counters. The average length of service and the interval of arrival of workers is 125.1 seconds and 67.5 seconds. Preliminary observational data show the average service time counters 1, 2 and 3 are 111, 136, 138 seconds. This fact shows that there is a long queue, the number of customers served at relatively low time units and work imbalances between operators. The results of the best queue simulation are proven to have shorter queues and the highest number of customers served per unit time. Based on the simulation results obtained by replacing 3 queue lines into 1 queue line and increasing the number of customers served is 74 people from 129 customers with the percentage of customers at each counter that is 28.38% for counter 1, 27.03% for counters 2 and 44, 59% for counter 3..

1. Introduction

Increased public interest in consuming fast food can cause its own problems for fast food businesses [1]. The problem that can be investigated from the increased interest of the community lies in its queuing system. The process of queuing into something is less favored by the people of Indonesia. This is also an obstacle experienced at XYZ fast food restaurant. The number of fast food enthusiasts cause long queues [2]. This condition causes a condition where the productivity of the fast food restaurant decreases. In the XYZ fast food store queue system there are 3 queue lines with a total of 3 service counters. Based on observations, the average service time at counter 1 is around 111 seconds, counter 2 is around 136 seconds and counter 3 is around 138 seconds. The limitation of this study is the amount of data as much as 100 data where at the time of observation counter 1 served 45 customers, counter 2 served 39 customers and counter 3 served 16 customers with a queue of about 3-7 people each. The queue itself is a line formed by the process of waiting in a system to get a thing or service [3]. Some queuing models commonly used in fast food restaurant are first in first out, last in first out, and random order in service [4]. Under these conditions it is necessary to design improvements to optimize the queue system in order to reduce the number of queues and also the idle time of each counter. Repairing the queuing system is done by identifying the type of arrival time distribution and also the length of service. With known types of arrival time distribution and length of service, we can determine the optimal queuing system model by simulating it with available simulation software.

Simulation is a technique that uses the help of computer devices to mimic the processes contained in a system [5][10]. Simulation determines whether a system is in accordance with systems that exist in the real world. Simulation can be a prediction of events that occur and can be collected data needed.

Many studies of the queuing system have been done. But there is rarely research on queuing systems using anylogic software. Different from other software Anylogic software can show the number of customers served at a certain time unit. In this study the simulation software used is Anylogic Software. Anylogic is a multimethod simulation modeling tool developed by The Anylogic Company that can be used for dynamic systems, discrete events and agent based [6]. This research was also assisted by the use of easyfit software in identifying the distribution of service length and arrival time at XYZ food restaurant. The use of these two software allows the determination of an optimal queue model for XYZ fast food restaurant

2. Methodology

The stages of research conducted at XYZ fast food restaurant are as follows:

2. 1 Data Collection

Data collection is used to identify the type of distribution that exists in the queuing system at the XYZ food restaurant that are used as important data so that it can be simulated in Anylogic Software. The data used is the arrival of customers, the length of time of service and also the length of time the customer moves to the service. Data was collected by direct observation at XYZ food stalls during work time. Data to be taken is the average time of service / process.

2.2 Creating a System Description

The system is designed using Easyfit Software and Anylogic Software so that the system designed can reduce the number of queues and idle time per counter.

2.3 System Design

The stages in designing this system consist of:

1.Distribution Testing

Distribution testing is performed on the old service data and the time interval of arrival that has been collected during the observation using Easyfit Software.

2.Making a Queue System Simulation

At this stage the most optimal queuing system simulation is made using Anylogic Software according to the type of arrival time distribution and customer service length obtained from Easyfit Software.

3. Result and Discussion

3.1. Service / Process Average Time Data

Data on average time of service at XYZ food restaurant can be seen in Table 1.

Table 1. Data on Average Time of Service at XYZ's Fast Food Stalls

Process	Average Time (Detik)
Time of arrival	67.5
Duration of Service	111

Counter1	
Duration of Service Counter 2	136
Duration of Service Counter 3	138

In table1 based on the observation data that has been collected on XYZ Fast Food Restaurant in total of 100 sampling data, the average length of service and the interval of arrival of workers is 1267.5 seconds. Preliminary observational data shows the average service time counters 1, 2 and 3 are 111, 136, 138 seconds.

3.2. Distribution Testing

The following is a distribution test carried out on the old service data and arrival time intervals that have been collected during the observation using Easyfit Software which can be seen in Figure 1 and Figure 2.

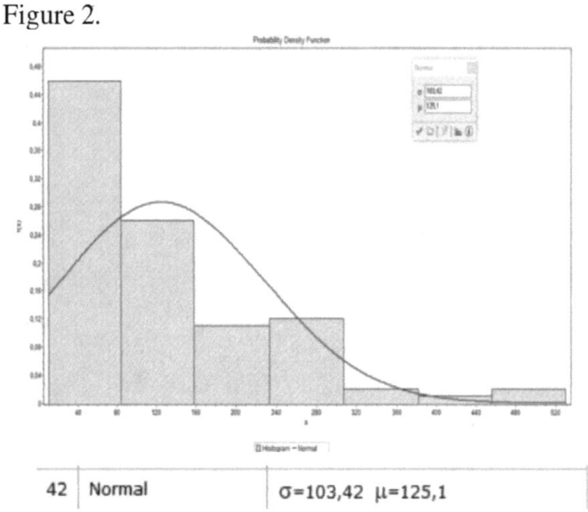

Figure 1. Testing of Distribution of Duration of Service

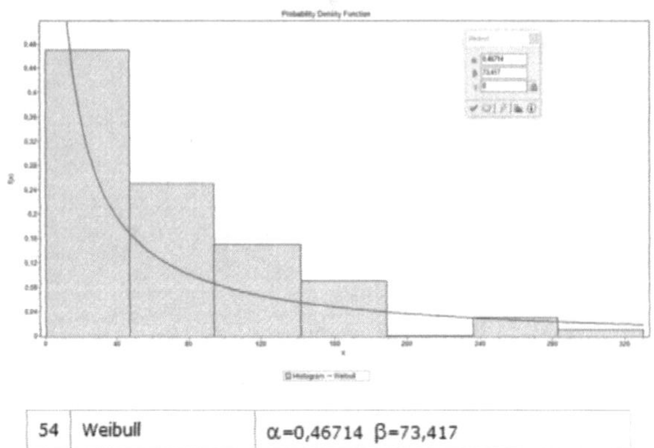

Figure 2. Testing the Distribution of Arrival Times

In figure 1 we can see that the distribution of the duration service is Normal distribution. As in figure 2 we can see that the distribution of the customer arrival time is Weibull distribution. This later will be used to determine the model of the queue system.

3.3. Making a Queue System Simulation

IOP Conf. Series: Materials Science and Engineering **851** (2020) 012028 doi:10.1088/1757-899X/851/1/012028

At this stage the most optimal queuing system simulation is made using Anylogic Software according to the type of arrival time distribution and customer service length obtained from Easyfit Software. Simulation results with 2D and 3D Anylogic Software can be seen in Figure 3 and Figure 4.

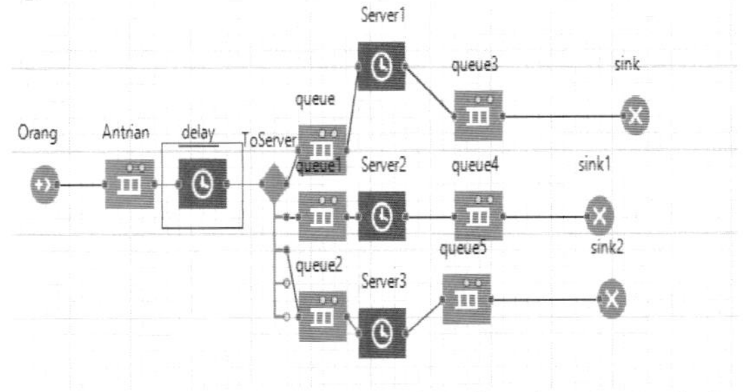

Figure 3. Simulation Results with 2D Anylogic Software

Figure 4. Simulation Results with 3D Anylogic Software

3.3.1. Test the System

After simulation with Anylogic Software with 3 queue lines and 3 counters (can be seen in Figure 5), the result is that the total number of customers served is 5 customers with 40% of customers served at counter 1, 40% are served at counter 2 and 20% is served at counter 3.

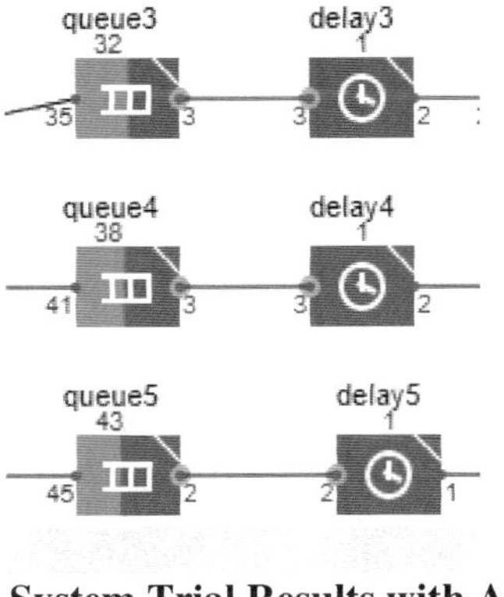

Figure 5. Current System Trial Results with Anylogic Software

IOP Conf. Series: Materials Science and Engineering **851** (2020) 012028 doi:10.1088/1757-899X/851/1/012028

Based on the results of the current system simulation with Anylogic Software, it can be seen that there are a small number of customers served, causing long queues. This shows that the XYZ fast food restaurant have long queues.

3.3.2. Testing of Simulation Systems 1

The simulation system trial 1 was conducted by changing 3 queue lines to 1 queue line and reducing the number of counters from 3 to 2. The simulation results can be seen in Figure 6.

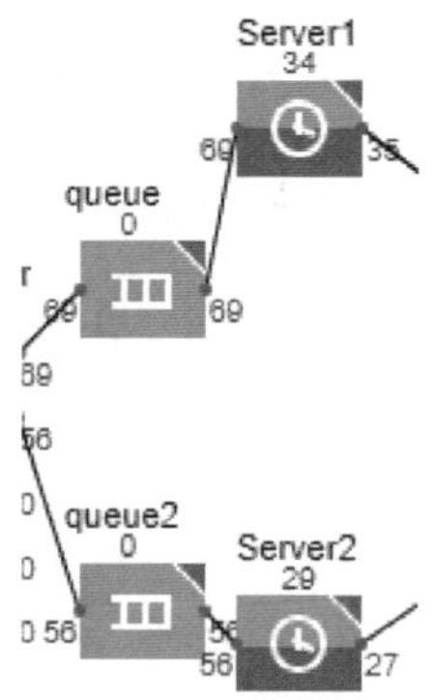

Figure 6. Simulation Results of Test 1 with Anylogic Software

According to the simulation results obtained using Anylogic Software that an increase in the number of customers served by 62 people from 125 customers with a percentage of customers at each counter is 56.45% for counter 1 and 43.55% for counter 2.

3.3.3. Testing of Simulation Systems 2

Trial simulation 2 is carried out by means of 3 queue lines being changed to 1 queue line with 3 counters. Simulation results can be seen in Figure 7.

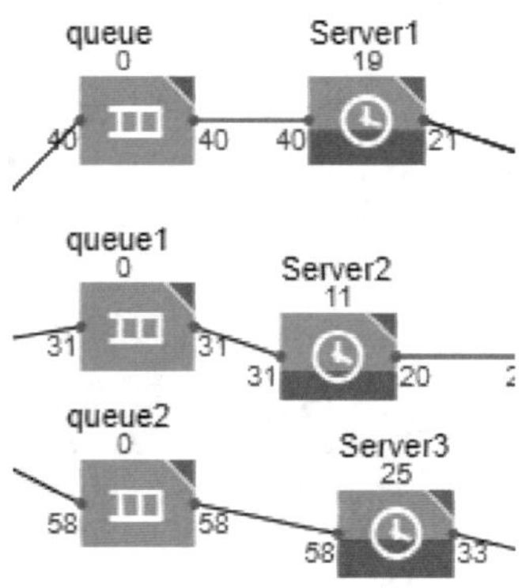

Figure 7. Simulation Results of Test 2 with Anylogic Software

According to the simulation results obtained using Anylogic Software that an increase in the number of customers served is as many as 74 people from 129 customers with the percentage of customers at each counter that is 28.38% for counter 1, 27.03% for counter 2 and 44.59% for counter 3.

4. Conclusion

1) Based on the results of the current system simulation with Anylogic Software, it can be seen that there are a small number of customers served, causing long queues. This shows that customers of XYZ fast food restaurant have long and long queues.
2) According to the simulation results obtained using Anylogic Software, simulation model is shown to have an increase in the number of customers served by 62 people from 125 customers with a percentage of customers at each counter is 56.45% for counter 1 and 43.55% for counter 2. Simulation model 2 is shown to have ane increase in the number of customers served is as many as 74 people from 129 customers with the percentage of customers at each counter that is 28.38% for counter 1, 27.03% for counter 2 and 44.59% for counter 3.
3) Based on the Anylogic software it is shown that simulation model 2 using 1 queue line and 3 counters is the most optimize model of queue system for the fast food restaurant.

Acknowledgments

The Universitas Sumatera Utara has provided financial support in completing this paper. The authors thank you for your help and support in completing this paper.

References

[1]. Nelson B. L, *Discrete event system simulation* (Prentice hall, 2001).

[2] Levine A. I., 2013 Healthcare Simulation: From "Best Secret" to "Best Practice". *The Comprehensive Textbook of Healthcare Simulation*, pp. 3-4..

[3] A. Law, Simulation Modeling and Analysis Fourth Edition. (McGraw-Hill, 2007)

[4] Agrawal G. A. 2010 Queuing Model for Health Centre *International Conference on Electronics and Information Engineering (ICEIE 2010)* pp. 542-547

[5] Zhu H., Gong J., and Tang J. A., 2013 Queuing Network Analysis Model in Emergency Departments. 25th Chinese Control and Decision Conference (CCDC) pp. 1829-1834.

[6] Gang F.Y. and Hong F., 2014 A Simple Simulation Method for Queue System. S*ixth International Conference on Intelligent Human-Machine System and Cybernetics*, Hangzhou, pp. 177-180

[7] Jiang S.H and Fu X.L., 2008 Under Queuing Theory to Explore Supermarkets' Service Model. *Logistics Technology* pp. 141-142

[8] Ratnasari S., Rahadian N., and Liquidannu E 2018. Pemodelan dan Simulasi Sistem Antrian Pelayanan Konsumen Gerai MCD Solo Grand Mall dengan Arena," *Pros. Semin. dan Konf. Nas. IDEC*, pp. 7–8,

[9] Hasian D P and Putra A K 2010 Simulasi pelayanan pengisian bahan bakar di SPBU Gunung Pangilun **9**(1), pp. 31–36, 2010.

[10] Rahmat R F, Anthonius, Muchtar M A, Hizriadi A, Syahputra M F 2018 Virtual reality interactive media for universitas sumatera utara - A campus introduction and simulation *Journal of Physics: Conference Series* **978**(1).

[11] Darmansyah S 2012. "*Analisis Stasiun MRT Jakarta Dengan Pemodelan Berbasis Agen*" Fakultas Teknik Universitas Indonesia. Depok.

[12] Hasugian I, Sidauruk P H, and Zalukhu A 2020 Evaluasi Ekonomi Teknik Pada Usaha Bengkel Las (Studi Kasus Bengkel Las XYZ, Medan, Sumatera Utara)", Buletin Utama Teknik,**15**(2), pp. 190-195

IOP Conf. Series: Materials Science and Engineering **851** (2020) 012029 doi:10.1088/1757-899X/851/1/012029

Analysis of Defective Causes in Coffee Product Using Decision Tree Approach

I Rizkya*, K Syahputri, RM Sari, S Christin
Departement of Industrial Engineering, Faculty of Engineering, Universitas Sumatera Utara

*e-mail: indahrizkya@usu.ac.id

Abstract. The development of Indonesian coffee through exploration of regional wealth needs to be done to add coffee products when expanding the domestic and overseas coffee market. But there are still many defective products found in domestic coffee, where defective products are products produced in the production process, where the product resulted is not in accordance with the applied quality standards, but it can still be improved by spending certain costs. The method used to identify defects in the product is assessed of space, workers and machinery aspects, and then a decision tree is made to determine the critical point. The results of the research conducted found that most defects are in coffee that does not has epidermis about 21% of the total sample of 23.000 grams. With fishbone diagrams and the decision tree approach the cause of the defect is known.

1. Introduction

Coffee is a potential commodity that is widely cultivated by smallholders and large plantation. Reviewing from its economic activities, coffee is seen as a commodity that has a high economic value and strategic for equitable income distribution so that it gives a high contribution to improve the welfare of farmers in secluded areas, providing employment opportunities, and providing foreign exchange income. Thus, coffee can be valued as a plantation commodity provides a multiplier effect for several other economic sectors [1].

Market globalization and the competitiveness growth in the manufacturing industry demand that product quality be a key of success factor [2]. Quality control of the production process is an activity of planning and control the production process starting from raw materials until the materials are turned into finished products in accordance with company standards. If the product produced by the company does not meet the established standards, then the product categorized into the defective product group. The defective product is analyzed for the defect causes so that the company make a improvements to the production process and the processed raw materials. The goal is that the company does not make the same mistakes in the future and get a quality product [3].

The quality of coffee is affected by many factors, depending on the quantity of time coffee is stored in the warehouse [4]. Good condition, proper appearance, size and coffee beans are also criteria for assessing the coffee quality [5]. Much research has been done to assess the coffee quality. Some of these studies see the effect of coffee bean composition on the coffee quality [6,7,8]. Other studies have also examined the quality of coffee beans [9,10].

This study examines the defects that often occur in coffee bean processing. Then describe the defect causes found in the processed coffee beans. In one of the coffee bean processing industries in Indonesia, defects are often found due to errors during the production process and raw materials used. Defects in coffee beans produced include spindles beans, beans which are still have epidermis in the dry process, grade 1 of hole coffee beans, grade 2 of hole coffee beans, broken beans, speckled beans, and black beans.

2. Methodology

Defect analysis is done by collecting information about existing defects. Then brainstorming was carried out with the whole team to find ideas, opinions about coffee product defects. Brainstorming is a technique used to obtain a large quantity of ideas from a team using its collective power [11]. The opinions obtained are be set out in the Cause and Effect Diagram or often called the Fishbone Diagram or Ishikawa diagram which was introduced by Prof. Kaoru Ishikawa [12]. Cause and Effect Diagrams are tools used to systematically place the root cause of a problem [13].

Data analyzed by cause and effect diagrams usually come from brainstorming sessions. This diagram is also used to identify and analyze a process and to find possible causes of a problem. Cause and effect diagram (fishbone) for the biggest defect problem is the coffee beans that still have epidermis. To find out the specific cause at each defect, decision tree tools are used so that the root cause of the problem is obtained from each work station [14]. The results of the product analysis, space, equipment, and workers then made a decision tree to describe the specific causes of defect at each work station. Furthermore, suggestions are given to prevent and reduce defects in coffee beans produced in the future.

3. Result and Discussion

In the initial stage, a product analysis that have the most defects are do not have epidermis, the defective results at each product can be seen in Table 1.

Table 1. Types and Quantity of Coffee Beans Defect

No.	Types of Coffee Bean Defects	Quantity of Defective Products (gram)	Total Sample (gram)
1	Broken	1.050	
2	Hole of Grade 1	410	
3	Hole of Grade 2	20	
4	Spotting beans	10	23.000
5	Black beans	5	
6	Have epidermis	4.830	
7	Spindle	450	

There are 7 types of defects that most often occur in coffee beans, namely broken bean, hole of grade 1, hole of grade 2, spotting beans, black beans, have epidermis, and spindle beans. Of the 23,000 grams of sample taken, coffee beans have epidermal defects that occur most frequently at 4,830 grams (21%). Broken bean defects were found as much as 1,050 grams (4.57%), spindles were found as many as 450 grams (1.96%), and hole of grade 1 was found as much as 410 grams (1.78%). Furthermore, a fishbone diagram was made to find out the cause of defects in coffee beans are still have epidermis as shown in Figure 1:

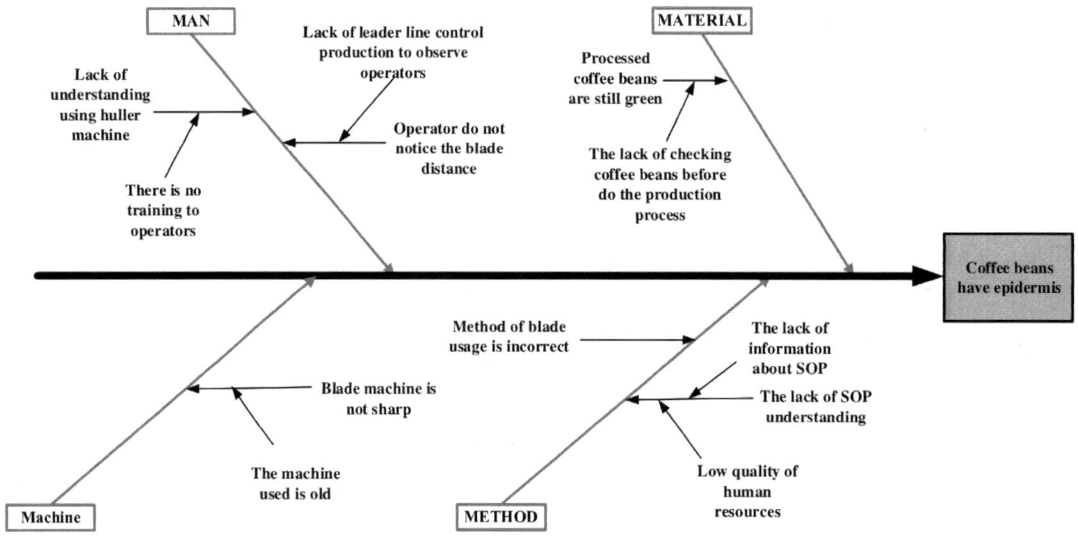

Figure 1. Fishbone Diagram of Epidermis Coffee Beans Defect Causes

To find out the specific causes of each defect, decision tree tools are used so that the root causes of the problems are obtained from each work station.

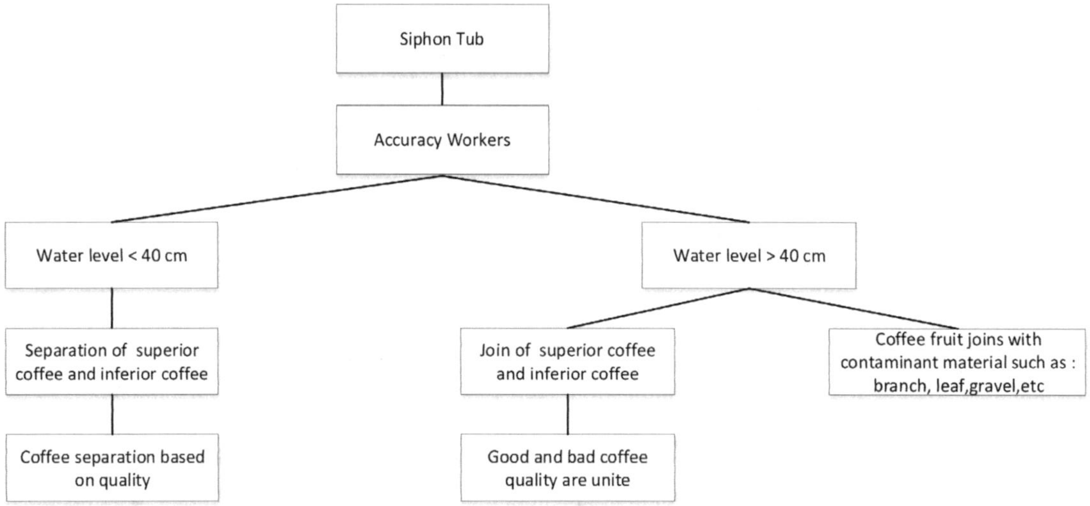

Figure 2. Decision Tree at Siphon Tub Station

From the decision tree above, it can be seen that the causes of defect is sourced from workers who are not careful in observing the water level in the siphon tub which is sometimes exceeds 40 cm so that the superior and inferior coffee beans join.

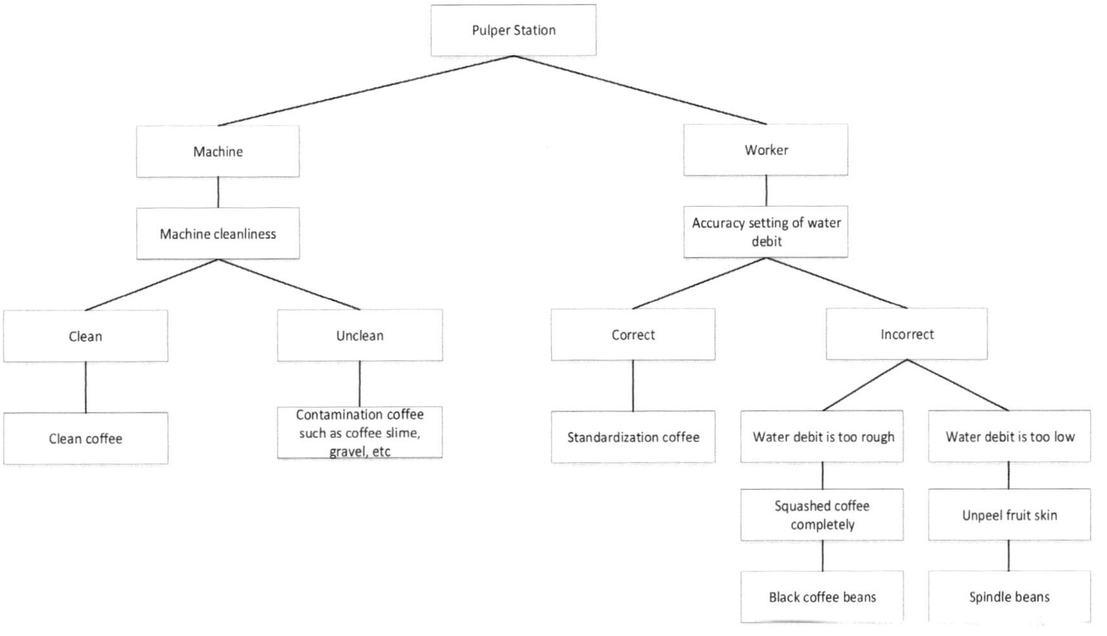

Figure 3. Decision Tree at Raung Pulper Station

From the decision tree above, it can be seen that the causes of defect are sourced from the worker and also the machine used. Inaccuracy in workers in regulating water debit will cause the coffee fruit to be squashed completely which if further processed will cause black coffee beans and also coffee beans that do not experience skin removal causing coffee beans spindle.

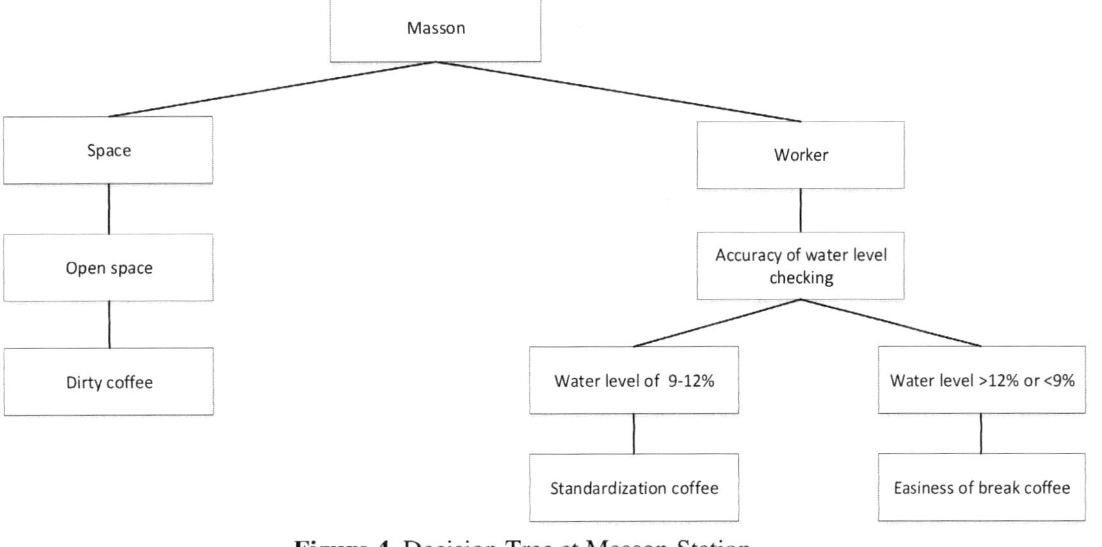

Figure 4. Decision Tree at Masson Station

From the decision tree above, it can be determined that the cause of the defect is sourced from the space, where at the Masson Station the finished coffee beans will be accommodated with burlap and the rest will be poured directly into the production floor, so that with the open space, it will cause contaminated coffee beans with dust. The workers are also a cause in checking the water content of coffee beans because the incompatibility will cause the coffee to break.

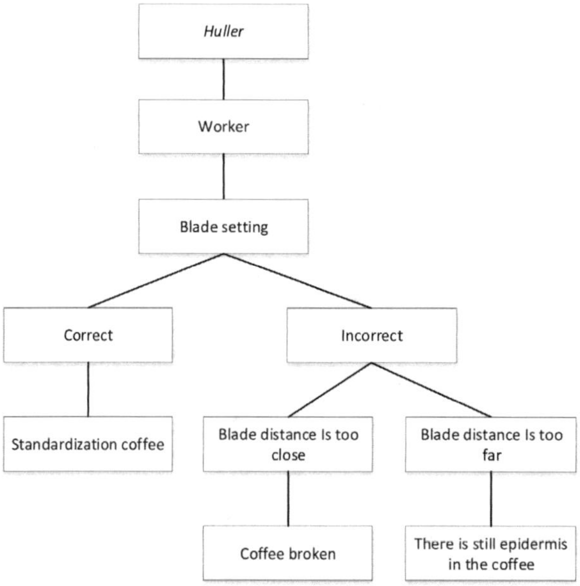

Figure 5. Decision Tree at Huller Station

From the decision tree above, it can be seen that the causes of defect are sourced from workers in setting the blade. Worker errors in adjusting the blade cause the epidermis do not to peel completely, which causes many epidermis on the coffee beans and also cause the coffee beans to break.

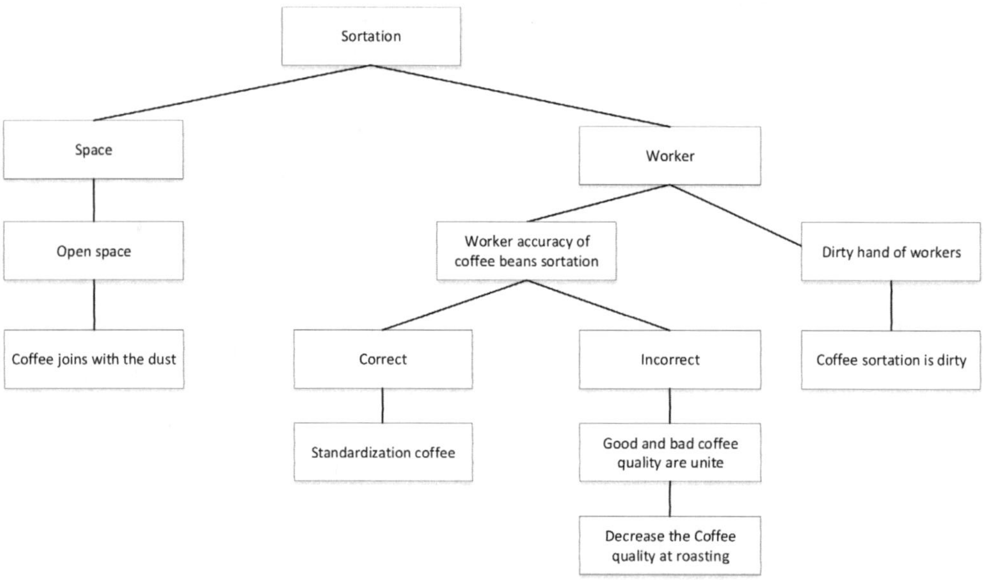

Figure 6. Decision Tree at Sortation Station

From the decision tree above, it can be concluded that the causes of defect is sourced from open spaces and also workers in the sorting process. Worker error in the coffee inspection of coffee beans done manually is the inaccuracy of workers in differentiating coffee beans quality 1, quality 4, and local. The mistakes made will reduce the quality of the next process which is roasting. Unclean workers hands will also cause coffee to become dirty.

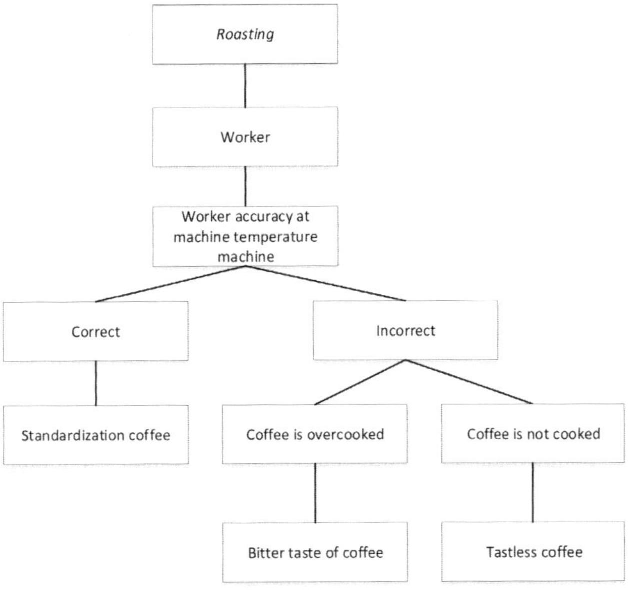

Figure 7. Decision Tree at Roasting Station

From the decision tree above, it can be seen that the causes of the defect are the worker sourced in the machine temperature regulation. Worker errors in regulating the machine temperature cause coffee taste too bitter or the tasteless of coffee produced.

4. Conclusion

Based on the decision tree analysis above, the improvement obtained at siphon tub Station is that workers are more careful in observing the water level in the siphon tub so it does not exceed 40 cm. At the Pulper Station, workers must be more careful in regulating the water flow so as not to find the crushed coffee fruit and skin removal in the coffee fruit. At the Masson station, improvements are made to an open space so that the coffee is not contaminated with dust in the area and workers must be more careful in checking the water content of the coffee beans. At huller station, workers must be more careful in blade managing. Inaccuracy in the blade setting causes there are so many epidermis on the coffee beans and also cause coffee beans to burst. At the sorting station, workers must be more careful in the inspection activity of coffee beans which are carried out manually. Inaccuracy of workers occurs in categorizing coffee beans of quality 1, quality 2, and local. At the roasting Station, workers must be more careful in regulating the machine temperature which is cause coffee tastes are too bitter or tasteless. To improve the workers accuracy and carry out their activities according to Standard Operational Procedures (SOP) it is necessary to conduct related training, observating, and controlling from management.

Acknowledgement

Authors would like to thank to Universitas Sumatera Utara for its financial support so that this paper can be published. Authors also thanks one of the coffee industries in Indonesia for allowing this research to be carried out.

References

[1] M Malahayati 2018 Open Agriculture **3**, pp 171-179
[2] M Colledani and T Tolio 2006 CIRP Annals 55 1, pp 453–456
[3] M Sharabi 2010 International Journal of Quality and Service Sciences **2** 2, pp 189-205
[4] A Illy and R Viani 2005 Second Edition Espresso Coffee The Science of Quality Elsevier Academic Press: United Kingdom

[5] J Rodríguez, C Durán, and A Reyes 2010 Sensors Journal **10**, pp 36-46
[6] P Mazzafera P, K V Gonçalves and M M Shimizu 2002 Scien. Agric **59**, pp 695-700.
[7] P Montavon, E Duruz, G Rumo and G Pratz 2003 J Agric. Food Chem **51**, pp 2328-2334.
[8] M K V Carr 2001 Expl Agric **37**, pp 1-36.
[9] F M Borem et al 2013 Journal of Stored ProductS Research **52**, pp 1-6
[10] L S Oliveira 2006 LWT-Food Science and Technology **39** 3, 235-239
[11] H S Gitlow and D M Levine 2009 *Six sigma for green belts and champions. 7th Edition* FT Press
[12] K Ishikawa 1985 *What is total quality control* New Jersey: Prentice Hall.
[13] R E Devor, T Chang and J W Sutherland 2007 *Statistical quality design and control: Contemporary concepts and methods. 2nd Edition* Pearson Prentice Hall.
[14] J K Kim et al 2007 Expert System **22** 4, pp 193-205

IOP Conf. Series: Materials Science and Engineering **851** (2020) 012030 doi:10.1088/1757-899X/851/1/012030

Simulation of Queuing System for Customer Service Improvement: A Case Study

Ivo Andika Hasugian[1*], Fernando[1], Supriadi[1]

[1] Department of Industrial Engineering, Faculty of Engineering, Universitas Sumatera Utara, Indonesia

*Email: ivo.andika@usu.ac.id

Abstract. This paper describes the queuing system for buying tickets in a cinema. Queues can be interpreted as a long line of customers who need service from one or more service facilities. Queues occur because of an imbalance between those served and their services. Some cinema consumers are annoyed because of the long queue to buy tickets and sometimes cancel them. One cinema that does not escape the length of the queue waiting time is the XYZ cinema. The purpose of this study is to optimize the queue system in buying tickets in the cinema to avoid long queues. The queuing system is optimized with a simulation approach using discrete event simulation. Simulation results show the best value obtained in the proposed scenario simulation model is to add a queue path to the queuing system. The results obtained are by increasing the queue path from 1 to 2 so that the length of the queue decreases from 91.2068 people per hour to 54.1520 and 50.8887 people per hour. Service time at counter 1 is 4.2201 minutes with an average number of 58.5861 people per hour. Service time at counter 2 is 3.9701 minutes with an average number of 51.1310 people per hour.

1. Introduction

The development of the world film industry is developing rapidly in the digital age and producing quality films. Indonesia is known as one of the countries that has a large number of citizens [1]. This certainly provides enormous potential in the human resources sector which can have massive effects on various industries in Indonesia, one of which is cinema. The number of cinema viewers experiencing high growth makes the number of cinemas increase each year [2]. The number of cinemas in Indonesia in 2019 is 2000 cinemas [3]. Comparison of the number of cinemas and the large audience still makes a long queue in several theaters. Queues can be interpreted as a long line of customers who need service from one or more service facilities [4]. Queues occur because of an imbalance between those served and their services [5]. Based on the results of interviews conducted with several cinema managers, many viewers were annoyed with the long queues when buying tickets and made some viewers cancel to watch. The cancellation made the cinema revenue reduced because customers chose to go to other theaters in the hope that the queue for more ticket purchases a little. One of the famous cinemas in Medan is the XYZ cinema, this cinema experiences a long queue in purchasing tickets, especially holidays. From interviews with cinema managers, analysis and simulation approaches are needed to illustrate and optimize the ticket purchase queue system so that it can determine the number of lines to be made in ticket purchases. The purpose of this study is to optimize the queue system in buying tickets at the XYZ cinema to avoid long queues. The basic process components of the queue are:

a. Arrival

IOP Conf. Series: Materials Science and Engineering **851** (2020) 012030 doi:10.1088/1757-899X/851/1/012030

Every queue problem involves arrival, for example cars, people, phone calls to be served, and others. This element is often called the input process. The input process includes the source of arrival or commonly called the calling population, and the manner of arrival which is generally a random variable. A random variable is a variable whose value can be anything as a result of a random experiment. Random variables can be either discrete or continuous. If a random variable is only possible to have a few values, it is a discrete random variable. Conversely, if the value is possible to vary over a certain range, it is known as a continuous random variable.

b. Service

A service or service mechanism can consist of one or more servants, or one or more service facilities. Each service facility is sometimes referred to as a channel. For example, a toll road can have several toll booths. The service mechanism can only consist of one service in one service facility found at a ticket window such as ticket sales at a movie theatre.

c. The queue

The essence of queuing analysis is the queue itself. The emergence of the queue depends on the nature of the coming and the service process. If there is no queue, it means that the servants are unemployed or have excess service facilities [2]. In this case, we request a simulation to solve the queue problem. Simulation model is one model that asks anyone involved in this strategy to consider themselves as other people who support to learn how others do it and feel [3][14]. The software used is Arena Software. Arena Software is software developed and systems developed by System Modeling. This software is able to build experimental models using modules that state processes or logic. Furthermore, this software has a connecting line that is used to declare the relationship between modules and or express the flow of entities.

2. Research Methodology

In conducting research, there are several stages that are used, namely the stage of data collection, initial identification of problems, setting goals, processing data, then conclusions.

2.1. Data Collection

At this stage the data obtained from direct observation to the field. Data needed is data between arrivals [4], length of service.

2.2. Distribution Testing

Distribution testing for a long time to service [5] is done with Easyfit Software. From the distribution, the distribution parameters taken to be included in the Arena Software are the distribution of time between arrivals and the distribution of service time.

2.3. Initial System Identification

In the initial identification the problem will be sought in the XYZ Cinema queuing system.

2.4. Analysis of the Arena Initial Model

At this stage the simulation is made in the form of modules using Software arena.

2.5. Arena Proposed Model

At this stage the proposal is made in the form of module simulations using the Arena Software. Alternative queuing system solutions are proposed to reduce queue length at XYZ Cinema consumers. Several previous studies using different methods that the authors used as a reference [7] such as the PLN payment counter queue [8]; simulation of vehicle movement in two-dimensional space [9]; queue at the supermarket cashier [10]; optimization of production processes in fish canning companies [11]; queues at pharmacies [12]; XYZ skin care consumer queues [13].

IOP Conf. Series: Materials Science and Engineering **851** (2020) 012030 doi:10.1088/1757-899X/851/1/012030

3. Result and Discussion

3.1. Data Collection

At this stage the data is obtained from direct observation to the field, on Saturday, October 26, 2019 for 8 hours, starting at 14.00-22.00 WIB. The data needed is data between arrivals, length of service. Data obtained during observation can be seen in the Appendix.

3.2. Distribution Testing

The following is a distribution test carried out on the data between arrival time and service length that has been collected during the observation using Easyfit Software which can be seen in Figure 1 and Figure 2.

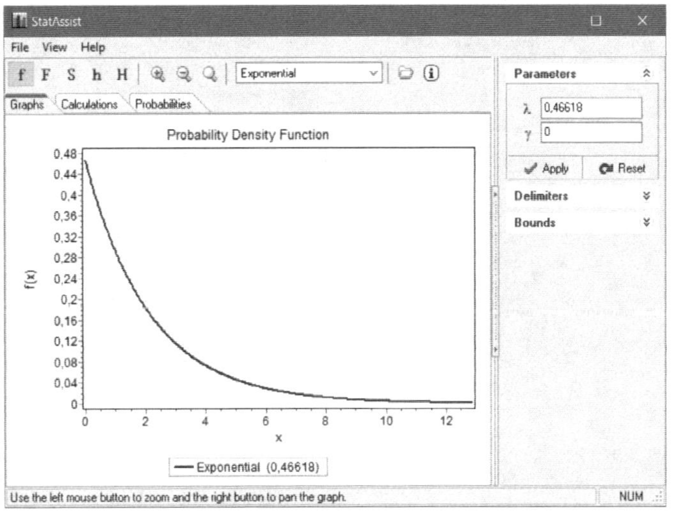

Figure 1. Testing the Distribution Of Time Between Arrivals

Based on Figure 1. The distribution of time between arrivals is Exponensial Distribution with α = 0.46618.

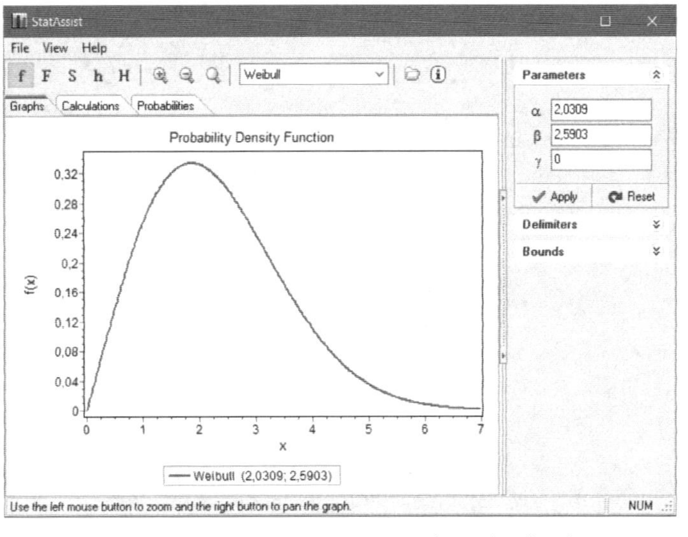

Figure 2. Testing the Long Service Distribution

Based on Figure 1. The distribution of time between arrivals is Exponensial Distribution with α = 0.46618.

3.3. Initial System Identification

The system in the object of observation consists of 4 stages, namely the arrival of consumers, queues, services and consumers out. The queuing system in XYZ Cinema has several problems namely the time consumers wait very long because there is only 1 queue line even though there are 2 counters. Therefore, it is necessary to improve the XYZ Cinema queue system.

3.4. Analysis of the Arena Initial Model

At this stage the simulation is made in the form of modules using Software arena. Following is a module in activity simulation at XYZ Cinema which can be seen in Figure 3.

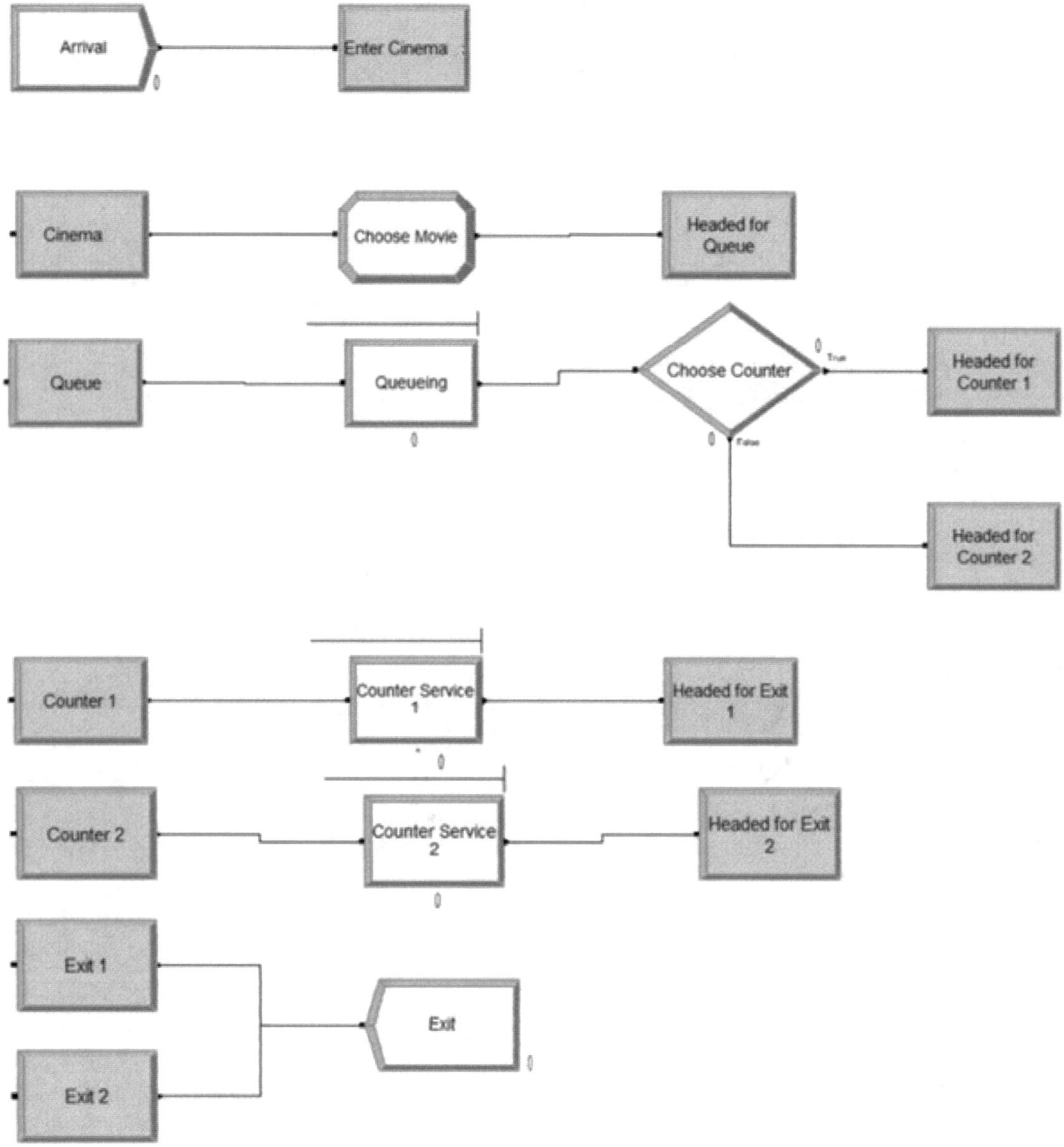

Figure 3. The initial model simulation arena

Based on Figure 3. The initial model explain than when the customer get in to XYZ Cinema, They get into the queue without choosing the queue, because there is only 1 line, After finished queueing, they pick the counter that are available to be served then the costumers leave.

Category Overview

Unnamed Project

Replications: 1 Time Units: Hours

Queue

Time

Waiting Time	Average	Half Width	Minimum Value	Maximum Value
Counter Service 1.Queue	4.3553	(Correlated)	0.00	12.8841
Counter Service 2.Queue	4.3280	(Correlated)	0.00	12.7578
Queueing.Queue	3.3962	(Correlated)	0.00	8.7660

Other

Number Waiting	Average	Half Width	Minimum Value	Maximum Value
Counter Service 1.Queue	59.3596	(Correlated)	0.00	177.00
Counter Service 2.Queue	57.2442	(Correlated)	0.00	174.00
Queueing.Queue	91.2068	(Correlated)	0.00	352.00

Figure 4. Results of Initial Model Arena Simulation

Based on Figure 4. it is obtained that the average queue time of simulation is 3.3962 minutes with an average number of 91.2068 people per hour. Service time at counter 1 is 4.3553 minutes with an average number of 59.3596 per hour. The service time at counter 2 is 4.3280 minutes with an average amount of 57.2442 per hour.

3.5. Arena Proposed Model

At this stage the proposal is made in the form of module simulations using the Arena Software. An alternative queuing system solution is proposed to reduce the queue length for XYZ Cinema consumers, namely to increase the queue path in the XYZ Cinema queue system. Following is an alternative module offered solution which can be seen in Figure 5.

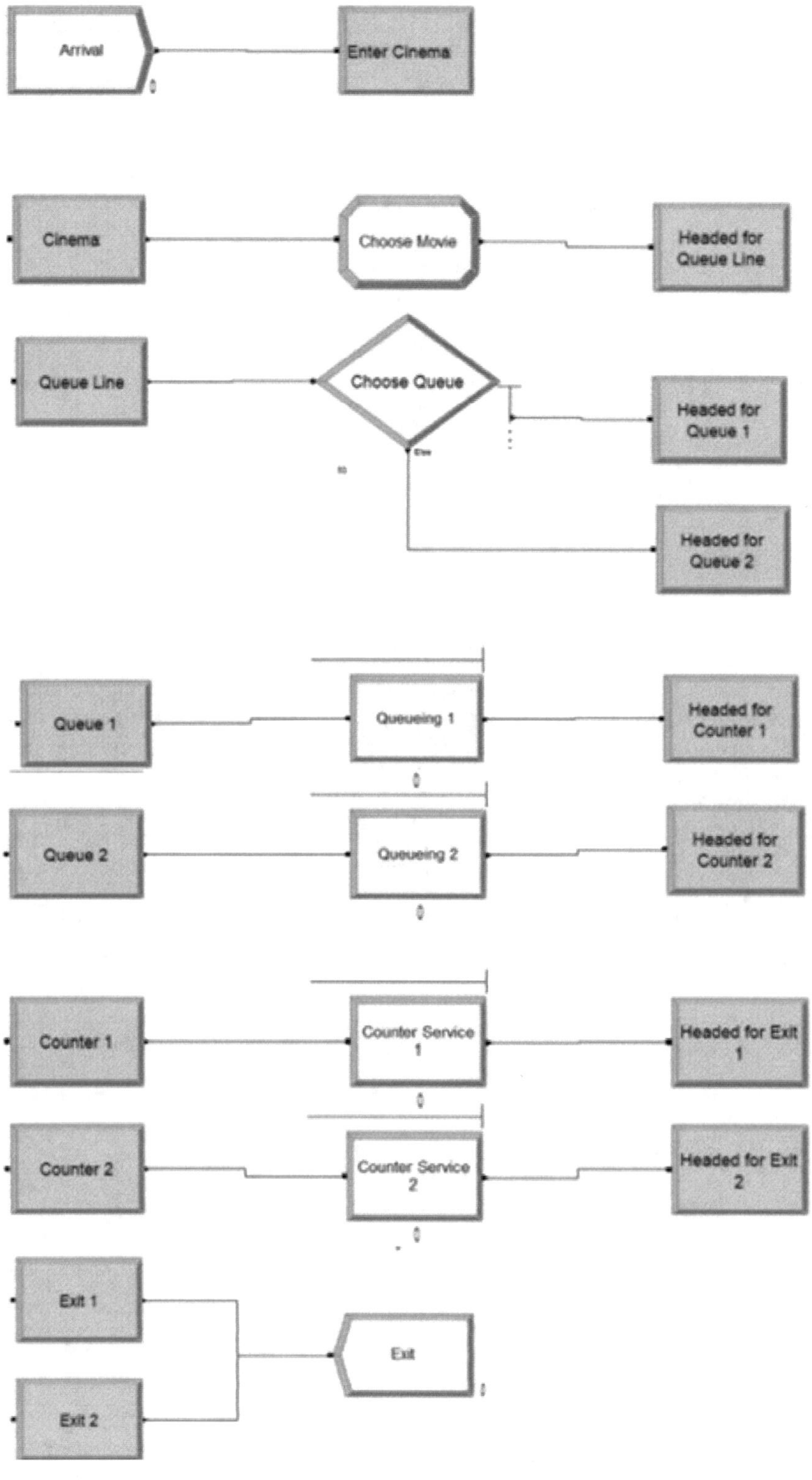

Figure 5. Results of Initial Model Arena Simulation

Based on Figure 5. the proposed arena simulation model explain that when the customer get into XYZ Cinema, They get into the queue and they choos the queue and they will wait. After finished queueing, they headed to the counter in front of their queue, after being served, the customer left.

Category Overview

Unnamed Project

Replications: 1 Time Units: Hours

Queue

Time

Waiting Time	Average	Half Width	Minimum Value	Maximum Value
Counter Service 1.Queue	4.2201	(Correlated)	0.00	10.9368
Counter Service 2.Queue	3.9701	(Correlated)	0.00	10.5957
Queueing 1.Queue	3.9007	(Correlated)	0.00	10.8107
Queueing 2.Queue	3.9513	(Correlated)	0.00298124	10.9262

Other

Number Waiting	Average	Half Width	Minimum Value	Maximum Value
Counter Service 1.Queue	58.5861	(Correlated)	0.00	152.00
Counter Service 2.Queue	51.1310	(Correlated)	0.00	142.00
Queueing 1.Queue	54.1520	(Correlated)	0.00	152.00
Queueing 2.Queue	50.8887	(Correlated)	0.00	142.00

Figure 6. Simulation Results of the Proposed Model Arena

Based on Figure 6. it is obtained that the average queue time of simulation is 3.9007 minutes and 3.9513 with an average number of 54.1520 people per hour on line 1 and 50.8887 on line 2. Service time at counter 1 is 4.2201 minutes with an average number of 58.5861 per hour. The service time at counter 2 is 3.9701 minutes with an average amount of 51.1310 per hour.

4. Conclusion

The simulation results show that the best value is obtained in the arena simulation of the proposed scenario model, namely by adding a queue path to the XYZ Cinema queuing system. Based on the simulation of the proposed model arena, it was obtained that the average queuing time of the simulation was 3,9007 minutes and 3.9513 minutes with an average number of 54,1520 people per hour and 50,8887 people per hour. Service time at counter 1 is 4.2201 minutes with an average amount of 58.5861 per hour. Service time at counter 2 is 3.9701 minutes with an average amount of 51.1310 per hour the length of the queue is reduced, and it will reduce invalidation and will increase the turnover of XYZ Cinema.

Acknowledgments

We would like to thank the Universitas Sumatera Utara for supporting and providing funding for this paper, we also want to thank all our participants who helped us complete this paper.

References

[1] Nelson B J, 2001 *Discrete event system simulation* (Prentice hall,)

[2] Xing W J, Li S B, He L H, 2015 Simulation Model of Supermarket Queuing System. *34th Chinese Control Conference*, Hangzhou, , pp. 8819-8823.

[3] Ghaleb M A and Suryahatmaja 2015 U.S. Modeling and Simulation of Queuing System Using Arena Software: A Case Study. *Proceeding of the 2015 International Conference on Industrial Engineering and Operation Management*. Dubai,

[4] Madadi N, Roudsari A H, Wong K Y, and Galankashi M R, 2013 Modeling and Simulation of a Bank Queuing System. *Fifth International Conference on Computational Intelligence, Modelling and Simulation*, , pp. 209-215.

[5] Chu L, 2013 Simulation of Theme Park Queuing System by using Arena. *Ninth International Conference on Intelligent Information Hiding and Multimedia Signal Processing*, Taiwan, , pp. 17-20.

[6] Sekar A K and Mulyati H 2011 Analisis Sistem Antrian pada Proses Penyelesaian Klaim di PT Taspen (Persero) Kantor Cabang Bogor. *Jurnal Manajemen*, 1-12..

[7] Ekoanindiyo F A, 2011 Pemodelan Sistem Antrian Dengan Menggunakan Simulasi *Din. Tek.*, vol. V, no. 1, pp. 72–85,.

[8] Praba Martha N G, Sukarsa K G, And Nila Kencana I P E, 2012 Analisis Sistem Antrian Pada Loket Pembayaran PT. PLN (Persero) Area Bali Selatan Rayon Kuta *E-Jurnal Mat.*, vol. 1, no. September, pp. 6–11.

[9] Saputri T, Nugraha C, and Amila K, 2014 Model Simulasi Untuk Pergerakan Kendaraan Pada Ruang Dua Dimensi Kontinu Dengan Pendekatan Pemodelan Berbasis Agen Tari Saputri, Cahyadi Nugraha, Khuria Amila *J. Online Inst. Teknol. Nas.* Oktober, vol. 2, no. 4, pp. 2338–5081

[10] Rahmadani D, and Julasmasari F, 2010 Simulasi Pelayanan Kasir Swalayan Citra Di Bandar Buat, Padang, **9**(1), pp. 19–24

[11] Sugiarto F and Buliali J L, 2012 Implementasi Simulasi Sistem untuk Optimasi Proses Produksi pada Perusahaan Pengalengan Ikan *J. Tek. ITS*, vol. 1.

[12] Aji S P and Bodroastuti T, 2013 Penerapan Model Simulasi Antrian Multi Channel Single Phase Pada Antrian Di Apotek Purnama Semarang *J. Kaji. Akunt. dan Bisnis*, **1**(1), pp. 1–16,

[13] Nursanti, Ida., Lisa, Anandistya., Qoyyiimah, Milati., Hannas, M. 2015 Analisis Perbaikan Sistem Antrian Konsumen Skin Care XYZ dengan Menggunakan Model Simulasi. Seminar Nasional *IENACO (2015)* ISSN: 2337 – 4349.

[14] Rahmat R F, Anthonius, Muchtar M A, Hizriadi A, Syahputra M F 2018 Virtual reality interactive media for universitas sumatera utara - A campus introduction and simulation *Journal of Physics: Conference Series* **978**(1).

Inventory Optimization in the Cigarette Industry

I Siregar[1] , I Rizkya[1], Rahmi M Sari[1], Khalida Syahputri[1]
[1]Department of Industrial Engineering, Faculty of Engineering, Universitas Sumatera Utara

E-mail: ikhsan.siregar@usu.ac.id

Abstract. The increasing cigarette industry makes cigarette-producing companies continue to make improvements in inventory planning determination continuously in order to improve the effectiveness and efficiency of the company. The company sets the number of production based on a rough estimate of consumer demand in previous periods. This causes an imbalance between the company's production capacity and the number of demands for consumers. A large accumulation of inventory causes a high carrying cost. The incompatibility in the number of demands with the quantity produced results in the emergence of the other costs so that the company loses the opportunity to get maximum profit. Therefore, it is necessary to analyse the inventory quantity to reduce carrying costs. The company optimizes the inventory quantity based on the concept of the Economic Order Quantity using the quantity of safety stock as a minimum inventory point so that the company can plan the production number in the next periods. The results showed that with the proposed inventory system a total cost savings carrying cost of IDR 722.795.625 obtained.

1. Introduction

Company competition in the industry world requires companies to win competition among similar companies. In today global market, companies are always observing for development and chances to decrease their investment with the lowest cost. The achievement of several business is related of their capability to offer goods and services at correct period and in correct place. Dissimilar organization implement their own strategy to compete with the other companies. At the beginning of the 19th century, companies had to keep large amounts of inventory to survive. During subsequent times, disadvantages of over-inventory were clearly understood by companies. In the 20th century, the importance of the cost inventory was understood. Therefore, companies-implemented based inventory control methods to manufacture with low quantity inventories [1].

Inventory in question is the source of raw materials, semi-finished goods named goods in process and finished goods, which are managed by the organization to meet their operational needs. It has considerable asset and possible waste sources that need to be controlled prudently. The amount of inventory that must fall to signal that an order must be made to refill an item [2]. Inventory of raw materials in a company is an important requirement in carrying out a goods production. If raw materials are unavailable, the production process cannot be carried out. Therefore, it is necessary to provide raw materials by the company. The quantity of raw material needed by the company is influenced by the size of the company's production. While the company's production costs are influenced by the level of demand for a product produced by the company so that in the company's inventory determination must always consider the quantity of raw materials used in a period. In order to inventory cost to be as much as possible, in a period the quantity of raw materials supply must be in accordance with the quantity of raw materials usage[3][4]. Inventory has been considered as one of the major cost drivers in manufacturing operation [5]. Inventory management is one of the most significant factors in organization and the percentage of inventories to whole advantage largely varies between 15 to 25%. Inventory management system has mainly two concerns, one is level of customer service in order to have right goods, in right place and at right time and other is cost of ordering and carrying inventories [6]. Inventory is a major asset that sell or manufacture products. In inventory management involves balance between customer service, or product availability and inventory cost. There are a number of factors that influence inventory-decision making. In addition, inventory management manufacturing, wholesaling, retailing, importing/exporting, and other fields, inventory constitutes one of the largest controllable assets of business [7].

Companies engaged in cigarette production are still experiencing problems in the optimal inventory quantity determination. The company applies inventory roughly so that there is an imbalance between production capacity and total inventory. This results in a high carrying cost. Each inventory cost is expected to be less than 15% while in reality the total inventory storage costs reach 40% of the total inventory cost. One approach to determine the optimal quantity of inventory is by the Economic Order Quantity (EOQ) to minimize carrying costs based on safety stock. EOQ on paper by James A. Cargal obviously describes the fundamental theory of the Economic Order Quantity. EOQ model adopts that demand is constant, and that deprived inventory at a fixed rate until it ranges zero [8]. EOQ has been a well-known formula that calculates the most favourable Economic Order Quantity. Engineers study the EOQ formula in engineering economics and industrial engineering courses. On the other hand, business discipline studies the EOQ in both operational and financial courses. In both, EOQ formulas have practical and exact applications in defining concepts of cost trade-offs; as well as specific application in inventory [9]. The quantity of an economic order is the quantity at which noted costs are low. Registered costs are costs associated with costs incurred for storing inventory items in the store. That be contingent on the amount and period of time until the inventory will be warehoused. This contains storage costs, damage costs, depreciation, handling fees, insurance costs, etc. [10]. Carrying costs and EOQ are influenced by the value of safety stock. Safety stock is inventory that is reserved as a safeguard from the continuity of the company's production process. Safety stock is needed because in reality the quantity of raw material needed for the production process is not always as planned. Safety stocks are used to maintain uncertainty, improve service levels and increase the stability in planning and controlling systems [11][12].

Research to overcome the problem of inventory planning has been done before [16]. The study was conducted by Eka Febryan Prayitno et al. 2016 in a fertilizer industry company which produces two types of products phosphate and dolomite. However, in raw manufacturing process still has disadvantage like there is no control in specific inventory of raw materials which could make disruption production process in financial loses. EOQ method is used to provide the quantity of raw materials. The data showed the EOQ value, the Safety Stock and the Reorder Point on each form automatically. It was made the raw material inventory control process scheduled without interrupting the production process [13]. Other studies presented a research using EOQ method to determine the inventory management in Automotive Service Industry in Malaysia. This method is planned to be applied in the independence automotive services workshop as an alternative to the high cost investment of the system in the selected industry. Based on the results, the data that predicted by Simple Moving Average technique gives the most accurate data and this EOQ method is selected in the determination of optimal number inventory model [14][15]. On this research we try to applied the method in cigarette industry, that is the different between this research compared to previous research.

2. Methodology

The research was conducted in one of the industries in the Medan city that produce cigarettes. The object examined in this study is the accumulation of high inventory quantity. Research begins with observations to observe and see the state of the industry. After observation, the topic and purpose of the study are determined according to the conditions on the production floor. Then data collection is needed to overcome the problem of the inventory quantity that occurs. Data collected in the form of production capacity data, the number of demands and the inventory quantity. The data collected is then processed using the EOQ approach based on safety stock and carrying cost. EOQ calculation is done with the following formula:

$$\text{EOQ} = \sqrt{\frac{2C_0\,D}{CiU}} \quad (1)$$

Where, C_0 = ordering cost, Ci= Carrying cost per year, D=number of demands per year and U= production cost. Safety stock value is obtained by the following formula.

$$\text{Safety Stock} = \text{k x } \sigma_c \quad (2)$$

Where k is the factor that corresponds with f(k) and σ_c is deviation standard combination. Calculation of σ_c by the following formula.

$$\text{f (k)} = \text{(1-SL) x } \frac{Q}{\sigma_c} \quad (3)$$

Where k= a function of the normal curve shows the right area of the normal distribution, SL= level of product availability (assumption= 99%), σ_c = deviation standard combination and Q= procurement quantity (obtained by EOQ formula). After obtaining the EOQ, Safety Stock and Corresponding Fator to the final calculation is carrying cost. Carrying costs can be calculated with the following formula.

$$\text{Carrying Cost} = \sum_{i=1}^{n} Xi\text{(1-SL)}\,\frac{15\%}{12\ month}\,\text{P} \quad (4)$$

Where Xi= Inventory quantity per month, 15% is carrying cost condition for 1 year and P= cost of finished goods. The last step is to compare the actual carrying cost and after the EOQ approach is carried out.

3. Result and Discussion

3.1. Actual Inventory

Actual company inventory in graphical form can be seen in Figure 1.

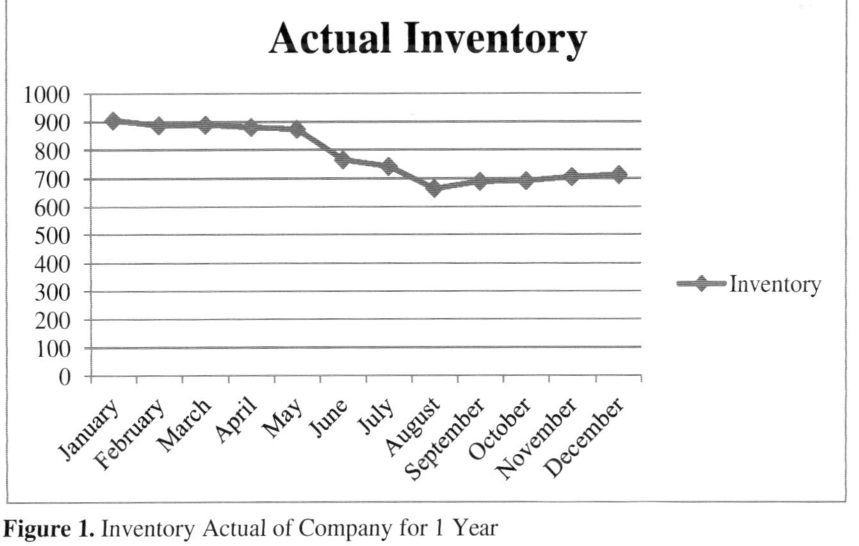

Figure 1. Inventory Actual of Company for 1 Year

Based on the graph above, the company's actual inventory fluctuates with a high quantity of inventory.

3.2. Proposed Inventory
Proposed company inventory in graphical form can be seen in Figure 2.

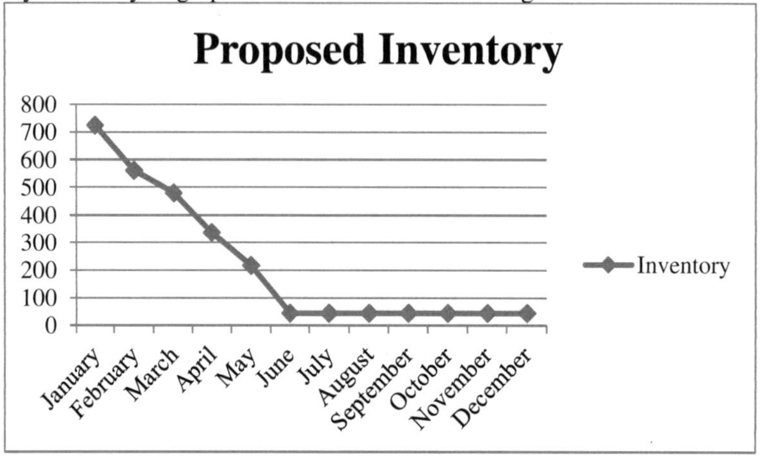

Figure 2. Inventory Proposed of Company for 1 Year

Inventory in the chart above is obtained based on EOQ and Safety Stock calculations. Based on the graph above, the company's proposed inventory experiences a significant inventory quantity reduction. Inventory proposed in the form of existing inventory utilization without production again.

3.3. Calculation of Carrying Cost in Actual Inventory
Recapitulation of the calculation results in actual carrying cost of inventory shown in Table 1.

Table 1. Carrying Cost Recapitulation of Actual Inventory

Month	Inventory (ton)	Carrying Cost (IDR)
January	906	96.828.750
February	888	94.905.000
March	890	95.118.750
April	881	94.156.875
May	874	93.408.750
June	765	81.759.375
July	742	79.301.250.
August	663	70.858.125
September	688	73.530.000
October	690	73.745.750
November	705	75.346.875
December	711	75.988.125
Total		1.004.945.625

The table above shows the results of the calculation is high carrying cost of IDR 1.004.945.625 which is resulted in company losses. This is because the quantity of actual company inventories is very high.

3.4. Calculation of Carrying Cost in Proposed Inventory
Recapitulation of the calculation results in carrying cost of the proposed inventory can be seen in Table 2.

Table 2. Carrying Cost Recapitulation of Proposed Inventory

Month	Inventory (ton)	Carrying Cost (IDR)
January	726	77.591.250
February	563	60.170.625
March	481	51.406.875
April	337	36.016.875
May	218	23.298.750
June	45	4.809.375
July	45	4.809.375
August	45	4.809.375
September	45	4.809.375
October	45	4.809.375
November	45	4.809.375
December	45	4.809.375
Total		282.150.000

Based on the calculation of safety stock, the total inventory for June to December is 45. The EOQ calculation in inventory can minimize the cost of carrying costs from IDR 1.004.945.625 to IDR 282.150.000. By calculating the EOQ by the company experienced a carrying cost savings of IDR 722.795.625.

4. Conclusion

EOQ calculation based on safety stock and carrying cost calculation is used by the company to determine the optimal inventory quantity by maximizing the quantity of available inventory without production again. Based on EOQ calculation and safety stock the inventories quantity in June to December changed to 45. Calculation of carrying cost with the quantity of proposed inventory was IDR 282.150.000 while the calculation of carrying cost with actual inventory cost IDR 1.004.945.625. Cost savings with carrying costs based on the quantity of proposed inventory is IDR 722.795.625.

References

[1] R Yildiz, and R Yaman 2018 *Management and Production Engineering Review,* **9** (3) 23-32
[2] N Sohail 2018 *Journal of Advance Research in Dynamical & Control System,* **10** (10) 1176-1190
[3] J. Heizer and B. Render 2008 *Manajemen Operasi* (Jakarta: Salemba Empat)
[4] A Maulana, and A Kusumawardhani 2015 *Dipenogoro Journal of Management,* **4** (2) 1-14
[5] R A Hadiguna, H S Jaafar, and S Mohamad 2011 *International Journal Enterprise Network Management,* **4** (4) 354-366
[6] I Nishad, and Arunkumar 2018 *International Journal for Research in Applied Science & Engineering Technology,* **6** (6) 309-315
[7] H Shen., Q Deng., R Lao., S Wu 2016 *Nang Yan Business Journal* **5** (1) 20-40
[8] E Guga, and O Musa 2015 *International Journal of Economics, Commerce and Management,* **3** (12) 174-182
[9] A S Sunhal., D Mangal 2017 *International Journal of Engineering Sciences & Research Technology* **6** (10) 303-309
[10] P P Gokhale., and M B Kaloji 2018 *International Journal of Latest Technology in Engineering, Management & Applied Science,* 7 (9) 28-31

[11] E Y Kristyaningrum., T Ekowati., A Setiyadi 2016 *Jurnal UGM* **27** (1)
[12] Ginting, R., Hidayati, J., & Siregar, I. (2018, March). Integrating Kano's model into quality function deployment for product design: A comprehensive review. In *IOP Conference Series: Materials Science and Engineering* (Vol. 319, No. 1, p. 012043). IOP Publishing.
[13] E F Prayitno., S H E Wulandari., V R Hananto 2016 *JSIKA* **5** (11) 1-8
[14] E N Roslin., S N A Razak., M Z Bahrom., M A A Rahman 2016 *Journal of Science & Engineering Technology* **2** (2)
[15] Syahputri, K., Sari, R. M., Tarigan, I. R., & Siregar, I. (2018, February). Application of lean six sigma to waste minimization in cigarette paper industry. In *IOP conference series: materials science and engineering* (Vol. 309, No. 1, p. 012027). IOP Publishing.
[16] Siregar I, Muchtar M A, Rahmat R F, Andayani U, Nasution T H, and Sari R M 2018 Method of calculation overall equipment effectiveness in fertilizer factory. *IOP Conference Series: Materials Science and Engineering* **308**(1), p. 012053

IOP Conf. Series: Materials Science and Engineering **851** (2020) 012032 doi:10.1088/1757-899X/851/1/012032

Production floor layout design for vise manufacturing using CRAFT algorithm

Muhammad Haikal Sitepu*, Tania Alda, Meilita Tryana Sembiring, Andri Nasution, Mutiara Sani Lubis and Andrian Jonathan

Industrial Engineering Department, Faculty of Engineering, Universitas Sumatera Utara, Medan, Indonesia, 20155

*E-mail: mhd_haikalkarana@usu.ac.id

Abstract. Production floor is the core for manufacturing company. The value added activities are mostly performed in the production floor. Several factors influence a productivity of production including technology, skill of labor and production floor layout. The manufacturing company with optimal production floor layout will produce lower material flow which in turn reduce production cycle time and material handling cost. As the result, production floor layout design is necessary particularly for manufacturing with assembly product such as vise manufacturing. Several parts with different processes and materials will require high material flow between machines. If the machines are not arranged properly in production floor, it will increase material flows. This paper aims to demonstrate the design of production floor layout for vise manufacturing using CRAFT (Computerized Relative Allocation Facilities Technique) algorithm. This study produces alternative of production floor layouts by minimizing the distance between machines.

1. Introduction

Manufacturing companies focuses on producing product by changing raw materials into finished goods. Several processes might require to produce finished goods. Within these processes, more values are added to the product. These processes are performed in production floor. Hence, it is necessary to maintain the productivity in production floor in order to get high value added. However, this productivity depends on several factors such as technology, skill of labor and production floor layout. Production floor layout design refers to a plan of an optimum arrangement of machines, operating equipment, storage space, material handling equipment and all other supporting services along with the design of the best structure to contain all these facilities [1].

The manufacturing company with optimal production floor layout will produce lower material flow which in turn reduce production cycle time and material handling cost. As the result, production floor layout design is necessary particularly for manufacturing with assembly product such as vise manufacturing. Several parts with different processes and materials will require high material flow between machines. If the machines are not arranged properly in production floor, it will increase material flows. This paper aims to demonstrate the design of production floor layout for vise manufacturing.

CRAFT (Computerized Relative Allocation Facilities Technique) is one approach for designing production floor layout. CRAFT evaluates the layout by exchanging the location of departments.

Inputs for the CRAFT algorithm include initial layout, data flow or frequency of displacement, cost data per unit distance, and the number of departments that do not change or remain. For the make to order manufacturing company such as vise manufacturing, the flow of material from initial process to final process is not regular because vise consists of several parts that require different processes and materials. This certainly has an impact on the magnitude of the moment of distance transfer and may have an impact on the productivity level of the company. Based on this, the production floor layout design needs to do carefully [2]. This paper demonstrates the use of CRAFT to design production floor layouts for vise manufacturing.

2. Method

This research uses experiment as research methodology. Experiment is used because this research focuses on determining the optimal production floor layouts by evaluating the impact from arranging the machines in the layout on the distance between facilities. This research uses CRAFT algorithm to support the experiment and to find the optimal production floor layouts.

2.1. Research process

Research is started with preparing input data such as type of machines required, number of machine and from to chart. This is continued by designing initial layout and determining the centroids of the departments in the initial layout. The initial layout is developed based on the information in the from to chart.

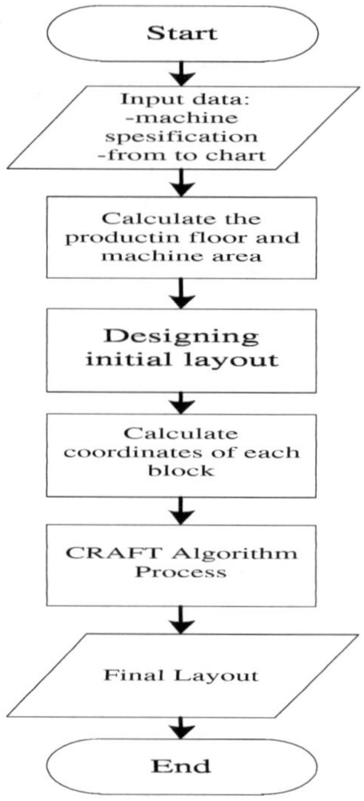

Figure 1. Research process

This is followed by calculating the rectilinear distance between pairs of department centroids and stores the value in a distance matrix. Then, CRAFT updates the layout according to the best exchange and compute the new department centroids as well as the new layout cost to complete the first iteration. Next, CRAFT considers all possible two ways (pairwise) exchange and identify the best exchange which is the one that yields the largest reduction in the layout cost. The cost is determined by multiplying total movements with unit cost matrix and distance matrix. The iteration will continue until no further reduction in layout cost can be obtained [3][4][5]. Research process is shown in Figure 1.

2.2. *Input Data*
Input data used in the design of the production floor layout with the CRAFT algorithm are:

2.2.1. Machine Specification Data
The machine specification data containing the dimensions of the machine used in the production process of vise can be seen in Table 1.

Table 1. Machine Data

No	Symbol	Machine	Amount	Machine Specifications (cm)	Number of Machines	Number of operators
1	PM	Entrance	1	-	-	-
2	G	Grinding	1	37 × 37	7	7
3	D	Drilling	1	53 × 65	2	2
4	S	Scrap	1	220 × 135	4	4
5	M	Milling	1	220 × 106	3	3
6	T	Tap and Dies	1	55 × 47	2	4
7	B	Lathe	2	172 × 84	3	3
8	A	Assembly	4	160 × 160	-	-
9	PK	Exit	1	-	-	-

Source: Data Collection

2.2.2. From to Chart Data
From to chart data can be seen in Table 2.

Table 2. From to Chart

	PM	M	D	S	TD	G	B	A	PK	Total
PM	0	0	0	20,40	0	50	19,20	0	0	89,60
M	0	0	0	0	8,40	0	0	48	0	56,40
D	0	21,60	0	9	8	8,40	0	0	0	47
S	0	0	31,60	0	0	2	0	0	0	33,60
TD	0	0	0	0	0	6,80	0	0	0	6,80
G	0	0	54	0	0	0	0	24	0	78
B	0	0	14,40	0	0	2,80	0	0	0	17,20
A	0	0	0	0	0	0	0	0	20	20
PK	0	0	0	0	0	0	0	0	0	0
				Total						**348,60**

Source: Data Collection

3. Result and Discussion

3.1. Initial Production Floor Layout Design
The calculation of the station area is used by the formulas below.

Area = (Length × Scale)(Width × Scale) (1)

Total Area = Area × Number of Machine (2)

Total Station Area $= \frac{\text{Total Area}}{10000}$ (3)

Selected Area = the Largest Total Station Area

Calculation of station area can be seen in table 3.

Table 3. Calculation of machines area

Machine	Length (cm)	Width (cm)	Area (cm^2)	Number of Machine	Total Area (cm^2)	Total Station Area (m^2)	Selected Area (m^2)
Drilling	220	180	39600	7	277200	27,72	29,64
Milling	420	220	92400	2	184800	18,48	29,64
Grinding	260	180	46800	4	187200	18,72	29,64
Tap and Dies	220	170	37400	3	112200	11,22	29,64
Lathe	350	300	105000	2	210000	21,00	29,64
Scrap	380	260	98800	3	296400	29,64	29,64

Source: Data Processing

Next is the calculation of the actual production floor area. Allowance used is 80%. The calculation of the actual production floor area is shown as follows.

Allowance = 80%

Machine Area $= \sqrt{\text{Selected Area (1+Allowance)}}$ (4)

$= \sqrt{29{,}64\ (1+0{,}8)}$

$= \sqrt{53{,}35}$

$= 8\ m^2$

Station Area Added Allowance = Selected Area (1+Allowance) (5)

= 53,35 (1 + 0,8)

$= 96{,}03\ m^2$

Production Floor Area = Station Area × Total Station (6)

= 96,03 × 9 Station

$= 865\ m^2$

Theoretical Production Floor Area $= \sqrt{\text{Production Floor Area}}$ (7)

$= \sqrt{865}$

$= 29{,}41\ m^2$

Actual Production Floor Area = Rounding Up Theoretical Production Floor Area (8)

$= 30\ m^2$

From above calculation results, the initial layout of each machine area is designed and the blocks drawn are shown in Figure 2

IOP Conf. Series: Materials Science and Engineering **851** (2020) 012032 doi:10.1088/1757-899X/851/1/012032

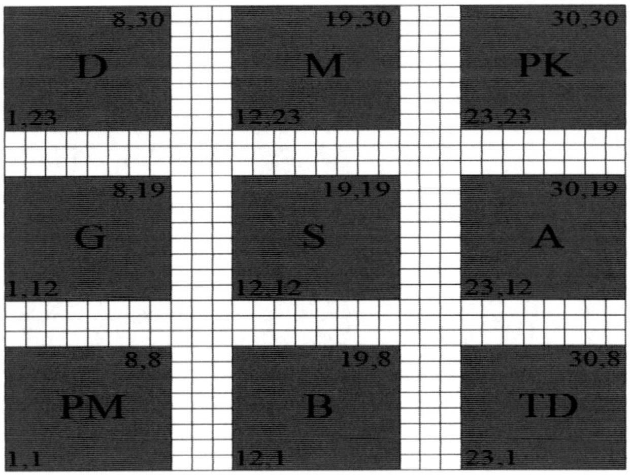

Figure 2. Initial Production Floor Layout Design

3.2. Production Floor Layout Design using CRAFT Algorithm

The following is one step of the CRAFT method using WinQsb software.

1. Input From To Chart and coordinates data to the following table.

Functional Layout Information for CRAFT Kelompok V Selasa

1 : Department Name: PM

Department Number	Department Name	Location Fixed	To Dep. 1 Flow/Unit Cost	To Dep. 2 Flow/Unit Cost	To Dep. 3 Flow/Unit Cost	To Dep. 4 Flow/Unit Cost	To Dep. 5 Flow/Unit Cost	To Dep. 6 Flow/Unit Cost	To Dep. 7 Flow/Unit Cost	To Dep. 8 Flow/Unit Cost	To Dep. 9 Flow/Unit Cost	Initial Layout in Cell Locations [e.g., (3,5), (1,1)-(2,4)]
1	PM	Yes	0	0	0	20,4	0	50	19,2	0	0	(1,1)-(8,8)
2	M	No	0	0	0	0	8,4	0	0	48	0	(12,23)-(19,30)
3	D	No	0	21,6	0	9	8	8,4	0	0	0	(1,23)-(8,30)
4	S	No	0	0	31,6	0	0	2	0	0	0	(12,12)-(19,19)
5	TD	No	0	0	0	0	0	6,8	0	0	0	(23,1)-(30,8)
6	G	No	0	0	54	0	0	0	0	24	0	(1,12)-(8,19)
7	B	No	0	0	14,4	0	0	2,8	0	0	0	(12,1)-(19,8)
8	A	No	0	0	0	0	0	0	0	0	20	(23,12)-(30,19)
9	PK	Yes	0	0	0	0	0	0	0	0	0	(23,23)-(30,30)

Figure 3. From to Chart and Coordinate Data

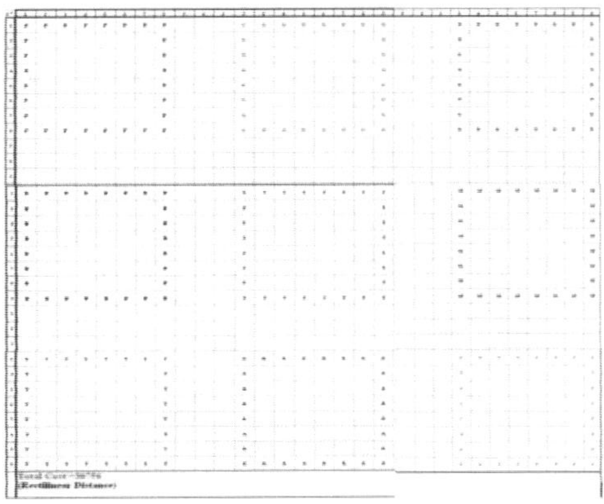

Figure 4. CRAFT Method Layout Results

Here is the final layout for production floor using the CRAFT algorithm Figure 4 and 5. The design of the production floor layout obtained using the CRAFT algorithm and WinQSB software with an allowance of 80% of the station area of 8 m x 8 m and the production floor area of 30 m x 30 m is optimal because the distance between work stations obtained using the aisle distance method is 660 m.

IOP Conf. Series: Materials Science and Engineering **851** (2020) 012032 doi:10.1088/1757-899X/851/1/012032

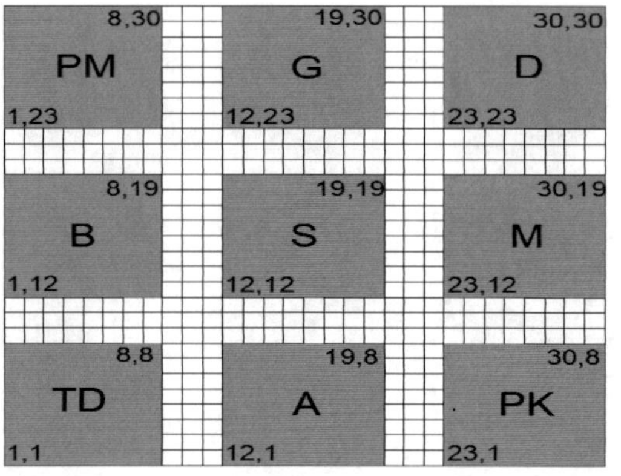

Figure 5. Final layout for production floor of Vise Manufacturing (redraw using AUTOCAD)

4. Conclusion

This paper demonstrates the use of CRAFT algorithm in designing production floor layout for vise manufacturing. CRAFT algorithm is used to evaluate initial layout based on the distance between work centers/stations and material flow. Further work is widely open in combining CRAFT algorithm with other approach so it will be able to consider more factors in designing production floor layout. For example, to consider sustainability aspects (environmental, social and economic) in production layout design, it requires an approach that is able to use more than two factors.

References

[1] Lekan O K, *et al* 2017 Analysis of Plant Layout Design for Operational Efficiency with Craft Algorithms 13 (4)

[2] Wignjosoebroto S 2009 Tata Letak Pabrik dan Pemindahan Bahan Edisi Ketiga Cetakan Keempat Guna Widya Surabaya

[3] Bunterngchit C 2018 The Application of CRAFT Algorithm for Increasing Material Flow Efficiency: A Case Study of Wooden Door Panels Manufacturing Factory *Proceedings of International Conference on Technology and Social Science*

[4] Tahir S, dkk 2015 Usulan Perbaikan Tata Letak Fasilitas dengan Menggunakan Algoritma CRAFT 4 (2)

[5] Wattimena E and Nil E M 2015 Usulan Perbaikan Tata Letak Fasilitas Produksi Gudang Tujuh PT. Mulchido dengan Menggunakan Metode Craft 9 (1)

[6] Siregar I, Muchtar M A, Rahmat R F, Andayani U, Nasution T H, and Sari R M 2018 Method of calculation overall equipment effectiveness in fertilizer factory. *IOP Conference Series: Materials Science and Engineering* **308**(1), p. 012053

IOP Conf. Series: Materials Science and Engineering **851** (2020) 012033 doi:10.1088/1757-899X/851/1/012033

Effect of Tip Speed Ratio on the Performance of H-Darrieus Wind Turbine with NACA 0018 Airfoil

Wahyu Hamdani[1], Hendrik V. Sihombing[2], Himsar Ambarita[1,2,*], and Koki Kishinami[3]

[1]Mechanical Engineering Department, Universitas Sumatera Utara, Jl. Almamater, Kampus USU, Medan 20155, Indonesia
[2]Sustainable Energy Research Centre, Universitas Sumatera Utara, Jl. Almamater, Kampus USU, Medan 20155, Indonesia
[3]Mechanical Engineering Department, Muroran Institute of Technology, 27-1 Mizumoto, Muroran 8585, Hokkaido, Japan

Email: himsar@usu.ac.id

Abstract. The effect of Tip Speed Ratio (TSR) to the performance of an H-Darrieus Vertical Axis Wind Turbine was studied using numerical method. The blade of the turbine was formed by using NACA 0018 mode. Three-dimensional governing equations are developed. The turbulent flow was modelled using $k-epsilon$ equation. The commercial code Computational Fluid Dynamic is employed to solve the problem. The focus in here is to explore effect of the TSR. The tip speed ratio is varied from 1.65 to 1.8. As a results, the contour velocity was generated and discussed. The average velocity of the wind leaving from the turbine is employed to estimate the power coefficient. The power coefficient for tip speed ratio 1.65 and 1.8 are discussed. The conclusion here is that tip speed ratio and the power coefficient strongly affected each other. Increasing the tip speed ratio will increase the coefficient power.

1. Introduction

Global warming catastrophe, critical fossil fuel sources, increases of oil price, climate change, etc. have gained the interest of many researchers to generate energy from renewable energy sources. One of the valuable renewable energy sources is wind power. A wind turbine is a structure that can be used to harvest the wind kinetic energy into rotational mechanical energy. This can be transformed into electric energy by using generator. There are several types of wind turbine that can be found in the wind farm. Those can be divided into Vertical Axis Wind Turbines and Horizontal Axis Wind Turbines. Mollerstrom [1] reviewed the performance vertical axis wind turbines that are found in the field with output power similar and more than 100 kW. The performances of the vertical axis wind turbines that have being operated were examined. It was revealed that many problems on development the vertical axis wind turbines mainly on metal fatigue and durability.

In the present study, the paper focuses on the study on the characteristics of H-Darrieus wind turbine. Some studies which focused on investigation of H-Darrieus wind turbine were reported in literature. Padmanabhan and Saravanan [2] compared NACA and NREL air foils of a wind turbine rotors with different tip speed ratio values. The two turbines showed the highest Cp values at an optimum tip speed ratio of 10 than the NREL at Cp=0.5, NREL has more Cp values for a bigger range of tip speed ratio. In addition, NREL is more efficient than NACA in order to get energy from the wind flow. Mc Cosker

[3] reported a design of three-blade horizontal axis wind turbine with a rotor radius of 2.5 m using turbine blade model NACA 4412 and NACA 23012, which follow Schmitz formula. In the study six different rotor designs are compared at different tip speed ratio. The comparisons are made by using BEM theory. The results reveals that rotor with NACA 23012 shows the bigger power coefficient at lower tip speed ratio. On the other hand, the rotor with NACA 4412 design shows the higher power coefficient for wide range of tip speed ratio especially for bigger tip speed ratio. The optimum is found to be at tip speed ratio of 10 for rotor with NACA 4412 with design tip speed ratio of 8. In addition, for rotor with NACA 23012 model the optimum tip speed ratio is 11 with design tip speed ratio of 8. It reveals that Schmitz formula and BEM theory can be employed to design the blade model with a proposed design tip speed ratio in order to get maximum power coefficient.

Vijayaragavan and Vijay [4] design of wind turbine blade using computer aided three-dimensional interactive application and analysis structural and flow characteristics using ANSYS. It shown the strain distribution, the stress distribution of blade for various materials with and without a notch. Salman Siddiqui et al. [5] investigated the tip speed ratio on wake stream specifications at operating conditions. It was plotted the vorticity and speed divisions in different locations in the downstream areas. Arian Hosseini and Navid Goudarzi [6] had designed, simulated and evaluated the characteristics of an innovative hybrid vertical axis wind turbine. In the study, a commercial computational fluid dynamics software was employed to estimate the performance parameter of the blade and structure system. It reveals an indication that when a Darrieus rotor shows that the highest power performance (Cp) it is loaded with high start-up torque demands. Pramono et al. [7] investigated the effect of chord length to the characteristics of an H-Darrieus Vertical Axis Wind Turbine (VAWT). The commercial computational fluid dynamic code is used to solve the problem. The results revealed that non-dimensional chord length strongly affects the power and adding the cord length will increase the power output of the turbine. Jerson and David [8] analyzed wind turbin performance at low tip speed ratio using the Betz-Goldstein model. Chaianant et al. [8] analyzed design factors like a material type, a number of blades height to radius ratios, design modifications, and turbine patterns by using CFD-based simulation. The number of blades and height to radius ratio has a significant effect on operating conditions of rotation due to inertia. Sobhani [10] compared NACA 0015 and J-NACA 0015 and investigated dimple effects on Darrieus. Computational Fluid Dynamic was employed to estimate the forces employed on the wind rotor under turbulent flow regime conditions. It was shown that the averaged efficiencies of the turbine were enhanced for an optimal tip speed ratio which is varies from 18% to 25 % if the airfoil with a cavity was used.

Those reviewed literatures reveal that the performance characteristics of an H-Darrieus vertical axis wind turbine strongly affected on several design parameters. In addition, commercial Computational Fluid Dynamics code has mainly applied to examine the effect of tip speed ratio to the performance. The aerator is powered by H-Darrieus VAWT comprising NACA 0018 airfoil. In specific, the effect of the tip speed ratio on performance will be explored here.

2. Method and Governing Equations

In the previous section, it was mentioned that the H-Darrieus VAWT has been used in the developed aerator. The used H-Darrieus vertical axis wind turbine will be drawn as air foil design to be studied. The model of the H-Darrieus vertical axis wind turbine is depicted in Fig. 1 (a).

The structure is employed as a wind energy harvester to operate the rotor of the aerator. Figure 2(b) depicts the diagram of the H-Darrieus Vertical Axis Wind Turbine. The location of the blade in the rotating area is depicted in Figure 1(c). The NACA 0018 is used as model to develop the blade of the turbine as depicted in the Figure 4(d). The characteristic parameter design that is explored here is Tip speed ratio and chord length of the turbine blade. The tip speed ratio is outlined as comparison of the tangential velocity of the tip of an air foil and the actual velocity of the stream flow. It is calculated by the below equation.

$$\lambda = \frac{\omega . R}{v} \tag{1}$$

In this study, a commercial CFD code is employed to carry the examination. A two-dimensional computation domain is employed and the turbulent regime flow is taken into consideration. The continuity equation and momentum equations are shown below [9].

$$\frac{\partial \rho}{\partial t}+\frac{\partial}{\partial x_i}(\rho u_i)=0 \tag{2}$$

$$\rho\frac{Du_i}{Dt}=-\frac{\partial p}{\partial x_i}+\frac{\partial}{\partial x_j}\left[\mu\left(\frac{\partial u_i}{\partial x_j}+\frac{\partial u_j}{\partial x_i}-\frac{2}{3}\delta_{ij}\frac{\partial u_i}{\partial x_i}\right)\right]+\frac{\partial}{\partial x_j}\left(-\rho\overline{u_i'u_j'}\right) \tag{3}$$

In the equation (3) the last term is known as the Reynolds-stresses and it will be given by

$$-\rho\overline{v_i'v_j'}=-\rho\left(\overline{v_i'v_j'}-\overline{v}_i\overline{v}_j\right) \tag{4}$$

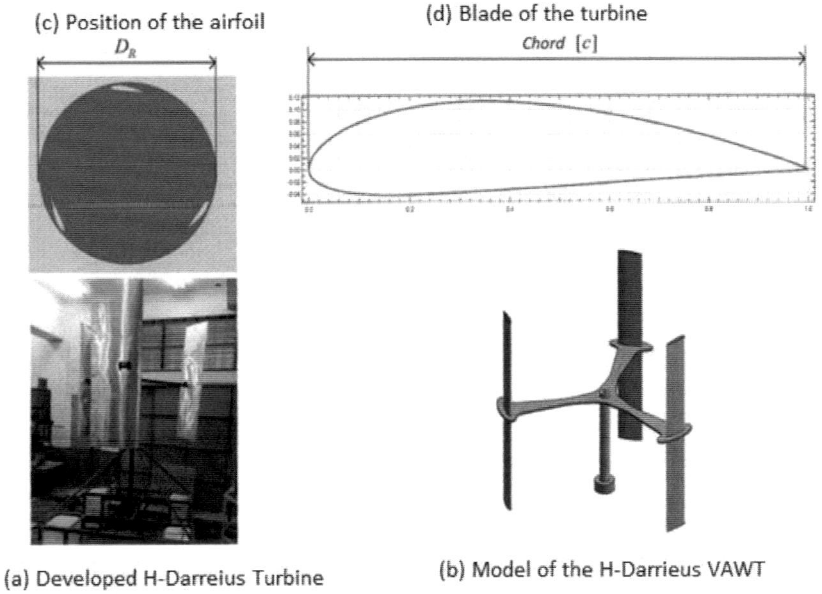

Figure 1 The model of H-Darrieus Wind Turbine used in this study

The stresses and the mean velocity gradients is combined by the Boussinesq hypothesis. Thus, the Reynolds-stresses ca be calculated by the equation shown in equation (5).

$$-\rho\overline{u_i'u_j'}=\mu_t\left(\frac{\partial u_i}{\partial x_j}+\frac{\partial u_j}{\partial x_i}\right)-\frac{2}{3}\left(\rho k+\mu_t\frac{\partial u_i}{\partial x_i}\right)\delta_{ij} \tag{5}$$

To close the equations above, several complementary governing equations are needed. In this method, the standard $k-\varepsilon$ turbulence formulation suggested by Lauder and Spalding [10] is employed. Furthermore, the two complementary governing equations are used in here. Those equations are the turbulent kinetic energy (k) equation and the turbulent dissipation rate (ε) equation. The mentioned equations are given in the below equations.

$$\rho\frac{Dk}{Dt}=\frac{\partial}{\partial x_i}\left[\left(\mu+\frac{\mu_t}{\sigma_k}\right)\frac{\partial k}{\partial x_i}\right]+G_k+G_b-\rho\varepsilon-Y_M \tag{6}$$

$$\rho\frac{D\varepsilon}{Dt}=\frac{\partial}{\partial x_i}\left[\left(\mu+\frac{\mu_t}{\sigma_\varepsilon}\right)\frac{\partial \varepsilon}{\partial x_i}\right]+C_{1\varepsilon}\frac{\varepsilon}{k}(G_k+C_{3\varepsilon}G_b)-C_{2\varepsilon}\rho\frac{\varepsilon^2}{k} \tag{7}$$

In those equations content two type of viscosities, they are the real dynamic viscosity and the proposed turbulent viscosity μ_t . The proposed turbulent viscosity can be estimated as follow.

$$\mu_t = \rho C_\mu \frac{k^2}{\varepsilon} \tag{8}$$

Where G_k and G_k is known as generation of turbulent kinetic incorporated to mean velocity gradient, generation of turbulent kinetic energy incorporated to buoyancy. And Y_M is defined as the contribution of fluctuating dilation in compressible turbulence to the overall dissipation rate. In addition, the symbol equation of $C_{1\varepsilon}$, $C_{2\varepsilon}$, $C_{3\varepsilon}$ and C_μ are defined as constants. In this simulation, the original number from commercial Computational Fluid Dynamics code are employed. The mathematical symbol of σ_k and σ_ε are defined as the turbulent Prandtl numbers for k and ε equation, respectively.

The computational method used in this study, in the simple explanation, can be said as follows. The finite volume method is applied to discretize the developed governing equations. The SIMPLE algorithm is used to merge the speed field and other fields, iteratively. The errors of the continuity and k and ε equations are plotted. The trial and error process will be terminated if the highest total error reached a value of less than 10^{-3} or the total error seems to be constant.

3. The Results and Discussions

The numerical simulations airfoil NACA 0018 have been executed for 3 number of the blade. Here, the wind velocity at the inlet and the diameter of the rotation region is 5 m/s and 150 cm, respectively.

3.1. Validation of numerical method

A validation to the used numerical method has been carried out. The used parameter in the validation is Lift coefficient (C_L) of an air foil NACA 0018 that is tried in experimental work in a wind tunnel at different angle of attack (α). Figure 2 shows the comparison. In the figure parameter resulted in the experiment shown by the blue color. The present CFD method was also employed to calculate the Lift coefficient at different angle of attacks. The results are also presented in Figure 2 by read color. The comparison reveals that the numerical result in the proposed method do agree with the experimental result. Based on this fact, the proposed numerical method can be employed to explore the effect of tip speed ratio to the performance of the wind turbine.

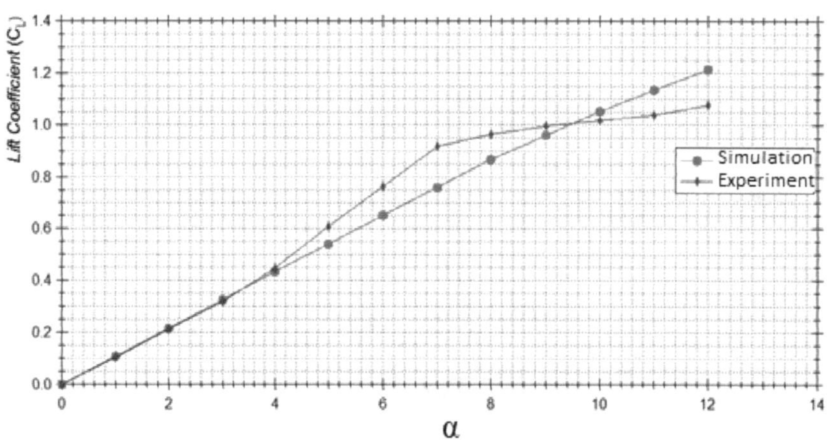

Figure 2 Numerical validation of NACA 0018

3.2. Velocity field

Figure 3 depicts the speed value in the computational domain given by the CFD results for all cases. Figure 3 and Figure 4 shows the contour velocity. To be noted, the red colour related to big speed and the blue colour shows low velocity. It shows that the wind speed lower in the inner region of rotating areas. Furthermore, on the inner regions of the turbine blade the speed almost zero (shown by blue colour). In contrast, the outer surface the velocity higher (shown by red colour). The difference of speed from the two surfaces results in power of the turbine. The effect of the tip speed ratio can be shown by making comparison of Figure 3 case with the velocity with 1.65 of tip speed ratio and Figure 4 with the speed with 1.8 of tip speed ratio.

The influence of tip speed ratio to the performance is examined by using comparing those two figures. The figure reveals that if the of tip speed ratio increase, the performance will increase. This is related to the energy of the wind flow drawn by the air foil to become mechanical kinetic energy in the rotor.

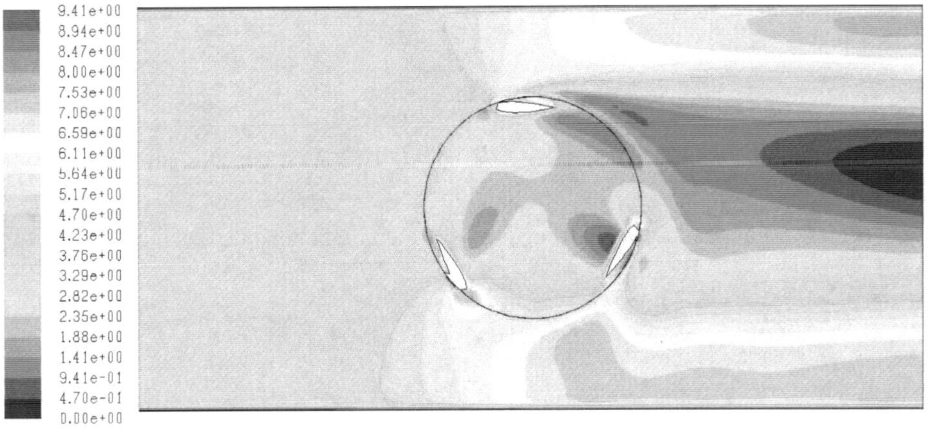

Figure 3 Contour velocity with 1.65 tip speed ratio

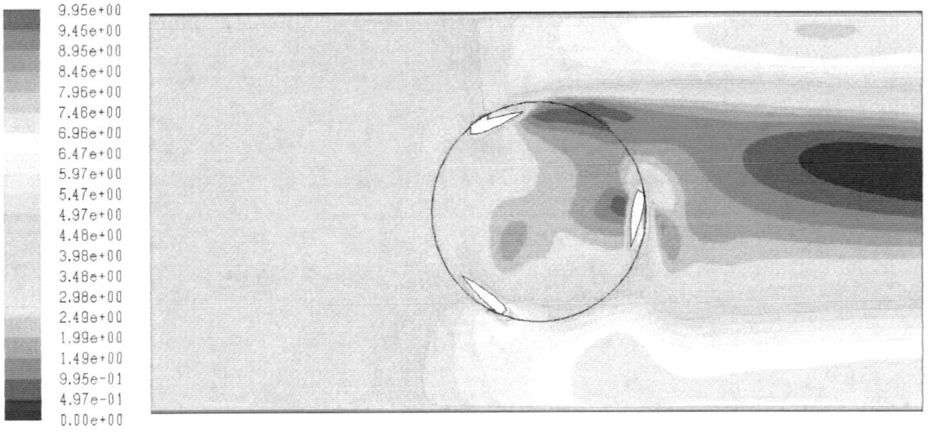

Figure 4 Contour velocity of 3 blade with 1.8 tip speed ratio

Figure 5 shown air flow inlet and outlet. It has shown contour of velocity wind turbine. The blue one was the lower flow and the red one was the highest flow. It shown the air flow with tip speed ratio 1.65 had higher velocity than turbine with 1.8 tip speed ratio.

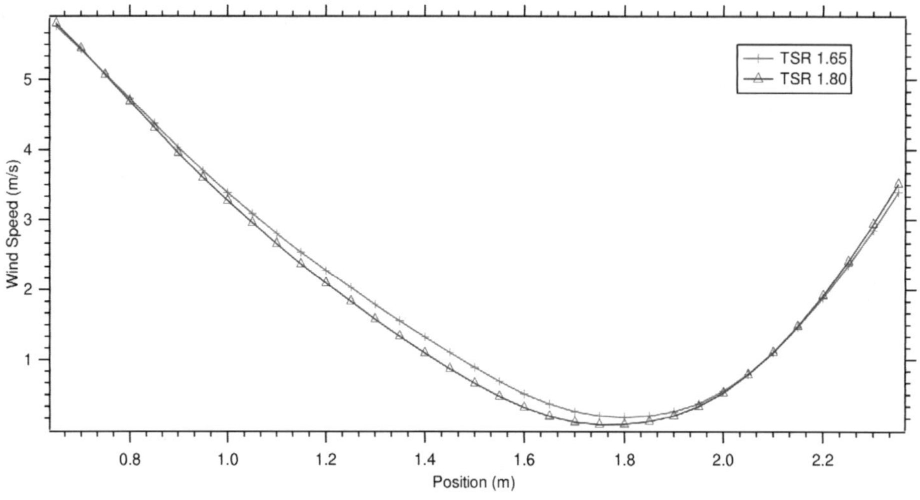

Figure 5 Correlation of velocity and tip speed ratio

3.3. Power coefficient

The characteristic of the velocity of the wind when it is leaving the air foil will be examined to calculate the power coefficient. The coefficient of the turbine power (c_p) is calculated by the equation formulated in the following.

$$c_p = \frac{P}{P_0} = \tfrac{1}{2}\left[1-\left(\frac{v_2}{v_1}\right)^2\right]\left(1+\frac{v_2}{v_1}\right) \tag{9}$$

Here, the parameter P [Watt] and P_0 [Watt] is the mechanical power drawn from the stream of the air flow and power in the air stream flow, respectively. In addition, the speed v_2 and v_1 is the air flow velocity before entering the turbine and the wind velocity after leaving the turbine, respectively. The average velocity leaving the turbine is given in Table 1. It show s that the average speed of the wind lower as tip speed ratio higher. In addition, as expected, the Power Coefficient higher as tip speed ratio higher.

Table 1 The effect of average velocity on Power coefficient

Tip Speed Ratio	Average velocity v_1 [m/s]	Power coefficient c_p	Power Pt [watt]	Efficiency [%]
1.65	2.095	0.553	50.778	55.284
1.8	2.005	0.562	51.608	56.203

4. Conclusions

The aim of this study was to explore a numerical simulation in order to explore the influence of tip speed ratio on the performance of the H-Darrieus Vertical Axis Wind Turbine. The turbine has the following specification: diameter of rotating region 1.5 m, the length of the blade is 1.55 m and number of blade is 3. The tip speed ratio was varied from 1.65 and 1.8. The velocity contour and the velocity profile of the wind leaving the turbine are shown. The conclusion can be drawn here is that the non-dimensional tip speed ratio strongly affects power. In addition, increasing the tip speed ratio will increase the coefficient of power.

References

[1] Mollerstrom E, Gipe P, Beurskens J and Ottermo F 2019 *Renewable and Sustainable Energy Reviews* **105** 1-13.
[2] Padmanabhan K.K, Saravanan R. 2012. European Journal of Scientific Research 440-446
[3] Mc Cosker J. 2015. Design and Optimization of Small Wind Turbine. Connecticut, USA: Requirements for MSc of Mechanical Engineering Renselaer Polytechnic Institute Hartford.
[4] Vijayaragavan E, Vijay V. 2015. International Journal of Applied Engineering Research. Vol.10 No. 68
[5] Siddiqui M. S, Rasheed A, Kvamsdal T, Tabib M. 2017. IOP Conf. Series: Journal of Physics: Conf. Series **854** 012043.
[6] Hosseini A, Goudarzi N. 2019. Energy Conversion and Management **189** 49-59
[7] Pramono I, Ambarita H, Kishinami K. 2019. IOP Conf. Series: Materials Science and Engineering **648** 012030
[8] Vaz J.R.P, Wood D.H. 2016. Energy Conversion and Management **126** 662-672
[9] Sranpat C, Unsakul S, Choljararux P, Leephakpreeda T. 2017. Energy Procedia **138** 500-505
[10] Zamani M, Nazar S, Moshizi S.A, Maghrebi M.J. 2016. Energy **116** 1243-1255

IOP Conf. Series: Materials Science and Engineering **851** (2020) 012034 doi:10.1088/1757-899X/851/1/012034

Analysis of the use of CMUcam5 Pixy camera in wheeled soccer robots

T H Nasution[1*], A Simon[1], F Fahmi[1], K Tanjung[1], M Zarlis[2] and N Noer[3]

[1]Department of Electrical Engineering, Universitas Sumatera Utara, Medan Indonesia
[2]Department of Computer Science, Universitas Sumatera Utara, Medan Indonesia
[3]Department of Physics, Universitas Sumatera Utara, Medan Indonesia

Correspondent Author: tigor.nasution@usu.ac.id

Abstract. Wheeled soccer robot is a robot that can play football like humans but maneuvers using wheels. Wheeled soccer robot has an event that is held routinely every year in Indonesia, namely in the Indonesian Robot Contest. To play well, the robot is certainly equipped with sensors and actuators that are qualified. One of the sensors used is the camera. The camera on the soccer robot functions to capture images around the robot that will later be processed on a computer. In this study, we analyzed the use of CMUcam5 Pixy for wheeled soccer robots. testing is carried out as much as 2 stages, namely testing without noise and testing with noise. From the results of tests that have been carried out, the CMUcam5 Pixy camera has a certain limit for detecting the ball. In testing with CMUcam5 Pixy camera noise can only detect up to 490 cm. Whereas in testing with CMUcam5 Pixy camera noise can only detect up to 420 cm.

1. Introduction

Wheeled soccer robot is a robot that can play football like humans but maneuvers using wheels. Wheeled soccer robot has an event that is held routinely every year in Indonesia, namely in the Indonesian Robot Contest. Indonesian Robot Contest has 5 branches of games, namely: Abu Robocon Indonesian Robot Contest, Indonesian Fire Extinguisher Robot Contest, Indonesian Humanoid Football Robot Contest, Indonesian Wheeled Robot Soccer Contest, Indonesian Dance Robot Contest[1]. Among these 5 branches, soccer robots can be included in the Wheeled Robot Soccer Contest branch.

To play well, the robot is certainly equipped with sensors and actuators that are qualified[2]. One of the sensors used is the camera. The camera on the soccer robot functions to capture images around the robot that will later be processed on a computer. The processed image will produce output in the form of objects that have been segmented from other objects. This segmented object will be created as a reference for the movement or activity that will be carried out by the robot.

The practical technology commonly used to insert cameras into robots is CMUcam5 Pixy to detect ball objects[3,4]. CMUcam5 Pixy is equipped with a Graphic User Interface that makes it easy to select objects to track. It also has firmware from the developer so that it can be operated easily even for beginners. In this study, we will analyze the use of CMUcam5 Pixy for wheeled soccer robots.

2. Methods

2.1. Research procedure

The first step of this research is the study of literature based on existing research. We did some research literacy about soccer competitions[5–8]. this research is also based on our previous research and several other researchers about image processing[9–11]. Then proceed to the testing stage using a CMUcam5 Pixy camera.

The flow of the test can be seen in Figure 1. Based on Figure 1, the first step carried out in the system design is to prepare objects in the form of futsal balls. Because the authors found it difficult to find an orange futsal ball in accordance with the rules of the competition in the Indonesian Wheeled Robot Soccer Competition in the market, the futsal ball was replaced with red. The colored futsal balls that were replaced did not cause significant problems in this study. Furthermore, testing is carried out as much as 2 stages, namely testing without noise and testing with noise.

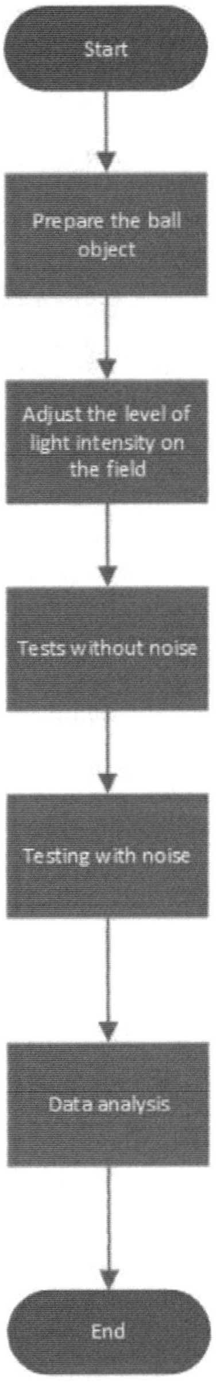

Figure 1. Stages of testing

IOP Conf. Series: Materials Science and Engineering **851** (2020) 012034 doi:10.1088/1757-899X/851/1/012034

2.2. Testing

The testing place is done indoors with enough light intensity. The testing site is carried out on a green carpet that has been formed into a wheeled soccer robot match field by the rules of the Wheeled Soccer Robot Contest. The test will be carried out by looking at the detection results with the Cmucam 5 Pixy camera module which already has firmware that can detect colors through a GUI (Graphical User Interface). Pixel 5 Pixy will be calibrated by entering the orange ball signature that will be used. Cmucam 5 Pixy configuration can be seen in Figure 2.

Figure 2. Configuring a CMUcam5 Pixy camera

After the object to be has been prepared, the next step is to adjust the test environment. The thing to note is the level of intensity that is given in the testing room. This is so that testing can be done with valid conditions in accordance with the conditions listed in the Wheeled Robot Soccer Competition rules. The intensity of the light will also affect the quality of tracking on objects and therefore, testing will be done indoors (indoor).

The intensity of the light provided in the room must be bright enough so that the image can be seen clearly to the end of the room or the test field. In the testing room, light from the sun is not arranged to enter the room so that the light intensity from outside the testing room does not interfere during the testing process. This is because the light from outside the room will continue to change along with the process of testing. Light from outside which changes the intensity of light in the room will also change the calibration process that will be done.

After the configuration of Cmucam 5 Pixy is finished, the layer of Cooked Video in the Pixymon software will display the output in the form of objects that have been marked with a box that says the signatures. The output of Cmucam 5 Pixy can be seen in Figure 3.

Figure 3. Display layer from Pixymon software output after configuration

3. Results and discussion

The first test was carried out 15 times without noise testing with distances ranging from 50 cm to 550 cm. Data from the test results can be seen in table 1. From the test results the CMUcam5 Pixy camera can only detect balls up to 490 cm.

Table 1. Tests without noise

Distance (cm)	Results
50	Detected
100	Detected
150	Detected
200	Detected
250	Detected
300	Detected
350	Detected
400	Detected
450	Detected
460	Detected
470	Detected
480	Detected
490	Detected
500	Not Detected
550	Not Detected

Furthermore, the experiment was conducted using 15 times of noise from 50 cm to 450 cm. The test results can be seen in Table 2. From Table 2 it appears that the CMUcam camera can only detect up to 420 cm.

Table 2. Tests with noise

Distance (cm)	Results
50	Detected
100	Detected
150	Detected
200	Detected
250	Detected
300	Detected
350	Detected
400	Detected
410	Detected
420	Detected
430	Not Detected
440	Not Detected
450	Not Detected
500	Not Detected
550	Not Detected
600	Not Detected
650	Not Detected
700	Not Detected

4. Conclusion
From the results of tests that have been carried out, the CMUcam5 Pixy camera has a certain limit for detecting the ball. In testing with CMUcam5 Pixy camera noise can only detect up to 490 cm. Whereas in testing with CMUcam5 Pixy camera noise can only detect up to 420 cm.

Acknowledgment
We thank the Ministry of Education and Culture and Universitas Sumatera Utara for supporting the implementation of this research.

References
[1] Pitowarno E and Nasrullah H 2019 Buku Panduan KRSBI Humanoid 2019 versi 27.12.18 *ATURAN PERTANDINGAN KONTES ROBOT SEPAK BOLA INDONESIA (KRSBI) Humanoid 2019*

[2] Cavalcanti A F, Araujo M V. and Pedrosa D P F 2015 Small Wheeled autonomous mobile soccer robot *Proceedings - 2nd SBR Brazilian Robotics Symposium, 11th LARS Latin American Robotics Symposium and 6th Robocontrol Workshop on Applied Robotics and Automation, SBR LARS Robocontrol 2014 - Part of the Joint Conference on Robotics and Intelligent Systems, JCRIS 2014*

[3] Ahmad M F, Alhady S S N, Rahiman W, Othman W A F W and Zahir A A M 2018 Visual based distance recognition technique by using pixy CMUcam5 *Lecture Notes in Mechanical Engineering*

[4] Yusof A A, Asohkumaran S K, Ramli M A A and Yusof A A 2020 Non-static vision guided pneumatic pick and place robotic system with shared supervisory control *Lecture Notes in Mechanical Engineering*

[5] Al Mamun M A, Nasir M T and Khayyat A 2018 Embedded system for motion control of an omnidirectional mobile robot *IEEE Access*

[6] Sugandi B, Prayoga S, Riandi I A and Tinambunan D G 2018 Goal Detection and Opponent Avoidance Algorithm for Wheeled Robot Soccer using Color Filtering and Contour Extraction *Proceedings of the 2018 International Conference on Applied Engineering, ICAE 2018*

[7] Xiong X 2019 Artificial Intelligence Control Algorithm for the Steering Motion of Wheeled Soccer Robot *Int. J. Pattern Recognit. Artif. Intell.*

[8] Nugraha R and Ibrahim N 2018 Sistem Kontrol Robot Sepak Bola berbasis Deteksi Multi Warna dengan PID Controller *TESLA J. Tek. Elektro*

[9] Nasution T H and Andayani U 2017 Recognition of roasted coffee bean levels using image processing and neural network *IOP Conference Series: Materials Science and Engineering*

[10] Fahmi F, Nasution T H and Anggreiny 2017 Smart cloud system with image processing server in diagnosing brain diseases dedicated for hospitals with limited resources *Technol. Heal. Care* **25**

[11] Siahaan E and Nababan E 2017 Subject Bias in Image Aesthetic Appeal Ratings *Data Sci. J. Comput. Appl. Informatics* **1**

IOP Conf. Series: Materials Science and Engineering **851** (2020) 012035 doi:10.1088/1757-899X/851/1/012035

The Implementation of Deep Learning Using Convolutional Neural Network to Classify Based on Stomata Microscopic Image of *Curcuma* Herbal Plants

U Andayani[1], I B Sumantri[2], Andes Pahala[1], M A Muchtar[1]

[1]Department of Information Technology, Faculty of Computer Science and Information Technology, Universitas Sumatera Utara, Medan, Indonesia
[2] Department of Pharmacy, Faculty of Pharmacy, Universitas Sumatera Utara,

ulfi.andayani@usu.ac.id

Abstract. There are so many types of herbal plants that come from the same genus. The similarity of features in herbal plants makes it difficult to distinguish. Especially in the pharmaceutical field, which is very risky for making mistakes. The part of plants that is often used as ingredients for herbal medicines is the leaves, so research on leaves and their constituent organelles is very important for the pharmaceutical world. Therefore, an approach is needed to identify the types of organelles present in the leaves, where the organelles most frequently studied are the stomata. So, a neural network approach is needed to distinguish plant characteristics in the same genus. In this research there are 2 species of plants in the *Curcuma* genus, namely turmeric and ginger. This research was conducted by implementing Deep Learning using Convolutional Neural Network (CNN) with the help of the Gabor filter process and feature extraction using the Gray Level Co-occurrence Matrix (GLCM). The study was conducted using 160 microscopic images of Curcuma herbal plants as training data with training accuracy of 93.1% and test data of 40 images with an accuracy of 92.5%.

1.Introduction

The World Health Organization (WHO) defines herbal plants as plants that are used for medicinal purposes and are original ingredients in making herbal medicines [1]. Herbal plants themselves have thousands of species with a total of around 40,000 types of herbal plants that have been known in the world, where around 30,000 species are allegedly located in Indonesia. This amount represents 90% of medicinal plants in Asia. Of these, 25% of them or about 7,500 species are known to have herbal or medicinal properties. However, only 1,200 types of plants have been used as ingredients for herbal medicines.

Based on the Decree of the Head of BPOM Number HK.00.05.4.2411 of 2004 Concerning the Basic Provisions for Grouping and Marking Indonesian Natural Medicines, Indonesian herbal medicines can be grouped into herbs, Standardized Herbal Medicines (OHT) and Fitofarmaka. Jamu only has traditional evidentiary claims, while OHT is a herbal medicine that has been scientifically proven for its quality, safety, and benefits and uses raw materials that have met the standards and are usually pre-clinically tested. While fitofarmaka is a standardized herbal medicine that has been carried out scientifically higher proof, because it has been carried out clinical trials [2].

The various types of herbal plants available, more is used as raw material for herbal medicine, even though it is not curative or healing, but only prevention. Only about 5% of species used as phytopharmaca ingredients, which have been clinically tested in humans can increase curative levels or can heal. This is caused by one of them due to the lack of further research carried out and the technology base.

One of the most used herbs is from the genus *Curcuma,* with examples of species such as kunyit (*Curcuma longa*) and temulawak (*Curcuma zanthorrhiza*). Both of these species are included in the 10 species of herbs that are most widely used, especially turmeric occupies the second position after ginger of the genus *Zingiber*. The terms genus and species are studied in taxonomy. The taxonomy itself is a grouping of individuals into species, arranging species into larger groups, and naming these groups, so as to produce a classification [2]. The level of classification from the most unique to the level of greater similarity, namely species, genus, family, order, class, division, and kingdom. The similarity of plants is also one of the factors that inhibit the use of herbal plants, which will be very risky if there is an error in choosing the type of plant that is used as medicinal material. The difference can be seen in terms of plant morphology or anatomy.

Plants morphologically or externally have 3 main parts, namely roots, stems and leaves. The leaves are the most important organ for plants in carrying out their lives because plants are *obligate autotrophic* organisms which must supply their own energy needs through the conversion of light energy into chemical energy. The leaves are the most frequently studied parts of plants in biology and pharmacy, especially their microscopic constituents.

Anatomically, the leaves have 3 tissues, namely epidermal tissue (outer layer), mesophyll tissue (inner layer) and transport tissue. In the leaf epidermis tissue, there are stomata. Stomata are easier to study because they are found in the leaf epidermis layer which is the outer layer of leaves, making it possible to be seen directly from a microscope without special treatment such as observation of mesophyll tissue.

Some previous studies on plant leaves have been carried out to resolve identification problems including those carried out by [3] to identify and diagnose diseases in cotton leaves using the *fuzzy* method, *Artificial Neural Network* (ANN), and *Support Vector* (SVM). The researcher also used color and texture extraction in this study. Researchers used 20 diseased leaf images and 25 normal leaf images. Using color segmentation and good image quality will improve diagnosis results. In this study an accuracy of 85% was obtained.

Furthermore, [4] conducted a study to classify the leaves of healthy or affected cocoa plants using *Backpropagation*. In this study an accuracy of 86% was obtained from a total of 90 images data. Subsequent research by [5] conducted a study using *Convolutional Neural Network* (CNN) to detect disease in plant leaves through the detection of artificial neural networks. In addition, researchers also use *invariant moments* as feature extractions. In this study an accuracy of 88% was obtained.

2. Data and Methods Used

2.1. Data Used

The data used in this study were microscopic images of kunyit (*Curcuma longa*) and temulawak (*Curcuma zanthorrhiza*) *plants*. Microscopic data in this study were obtained from research laboratories at the Faculty of Pharmacy, Universitas Sumatera Utara. The image was taken using an electron microscope with a magnification of 10 x 40 magnification scale. As for the training data using 80 microscopic images for each plant, so the total training data there were 160 images, while for the test data using 20 microscopic images for each plant, so the total training data there were 40 images.

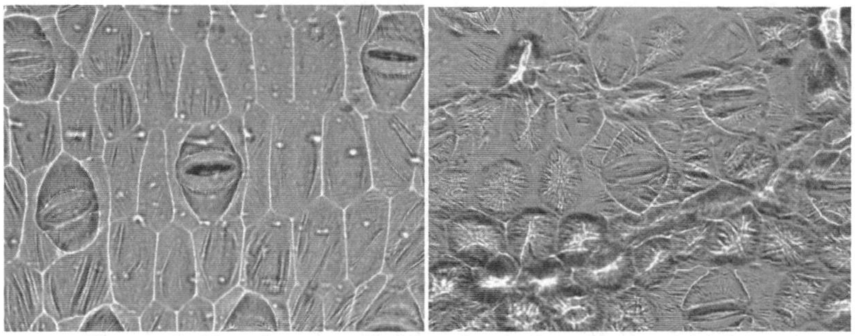

Figure 1. Kunyit stoma (left) and temulawak stoma (right).

2.2. *Method Used*

The stomata classification system in this study was carried out with several steps, starting from the determination of the dataset, processing, to the resulting output. Here is a general description of architecture.

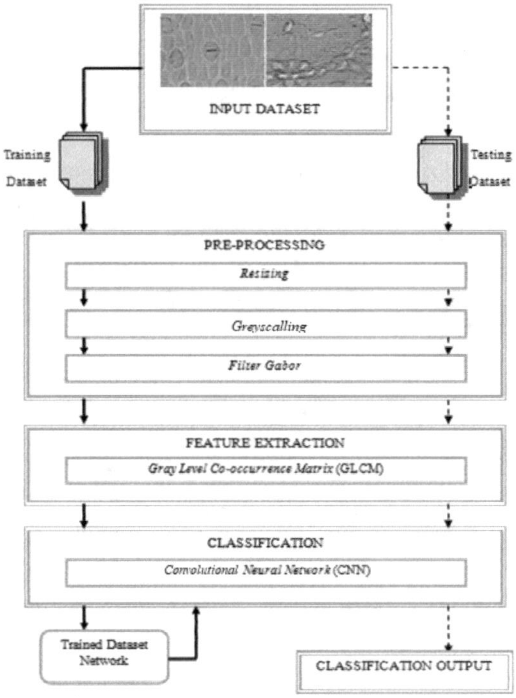

Figure 2. General description of architecture.

2.2.1. *Input*

Input in the form of stomata microscopic image data from Faculty of Pharmacy, Universitas Sumatera Utara, which was taken using a microscope with a magnification of 10x40.

2.2.2. *Preprocessing*

Pre-processing stages allow to produce a better image, so that it will be more easily processed and produce a more optimal *output* at later stages.

1) *Resizing*

The inputted image will change in size (resizing) to 100x100. This process aims to speed up the classification process and generalize the image size of each data.

2) *Greyscale*

At this stage, images that have been resized will be *grayscale* processed into gray images. Since this stomata research require images texture data and not color images, *grayscale* can be done.

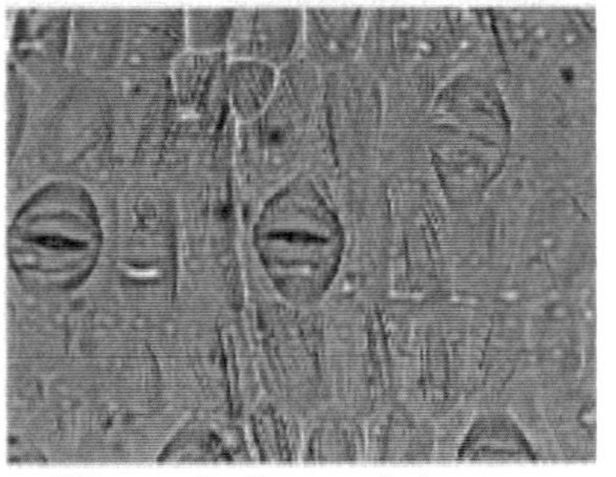

Figure 3. Grayscale image.

3) Gabor Filters

This stage is done by initializing the following variables:

- Determine the parameters *u, v, m, n* to be used in calculating *the bank filter Gabor.*
  The variable *u* is the number of scales (small to large), with a default value of 5, variable *v* is the number of orientations (horizontal, vertical, diagonal), with a default value of 8. Variable *m* is the number of rows in the 2-dimensional filter Gabor matrix, its value must be odd integer, with the default value is 39, and *n* is the number of columns in the 2-dimensional filter Gabor matrix, the value must be an odd integer, with the default value is 39. Thus, the author sets the following values in the program code:

```
gaborFB.u = 6
gaborFB.v = 9
gaborFB.m = 39
gaborFB.n = 39
```

- Specify the parameters *d1, d2* to be used in calculating *Gabor features.*
  The *d1* and *d2* parameters are *down sampling* of matrix rows and columns. Its function is to reduce the size of the matrix by taking important points from the matrix. The greater the down sampling value, the smaller the size of the resulting matrix, so that calculations can be done faster, but the level of compatibility will be weaker, and vice versa. Following are the values set by the author for the *down sampling* parameter:

```
gaborFE.d1 = 3
gaborFE.d2 = 3
```

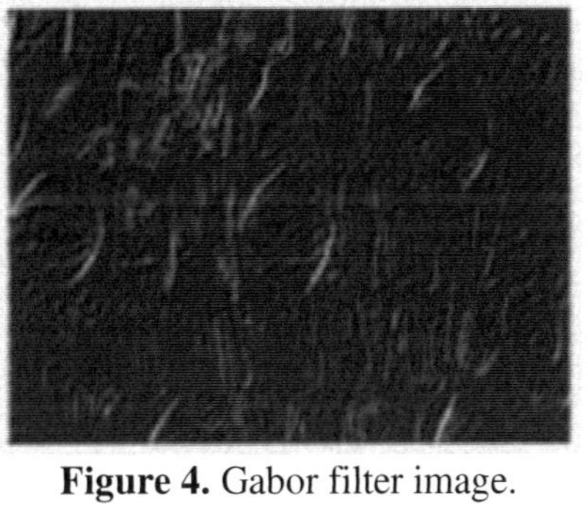

Figure 4. Gabor filter image.

2.2.3. Feature Extraction

Feature extraction is a fundamental part of images analysis. A feature is a characteristic or unique characteristic of an object [6]. Feature extraction used in this study is the *Gray Level Co-Occurrence Matrix* (GLCM). The GLCM feature extraction method is one of the second order extractions of the texture statistical features. Second order extraction shows a statistical relationship between 2 pixels. GLCM is a matrix with the number of rows and columns proportional to the number of *gray levels* (G) in an image.

The GLCM method can produce at least 4 feature extractions from a digital image per pixel neighboring angle. These quantities include *contrast, correlation, energy,* and *homogeneity.*

1) *Contrast*

Contrast is a measure of the existence of gray level variations between image pixels with relative locations. *Contrast* has a limit value from 0 to rank 2 of the symmetric GLCM matrix length. In images with pixel elements that have the same overall value, *contrast* is 0. To find the extraction of *contrast* features, the following formula can be used:

$$Contrast = \sum_{i=1}^{L} \sum_{j=1}^{L} (i-j)^2 GLCM(i,j) \quad (1)$$

2) *Correlation*

Correlation is a measure of linear dependence between gray levels in the image. *Correlation* in GLCM serves to regulate the linear dependence of the gray level in the neighboring pixel image. To find the extraction of features, the following formula can be used:

$$Correlation = \sum_{i=1}^{L} \sum_{j=1}^{L} \frac{(i-\mu_{i'})*(i-\mu_{j'})*(GLCM(i,j)}{\sigma_{i'} * \sigma_{j'}} \quad (2)$$

3) *Energy*

Energy is the value of the sum of squares on the elements of the GLCM matrix. *Energy* has a high value when the image has good homogeneity or almost the same pixel value. To find the extraction values, the following formula can be used:

$$Energy \sum_{i=1}^{L} \sum_{j=1}^{L} (GLCM(i,j))^2 \quad (3)$$

4) *Homogeneity*

Used to measure homogeneity, which is a measure of the proximity of the distribution of each element in the GLCM matrix to the diagonal GLCM matrix. To find the extraction of the features, the following formula can be used:

$$Homogeneity \sum_{i=1}^{L} \sum_{j=1}^{L} \frac{GLCM(i,j)}{1+(i-j)^2} \quad (4)$$

2.2.4.Classification with CNN

Convolutional Neural Network (CNN) is a development of the *Multilayer Perceptron* (MLP) which is designed to process two-dimensional data [7][8]. At this stage the images data will be classified into 2 types of output, namely kunyit or temulawak, from the stomata images. Total data used are 300 images that will be divided into training data and test data with a ratio of 250: 50, where 250 training data contains 125 data from each category, while 50 test data contains 25 data from each category.

The initial stage of this image processing is to change the size of the image that was originally sized 100x100 pixels to 32x32 pixels. It is recommended to use images with a power value of 2, such as 16, 32, 64, and so on. For sizes 16x16 and under will make a lot of pixel information disappear, whereas for sizes 64x64 or larger will make the processing rate slower. So the author makes the size value used is 32x32.

In the previous chapter it was explained that the general process on CNN consists of the convolution process, activation function, and pooling. The number of processes in this stage is adjusted to the research needs.

The convolution process is carried out 4 times, or there are 4 *convolution layers* used. In general, already quite two layers are used for the classification model, but in this study used a network of more *deep* as the implementation of *deep learning* to train the existing models and see how the performance of the model process.

The activation function used is Rectified Linear Unit *(ReLU)* to make the training process faster. The size of the kernel / filter used for each convolution *layer* is 3x3. While the *pooling* size used is 2x2. *Pooling is* done twice, namely after the first two convolution processes and after the next two convolution processes. This is done so that the size of the input is not drastically reduced in each process carried out, so that the input image information that is owned is still useful and can be used in the classification process. While the number of filters / kernels used is also varied, namely as many as 32 filters for the convolution *layers* 1 and 2 as well as 64 filters for the convolution *layers* 3 and 4. The use of more filters in the last two convolution *layers* is because the size of the input at the two layers is smaller, so more filters are needed to extract image information. In the final process of classification, *SoftMax classifier is* used to provide more intuitive results, making it easier to do the classification of probabilistic interpretations for all labels produced.

The process that occurs from the model above is started by "encoding "an image into *features* in the form of numbers that represent that image. In the first convolution process, the image as an *input* size of 32x32 pixels is actually a multidimensional *array* with a size of 32x32x3 (3 is the number of RGB *channels)*. This image will use several processes as mentioned in the model. The filter will be moved by *stride* 1 to all parts of the image input, starting from the upper left corner to the lower right. Every filter shift on the image input is carried out a "dot" operation or mathematical calculation. In this study *zero padding* was not added, so the input size used was equal to 32x32 pixels. In other words, the *output* produced from the convolution process has a smaller size.

The size of the image generated from the first convolution process is 30x30 pixels. The results of this measure can also be obtained using the calculation using equation below:

$$(W\text{-}F+2P)/S+1 = 32\text{-}3+2(0)/1 + 1 = 30 \qquad (5)$$

And so on to do the calculation of the size of the image formed. The results of this operation are then subject to the activation functions, namely ReLU and pooling. The input for the activation function is the real value and the output of the function is a value between 0 and 1. If the input is very negative, the output obtained is 0, whereas if the input is very positive then the output value obtained is 1. In principle, the *pooling layer* consists of a filter with a certain size and *stride* will shift to the entire *feature map* area. *Pooling* used is *Max Pooling*. The purpose of using the *pooling layer* is to reduce the dimensions of the *feature map (down sampling)*.

By using a 2x2 size filter and *stride* 2 in the *max pooling* operation, the size of the image formed after the first convolution process is carried out, is subject to an activation function, then the *max pooling* operation is 14x14 pixels. The result of the convolution process is a *feature map* that is reused as *input* for the next convolution process. The process runs until the convolution process ends. In addition, the regularization method used is a *dropout, in* which several neurons will be randomly selected and not used during training. These neurons can be said to be disposed of at random. This means that the contribution of disposed neurons will be temporarily stopped while new tissue and weights are also not applied to the neurons when doing *backpropagation*.

2.2.5 Output
The classification results will display the prediction of plant species in which category based on the image that has been inputted, in this case turmeric or ginger.

3. Research Results and Discussion
This study used training data of 160 images and test data of 40 images of kunyit and temulawak. In this study there are parameters that greatly affect the accuracy of the classification and also the speed of image data processing, such as *Max Epochs* and *Mini Batch*.

IOP Conf. Series: Materials Science and Engineering **851** (2020) 012035 doi:10.1088/1757-899X/851/1/012035

Obtaining the accuracy obtained from the testing of the stomata images classification system can be calculated by the percentage of the correct testing data divided by the total amount of testing data:

$$Accuracy = \frac{Total\ True\ Testing\ Data}{Total\ All\ of\ Data}\ x\ 100\%$$

$$= 37/40 \text{ x } 100\%$$

$$= 92.5\%$$

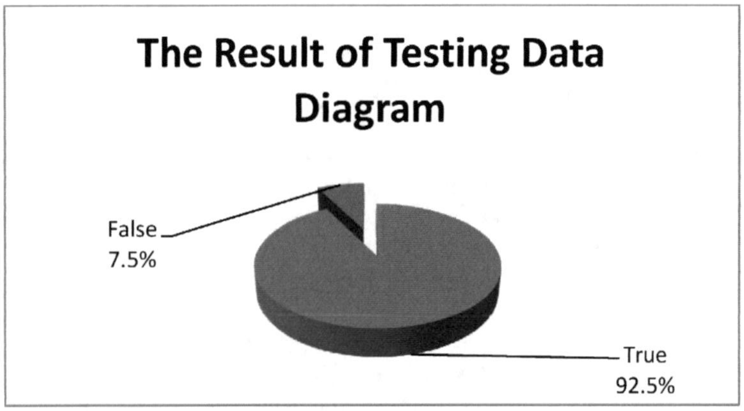

Figure 5. Data result.

From the research conducted, that the accuracy obtained from the training data reached 9 3.1 % accuracy with a total data of 16 0 images, *Max Epochs* 30, *Mini Batch* of 128 and accuracy of test data with an accuracy of 9 2.5 % of the total 4 0 test images shows that the implementation of the *Deep Learning* method using *Convolutional Neural Network* is good enough to classify the microscopic image of herbal stomata from the genus *Curcuma.*

4. Conclusions and Suggestions

4.1. Conclusion

The conclusions that can be drawn from the results of system testing using the *Convolutional Neural Network* (CNN) method for the classification of *Curcuma* plant stomata are as follows:

1. CNN method is able to classify *Curcuma* stomata microscopic image of herbs with an accuracy rate of 93.1% in the training data, and an accuracy of 92.5% in the test data
2. The use of more training data can improve accuracy.
3. Determination of filter size on CNN greatly affects the accuracy results. The smaller the filter size used, the better information obtained.
4. Determination of *Max Epochs* and *Mini Batch* parameters that are too high or too low can affect the accuracy results.

4.2. Suggestions

Suggestions that the author can give for subsequent research are as follows:

1. Establish a broader classification system of herbal plants.
2. Use other classification methods to get higher accuracy.
3. Trying to classify stomata images with other parameters such as stomata type shapes and neighboring cells.
4. Try classification of other microscopic organelles, such as those in stems or roots

IOP Conf. Series: Materials Science and Engineering **851** (2020) 012035 doi:10.1088/1757-899X/851/1/012035

References

[1] WHO. 1998. *Guidelines for the Aooripriate use of Herba Medicines*. Manila: WHO Regional Publications, Western Pacific Series No. 23.

[2] Buletin Infarkes, Kementerian Kesehatan. 2015. *Bugar dengan Jamu*. Direktorat Jenderal Kefarmasian dan Alat Kesehatan, Kementerian Kesehatan. Jakarta.

[3] Judd, W. S.; Campbell, C. S.; Kellogg, E. A.; Stevens, P. F.; Donoghue, M. J. (2007). *"Taxonomy". Plant Systematics: A Phylogenetic Approach (3rd edition)*. Sunderland: Sinauer Associates.

[4] Permata, E., Suherman, A. & Maulana, A. 2014. *Klasifikasi Daun Tanaman Theobroma Cacao menggunakan metode Neural Network.*

[5] Gharat, A., Bhatt, K. & Kanase, Bhavesh. 2017. *Leaf Disease Detection Using Image Processing*. The Imperial Journals.

[6] Sutoyo, T. 2009. *Teori Pengolahan Citra Digital*. Yogyakarta: Andi

[7] LeCun, Y., Bottou, L., Bengio, Y., and Haffner, P. 1998. *Gradient-based learning applied to document recognition*. Proceedings of the IEEE, 86(11), pp. 2278–2324

[8] Rahmat R F, Dennis, Sitompul O S, Purnamawati S, Budiarto R, 2019 Advertisement billboard detection and geotagging system with inductive transfer learning in deep convolutional neural network *Telkomnika* **17**(5), pp 2659-2666.

IOP Conf. Series: Materials Science and Engineering **851** (2020) 012036 doi:10.1088/1757-899X/851/1/012036

An Analysis of One-Roof Application Services Quality at the Faculty of Social and Political Sciences, University of North Sumatra

Muhammad Arifin Nasution [1] **, Surya Yudha Regif** [2]
[1]Public Administration, Faculty of Social and Political Sciences, Universitas Sumatera Utara

[2]Government Science, Faculty of Social and Political Sciences, University of Timor

Abstract. The University of North Sumatra is a government agency that organizes education, research and community service. It also includes efforts to enforce effective and efficient administrative governance. Integrated administration services in the world of collegeto serve student administrative affairs have been carried out by the Faculty of Social and Political Sciences, The University of North Sumatra through online services. The online-based service is called the One-Roof Application or abbreviated as ASA to facilitate the process of correspondence between students and the administration of the faculty in the orderly administration of the academic world which is provided under one roof of service. This study uses a questionnaire as one of the data collection tools, then the study used research respondents who used the one-roof application. Based on information received from the One-Stop Application Service Unit of the Faculty of Social and Political Sciences, that every day in the customer hall serves approximately 20 people using one roof with a range of service days in 1 month, Monday - Friday totaling 20 x 20 days = 400 population with a sample size of 196 people. This is based on the formulation of the creative and organic tables set out in the Regulation of the Minister of Administrative Reform and Bureaucratic Reform of the Republic of Indonesia Number 14 of 2017 Concerning Guidelines for the Preparation of Public Satisfaction Survey Public Service Provider Units. The results of the service satisfaction study are categorized GOOD.

1. Introduction

As demands and challenges in the development of education services in higher education, institutions must be able to give birth to new approaches in the implementation of governance in which one of them is the innovation of public service delivery, taking into account the strengthening of public development to obtain excellent service quality in Good University Governance which ultimately provides a sense of trust in students towards the educational institution that cover it. Furthermore, the picture of University governance itself is interpreted as a form as well as a constitutional process while the university manages its affairs independently [3]. In a statement from the Head of the Republic of Indonesia's State Administrative Institution (LAN), Prof. Dr. Agus Dwiyanto that "Innovation is, now, a must either sooner or later[1]. We believe that there is always room for innovation. Do not let the people who force to make changes because the community is now very aware of their rights and demanded their rights well ", (Seminar and Workshop on Innovation in the Implementation of Regional Public Services, at GSP (Grha Sabha Pramana) Gadjah Mada University, following the paradigm of Reinventing Government and Good Governance emphasized that the delegation of some authorities must be able to maximize the 4E principle, namely effectiveness, efficiency, equity/fairness and economy [5].

Whereas in the Open government perspective it has also been carried out. Yang, Lo, & Shiang [7] which specifically found influential factors and their impact on open government initiatives were investigated from four perspectives namely legislation and policy, technology, organization, and environment. Legislation and policies were found to have the most significant impact while existing regulatory and policy institutions acted as constraints. Technological factors provide constraints but are considered relatively more easily resolved with adequate support. Factors that are in the perspective of the organization and the environment follow secondary impacts.

The means to bring the government closer to the public in the field of public services in the era of the industrial revolution 4.0 is e-government [2].The first principle that must be done is to focus on improving government services to the public[6]. Because there are so many types of services provided by the government to its people, it must be considered which services are the priority. The University of North Sumatra has a commitment that shields itself from the effects or influences or challenges of the internal or external environment that is destructive or has bad content that can undermine good governance in universities. the value of service as one of the values needed to support the implementation of administrative governance formulations in it so as not to nearly the review of the development of demands and changing challenges in the academic world. It was realized that the role as a center of knowledge in the province, North Sumatra University, in this case, the Faculty of Social and Political Sciences was demanded to balance its functions in addition to the three main services of the university as well as maintaining the condition of social satisfaction in its faculty environment as a way to deal with changes and developments in technology/information.

With the three main services of the University, indirectly college is a small form of simple state management based on rules and values based on the noble value of education. This also underlies the presence of Good University Governance in organizing education in an educational institution in North Sumatra. Therefore, the application of the principles of good governance in the management of systems and processes within the Faculty of Social and Political Sciences must be sensitive to the development of science and technology as a result of dynamic knowledge.

In North Sumatra itself in general, college has challenges in the future, namely:

1. Optimizing the quality of education, the ratio of the number of educated people to employment, the product capacity produced, appropriate and good university management.
2. Independence institutions that have the responsibility for anticipating external socio-cultural developments that affect the corridors and direction of the development of a nation.
3. the development of science that produces Science and Technology that affects the demands and interests of education in the area of North Sumatra.

The implementation of the One-Roof Application Program (ASA) is one part of the challenges ahead for tertiary institutions, namely good university governance as stipulated in the CollegeLaw No. 12/2012 Article 64 which covers the autonomous management in the non-academic field as referred to in paragraph (1) includes operational norms and policies and implementing settings:

a. organization;
b. finance;
c. student affairs;
d. power; and
e. infrastructure.

From this implementation, the principles of good university governance are upheld, including:

1. Organize the organization & make their own decisions;
2. Hunan Research management;
3. Asset management;
4. Financial management.

Therefore, it must be analyzed about the implementation of the ASA (One-Roof Application) at the Faculty of Social and Political Sciences, University of North Sumatra following the basis and objectives as well as the ability of universities to carry out autonomy as referred to in paragraph (2) of the 12-year Law 2012 which was evaluated independently by the University. Evaluation of the implementation is to measure the success of faculty policies at tertiary institutions so that the evaluation of this scientific work will become one of the benchmarks of achievement and

IOP Conf. Series: Materials Science and Engineering **851** (2020) 012036 doi:10.1088/1757-899X/851/1/012036

improvement of the performance of academic services at the University of North Sumatra in improving the quality of services for students who use services (One-Roof Application) at the Faculty of Social and Political Sciences, University of North Sumatra.

2. Research Method

In this study, the research location is in Building A, Faculty of Social and Political Sciences, located in the University of North Sumatra Complex. The design of this study is a combination of qualitative and quantitative approaches with a dominant-less model. There are several considerations for using this method. The first is easy to switch if double facts are found. The second directly provides the relationship between researchers and respondents. The third reason is that it is more responsive and able to adapt to the many influences of sharpness and value tendencies encountered (Moelong, 2006). Data collected and analyzed using descriptive-analytic methods.

3. Data Collection

The primary data source in this study is the respondents/service users (students) who use the service (ASA) One-Roof Applications at the Faculty of Social and Political Sciences.

The population is a generalization area consisting of objects/subjects that have certain qualities and characteristics [4]. The population in this study are students who have received one-stop application (ASA) services at the Faculty of Social and Political Sciences. Because in this study using a questionnaire as a data collection tool, this study also used research respondents. Based on information received from the One-Stop Application (ASA) unit at the Faculty of Social and Political Sciences. serving less than 20 people per day who need the One-Stop Application at the Faculty of Social and Political Sciences (ASA). in the range of working days in 1 month, Monday - Friday totaling 20 x 20 working days = 200 populations / 196 samples.

Table 1. Krejcie and Morgan

Population (N)	**Sample (n)**
400	196
420	201

Sampling techniques can be adapted to the type of service, survey objectives, and data to be obtained. Respondents were randomly selected according to the coverage area of each service unit. As for the sample size and population using a sample table from Krejcie and Morgan or calculated using the formula:

$$S = \{ \lambda^2 . N. P. Q\}/ \{d^2 (N-1) + \lambda^2 . P. Q \}$$

Where:

S = number of samples
λ2 = lambda (multiplier factor) with
dk = 1, the error rate can be 1%, 5%, 10%
N = population
P (population spread normally) = Q = 0.5 d = 0.0

Sampling-based on the limitations of observing the entire sample, for time and cost efficiency to produce generalizations to the population and reduce research errors in sampling. So that the sample was taken in this study can represent the population, it can be determined the number of samples calculated using the Morgan and Krejcie regulated formula in the Minister of Administrative Reform and Bureaucratic Reform of the Republic of Indonesia Number 14 of 2017 Concerning Guidelines for Preparation of Public Satisfaction Survey Public Service Provider Unit.

4. Data Analysis

The analysis technique used is in making the frequency distribution table classifications of answers set by 4 classes, with the consideration that the determination of the classification of answers is adjusted to the number of categories/classes used in making the questionnaire based on a Likert scale, namely:

Table 2. Likert Scale

Perceptional values	Interval values (NI)	Interval values of conversion (NIK)	Service qualities (x)	Performance of service unit (y)
1	1,00 – 2,5996	25,00 – 64,99	D	Not good
2	2,60 – 3,064	65,00 – 76,60	C	Less good
3	3,0644 – 3,532	76,61 – 88,30	B	Good
4	3,5324 – 4,00	88,31 – 100,00	A	Very good

With size indicators through the Public / Student Satisfaction Survey Elements include:

1. Requirements
2. Systems, Mechanisms, and Procedures
3. Settlement Time
4. Fees / Rates *)
5. Product Specifications Service Type
6. Implementing Competencies **)
7. Implementing Behavior **)
8. Handling Complaints, Suggestions, and Feedback
9. Facilities and infrastructure

5. Results and Discussion

Based on the assessment of the elements of Student Satisfaction in the Unit (ASA) One-Roof Application at the Faculty of Social and Political Sciences, University of North Sumatra include, among others:

5.1 Requiremens

Requirements are conditions that must be met in the maintenance of a type of service, both technical and administrative requirements. From the analysis results obtained an average score of 3.07 is in the interval score of 3.0644 - 3.532 categories "GOOD". Thus student satisfaction from the requirements element is in the "GOOD" category. The results of the requirements elements are presented in the following table.

Table 3. Requirements

No	Answer	Score	Frequency		
			F	f.x	%
1.	Very Good	4	74	296	38
2.	Good	3	78	234	39
3.	Less good	2	29	58	15
4.	Not good	1	15	15	8
	Total		196	603	100
				603/196=3,07	

In the initial requirements for registering through:

1. Students can register to the ASA Unit (One-Roof Application) to get a username and password to enter the One-Roof Application service website at https://asa-fisip.usu.ac.id/
2. After a student using the One Roof Application Login service on the student website will be given two choices of correspondence services, namely the Application Form in which to make a request for making a letter, it is better to understand the following sections:
   a. ACADEMIC
      Print Letters Related to Academic Areas
   b. STUDENTS
      Printing Letters relating to Student Activities. With the following instructions:
      1. Select the type of letter to be requested

2. Fill in the Data Form according to the Type of Letter
3. Click Send Then wait for the next notification
4. Check the status of your letter, whether pending, accepted, or rejected
5. If pending, then the letter has not been printed
6. If rejected, please make a new application, maybe your data is incomplete/appropriate
7. If accepted, save / print receipt then take the letter to the Administration.

5.2 Systems, Mechanisms, and Procedures

The procedure is a standardized service procedure for service providers and recipients, including complaints. From the results of the analysis obtained an average score of 3.12 in the interval score of 3.0644 - 3.532 in the category "GOOD". Thus the satisfaction of students in services from the elements of the System, Mechanism, and Procedure is in the category of "GOOD". The results of the requirements elements are presented in the following table.

Table 4. Systems, Mechanisms, and Procedures

No	Answer	Score	Frequency		
			F	f.x	%
1.	Very Good	4	85	340	43
2.	Good	3	73	219	37
3.	Less Good	2	22	42	12
4.	Not Good	1	15	15	8
	Total		196	612	100
			612/196=3,12		

Service systems, mechanisms and procedures are considered to be good and running according to the existing service systems and mechanisms. Systems, Mechanisms, and Procedures of administrative services needed by students who use services in (ASA) One-Roof Applications at the Faculty of Social and Political Sciences, University of North Sumatra engages in the system and the use of service procedures mechanism does not interfere with student activities both in the classroom and outside the classroom with a simple mechanism for students doing service rights activities.

5.3. Service Time

Service time is the period needed to complete the entire service process of each type of service. From the results of the analysis obtained an average score of 3.19 in the interval score of 3.0644 - 3.532 in the "GOOD" category. Thus the satisfaction of students receiving service (ASA) One-Stop Application at the Faculty of Social and Political Sciences, University of North Sumatra. the service time element is in the "GOOD" category, with the following table:

Table 5. Service Time

No	Answer	Score	Frequency		
			F	f.x	%
1.	Very Good	4	85	340	43
2.	Good	3	75	225	38
3.	Less Good	2	26	52	13
4.	Not Good	1	10	10	6
	Total		196	627	100
			612/196=3,19		

IOP Conf. Series: Materials Science and Engineering **851** (2020) 012036 doi:10.1088/1757-899X/851/1/012036

5.4 Fees / Rates

Cost / Rates is the fee charged to students through the One-Stop Application (ASA) service at the Faculty of Social and Political Sciences, the University of North Sumatra, which does not incur any costs so that the satisfaction of students assessing with the category of analysis results obtained a mean score of 3.70 at the interval of scores 3, 5324 - 4.00 the category "VERY GOOD".

Table 6. Costs / Rates

No	Answer	Score	Frequency		
			F	f.x	%
1.	Very good	4	135	540	69
2.	Good	3	61	183	31
3.	Less good	2	-	-	-
4.	Not good	1	-	-	-
	Total		196	723	100
			612/196=3,70		

5.5 Product Specifications Service Type

Product type service specifications are the results of services provided and received by established conditions and accepted by service recipient students (ASA) One-Roof Applications at the Faculty of Social and Political Sciences, University of North Sumatra is the result of each type of service specification. From the analysis results, the average score of 3.19 is in the interval score of 3.0644 - 3.532 in the "GOOD" category.

Table 7. Product Specifications Types of Services

No	Answer	Score	Frequency		
			F	f.x	%
1.	Very good	4	79	316	39
2.	Good	3	85	255	43
3.	Less good	2	24	48	14
4.	Not good	1	8	8	4
	Total		196	196	627
			612/196=3,19		

From the specifications of the types of services, there are two types namely

1. Academic

a. Submission of Special Exams
b. Application for Practicum / PKL
c. Active Requests to Study Again (AKK)
d. Resignation Request
e. Request for Late Payment of SPP
f. Request for Postponement of Academic Activity (PKA)
g. Application for Guiding Thesis / Final Project
h. Application for Submission of a Thesis Title
i. Pre-Research Requests
j. Application for Research Permit
k. Certificate of Value
l. Application for Taking Diplomas / Value Transcripts
m. Application for Internship
n. Application for Invitation for Proposal Seminar / Research Results

o. Application for Thesis / Final Project Invitation
p. Application Entitled to Take a Thesis / Comprehensive Examination / TA
q. Application for Proof of Thesis Submission
r. Application for Alumni Information
s. Application for publishing scientific papers/journals

2. Students

a. Description of Study Program Accreditation
b. Application for Active Students
c. Lost KRS / KHS / KTM
d. Permission Not Following Lectures
e. Active Lecture for Family Dependents
f. Request for Scholarship Recommendations

After the service is completed and within a predetermined timeframe, feedback from submitting service requests will be replied to the follow-up matter through the personal account of each student registered in the one-stop application with the following information:

(Name)
(No.Resi and remarks)

(receipt location)

5.6 Implementing Competencies

Implementing Competence is an ability that must be possessed by implementers including knowledge, expertise, skills, and experience. From the analysis results obtained an average score of 3.12 is in the interval score of 3.0644 - 3.532 which is the category of "GOOD". Thus student satisfaction with (ASA) One-Roof Applications at the Faculty of Social and Political Sciences, University of North Sumatra, with the element of implementing competence in the category of "GOOD".

Table 8. Competency of Implementers

No	Answer	Score	Frequency		
			F	f.x	%
1.	Very good	4	79	340	43
2.	Good	3	85	219	37
3.	Less good	2	24	42	12
4.	Not good	1	8	15	8
	Total		196	612	100
			612/196=3,12		

5.7 Implementing Behavior

Implementing Behavior is the attitude of officers in providing services. From the results of the analysis obtained an average score of 3.0644 - 3.532 which is the category "GOOD" so that the elements of implementing behavior are in a good category. The results of the answers to the elements of implementing behavior are presented in the following table.

Table 9. Implementing Behavior

No	Answer	Score	Frequency		
			F	f.x	%

1.	Very good	4	81	324	40
2.	Good	3	92	276	50
3.	Less good	2	13	26	8
4.	Not good	1	10	10	2
	Total		196	636	100
			612/196=3,24		

5.8 Address Complaints, Suggestions and Feedback
Handling complaints, suggestions and input is the procedure for handling complaints and following upon them. From the results of the analysis obtained an average score of 3.01 is in the interval score of 3.0644 - 3.532 which is the category of "GOOD".

Table 10. Complaints, Suggestions, and Feedback

No	Answer	Score	Frequency		
			F	f.x	%
1.	Very good	4	78	312	38
2.	Good	3	74	222	39
3.	Less good	2	29	58	15
4.	Not good	1	15	15	8
	Total		196	607	100
			603/196=3,08		

Student satisfaction with (ASA) One-Stop Application at the Faculty of Social and Political Sciences, the University of North Sumatra from Handling complaints, suggestions and input is facilitated by questions about suggestions in the account feedback after online services.

5.9 Facilities and Infrastructure

Facilities and Infrastructure are media that are used or used as a place for supporting services. From the analysis, results obtained a score of 3.26 in the interval score of 3.0644 - 3.532 the category "GOOD". Thus the satisfaction of students of service users in (ASA) One-Roof Applications at the Faculty of Social and Political Sciences, University of North Sumatra, with elements of Facilities and Infrastructure are in a good category. The results of answers to the community satisfaction index through elements of facilities and infrastructure are presented in the following table

Table 11. Facilities and Infrastructure

NO	Answer	score	Frequency		
			F	f.x	%
1.	Very good	4	78	312	40
2.	Good	3	97	291	50
3.	Less good	2	16	32	8
4.	Not good	1	5	5	2
	Total		196	640	100
			640/196 = 3,26		

From the discussion of the questionnaire to primary data totaling 196 questionnaires to see the results of the student satisfaction survey of service users in (ASA) One-Roof Applications at the Faculty of

Social and Political Sciences, the University of North Sumatra in the "GOOD" category (at intervals of 3.0644 - 3,532) Namely 3.21 with the above results, consisting of nine size indicators namely:

Table 12. Community Satisfaction Index of 9 Service Elements

No.	Elements	Average Score	Category
1	. Requirements	3.07	Good
2	Systems, mechanisms, and procedures	3.12	Good
3	Good Service Time	3.19	Good
4	Cost / Tariff	3.70	Good
5	Product Specifications Type Service	3.19	Good
6	Implementing Competencies	3.12	Good
7	Implementing Behavior	3.24	Good
8	Handling Complaints, Suggestions and published	3.08	Good
9	Facilities and Infrastructure	3.26	Good
	Average	28.97 / 9 = 3.21	Good

6. Conclusion

Service Quality (ASA) One-Stop Application at the Faculty of Social and Political Sciences, University of North Sumatra shows the GOOD category with Type B. Satisfied University of North Sumatra. Quality of Service (ASA) One-Stop Application at the Faculty of Social and Political Sciences, University of North Sumatra can be concluded that:

1. Quality of Service (ASA) One-Stop Application at the Faculty of Social and Political Sciences, University of North Sumatra requires several things that need to be considered by the implementer of the One-Stop Application (ASA) service program. Like Optimizing the Capacity and Quality of Human Resources (Human Resources) that serve employees and employees, staff / employees support the implementation of this procedure in order to maximize the error or error in the delivery of data in filing,
2. Consultation of instruments that support and support service on-site and post-reporting must be immediately followed up through suggestions and attitude responses to the faculty so that it becomes an excellent service.
3. Impact, services from (ASA) One-Roof Applications at the Faculty of Social and Political Sciences, University of North Sumatra, provide interaction between services so as to contribute to the development of human resources that are responsive to the development of technology and information.

Based on the results of the study, the following suggestions can be given:

1. This process can make an example for universities who want to build Good University Governance in terms of service governance that cuts the flow of time and energy so that it can be effective and efficient in services for students.
2. Supervision and development of professional work culture can be applied through the Roadmap.
3. This also proves not only about world education but transmits a good reflection to students who receive service, so there needs to be a minimum value required by inspiring service standards.

References

[1] Dwiyanto A 2012 Proceeding Seminar & Lokakarya Inovasi Penyelenggaraan Pelayanan Publik Di Daerah, Fakultas Ilmu Sosial dan Ilmu Politik Universitas Gadjah Mada., Yogyakarta.
[2] Indrajit R 2016. Electronic Government. Jakarta: APTIKOM.
[3] Shattock, M 2006 Managing and Good Governance in Higher Education. Maidenhead, Berkshire: Open University Pres

[4] Sugiyono. 2012. Metode Penelitian Kuantitatif Kualitatif R&B. Bandung: Aflabeta.
[5] Wasistiono S 2012 Manajemen Pemerintahan Daerah. Bandung: Alqa Print
[6] Rahmat R F, Dennis, Sitompul O S, Purnamawati S, Budiarto R, 2019 Advertisement billboard detection and geotagging system with inductive transfer learning in deep convolutional neural network *Telkomnika* **17**(5), pp 2659-2666.
[7] Yang T M, Lo J, and Shiang J 2015 To open or not to open? Determinants of open government data. *Journal of Information Science*, 00(0), 1-17. doi:10.1177/0165551515586715

IOP Conf. Series: Materials Science and Engineering **851** (2020) 012037 doi:10.1088/1757-899X/851/1/012037

Facilities layout design for vise manufacturing using Blocplan

Muhammad Haikal Sitepu*, Tania Alda, Meilita Tryana Sembiring, Andri Nasution, Nurul Nadhirah Ayu and Muhammad Rizky Zein,

Industrial Engineering Department, Faculty of Engineering, Universitas Sumatera Utara, Medan, Indonesia, 20155

*Email: mhd_haikalkarana@usu.ac.id

Abstract. Production floor requires several supporting facilities such as warehouse, office, labor locker and shipping/receiving area. These facilities are used to do activities other than production such as storing raw materials, rest for labor and other administration works. These facilities must be designed around the production floor. Design facilities layout becomes important for production since it can influence the material and human flow between facilities and production floor. As the result the arrangement of supporting facilities is necessary to make production activities run smoothly. Mistake in arrangement of supporting facilities might increase material and human movements which in turn increase material handling costs. Vise manufacturing requires supporting facilities to support the production of vise. The purpose of this study is to demonstrate the use of Blocplan in arranging supporting facilities for vise manufacturing.

1. Introduction

In the industrial world, company productivity is influenced by the arrangement of production and supporting facilities. Material and human flow within the company is mainly depend on the arrangements of production and supporting facilities. As the result, to achieve good productivity, the company should have good production and supporting facilities layout. Different factors are required to consider in designing facilities layout such as interrelationship between facilities such as material flow and human flow between facilities, as well as hazard produced by the facilities such as noise and bad smells.

To consider these factors in designing supporting facilities layout, several approaches are required. BlocPlan is one approach available in literature to support the design of facilities layout. Block plan use qualitative and quantitative data in designing facilities layout [1][2]. Qualitative data is obtained from the opinion of expert about the importance of facilities to be brought near or far. Then, quantitative approach is used to convert the closeness opinion into facilities layout with minimum movements.

Vise is a support tool for many manufacturing processes mainly for finishing process such as cutting, drilling and grinding. Vise consists of several parts that are made from different types of materials and are needed different manufacturing processes. Similar with other manufacturing, to produce vise, several supporting facilities are required. Production floor requires several supporting facilities such as warehouse, office, labor locker and shipping/receiving area. These facilities are used to do activities other than production storing raw materials, rest for labor and other administration works. These facilities must be designed around the production floor. Design facilities layout becomes important for production since it can influence the material and human flow between facilities and production floor [3]. For example, warehouse facility is used to store raw material and finished goods from production. The size of warehouse mainly depends on the quantity of inventory. This quantity is affected by several factors such as holding cost, ordering cost and delivery time [4]. As the result the arrangement of supporting facilities is necessary to make production activities run smoothly. Mistake in arrangement of supporting facilities might increase material and human movements which in turn increase material handling costs [5]. This paper focuses to design facilities layout for vise manufacturing by using BlocPlan.

2. Methodology

This research uses experiment as research methodology. Experiment is good to use when the investigation focuses to evaluate the impact of alternatives into system performance. In this research, the impact of facilities layout design on material flow is evaluated. The experiment was conducted at Laboratory of Facility Layout and Material Handling, Faculty of Engineering, Universitas Sumatera Utara.

2.1. Research Process

This research is divided into three stages including identification facilities required to support vise manufacturing, identification the degree of closeness between facilities and arrangement the facilities position in the layout. In the first stage, the research focuses to identify the facilities required to support vise manufacturing. Literature review is used to identify essential facilities for vise manufacturing. Then, this is continued by identification the degree of closeness between facilities. Activity Relationship Chart (ARC) is used to identify the degree of closeness. ARC identify the closeness between facilities based on relationship between facilities such as material flow or human flow and based on the hazard triggered by facilities such as noise and bad smells. In the third stage, the research focuses to arrange the facilities in the layout by using the degree of closeness from stage 2. In this stage, Blocplan is used as tools to support facilities layout design. The research process can be observed in figure 1.

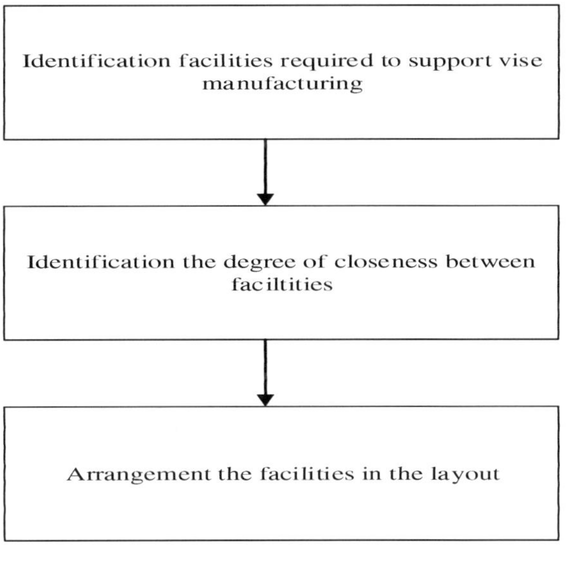

Figure 1. Research process

3. Result and Discussion

3.1. Facilities and degree of closeness between facilities

There are five types of facilities identified including production area, production service, personal service, general service and physical plant service [4]. Production service means that the facilities

IOP Conf. Series: Materials Science and Engineering **851** (2020) 012037 doi:10.1088/1757-899X/851/1/012037

required to support the production process. There are 4 facilities identified under production service including raw material warehouse, finished good warehouse, receiving area and shipping area. Personal service means that the facilities required to support the labor/ employee activities. There 4 facilities recognized under personal service including locker room/ toilet for employee, common room/ canteen/ rest area, Emergency room, and prayer room. General service is the facilities required for general purposes. There facilities have been identified under general service including office for administrative work, parking and security post. Furthermore, physical plant service is the facilities to support energy and maintenance for building and production. Two facilities have been identified under physical plant service including generator room and maintenance room.
After the facilities have been identified, the degree of closeness between these facilities are provided. Activity relationship chart were used to define the degree of closeness. This was identified based on the survey to students in Industrial Engineering Department, Faculty of Engineering, Universitas Sumatera Utara. Figure 2 shows the degree of closeness.

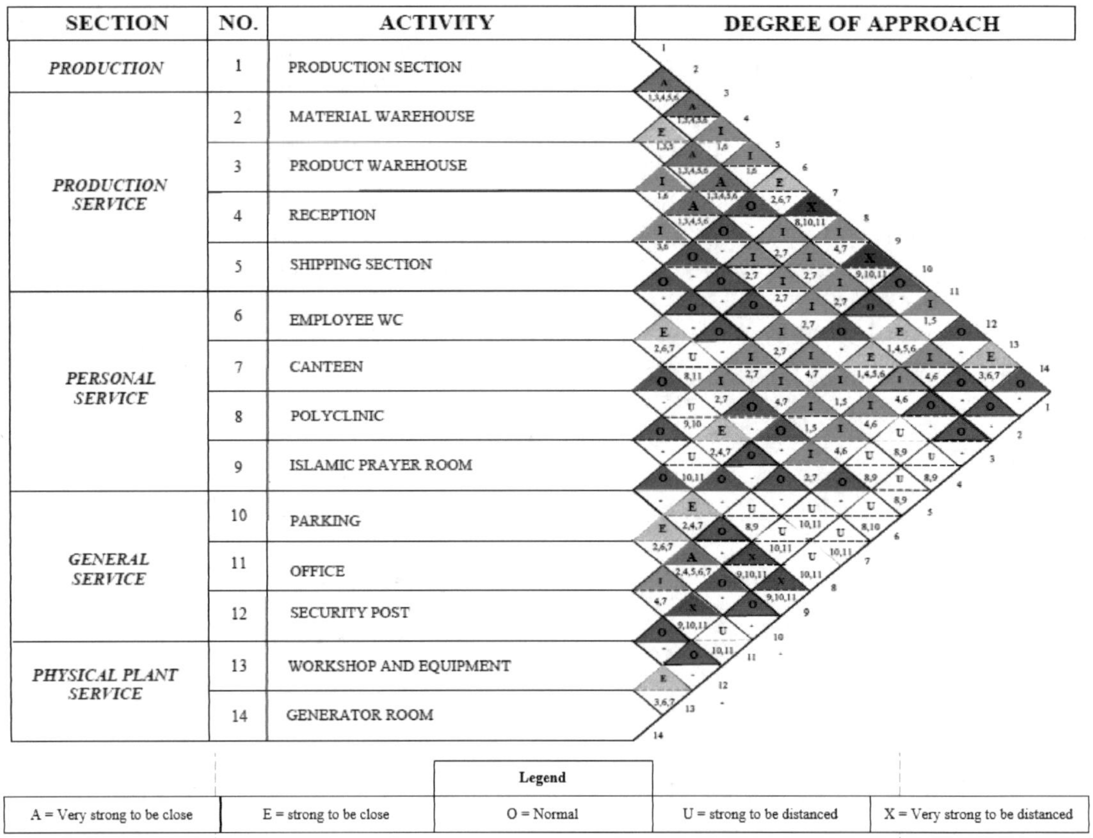

Figure 2. Degree of closeness

3.2. Facilities Layout Design

Blocplan software is used to support facilities layout design. It started by inputting total area for every facility and degree of closeness between facilities. This is shown in Figure 3 (a) and (b). Then, Blocplan changes the degree of closeness into numerical value. For example, degree closeness A is valued with 10 points, degree closeness E is valued with 5 points, and degree closeness X is valued with -10 points. Blocplan gives opportunity to change the value for degree closeness. Based on these value, Blocplan calculates the total score for every facility. Total score is calculated from multiplying facility's degree of closeness with the value. After this, Blocplan generates alternative layouts which have adjacency score, R-score and R-dist. The best alternative layout is the layout with highest adjacency score, highest R-score and lowest R-dist [6]. Alternative layout 19 is selected due to have highest adjacency and R-score. The Blocplan result and analysis can be observed in Figure 4.

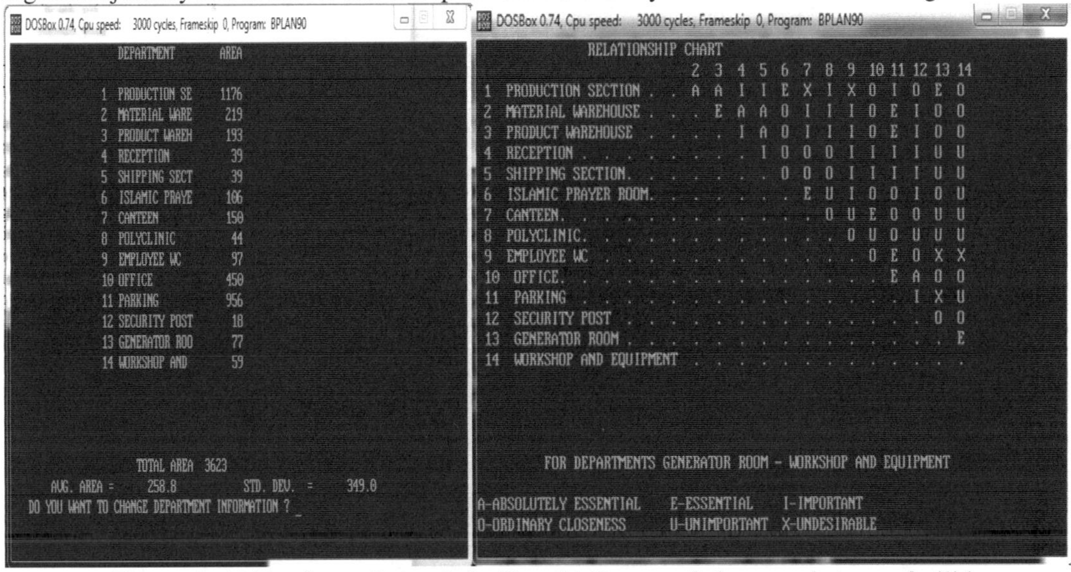

Figure 3. (a) Total area for each facility, (b) Degree of closeness between facilities

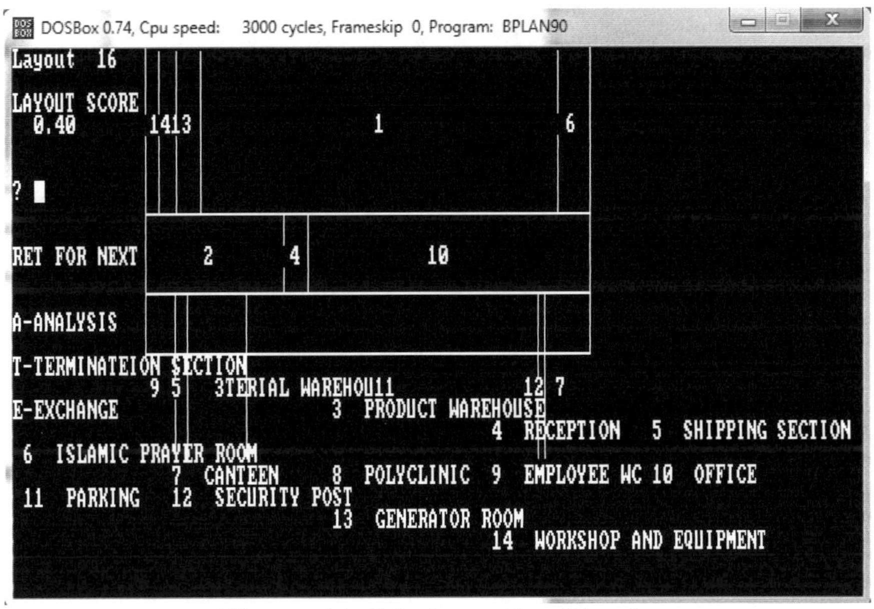

Figure 4. Facilities layout from Blocplan

The Blocplan software has certain limitations. Some of limitations are the layout produced by Blocplan is not reflected the real size of facilities and there is no space between facilities in the proposed layout. The final layout that follows the real size of facilities and bases on position produced by *BLOCPLAN Software* is shown by Figure 5 and 6.

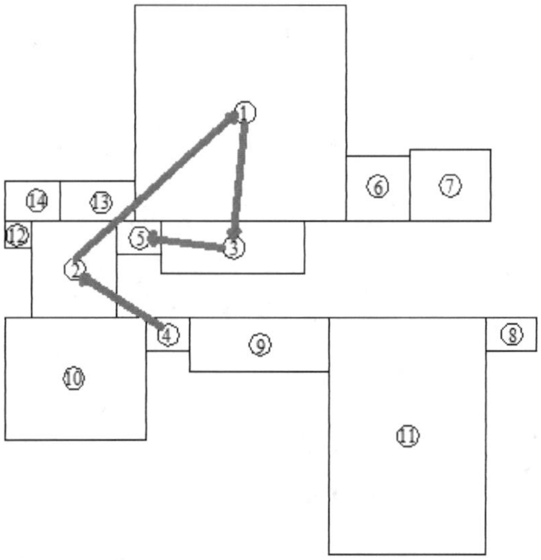

Figure 5. Area Allocation Diagram

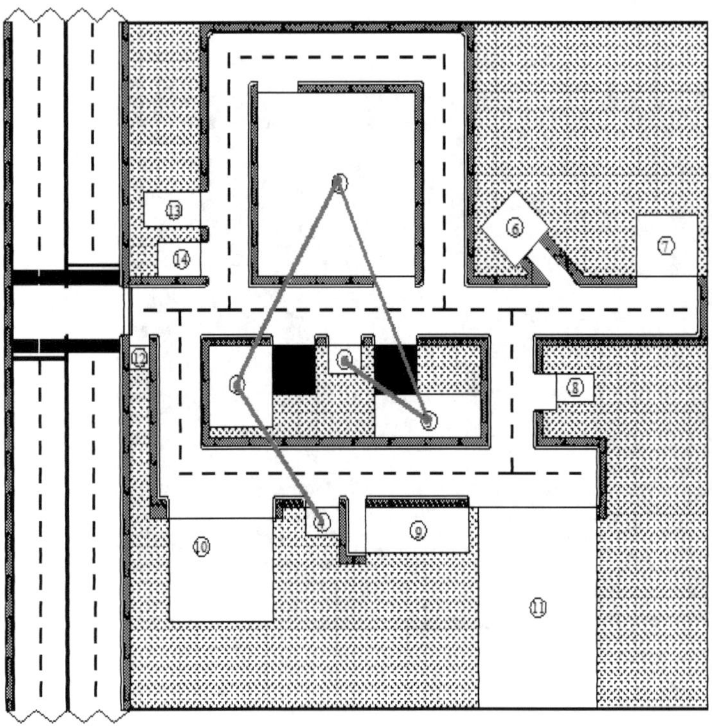

Legend	
1	Production
2	Raw Material Warehouse
3	Finished Goods Warehouse
4	Receiving area
5	Shipping area
6	Prayer room
7	Restroom/canteen
8	Emergency room/Clinic
9	Locker room/Toilet
10	Office
11	Parking
12	Security post
13	Generator room
14	Maintenance room

Figure 6. Block Layout of facilities for vise manufacturing

4. Conclusion

This paper demonstrates the use of Blocplan to design facilities layout for vise manufacturing. Based on the results of data processing using Blocplan software, there are 20 alternative facilities layouts,

Alternative 19 is selected as best layout due to the largest adjacency and *r-score* with a value of 0.74. The main limitation of Blocplan software lies on subjectivity on determining the degree of closeness. Further research can be focused on reducing the subjectivity by using different approach in determining the degree of closeness.

References

[1] Donaghey C 2008 Developing Facility Layouts with BLOCPLAN for Windows University of Houston Texas
[2] Heragu S S 2006 Facilities Design Second Edition University of Louisville United States of America
[3] Tompkins and White J A 1996 Facilities Planning 2nd ed. John Wiley and Sons Canada
[4] Sitepu M H, Situmorang D P, and Sembiring M T 2019 Inventory Planning at Painting Section in the Indonesian Automotive Industry *IOP Conf. Series: Materials Science and Engineering*
[5] Puspita I A, Iqbal M, Pratami D, and Pratomo A 2015 Production Facility Layout Design Using Blocplan Algorithm *Advanced Science Letters*
[6] Vaidya R. D., Shende, P. N., N. A. Ansari and S.M. Sorte 2013 Analysis Plant Layout for Effective Production *International Journal of Engineering and Advanced Technology (IJEAT)*. 2(3): 500-504
[7] Siregar I, Muchtar M A, Rahmat R F, Andayani U, Nasution T H, and Sari R M 2018 Method of calculation overall equipment effectiveness in fertilizer factory. *IOP Conference Series: Materials Science and Engineering* **308**(1), p. 012053)

IOP Conf. Series: Materials Science and Engineering **851** (2020) 012038 doi:10.1088/1757-899X/851/1/012038

Process and material flow design for vise manufacturing with routing sheet and from to chart

Muhammad Haikal Sitepu*[1], Alison McKay[2], Tania Alda[1], Talitha Almira Muharrami[1], Yohannes Tulus Martinez[1]

[1]Industrial Engineering Department, Faculty of Engineering, Universitas Sumatera Utara, Medan, Indonesia, 20155
[2]Institute Design, Robotic and Optimization, School of Mechanical Engineering, University of Leeds

*Email: mhd_haikalkarana@usu.ac.id

Abstract. Process design is important steps in production layout planning. At this step, processes and machines for making parts and products are identified. Then these processes and machines are arranged so the flow of materials can be minimized. Vise is a support tool for various manufacturing processes such as cutting, drilling and grinding. Vise consist of various parts such as platform base, body and clamping parts. To make these various parts several manufacturing processes are required. This paper aims to demonstrate the use of routing sheet to identify processes and machines to manufacture vise as well as to show the use of from to chart to arrange those machines in order to minimize material flows. Several machines configurations with minimum material flows are presented.

1. Introduction

Vise is an important support tool for manufacturing process particularly for finishing process. Vise is used to hold an item during the manufacturing process, for example during the cutting process, vise is beneficial to make the item motionless. Vise consists of several parts such as platform, clamping body and clamp jaw. Different manufacturing processes are required to manufacture each part of vise. Hence, these processes need to consider in designing production layout for vise manufacturing. This paper has an aim to demonstrate the process and material flow design for vise manufacturing

Process design focuses on configuring stage of process to manufacture a product. Routing sheet is one tool to support process design. Routing sheet accumulates information from production floor related to the process required to manufacture part of product. There are some information appeared in the routing sheet, namely: (a) The type of process needed for each part of product, (b) The processing times and (c) raw materials needed. These information are essential to support the arrangement of machines and to calculate the production capacity [1].

After manufacturing processes for each part identified, the arrangement of machines is performed. This step is critical in production layout planning because it determines the flow of materials in production floor [2]. Error in the arrangement of machine might increase the flow of materials which in turn incrementing material handling cost. From to chart is one support tool for setting the sequence of machines in production floor. From-to-chart is used to evaluate the initial sequence of machines by calculating the total movement of materials. This paper focuses on evaluating the proposed machines sequences using from to chart in order to determine better sequences with lower material movements [3].

2. Methodology

This research used experiment methodology to design the process and material flow for vise manufacturing. The experiment was performed in Laboratory of Facility Planning and material handling, Engineering Faculty, Universitas Sumatera Utara.

2.1. Research Process

This research is divided into four critical steps including manufacture the vise, develop the routing sheet, develop alternative of machine sequences and evaluate those alternative using from to chart. In the first step, the observation of the vise manufacturing process was conducted. At this step, manufacturing process and processing times for each part of vise were recording. This was followed by developing routing sheet by considering the data from observation. At this step, processes for each part of vise were ordering. Then based on the information from routing sheet, the alternative of machine sequences were developed. At the final step, alternative of machine sequences were evaluated using from to chart. The evaluation was implemented by considering quantity of part, quantity of part movements and weight of parts [4].

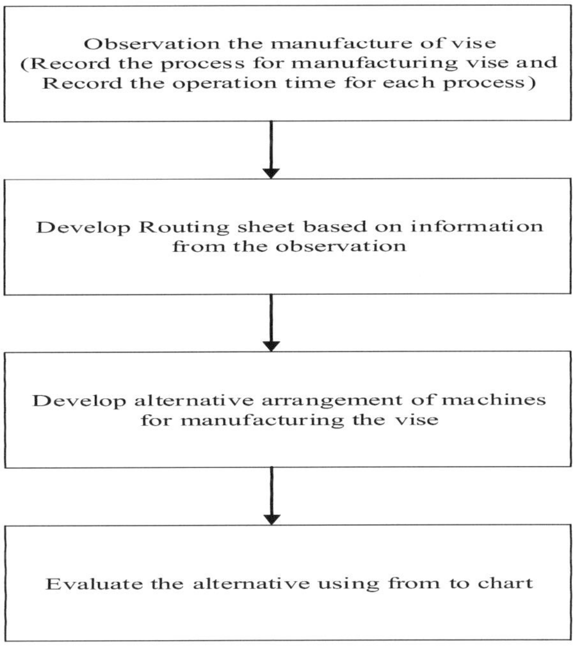

Figure 1. Research Process

3. Result and Discussion

3.1. Routing Sheet

Routing sheet is produced based on observation data. The observation was conducted on Laboratory of manufacturing process, Faculty of Engineering, Universitas Sumatera Utara. The main work in developing routing sheet is to determine the ordering process for making each part of vise. The following table displays routing sheet for part platform, clamp jaw and clamping body (right and left).

Table 1. Routing Sheets for Part of Vise

	Platform		**Left Clamping Body**		**Left Clamp Jaw**		**Right Clamping Body**	
No	**Machine Type**	**Operation Time (Seconds)**	**Machine Type**	**Operation Time (Seconds)**	**Machine Type**	**Operation Time (Seconds)**	**Machine Type**	**Operation Time (Seconds)**
1.	Grinding machine	600	Shredder	240	Grinding machine	300	Shredder	240
2.	Drilling Machines	360	Shredder	180	Drilling Machines	240	Shredder	180
3.	Drilling Machines	360	Shredder	180	Drilling Machines	240	Shredder	180
4.	Shredder	240	Drilling Machines	240	Shredder	120	Drilling Machines	300
5.	Shredder	240	Drilling Machines	240	Grinding machine	240	Drilling Machines	300
6.	Drilling Machines	240	Drilling Machines	240			Drilling Machines	360
7.	Drilling Machines	240	Drilling Machines	240			Tap and Dies	240
8.	Milling Machine	420	Tap and Dies	240			Tap and Dies	240
9.			Tap and Dies	240			Grinding machine	300
10.			Tap and Dies	240				
11.			Tap and Dies	240				
12.			Grinding machine	300				

3.2. Alternative of Machines Sequences

Based on information from routing sheet which is the order of process for each part, the alternative of machines sequences were developed. The following table shows the alternative of machine sequences.

Table 2. Alternative of Machines Sequences

Sequence Number	Alternative 1	Alternative 2	Symbol Meaning
1	PM	PM	PM = Entrance
2	G.	S.	G = Grinding Machine
3	S.	D	S = Scrap Machine
4	M	TD	M = Milling Machine
5	D	M	D = Drilling Machine
6	B	B	B = Lathe Machine
7	TD	G.	TD = Tap and Dies Machine
8	Assembly	Assembly	Assembly = Work Centre to assembly
9	PK	PK	PK = Exit

3.3. From to Chart

From to chart was used to evaluate the alternative of machines sequences. Two factors were used in evaluation including weight of part and quantity of part. To consider these factors, relative importance of part is calculated by diving total weight of vise with part with smallest weight. Equation 1 show the formulation. Table 3 shows the weight of part, quantity of part and relative importance of part.

$$Relative\ importance = \frac{Part\ weight * Quantity}{Smallest\ Total\ Part\ Weight} \quad (1)$$

Table 3. Calculation of relative importance

Part	Percentage	Part Weight	Quantity / Day	Total Weight / Part	Relative Importance
Base	40%	6	52	312	8
Left Clamping Body	10%	1.5	52	78	2
Left Clamp Jaw	5%	0.75	52	39	1
Right Clamp Jaw	5%	0.75	52	39	1
Right Clamping Body	10%	1.5	52	78	2
Rotating Board	14%	2,1	52	109.2	2.8
Lava Body	9%	1.35	52	70.2	1.8
Handle	7%	1.05	52	54.6	1,4

From to chart is developed by identifying the movement of each parts (from routing sheet). For example the movement from PM-G (entrance to grinding) are experienced by several parts including RI base, left and right clamp jaw. As the result, in from to chart, the colom PM-G is filled with 10 which is accumulated from 8 (relative importance from RI base), 1 (relative importance from left clamp jaw) and 1 (relative importance from right clamp jaw). Table 4 displays from to chart for alternative 1.

Table 4. From to chart for alternative 1

	PM	G.	S.	M	D	B	TD	Assembly	PK	Total
PM	0	10	6.8	0	0	3,2	0	0	0	20
G.	0	0	1	0	8	0	0	12	0	21
S.	0	1	0	0	15.8	0	0	0	0	16.8
M	0	0	0	0	0	0	2.8	8	0	10.8
D	0	2.8	9	10.8	0	0	4	0	0	26.6
B	0	1,4	0	0	1.8	0	0	0	0	3,2
TD	0	6.8	0	0	0	0	0	0	0	6.8
Assembly	0	0	0	0	0	0	0	0	0	0
PK	0	22	16.8	10.8	25.6	3,2	6.8	20	0	105.2
									Total	210.4

Then penalty point needs to assign to from to chart. Penalty point reflects number of processes or machines that is stepped over. For example, from PM to S (Entrance to scarp machine), there is one machine (grinding machine) that is stepped over. Penalty point is calculated by multiplying relative importance with penalty point and number of machine that is stepped over. For example, for PM to S total value is 13.6 which is accumulated by multiplying 6.8 with 2

Table 5. Penalty Point for Alternative 1

	PM	G.	S.	M	D	B	TD	Assembly	PK	Total
PM	0	10	13.6	0	0	16	0	0	0	39.6
G.	0	0	1	0	24	0	0	72	0	97
S.	0	2	0	0	31.6	0	0	0	0	33.6
M	0	0	0	0	0	0	8.4	32	0	40.4
D	0	16.8	36	21.6	0	0	8	0	0	82.4
B	0	11.2	0	0	3,6	0	0	0	0	14.8
TD	0	68	0	0	0	0	0	0	0	68
Assembly	0	0	0	0	0	0	0	0	0	0
PK	0	108	50.6	21.6	59.2	16	16.4	104	0	375,8
									Total	751.6

Using the similar steps, alternative 2 was evaluated using from to chart. Table 6 and 7 show the initial from to chart and penalty point for alternative 2.

Table 6. From to Chart for Alternative 2

	PM	S.	D	TD	M	B	G.	Assembly	PK	Total
PM	0	6.8	0	0	0	3,2	10	0	0	20
S.	0	0	15.8	0	0	0	1	0	0	16.8
D	0	9	0	4	10.8	0	2.8	0	0	26.6
TD	0	0	0	0	0	0	6.8	0	0	6.8
M	0	0	0	2.8	0	0	0	8	0	10.8
B	0	0	1.8	0	0	0	1,4	0	0	3,2
G.	0	0	9	0	0	0	0	12	0	21
Assembly	0	0	0	0	0	0	0	0	0	0
PK	0	15.8	26.6	6.8	10.8	3,2	22	20	0	105.2
									Total	210.4

Table 7. Penalty Point for Alternative 2

	PM	G.	S.	M	D	B	TD	Assembly	PK	Total
PM	0	6.8	0	0	0	16	60	0	0	82.8
G.	0	0	15.8	0	0	0	5	0	0	20.8
S.	0	18	0	4	21.6	0	11.2	0	0	54.8
M	0	0	0	0	0	0	20.4	0	0	20.4
D	0	0	0	5.6	0	0	0	24	0	29.6
B	0	0	10.8	0	0	0	1,4	0	0	12.2
TD	0	0	72	0	0	0	0	12	0	84
Assembly	0	0	0	0	0	0	0	0	0	0
PK	0	24.8	98.6	9.6	21.6	16	98	36	0	304.6
									Total	609.2

Based on the results of calculations, alternative 1 produces penalty point 751.6 while the penalty point for alternative 2 is 609.2. Based on this result, alternative 2 is selected as machine sequence due to lower penalty points.

4. Conclusion

This paper demonstrates the use of routing sheet to support process design and presents the use of from to chart to evaluate alternative of machines sequences. For product that consist of several parts and those parts required different manufacturing process, to arrange the machines in production floor requires further approach. Several factors such as weight and quantity of parts can be used as consideration in configuring the sequence of machines. However, the use of from to chart mainly depends on initial alternative for machines sequences. Further approach is required by considering more factors such as social aspects (safety of worker) and environmental aspects (emission) in the arrangement of machines.

References

[1] Setiabudi Y, et al 2018 ATV 12 Production Capacity Planning Using the Rough Cut Capacity Planning (RCCP) Method To Know Production Optimization Points (Case Study at PT Schneider Electric Manufacturing Batam). P-ISSN 2301-7244

[2] Rantung A M, et al 2018 Proposed Improvement of Production Floor Layout of PT. Superior Engineering Porters to Minimize Material Handling Costs and Production Time with Pairwise Exchange and Simulation Methods. ISSN: 2622-5131

[3] Winarno H 2015 Layout Analysis of Spatial Facilities at the Faculty of Engineering, University of Serang Raya Using the Activity Relationship Chart (Arc) Method. ISSN: 2407-1846

[4] Tanjung M 2017 Organizational Functions in Project Management *Penusa Mantik Journal* 1 (1) E-ISS: 2580-979741, p-ISSN: 2088-3943

[5] Siregar I, Muchtar M A, Rahmat R F, Andayani U, Nasution T H, and Sari R M 2018 Method of calculation overall equipment effectiveness in fertilizer factory. *IOP Conference Series: Materials Science and Engineering* **308**(1), p. 012053)

IOP Conf. Series: Materials Science and Engineering **851** (2020) 012039 doi:10.1088/1757-899X/851/1/012039

The effect of stage number on the performance of a vapor compression refrigeration cycle using refrigerant R32

A H Nasution[1], H Ambarita[1], H. V. Sihombing[1], E Y Setiawan[2], H. Kawai[3]

[1]Sustainable Energy and Biomaterial Centre of Excellent, Universitas Sumatera Utara, Jl. Almamater Kampus USU, Medan 20155, Indonesia
[2]Mechanical Engineering, National Institute of Technology Malang, Jl. Bendungan Sigura-gura No. 2 Malang 65145, Indonesia
[3]Mechanical Engineering Department, Muroran Institute of Technology, 27-1 Mizumoto, Muroran 8585, Hokkaido, Japan

Email: himsar@usu.ac.id

Abstract. In the refrigerant cycle of Air-Conditioning (AC) unit the mainly used cycle is vapor compression cycle. The performance of the refrigerant, named as Coefficient of Performance (COP), need to be improved by using modification. The modification by using multi stage is one of the potential solutions. In the present study, the effect of number of stage to the performance will be explored. A simple vapor compression cycle that is typically used in the AC unit is taken into study. IN the cycle the refrigerant is difluoromethane R32. Three different stages are examined, they are single stage, two-stages, and three stages. The numerical analyses are carried out using commercial Aspen One software. The COP, mass flow rate of the refrigerant and compressor power will be discussed. The result shows that increasing the number of stages related to increasing the COP.

1. Introduction

Climate change becomes a hot issue today. One of that was the effect of greenhouse gas (GHG). now, most of the people in the world consent to reduce GHG. In order to get a convenient air in a room, Air Conditioning (AC) unit is usually used. The AC is employed to maintain the temperature and humidity of the room. Due to increasing standard living, the AC system consumes high electrical energy. Since the energy is mainly resulted from fossil fuel the air conditioning system has significant impact on GHGs emission. Strategies and innovation to improve the performance of vapor compression cycle are extremely needed. There are several studies related to enhance energy efficiency of a vapor compression cycle such as using new refrigerant, inter-stages cycle, internal heat exchanger, etc. [1, 2]

In a single stage system, as compression ratio in the system increase. A high value of temperature drastically reduces the efficiency of the compressor and causing an additional load on both the condenser and compressor. One of the effective solution to these problems a inter stage system is used [3]. In the present work, we focus on the cycle comparison when it is operated as single stage, two-stages, and three-stages. Sihombing, et al. [4] has reported their study on investigation the effect of the multi-stage vapor compression cycle. It was shown the difference between performance single-stage and two-stages, it increases by 8.8 % than single-stage. Shuxue and Guoyuan [5] investigated the cooling performance of R32 two-stage vapor compression system with vapor injection for domestic air-conditioner. The results show that, the refrigerant R32 vapor injection system has a significant performance enhancement for cooling performance. The Coefficient of Performance of the two stage compression system can improve 5-15 % and 10-12 %, respectively. Nilesh et al [6] studied influence of inter stage pressure on the performance of two-stage refrigeration cycle using inter cooler. There were 6 refrigerants examined. They are R22, R134a and R134a as synthetic refrigerants and propane, carbon dioxide and nitrous oxide as natural refrigerants.

Cordin et al. [7] review effect of multi-temperature as heat pump. Their work review the highlights of the major benefits and challenges of mechanically driven heat pumps and refrigeration systems. Using thermodynamics simulations, they reveal that multi-stages compressor cycle have the biggest COP. The Coefficient of Performance of the multi-stages compressor cycles is on average 8 % bigger than single stage. Mazzei and Palombo [8] analyzed the potential using of multi-stage integrated evaporative HVAC systems. Energy saving are evaluated hourly, with the mass flow rate of the supply air being kept constant. Murat [9] compare the single-stage and cascade vapour-compression refrigeration systems using R134a as the refrigerant. The results show that, for the cascade system provides a lower evaporating temperature, lower compressor discharge temperature, lower ratio of discharge to suction pressure and higher compressor volumetric efficiency. Liu et al. [10] investigated the two-stage compression heat pump water heater system. In this paper showed that the two-stage rotary compressor can provide much better energy performance than the conventional heat pump water heater.

Those studies showed that the vapor compression refrigeration cycle has come under scrutiny. Thus, many of innovations have been proposed by researchers to increase the COP. In this study we propose to explore the characteristics of the cycles with different number of stages. They are single-stage, two-stages, and three-stages vapor compression refrigeration cycle with refrigerant R32. The commercial simulation code of ASPEN PLUS is used to carry out the analysis. The results are expected to supply the necessary information to develop high efficient vapor compression cycle for AC.

2. Method of solution

In this section the governing equations are explained. For convenient only the two-stage cycle is explained. The rate of energy in the first compressor ($\dot{W}_{c1}$) and the rate of energy of the second compressors ($\dot{W}_{c2}$) are calculated, respectively, by

$$\dot{W}_{c1} = \dot{m}_1 (h_2 - h_1) \quad (1)$$

$$\dot{W}_{c2} = \dot{m}_2 (h_4 - h_3) \quad (2)$$

In those equations the parameter h_1 and h_2 are named as enthalpy of the refrigerant at the inlet and the exit of first compressor. Furthermore, the parameter h_3 and h_4 are the enthalpy at the inlet and the exit of the second compressor, respectively. The total of the energy rate to the cycle is given by

$$\dot{W}_{tot} = \dot{W}_{c1} + \dot{W}_{c2} \quad (3)$$

The rate of exhaust heat by the system to ambient can be calculated by equation (4).

$$Q_c = m(h_4 - h_5) \quad (4)$$

In this equation h_4 and h_5 are enthalpy of the refrigerant at the inlet and the exit of the condenser. Here m is defined as the flow rate of the refrigerant entering the condenser. The rate of heat drawn by evaporator from the system can be calculate by the below equation.

$$Q_e = m(h_1 - h_9) \quad (5)$$

The next parameter is the effect of refrigeration (*ER*) which can be calculated by equation (6).

$$ER = h_1 - h_9 \quad (6)$$

Finally, the coefficient of performance of the vapor compression cycle (COP) is calculated by

$$COP = \frac{Q_e}{W_{tot}} \quad (7)$$

IOP Conf. Series: Materials Science and Engineering **851** (2020) 012039 doi:10.1088/1757-899X/851/1/012039

3. Results and Discussions

3.1. Description of the system

In order to perform the simulation, the commercial software of ASPEN Plus is carried out. The thermophysical properties of the refrigerant at every condition is estimated using REFROP which is embedded in the software. The used refrigerant here is R32. For all cases the evaporation temperature and condensation temperature is -10°C and 40°C, respectively. The cycle will be loaded with a cooling load of 1000 Ton of Refrigeration.

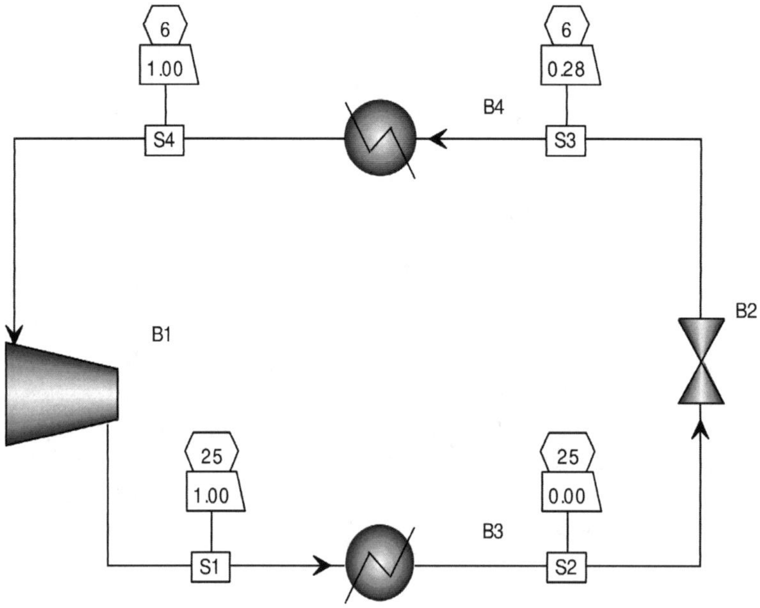

Figure 1. Single stage vapour compression cycle in Aspen-Plus

As a baseline as single stage vapour compression cycle is analysed. It consists an evaporator, a compressor, a condenser and an expansion valve. The schematic diagram of the single stage cycle in the Aspen is shown in Figure 1. In addition, the two-stage vapour compression refrigeration cycle is shown in Figure 2. The main components of the two-stage cycle are two compressors, a condenser, two expansion valves, and evaporator. The additional components in the two-cycles are a flash cooler and a mixing chamber. The position of all components in the cycle is shown in Figure 2. The liquid refrigerant from the condenser will be divided into liquid and vapour in the flash cooler. The liquid will be sent to the evaporator. On the other hand the vapour refrigerant will be sent to mixing chamber. The refrigerant from evaporator will compressed using first stage compressor. This refrigerant is in the vapour form and it will be collected with the refrigerant from the flash cooler in the mixing chamber. The total refrigerant from the mixing chamber will be compressed by the second stage compressor. This figure reveals that not of the refrigerant flow into the evaporator. This will affect the refrigerant effect of the system.

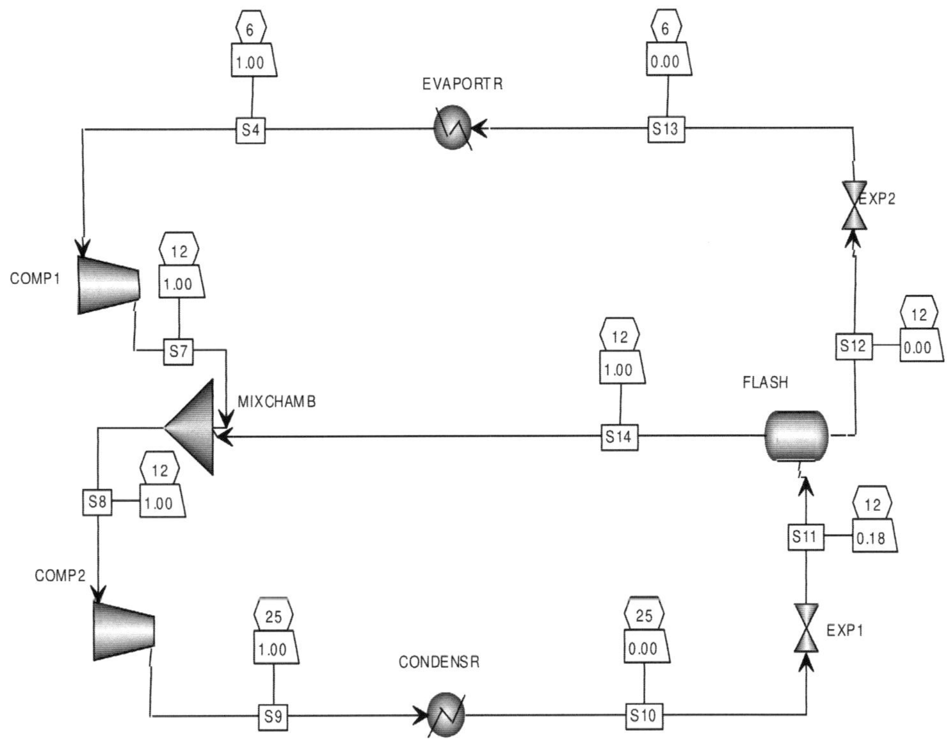

Figure 2. Two-stage vapour compression cycle in Aspen-Plus

Figure 3 shows the schematic diagram of the three-stage vapour compression cycle. It consists of three compressors, an evaporator, a condenser and two expansion valves. The additional components are two mixing chambers and two flash chamber. In the analysis the conservation of energy and conservation of mass will be employed in every condition.

3.2. Simulation result

The simulation for all cases have been carried out and the results are shown in Table 1. The parameter shown in the table are suction pressure, discharge pressure, compressor power, heat duty, the total mass flow rate, temperature before expansion and temperature after expansion. In the suction pressure, all of the lowest compressor of multi stages shows the similar value with the single stage. On the other hand, the discharge pressure of the upper compressor of the multi stages are similar. Only the intermediate pressure varies depend on the number of the stages.

The compressor power of the singe stage is 1048 kW. In the two-stage cycle the power in the first and the second compressor are 403.6 kW and 552.1 kW, respective. In total the total power in both compressors is 955.7 kW. This fact reveals that the total compressor power in the two-stage vapour compression cycle is lower than in the single stage. In the other word, the compressor power in the two-stage is lower 8.8% in comparison with single stage. Furthermore, for the three-stage cycle, the compressor power in the first, second and third compressors is 239 kW, 319 kW and 392 kW, respectively. In total the total power of compressor in the three-stage cycle is 950 kW. This fact reveals that the compressor power in the three-stage cycle is lower 9.35% in comparison with singe cycle. In can be said that the compressor power decrease as number of cycle increase. The mass flow rate of the refrigerant in the cycle also decreases as number of cycle increases.

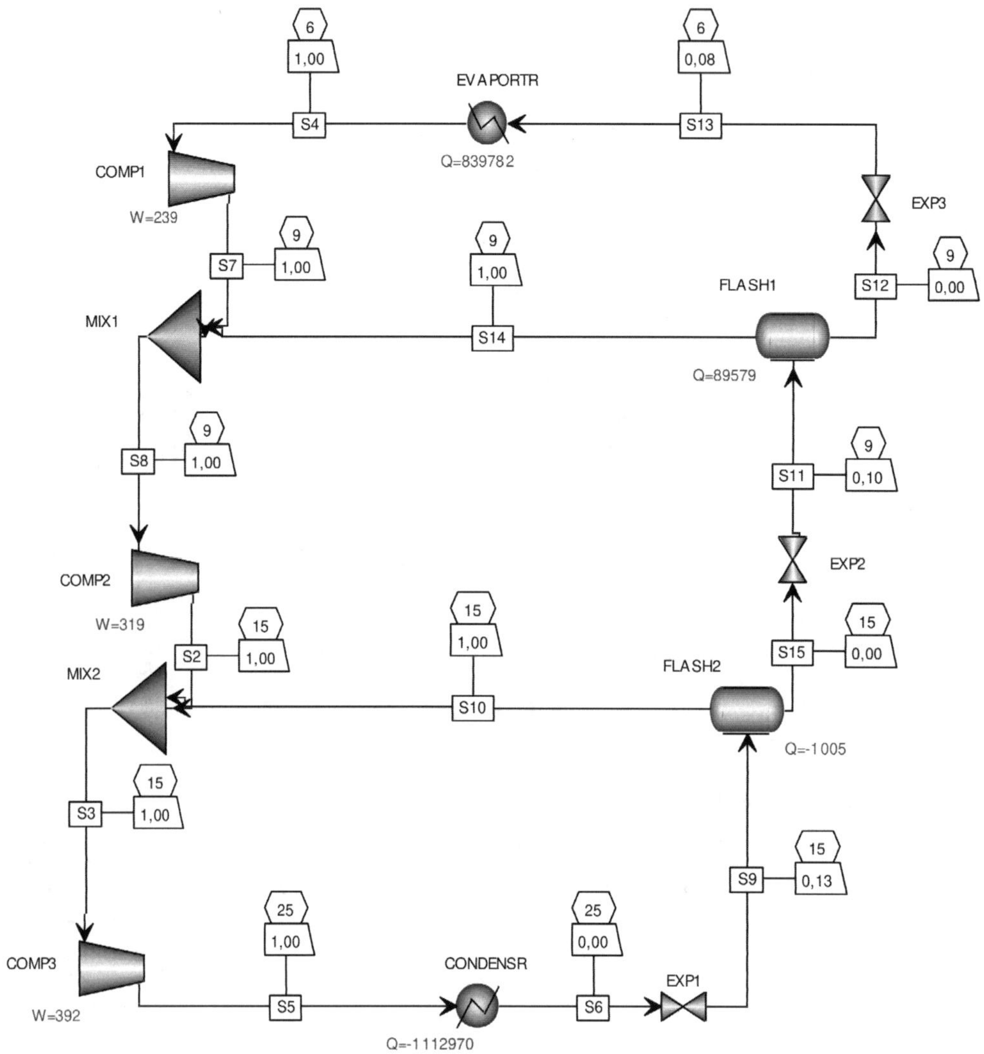

Figure 3. Three-stage vapour compression cycle in Aspen-Plus

Table 1. Result single-stage, two-stage and three-stage of the Aspen Plus simulation

	Single stage	Two-stages		Three-stages		
		First stage	Second stage	First stage	Second stage	Three stage
Suction pressure (bar)	6	6	12	6	9	15
Discharge pressure (bar)	25	12	25	9	15	25
Compressor power (kW)	1048	403.6	552.1	239	319	392
Refrigerant composition (mass fraction)	100 % Difluoromethane					
Heat duty (Cal/sec)	839224	839456		839782		
Total mass flowrate (kg/s)	14.8	12.11		11		
Temperature before expansion valve (K)	25	40	13	40	21	5

Temperature after expansion valve (K)	-10	13	-10	`21	5	-10

Finally the performance of the cycle is examined using the Coefficient of the Performance (COP). The COP for all stages are shown in Table 2. The COP for single stage, two-stage and three-stage are 3.35, 3.67, and 3.69, respectively. It was shown that the COP increases as number of stage increase. However, the difference from two-stage to three-stage is not as big as from single to two-stage. It is suggested that the two-stage is better than three-stage.

Table 2. Simulation results for refrigerant R32

Refrigerant	Coefficient of performance		
	Single Stage	Two Stage	Three Stage
R32	3.35	3.67	3.69

4. Conclusions

In this study the effect of number of stage in a vapor compression has been explored numerically using commercial code Aspen Plus. They are single-stage, two-stages, and three-stages vapor compression refrigeration cycle with refrigerant R32. The conclusion can be drawn here are as follows. The compressor power in the two-stage is lower 8.8% in comparison with single stage. The compressor power in the three-stage cycle is lower 9.35% in comparison with singe cycle. In can be said that the compressor power decrease as number of cycle increase. The mass flow rate of the refrigerant in the cycle also decreases as number of cycle increases.

References

[1] Ambarita H, Sihombing H. V. 2018. IOP Conf. Series: Journal of Physics: Conf. Series **978** 012098
[2] Ambarita H. 2018. IOP Conf. Series: Earth and Environmental Science **126** 012018
[3] Gupta V.K, Prasad M. 1983. International Journal of Refrigeration. Vol 6, Issue 2, 103-107.
[4] Sihombing H. V, Ambarita H, Naibaho W, Setiawan E.Y. 2020. IOP. Conf. Series: Materials Science and Engineering **725** 012017.
[5] Shuxue X, Guoyuan M. 2015. Energy and Building **93** 324-331.
[6] Purohit N, Gupta D.K, Dasgupta M.S. 2016. Energy Procedia **90** 171-178.
[7] Arpagaus C, Bless F, Schiffmann, Bertsch S.S. 2016. International Journal of Refrigeration **69** 437-465.
[8] Mazzei P, Palombo A. 2015. International Journal of Ambient Energy, 18:1, 3-12.
[9] Hosoz M. 2005. Turkish J. Eng. Env. Sci **29** 285-296
[10] Liu F, Tan J, Xiong J. 2010. International Refrigeration and Air Conditioning Conference 2213
[11] Khurmi R.S and Gupta J.K. 2006. Refrigeration and Air Conditioning. Ram Nagar, New Delhi.
[12] Arora R C 2010 *Refrigeration and Air Conditioning*, PHI, New Delhi.

IOP Conf. Series: Materials Science and Engineering **851** (2020) 012040 doi:10.1088/1757-899X/851/1/012040

Identification of Critical Waste on Outpatient Installation in Universitas Sumatera Utara (USU) Hospital

K Syahputri*, R M Sari, I Rizkya, J Leviza
Departement of Industrial Enginnering, Faculty of Engineering, Universitas Sumatera Utara, Medan, Indonesia

*Email : khalida@usu.ac.id

Abstract. Increasing the number of existing hospitals leads to increase a competition among hospitals. This condition causes hospital to become more competent by improving existing services, as well as providing continuous healthcare quality to every patient. Service system at outpatient installation of USU Hospital in BPJS patient often experiencing problem because there are still many waste at each part and patient service process become inefficient. Therefore, it is necessary to analyze critical waste identification to know the most dominant waste occurring in the process to do an action as soon as possible. The results obtained from the identification indicates that the most critical waste in BPJS patient service process is waiting, where the patient must wait for a long time starting from the registration process, inspection at polyclinic, and medicine- taking at the pharmacy.

1. Introduction

Hospital is a part of health resources are needed in supporting the implementation of health. Increasing the number of existing hospitals leads to increase a competition among hospitals. This condition causes hospital to become more competent by improving existing services, as well as providing continuous healthcare quality to every patient. Hospital services are now capital-intensive, labor-intensive, and technology-intensive in facing of global competition. Health care in hospitals is not only curative, but also rehabilitative which is both properties are implemented in an integrated manner through promotional and preventive efforts [1]. Thus, the goal of hospital health services is not only for individual patients, but also for the families of patients and general public.

The study was conducted on an outpatient installation at USU hospital located in Medan. Outpatient installation is one of the existing installations at USU Hospital and has a special role in treating patients who need the treatment. The process flow of the outpatient installation begins with the queue number-taking, registering at registration section, investigating by the doctor at the clinic, such as blood checks, CT scan, X-rays, and other investigations if necessary, and medicine-taking after medicine recipe by doctor at the pharmacy. Registration in the registration section is done by checking the patient card for the old patient while the new patient will perform the personal data-filling to input in medical record. After registering, patient moves to the clinic to check with the doctor. In this installation, there are 13 polyclinics consist of anesthesia poly, cardiac poly, ENT poly, eye poly, surgical poly, nerve poly, internist poly, pediatric poly, child development poly, pulmonary poly, obgyn poly, polyurium and genital poly, bone poly, and urology poly. After the investigation on one of the polyclinics is completed, the patient will receive a medicine recipe from the doctor if there is no investigation is required and go to the pharmacy to medicine-taking. If investigation is required, the patient will go to the laboratory for investigation and return to the polyclinic by bringing the investigation results after the investigation is complete. After that, the patient will receive the medicine recipe by the doctor and the patient go to the pharmacy to medicine-taking.

Outpatient installation at USU hospital treats general patients and BPJS insurance patients where the process of these two patients is only different at the end of the process. The general patient make a payment after medicine-taking at the pharmacy as doctor medicine recipe while the BPJS insurance patient does not need to make the payment. Patients BPJS insurance at USU hospital more than the general patient and BPJS insurance patients become the focus in this study. In the flow process of patient BPJS insurance, there are many problems due to a lot of waste on registration process, investigation by the doctor, and medicine-taking at the pharmacy. This is because there is still a lot of waste occurs at each section and patients must queue and wait for a long time at each process. This problem leads to patient satisfaction on the reduction service and critical waste is needed to identify each process in order to make repairs immediately.

Waste identification done by the activities description occurs of the service process into value added activities, non value added activities but necessary, and non value added activities lean approach [2]. Lean is a continuous effort to eliminate waste and increase the value added of products or services to provide value to costumer value [3]. Therefore, the lean concept applied in service-based fields. Lean concept in the field of service assumes the same principle, namely continuous improvementand waste. These principles are also currently applied in many service businesses for example call centers, health care, higher education, software development,and other professional services [4]. Lean healthcare is a strategy focused on eliminating inefficiencies and giving more time to patient service activities [5].

Many previous studies have been conducted in a medical center or hospital in the world to identify waste using a lean healthcare approach. One study was conducted at Virginia Mason Medical Center in Washington. The same study has also been conducted at the anatomical pathology lab using lean methods [6]. However, there is still many research with lean healthcare done in hospital in Indonesia especially in Medan City. This research is aimed to identify waste from service process to reduce waste happened in outpatient installation, and service quality and increase consumer satisfaction.

2. Methodology

In this study object of this study is the flow of outpatient installation process of USU hospital specially BPJS insurance patient. Research begins by making observations to see and observe the condition occurs in the outpatient installation. After that, determines the purpose of research in accordance to existing conditions on the outpatient installation. Based on the purpose of research, data collected as input in problem solving. The data collected for data processing in the form of waste occurs at each process, the patient processing time at each part, and waiting time of the patient. Based on these data, data processing done as a solution to problems that occurs in the outpatient installation. The research is done by using lean concept. This method is one of the best practices to identify waste that occurs in the work environment [7]. Lean is a production practice aims to minimize waste along the value streams mapping and create more value to customers. According to lean principles, many use of resources does not deliver consumer value is a target for change, reduction or elimination [8].

Problem solving with lean approach is done with several stages. The first stage is to map the service system of outpatient installation of USU hospital. The purpose of mapping is to know the process occurs in the outpatient installation of USU hospital at this time. Mapping is done by describing the service process of outpatient installation starting from the arrival of the patient until the patient returns. The result of this service installation system mapping is called the current state mapping service system and the flow of BPJS insurance patient service process on the outpatient installation of USU hospital. After obtaining big picture mapping, the next step is to identify the activity along the service system on the outpatient installation. This identification is done to know and calculate the percentage of activities including value added, non value added but necessary, and non value added categories. After the percentage at each activity is obtained, the next step is to identify the waste occurs in the service system of the outpatient installation. This waste identification aims to determine the waste occurs during the process and causing the patient service process be ineffective. Next stage is critical waste identification from the eight waste by distributing questionnaires to patients.

3. Result and Discussion

3.1. Current State Mapping

The process of describing the current state mapping service system on the outpatient installation of USU hospital is done from the arrival of the patient until the patient returns. Current state mapping of service system BPJS insurance patient on an outpatient installation in USU hospital can be seen in Figure 1.

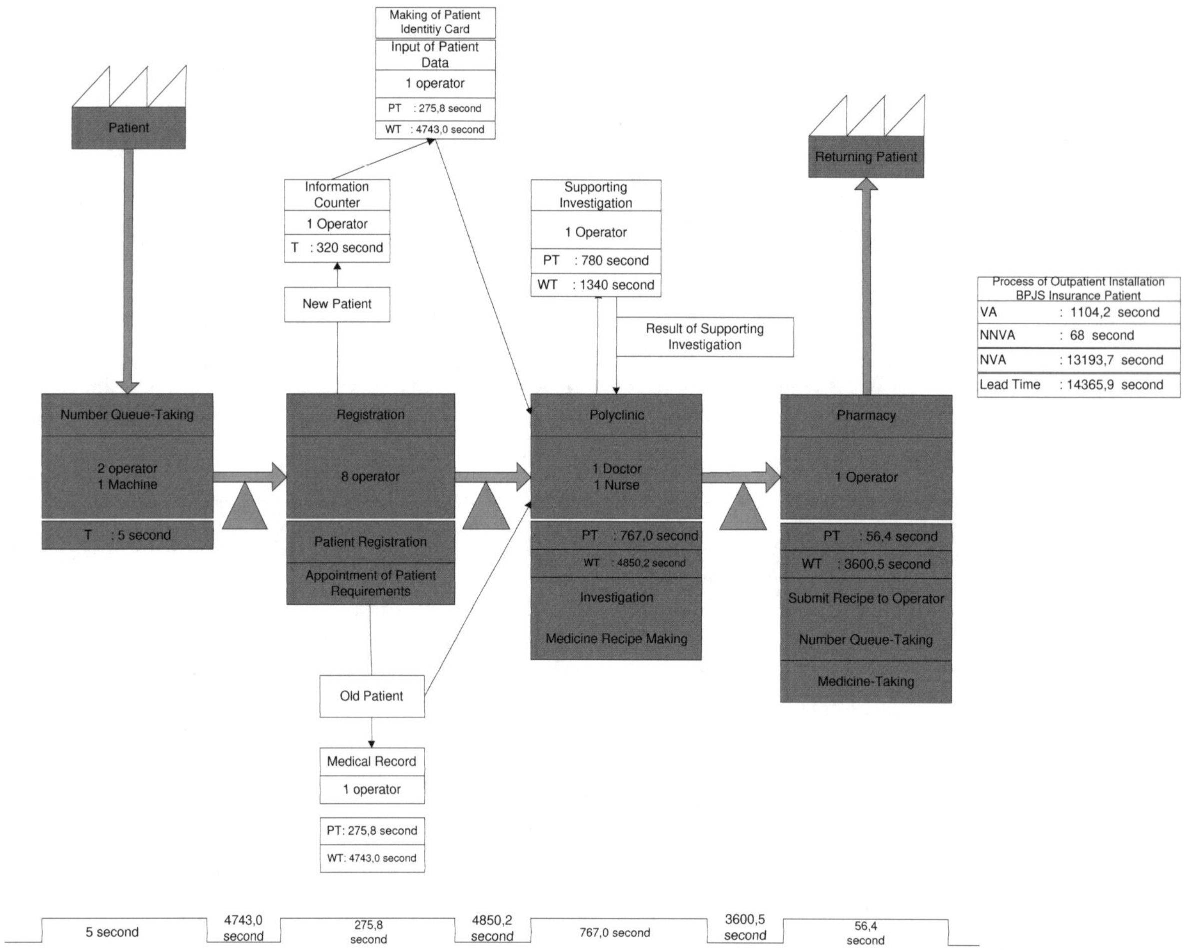

Figure 1. Current State Mapping of Service System BPJS Insurance Patient on an Outpatient Installation in USU hospital

Based on the figure above, it shows that the patient waiting for a long time at each part of the process flow. This is causing by several things such as a doctor has not arrive yet, the patient's medical record has not reached the clinic and doctor can not investigate the patient, and other causes makes the patient queue at each section.

3.2. Identification of Value Added, Non-Value Added but Necessary, and Non-Value Added Activities
Identification is done by calculating the percentage at each activity after the activity is categorized according to the mapping on the current state mapping. The classification of activities is based on the type of activity in the organization [10]. Percentage of outpatient service activity for BPJS patients insurance can be seen in Table 1.

Table 1. Budget of Job Item in Fruit and Boiling Stations at Basic Design, Alternative 1, and Alternative 2

Category	Percentage
Value added	12,73%
Necessary non value added	45,40%
Non value added	41,93%

Based on the table above obtaines that the outpatient installation service process for patients BPJS insurance producing necessary non-value added value activity of 45,40%. This result shows that there is still a lot of activity that necessary but it's not give value added because waste occurs in the service system and more identification is needed to identify the waste.

3.3. Identification of Waste in BPJS Insurance Patient Service Process
Before determining the critical waste occurs in the outpatient installation service system, first stage is identify the wastes occurs in the system. Waste identification during the BPJS insurance patient service process on an outpatient installation is waiting, overprocessing, and inventory. Waiting occurs when the patient waits at each process, such as queuing for the registration process, queuing for examination with the doctor, and queuing to take medication at the pharmacy. Overprocessing occurs in the medical record where the patient must wait for the results of medical records to doctor investigation because medical record process has not optimal yet and irregular. Inventory occurs when doctors have to replace the medicine recipe because medicine are not available at pharmaciy.

From these four wastes, determines the critical waste by distributing questionnaires. The results of the questionnaires distribution shows that from the four wastes occurrs during the outpatient installation service process, waste waiting is the most critical waste which is the condition causes waiting process in every patient service process on the outpatient installation starting from the registration process, the investigation on the polyclinic section, and the medicine-taking in the pharmacy.

4. Conclusion
The process of patient service has not optimal yet done by waste identification in the process. This waste identification aims to determine the waste occurs during the process and causes the service process to the patient be ineffective. The results obtaines that the identification indicates the most critical waste in the BPJS insurance patient service process is waiting, where the patient must wait for a long time starting from the registration process, investigation on the polyclinic, and medicine- taking at the pharmacy.

Acknowledgement
I would like to acknowledge to Universitas Sumatera Utara for funding this publication research. I would also like to expressed my gratitude to colleagues and all parties who helped complete this research.

References
[1] Undang-Undang Republik Indonesia No. 44 Tahun 2009 tentang Rumah Sakit.
[2] K syahputri 2017 IOP Conf. Ser.: Mater. Sci. Eng. 180 012119

[3] Gaspers, and A Fontana 2011 Lean Six Sigma for Manufacturing and Service Industries(Jakarta: PT. Gramedia Pustaka Utama
[4] Z F Ikatrinasari, E I Haryanto 2014 *Journal of Service and Management*, **7** 291-301
[5] Lestie, Hagood, Royer, Reece, and Maloney 2006 Aorn Journal, **84**(5).
[6] SJ Spear. 2005 *Harv Bus Rev.* **83**(9): 78-91.
[7] B H Ehrlich 2002 *Transactional Six Sigma Servicing, Leveraging Manufacturing Concept to Achieve World Class Service*(Boca Ratong: St. Lucie Press)
[8] D Arfmann, Federico, and G T Barbe 2014 *International Journal of Business and Social Science*, **5** (2).
[9] P Hines, and D Taylor 2000 *Going Lean*(Cardiff Business School: Lean Enterprise Research Center).

IOP Conf. Series: Materials Science and Engineering **851** (2020) 012041 doi:10.1088/1757-899X/851/1/012041

Availability and reliability assessment of production systems using simulations

Humala Napitupulu[1], Nurhayati[1]

[1] Department of Industrial Engineering, Universitas Sumatera Utara, Medan, Indonesia

e-mail: humala@usu.ac.id, nurhayatipandia68@yahoo.com

Abstract. Reliability assessments in the industry is needed to support production planning, facility maintenance, and other related matters. The reliability of an equipment is usually assessed based on the operating period of its components. Likewise the reliability of the system can be assessed based on the reliability of the subsystem. Assessment of the reliability of production systems as a combination of a number of subsystems can be done using analytical models suitable for series and parallel arrangements. For complex production systems, valuation can use a combination of series and parallel models. But for a production system that consists of a lot of equipment in a series arrangement, evaluations using analytical models will provide low reliability that does not reflect real conditions. In the industry studied, the reliability of the system with 14 subsystems in a series arrangement is around 0.2%. Therefore, reliability assessment needs to be done by simulation. Using the proposed simulation model, the reliability of the system obtained is around 90.4%. This assessment shows that the assessment of the reliability of the production system is more appropriate using simulations, and the proposed simulation model can be used in reliability assessments
Keywords : Reliability, Availability, Productivity, Production System

1. Introduction

A competitive world with increasing customer demand for reliable product shipments in recent years has made reliability engineering a more challenging task. The availability and effectiveness of production systems is also increasingly important in planning and optimization. In this case, reliability analysis is one of the main tools to ensure product quality and delivery deadlines that will maintain certainty in customer loyalty and company reputation. Therefore, evaluating equipment reliability is an important part of most production planning, product design and development, and maintenance program management. [1] This is increasingly challenging for systems consisting of various components, which have become global and essential in today's society. [2,3]

System reliability is usually assessed using mathematical and statistical equations that are appropriate for system configuration. The reliability of a system or sub-system can be assessed using analytic models for serial and parallel configurations. [4,5,6] However, the results of this assessment do not always show a real situation because the assessment of the reliability of a production system consisting of many subsystems in a long series arrangement will give very low results. In dealing with this problem, equipment reliability assessment needs to be done with simulation applications, by developing and using the appropriate simulation models as needed. [7,8,9].

IOP Conf. Series: Materials Science and Engineering **851** (2020) 012041 doi:10.1088/1757-899X/851/1/012041

2. Reliability

2.1 Definition
The reliability of a machine or a system is its ability to consistently perform the intended function or mission in accordance with the request, without degradation or failure. In manufacturing, reliability is the probability of failure-free performance over the useful life of an item or machine within a certain period of time under specified environmental conditions and duty cycles [10]. The analytical method used to calculate the reliability of a system will depend on the configuration of components. The two basic models of system configuration are series configuration and parallel configuration. [5,6]

2.2 Series Configuration
In a series configuration, failure of any component will cause a failure for the entire system, so the reliability of the system is determined by the probability of success of all units in the system. All units must succeed for the system to succeed. The reliability of the system is then expressed by the equation:

$$\begin{aligned} R_S &= P(X_1 \cap X_2 \cap \ldots \cap X_n) \\ &= P(X_1)P(X_2|X_1)P(X_3|X_1X_2)\cdots P(X_n|X_1X_2\ldots X_{n-1}) \end{aligned} \quad (1)$$

where: R_s is the reliability of the system
X_i is the event of unit i being operational
$P(X_i)$ is probability that unit i is operational

In the event that failure of any component will affect the failure rate of the other components, the conditional probabilities in the above equation must be considered. In the case of the independent component, the above equation becomes:

$$R_s = P(X_1)\,P(X_2)\,\ldots.\,P(X_n)$$

$$R_s = \prod_{i=1}^{n} P(X_i) \quad (2)$$

In a pure series system, the reliability of the system is the same as the product reliability of its constituent components, so the reliability of the series system is always lower than the reliability of the most unreliable components. The number of components determines the reliability of the system in a series arrangement. Increasing the number of components in series will reduce system reliability.

2.3 Parallel Configuration
In a system with n simple parallel units, at least one unit must be successful for the system to succeed. The units in a parallel arrangement are called redundant units, and redundancy is an important aspect in system design and reliability. Increased redundancy is one of several methods of increasing system reliability. The probability of failure for a system with statistically independent parallel components is the probability that 1 unit fails, 2 units fail, and so on until all units in the system fail. In this system, all n units must fail for the system to fail. The unreliable system is then provided by:

$$\begin{aligned} Q_s &= P(X_1 \cap X_2 \cap \ldots \cap X_n) \\ &= P(X_1)P(X_2|X_1)P(X_3|X_1X_2)\ldots P(X_n|X_1X_2\ldots X_{n-1}) \end{aligned} \quad (3)$$

where: Q_s is the unreliability of the system
X_i is the event of failure of unit i
P_i is the probability of failure of unit i

If the failure of one component affects the failure rate of the other components, then conditional probabilities in the above equation must be considered. In the case of the independent component, the above equation becomes:

$$Q_s = P(X_1)P(X_2)\ldots P(X_n)$$

$$Q_s = \prod_{i=1}^{n} P(X_i) \qquad Q_s = \prod_{i=1}^{n} Q_i \quad (4)$$

IOP Conf. Series: Materials Science and Engineering **851** (2020) 012041 doi:10.1088/1757-899X/851/1/012041

2.4 Reliability and Availability

The effectiveness of the production system is used as a standard to measure manufacturing productivity. This identifies the percentage of production time that is truly productive. A high effectiveness score means that the manufacturing system is reliable and produces a high percentage of good products. In this context, reliability determines availability and productivity because damaged equipment is not available for use but must stop during repair.[1,4]

Equipment failure and material shortages will result in loss of availability including unplanned outages such as replacement time and repair time. The use of equipment also generally requires time for preparation in the form of installation time or setup time. Likewise changeover time is included in the analysis of the effectiveness, because this determines the time that can be used for manufacturing.[1]

3. Method

System reliability assessment is carried out in two stages of research.

1) In the first stage, field research was carried out to obtain equipment reliability, sub-system reliability and system reliability. This reliability assessment uses an analytic model.
2) The second stage is reliability assessment by using simulations to obtain predictions of subsystem and system reliability. For this assessment, a simulation model is designed, and the generation of imitation data is carried out using the TTF and TTR distribution patterns obtained from field research.

Discrete system simulations are run using imitation data in the operation of simulation models. The simulation is carried out in the number of iterations and replications that meet the validation. The simulation results are then verified and validated by comparing them with the results of the analysis of the availability and productivity of the actual system.

4. Result and Discussion

4.1 Field research

To facilitate this research, one plant was chosen as the target. A palm oil extraction plant was chosen which is available to provide production and maintenace data needed in the reliability analysis. This plant consists of two parts, palm oil processing plant and palm kernel processing plant. The reliability assessment study was conducted only at palm oil processing plants.

At the palm oil processing factory studied, there are 32 units of equipment used in processing. All of this equipment can be grouped into 14 subsystems in a series arrangement, consisting of 6 subsystems with one unit of equipment and 8 subsystems with several units of equipment in a parallel configuration. By using the reliability assessment equation, the reliability of each unit of equipment can be obtained.

Historical data on failure time (TTF) and time to repair (TTR) of each equipment in the one year period, 2018, were obtained from the archives of the maintenance department. Observation and data collection is also carried out for production operations and elements of activities that use the available time. Next, the reliability assessment of each equipment and the operating time analysis were carried out to determine the availability of the system.

By grouping equipment into 14 subsystems, an analysis of the availability and reliability of systems and sub-systems can be carried out. The reliability of subsystems with multiple equipment is assessed using a parallel reliability assessment model. The analysis results are presented in table 1.

The configurations of all equipment that are grouped in 14 subsystems are series configurations. The reliability of the system can be defined as average reliability by considering that failure of certain equipment will affect the operation of the equipment or subsystem itself without causing termination of the entire plant.

By using analytic models for configuration series of 14 subsystems, the reliability of the production system is a product of the reliability of all equipment as follows:

$$Rs = 0.3845 \times 0.811 \times 0.2994 \times 0.6165 \times 0.6315 \times 0.4106 \times 0.4132 \times 0.7533 \times 0.5041 \times 0.9809 \times 0.9758 \times 0.9659 \times 0.9012 \times 0.9901$$

$$Rs = 0.00193 \text{ or } 0.193\%$$

Table 1. The configuration and reliability of 14 subsystems in factory of PT. XYZ

Order	Subsystem	Units	Configuration	Reliabilty	Availability
1	Loading Ramp	1	Single	0,3845	99.60
2	Sterilizer	2	Paralel	0,8110	98.79
3	Tippler	1	Single	0,2994	99.44
4	Fruit Bunch Conveyor	1	Single	0,6165	99.39
5	Thresher	2	Paralel	0,6315	99.06
6	Re-thresher	1	Single	0,4106	99.11
7	Main Fruit Conveyor	1	Single	0,4132	99.24
8	Fruit Elevator	2	Paralel	0,7533	99.60
9	Top Distribution Conveyor	1	Single	0,5041	99.27
10	Digester	4	Paralel	0,9809	98.98
11	Screw Press	4	Paralel	0,9758	98.63
12	Vibrating Screen	3	Paralel	0,9659	99.05
13	Oil Purifier	3	Paralel	0,9012	99.25
14	Sludge Centrifuge	6	Paralel	0,9901	99.16
					99.18

The reliability of the system as average reliability can be calculated as Rs.

$$Rs = (R_1+R_2 +......+ Rn)/n$$

$$Rs = (0.3845+0.811+0.2994+0.6165+0.6315+0.4106+0.4132+0.7533+0.5041+0.9809+0.9758+0.9659+0.9012+0.9901)/14= 0.6884$$

Average availability is 0,9918 or 99,18%

The reliability of the production system using an analytical model is very low under 1 percent, which is very difficult to accept as a result of an assessment that reflects the real situation. While the average reliability is 0.6884 or 68.84% with an availability rate of 99.18%. Thus the analytic model is not suitable for use in assessing the reliability of production systems. Therefore, reliability assessment needs to be carried out using simulations, and for this reason it is necessary to develop a simulation model that is suitable for the production system being assessed.

4.2 Simulation Results

Research in the laboratory is conducted to assess the reliability of subsystems and systems using system simulations. The simulation is run by operating the simulation model using imitation data. Imitation data for the time interval between damage is generated using random numbers and the pattern of time interval distribution between damage (TTF) as a reference model obtained from observational data analysis. In this case, the MTTF and time interval distribution patterns between damage are obtained from the analysis of historical data of each equipment using reliability analysis software. Reliability analysis was carried out for 32 equipment from 14 subsystems. Distribution patterns obtained from the analysis are 4 equipment with Normal patterns, 17 equipment with Lognormal patterns and 5 equipment with Weibull patterns.

The development of simulation models begins with the preparation of a reliability assessment model in the form of an algorithm or flowchart for determining the length of time used for equipment repair. The algorithm model is then translated into a series of formulas in an Excel worksheet.

Using imitation data, the imitation operation is performed to get a prediction of the total repair time for all equipment in a period of 1 year. The simulation is run in 287 working days in a year starting from the first day. The time between failures on each equipment and subsystem is predicted using random numbers as probabilities in the distribution pattern model, while the repair time is predicted randomly using uniform patterns. In accordance with this prediction, the simulation is executed by carrying out imitation operations of each subsystem on the system. The procedure for implementing the simulation in the form of an algorithm is presented in Figure 1.

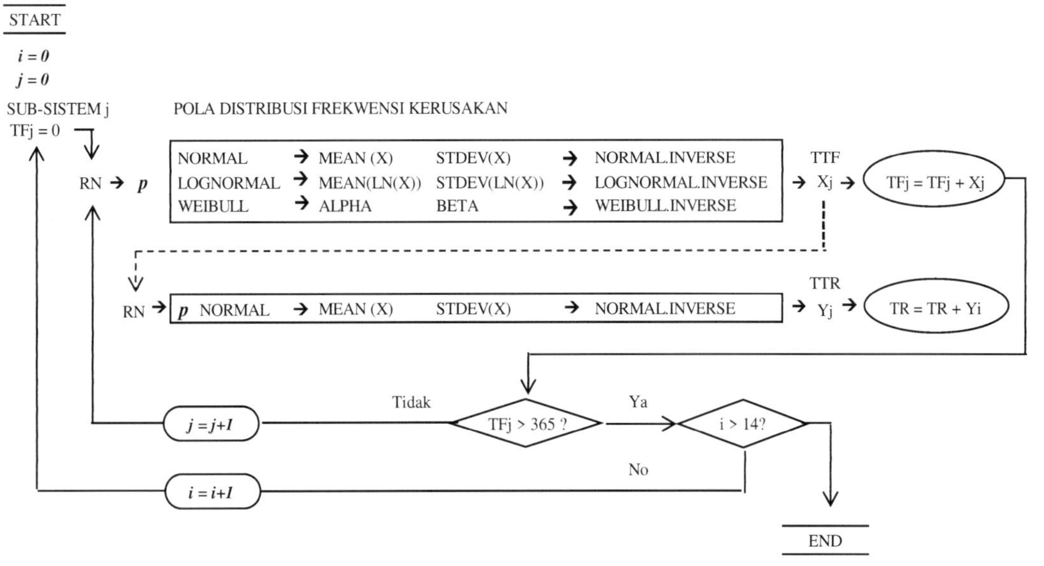

Figure 1. Simulation algorithm of system reliability assessment

The simulation is run for 365 calendar days in one year. The number of effective working days in one year is 287 working days after deducting 52 working days for Sunday as a free day every week, and 26 days for half a working day on Saturday. On Saturdays, the factory operates in half a day, between 8 and 12 hours. In this simulation, the time interval between damage (TTF) for each equipment is calculated in units of minutes with respect to the predicted repair time (TTR) based on field data obtained in units of minutes. From field information it is known that the factory operates an average of 18 hours per working day. Thus, the operating time in one year is 287 x 18 x 60 minutes = 309,960 minutes.

In accordance with company policy, equipment preparation must be carried out routinely every working day before production operations. The duration of preparation time, known as setup time, is determined by the factory manager. The length of setup time is not the same from day to day but varies depending on the situation at hand. The duration of this setup is not related to the length of time required to repair damaged equipment. From the field recording of the implementation of the setup in the past year, the total preparation time in one year is 48,852 minutes, so that the total time available for production operations in one year is 309,960 - 48,852 = 238,614 minutes.

By using the average time needed to repair damage to each unit of equipment, the simulation application provides a prediction of repair time for each equipment. These prediction results are obtained by taking into account the frequency of equipment damage in one year. The predicted total repair time for all equipment in one year is 25,350 minutes. Prediction of the total length of time to repair equipment, as shown in table 2, is the average value of the total time to repair equipment obtained from simulation applications with 10 replications, 10 iterations per each replication, and 30 cycles per iteration.

The effective time available in a one year period is 309,960 - 25,350 - 48,852 = 235,758 minutes. From the simulation, the total number of repairs is 430 events per year or around 1.5 repairs per day. Within 1 year, there were 57 days without repairs, and the highest number of repairs was 5 repairs in one day. With a total available time of 235,758 minutes, the availability of the system is 235,758 / 309,960 = 0.7607 or 76.07%. While the ratio of total processed materials (252/413 tons) to installed capacity (327/180 tons based on 30 tons / hour) is 0.7714 or 77.14%. The difference between availability and production ratio is quite small around 1%, so the simulation model can be used to determine system reliability. Based on this availability, the reliability of the system is 0.904 or 90.40%, which is higher than the analytical reliability value of 67%.

Table 2. Results of system simulation

Replication	Repairs	TTR (minutes)	Availability	Reliability
1	431	25525	0.7596	0.9053
2	433	25667	0.7604	0.9032
3	434	25040	0.7574	0.9055
4	429	25583	0.7602	0.9050
5	430	25677	0.7618	0.9029
6	426	25229	0.7614	0.9035
7	425	24942	0.7612	0.9042
8	432	25171	0.7633	0.9038
9	430	25304	0.7608	0.9029
10	426	25363	0.7613	0.9038
Average	**430**	**25350**	**0.7607**	**0.9040**

The average reliability of the production system as shown in table 1 is 99.18%, not so different from the simulation results which is 90.40%. Thus the assessment of system reliability, especially production systems that consist of many subsystems and equipment, is appropriate using simulations.

5. Conclusion

Discrete system simulation applications are suitable for evaluating the reliability of systems consisting of many equipments in long series settings. The results of the reliability assessment with the simulation application are more representative than the results of the assessment with the analytic model. Likewise in the case of availability assessment, the use of simulation provides more complete results including frequency and time to improve equipment in the system.

Acknowledgments

This research was funded by the Research Institute (LP) of the University of Sumatera Utara (USU) Medan, Indonesia, in accordance with the TALENTA Research Implementation Contract, Fiscal Year 2019, Number 363/UNS.2.3.1/PPM/KP-TALENTA USU/2019). The author is grateful for the approval and funding of the research provided by the Chancellor of USU and the Chairperson of USU's LP. The authors wish to acknowledge the contributions of the Chair of the Department of Industrial Engineering and the Dean of the Faculty of Engineering, as well as reviewers who gave their time and were willing to share their knowledge in carrying out the review.

References

[1] Borris Steven 2006 *Total Productive Maintenance*. (McGraw-Hill Companies, Inc.) p 448
[2] Dhillon B S 2006 *Maintainability, Maintenance, and Reliability for Engineers* 1st Edition CRC Press ISBN 9780849372438 London (Taylor & Francis eBooks) p 240
[3] Ramesh G and R Smith 2009 *Maintenance and Reliability Best Practices* New York (Industrial Press Inc.) p 416
[4] Leroy Alain 2018 *Production Availability and Reliability* 1st edition London (John Wiley & Sons, Inc.) ISBN-10: 1786301687 p 368
[5] Bloom Neil B 2006 *Reliability Centered Maintenance* New York (McGraw-Hill,Inc.) p 314
[6] Hoang Pham 2003 *Handbook of Reliability Engineering*. (Springer-Verlag London Ltd) p 663
[7] Guillaume Dubois 2018 *Modeling and Simulation: Challenges and Best Practices for Industry* 1st Edition London (CRC Press) ISBN 9780815374893 p 148
[8] Hans Dawei, Joachim B, Zimmer M and Buchhol D P 2013 *Modeling and Simulation : An Application-Oriented Introduction.* (Springer-Verlag Berlin Heidelberg) p 413
[9] Napitupulu Humala 2009 *Simulasi Sistem : pemodelan dan analisis* Medan (USU Press) p 282
[10] Kishor S T and Andrea B 2017 *Reliability and Availability Engineering: Modeling, Analysis, and Applications* Kindle Edition Cambridge (Cambridge University Press) p 726

IOP Conf. Series: Materials Science and Engineering **851** (2020) 012042 doi:10.1088/1757-899X/851/1/012042

Improving the work system with management engineering methods

I Siregar[1], T H Nasution , K Syahputri

[1]Department of Industrial Engineering, Faculty of Engineering, Universitas Sumatera Utara

Email : ikhsan.siregar@usu.ac.id

Abstract. Management engineering is often done in the industrial world, especially factories and manufacturing companies. In this research, management engineering will be applied in hospitals, especially in the emergency department, why is it done in this department, because in previous studies data collection has been carried out and it is concluded that nurses in this department have a high degree of flexibility, so can cause errors while working. Almost the same as the application of fatigue in the manufacturing industry that the level of fatigue on the operator can reduce work productivity. This research will apply a macro ergonomic approach, in which the model used is the System Engineering Initiative for Patient Safety (SEIPS) model at the Hospital. This approach uses interview and observation techniques. Through this approach the causal factors will be seen and, in the future, seen how the solution to overcome this fatigue factor. This method uses a survey system that involves a large state-owned hospital in the city of Medan by using 83 nurse respondents and 7 respondents from the hospital.

1. Introduction

Occupational health is a public health application in a place (company, factory, office, etc) and being a patient in public health is workers and the community around the company [1][15]. From an ergonomic point of view, each workload received by a person must be appropriate or balanced of the physical, cognitive and limitations of human ability to accept the workload [2]. Any work of any kind job requires muscle strength or thought is a workload for all the people do the work. Due to workloads are too heavy or weak physical abilities, it can result in a worker suffering from work-related disorders or diseases. All postures are not natural in working, such as attitudes to reach goods exceed the hand reach must be avoided. Symptoms of fatigue are experienced by employees when working is monotonous and repetitive. Occupational fatigue is often interpreted as a process of decreasing efficiency and reduced physical strength or endurance to continue the activities must be done [3].

Occupational fatigue is a condition which a decrease in work efficiency, skills and resulted by feelings of fatigue or boredom to continue activities or work. Fatigue cannot be underestimated because it affects the effectiveness, productivity, and safety of workers in general. High levels of workload cause workers to concentrate and focus when working and increase the risk of workplace accidents caused by human error[4].

Fatigue condition usually results from physical and spiritual disorders, stress, and overworks. That is why fatigue is also expalined as the inability to maintain and preserve the required or expected stamina and energy or the lack of energy in the working capacity. Working with fatigue is even equated to alcoholism [5] [6]. One profession that has a high level of fatigue is a nurse at the hospital. Hospital nurses who provide patient care service often experience temporal demand and heavier workload due to the shortage of staff and potential overtime. Nurses encounter nonstandard work schedules, long works hours and circadian adjustment to night shift which is physically, mentally and emotionally strenuous [7]. The nursing profession is one of the jobs that often uses work shifts, this is because working hours are relatively solid. Individuals who do night shift work in circulation will be exposed to fatigue which causes sleep deprivation. The results of fatigue include reduced memory, reduced reaction time, decreased speed of information processing, irritability, endangered problem solving and critical thinking, slipped in attention to details, decreased concentration, judgment, and motivation. Also, regarding the point that fatigue is associated with serious damage to nursing staff including muscle and cardiovascular damage, etc. Overall fatigue can cause a decrease in output, dissatisfaction, absence from work, increased sick leave, and high personnel turnover. improve the decision to leave work [8].

One method used in analyzing work fatigue problems due to fatigue is the macro ergonomic approach. Macro ergonomic is concerned with factors in the technological subsystem, subsystem personnel, external environment, and their interactions as their impact on work system design. The goal of work system design optimization in terms of its sociotechnical system characteristics is to obtain harmonized work systems fully [9]. The macro ergonomics approach aims to better understand and overcome the factors contribute to and/or prevent fatigue in the work system of nurses in hospitals. One of the macro ergonomics approaches is the Systems Engineering Initiative for Patient Safety (SEIPS) model. The SEIPS model was specifically developed to deal with patient and nurse safety. This model explains five components for the structure of work systems, namely person (person), organization, task or workload, work environment, and equipment and technology to pursue balance in the work system of hospital nurses by increasing positive factors when there are several negative factors cannot accept redesign or system intervention [10].

Work fatigue problems have been solved in previous studies. The research has been conducted in North Sumatra relates to these nurses, namely the influence of stress and leadership on the nurses performance in the general hospital in Pematang Siantar. Problems cause stress in work include the absence of a balance between nurses and patients in this case the quantity of beds as a comparison, and nurses feel exhausted, the long duration can be the cause of the emergence of stress experienced at work. The results of regression analysis indicates that the coefficient of determination (R2) is equal to 0,429, means that the independent variables (stress and leadership) together have an influence of 42,9% on nurse performance when the remaining 57,1% is influenced by other variables not examined [11]. Other studies on fatigue problems were also carried out by Shucisnigdha, et al. The study was conducted using fatigue alarms to improve operator performance. This study aims to look at the working conditions in one Intensive Care Unit (ICU) with the presence of a second unit officer and monitor for patients from other medical floors. physiological steps can be used together with mental workloads and affect to produce a complementary definition of alarm fatigue [12][13].

Based on the research that has been done before, this study uses a macro approach in analyzing the fatigue factors of nurses at Accident and Emergency Department.

2. Methodology

This study is conducted at one of the hospitals in the Medan city where the object examined in this study is the work system of nurses in hospitals. The study begins with observations to observe and see the condition of the company directly. From the results of observations, it is determined the formulation of the problem in accordance with the conditions occur in the company and the research objectives can be applied. The research objectives determined are a solution to existing problems. Furthermore, the data collection as the input in conducting this research. The data needed is in the form of questionnaires

IOP Conf. Series: Materials Science and Engineering **851** (2020) 012042 doi:10.1088/1757-899X/851/1/012042

data. The questionnaire is a quantity of written questions are used to obtain information from the respondent in the sense of reports about the person, or things. In research, the use of questionnaires is a very important thing in data collection. The main purpose of making a questionnaire is to obtain information is relevant to the goal by fillment in the question by the researcher towards the chosen respondents. Carayon (2001) has set five stages to develop a questionnaire survey, namely:

1. Conceptualization
   Determine what concepts will be measured between the elements of the work system that will be evaluated and the output elements that will be evaluated.
2. Operationalization
   Determine the goals to be achieved from the survey and how to match between the concepts used and the objectives to be achieved
3. Source of Questionnaire
   Determine the type of survey questionnaire that is used as a reference for research
4. Making of Questionnaire
   Reviewing the form of questionnaires, the rating scale of measurement and the question items determination, how to fill out questionnaires and etc.
5. Initial testing of the questionnaire who is the respondent will participate in this initial test, and determine the objectives of the initial testing (checking the clarity of the question items, the format and duration of questionnaire fillment).

The questionnaire are distributed to nurses and hospital management. The nurse is the operator with the highest level of fatigue. The management provides a work description. The scale used in the assessment of the questionnaire is the Likert scale. This study uses a Likert scale, Likert scale is a psychometric scale used in questionnaire-based research, and is a scale that is frequently cast-off in research in the method of surveys. When responding to queries on a Likert scale, respondents control their level of agreement with the declaration by choosing one of the obtainable options. Likert scale is used to portion attitudes, opinions and perceptions of a person or group of people about community phenomena.

Table 1. Likert Scale

	Preference		Preference		Preference
1	Very agree	1	Agree	1	Very Positive
2	Agree	2	Often	2	Positive
3	Doubt	3	Sometimes	3	Neutral
4	Disagree	4	Almost Never	4	Negative
5	Very disagree	5	Never	5	Very Negative

For the determinations of quantitative analysis, the answer is given a notch, for example: very agree / agree / very positive given a score of 5, then agree / often / positively given a score of 4 and so on [14]. The Systems Engineering Initiative for Patient Safety (SEIPS) macro ergonomic approach carried out after questionnaires are distributed. Some of the things discussed in macro ergonomics are organizational structures, interactions among people in the organization and aspects of workers inspiration. In other words, micro ergonomics only look at the level of work but macro ergonomics look at the level of work and also the level of the organization. Macro ergonomics has been known as a subdiscipline of ergonomics related to human, organizational and technological relations. Macro ergonomics is something that is integrated because it includes information, approaches, and equipment from socio-engineering systems, industrial psychology, systems engineering, physical ergonomics, and ergonomics theory. In its implementation, macro ergonomics presents a valuable niche that none of these areas has ignored. As a science, macro ergonomics directs to progress an empathetic of work systems, behavior, or personnel that interact with hardware or software in an internal physical environment, external environment, and organizational structure and processes to make it better.

IOP Conf. Series: Materials Science and Engineering **851** (2020) 012042 doi:10.1088/1757-899X/851/1/012042

3. Result and Discussion

3.1. *Stages of Ergonomic Approach Using Systems Engineering Initiative for Patient Safety* (SEIPS)

In macro ergonomics application there are several stages, namely:

1. Stage of Information Collection
   Information collection needed for making questionnaires about the work system and information about who the respondents will be involved in conducting direct observation in the hospital and conducting interviews with 7 respondents from the hospital management and 83 nurse respondents.
2. Stage of Objective Determination
   Determine the goals to be achieved in the study and the benefits will be obtained for the company. The goal to be achieved is to determine the factors affect the nurse mistakes and how to overcome them.
3. Stage of Implementation
   Determine when the survey will be conducted, the procedure to be used and the survey data collection method used by conducting an open questionnaire distribution based on a Likert scale.
4. Stage of Analysis and Interpretation
   Data processing is done with statistical software to process, analyze and draw conclusions from survey data, then connect with the objectives to be achieved.
5. Stage of Result Deliver
   Making reports on the results of the research conducted.
6. Stage of follow-up action is the final stage of a research to make the next activity plan / action, for example the making of a recommendation work system improvement or implementation of the results of the research conducted.

3.2. *Approach of Work System Design*

Each work system contains several or all work components, each interacting with each other. Ergonomics as both science and technology is always concerned with interfaces and interactions between operators and work components, and concentrates on the influence of interactions on work system performance. In a relationship table or interaction between operators and work components can be seen in Table 1.

Table 2. Components of Work System

Component	Design Area	Consideration
Hardware	Design and layout of components	Process, equipment, access
Operator	Physical Characteristics of Skills	Body characteristics, strength, work capacity, posture, fatigue, and endurance.
	Recipient of information and process	The five senses (vision, hearing, etc.), attention, memory, etc.
	Individual and social characteristics	Age, gender, cultural background, ethnicity, skills, training, motivation, job satisfaction, and interest, saturation, behavior, etc.
Organization	Personnel / production organization	Work-rest time, work rotation, rotating work, interest, satisfaction, responsibility, social interaction, and others.

From the table above it can be concluded that the factors that influence nurse fatigue are physical work environmental aspect, responsibilities and duties of workers aspect, equipment and technology used aspect, company organization aspect and personal nurses' aspect.

4. Conclusion

Macro ergonomics has been known as a subdiscipline of ergonomics related to human, organizational and technological relations. This research was conducted at one hospital in Medan City by using 83 respondents from the nurses and 7 respondents from the hospital to determine the factors that influence of nurse fatigue. Analysis of work systems to reduce nurse fatigue with the macro ergonomic method of the Systems Engineering Initiative for Patient Safety (SEIPS) model in hospitals created on the consequences of explanations and consultations of nurses and hospital management. The factors that influence the work system of nurses in hospitals are identified both physical work environmental aspect, responsibilities and duties of workers aspect, equipment and technology used aspect, company organization aspect and personal nurses' aspect.

Acknowledgement

Author would like to acknowledge to Universitas Sumatera Utara for subsidy the financial funding to issue this paper and the hospital which are willing to be the object of research. We appreciate the efforts of all those who have cooperated in conducting this study.

References

[1] S Notoatmodjo, 1997 *Ilmu Kesehatan Masyarakat* (Jakarta: Rineka Cipta).
[2] Tarwaka, 2004 *Ergonomi untuk Keselamatan, Kesehatan Kerja dan Produktivitas* (Surakarta: UNIBA PRESS).
[3] E Nurmianto, 1998 *Ergonomi Konsep Dasar dan Aplikasinya* (Surabaya: Guna Widya).
[4] D A D Pratiwi., D Setiawan 2017 Jurnal Jurusan Keperawatan 1-8.
[5] Z Pourmovahed., K Nasiriani 2016 *Jurnal Open Journal Women's Health* **3** (1) 1-7.
[6] P D Burnout 2011 *Health Program* **92** (4) 47-50.
[7] L M Steege, J G Rainbow 2017 *Journal of Nursing Studies*, **67** 20-28.
[8] D J Y Ross, C Burns 2014 *Workplace Health Saf*, **62** (6) 256-261.
[9] M Hartono 2008 *Journal of Logistics and Supply Chain Management*, **1** (2) 61-71.
[10] L M Steege 2016 *Applied Ergonomics*, **56** 19-26.
[11] M Fauzan 2017 *Jurnal Wawasan Manajemen*, **5** (1) 89-105.
[12] S Deb, D Claudio 2015 *IIE Transaction on Helathcare System Engineering*, **5** 183-196.
[13] Siregar, I., Nasution, A. A., & Sari, R. M. (2017, March). Effect of total quality management on the quality and productivity of human resources. In *IOP Conference Series: Materials Science and Engineering* (Vol. 180, No. 1, p. 012116). IOP Publishing.
[14] Y R Nasution 2018 *JISTech*, **3** (2) 20-35.
[15] Rahmat R F, Herly E T, Siregar B, Syahputra M F, Sitompul O S 2018 RFID presence monitoring system as an input to measure the workload of employee *Proceedings of the 2017 4th International Conference on Computer Applications and Information Processing Technology, CAIPT 2017* pp 1-6

IOP Conf. Series: Materials Science and Engineering **851** (2020) 012043 doi:10.1088/1757-899X/851/1/012043

Seismic micro-zonation framework for earthquake risk in Medan-North Sumatera Indonesia using geotechnical engineering survey

Teguh Rahayu[1,*1], Miswar Budi Mulya[1*], Ayun Ria Ainun[2], Marzuki Sinambela[2,3]

[1]Department of Environmental and Natural Resource Management, Universitas Sumatera Utara, Medan, North Sumatera, Indonesia

[2]Badan Meteorologi, Klimatologi dan Geofisika, Indonesia

[3]Department of Physics, Faculties of Mathematics and Natural Science, Universitas Sumatera Utara, Medan, North Sumatera-20155, Indonesia

Corresponding author: *ayugraha76@gmail.com, miswarbm_bio@usu.ac.id, sinambela.m@gmail.com

Abstract. The micro zonation is very essential to identify the tectonic and geological formation in the study area, and still fundamental for determining the seismic source and establishing a realistic earthquake hazard model. Indonesia is located in the Pacific Ring of Fires, it is constantly at risk of earthquakes, tsunamis, and the eruption of volcanic. In this case, we focused to review the previous study of micro zonation in Indonesia and making the framework of micro-zonation for earthquake risk in Medan-North Sumatra, Indonesia. The goal of this study to illustrate the diversity of methodologies presently in use for the preparation of seismic hazard maps and to evaluate basic principles of zonation for different purposes and at different scales in new planning in Medan, North Sumatera. We select the micro zonation methods and looking for the opportunity for planning the micro zonation based on the literature review. The result of this review, the plan and framework of micro zonation in Medan, North Sumatera will be important and needed for mapping the seismic hazards. The output of seismic micro-zonation map based on the geotechnical engineering survey with the Multichannel Analysis of Surface Waves (MASW), Spatial Auto Correlation (SPAC) and Horizontal to Vertical Ratio (HVSR) for dominant period.

1. Introduction

Seismic micro-zonation is very important not only in other countries but also in Indonesia. The potential of the earthquake and geothermal activity, making Indonesia mitigate seismic risk. Indonesia is located between the Ring of Fire. This location making why Indonesia has so many earthquakes and potential of geothermal and volcanic activity. In general, the Indonesia archipelago has the unique geological condition [1], [2]. It is located between two big plates, Indo-Australia Plate in the south and Eurasia Plate in the north [3], [4].

IOP Conf. Series: Materials Science and Engineering **851** (2020) 012043 doi:10.1088/1757-899X/851/1/012043

Seismic micro-zonation is one of the steps, in earthquake risk mitigation study and needs a multidisciplinary approach with major influences from the field of geology, seismology, geophysics and geotechnical engineering. The seismic micro-zonation is very imperative to identify the tectonic and geological formation in the study area and also for beginning a realistic earthquake hazard models and for investigation [5]. The micro-zonation studied attempts to identify geologic zones of an area of interest with similar seismic hazards at the local scale.

Seismic micro-zonation is the most current tool for earthquake risk assessment and creates a database for planning pre-earthquake disaster management [6]–[8]. It has included delineation of the homogenous zones is seismological and geological characteristics and description of zone characteristics by dynamic parameter [9]. The seismic micro-zonation had been studied and developed in two decades, [10] initiated a model of seismic micro-zonation of the Skim Himalaya by integrated the geological and strong motion feature using geographic information system. These output of maps, generally a better spatial representation of seismic hazard including the site-specific analysis. Based on [11] present the deterministic seismic micro-zonation of Kolakat city by the computing of spectral ground motion amplification. One of the natural disaster that have expressively impacted on risk and damage is an earthquake, [12] investigating the seismic zone and earthquake hazard micro-zonation by using Neo-Deterministic seismic hazard analysis.

In this case, we review some literature about micro-zonation in other countries to compare the micro zonation which had been done in Indonesia, and a potential plan of micro zonation in Medan, North Sumatera. The aim of this study to evaluate micro zonation in some areas in Indonesia and how seismic micro-zonation is very important in a new location of micro-zonation in Medan, North Sumatera, Indonesia.

2. Literature Review

It has long been known that the location of Indonesia in the ring of fire can greatly influence the activity of earthquake and impact effect and risk. In micro zonation in some areas in Indonesia had been done. Figure 1 shows the micro zonation in some areas in Indonesia, started from Padang, Jakarta, Bandung, Semarang, Yogyakarta, Banjarnegara, Singaraja and Palu.

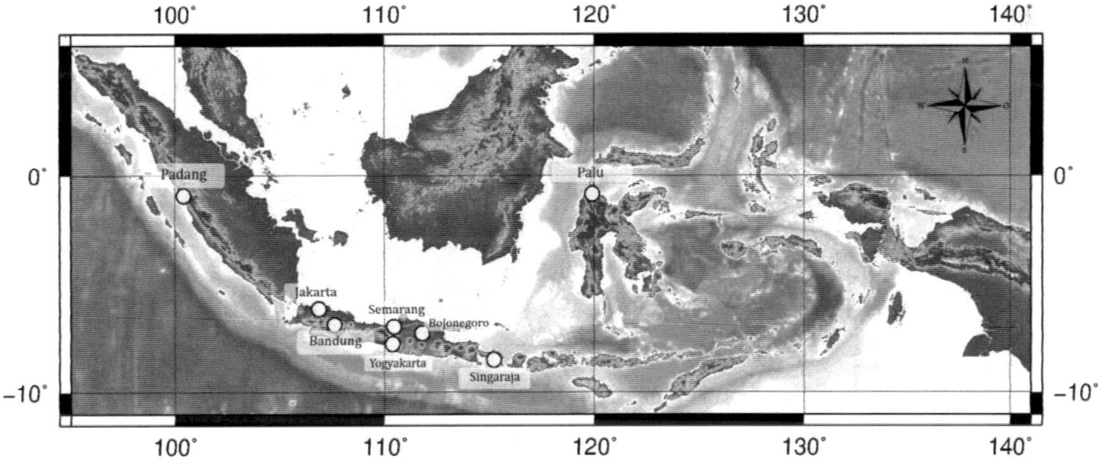

Figure. 1 A brief of history micro-zonation in some city in Indonesia

The micro-zonation in some areas in Indonesia had been done. The micro-zonation using some methods or models for plotting the map. In Table 1. show the methods of micro-zonation based on input data and be contingent on the scale of mapping in the level of zonation. The micro-zonation in Indonesia is very urgent based on the location of Indonesia at the Pacific in Ring of Fire. Generally, the histories of micro-zonation in Indonesia consisted of three grade show in the Table 1, grade I based on the historical earthquake and current information of geological and geomorphological maps. Grade 2, the micro-zonation according to the micro-tremor studies, and simplified geotechnical studies. Grade 3, based on the complete geotechnical investigations and ground response analysis.

Table 1. The reviews methods for micro-zonation in Indonesia

City/Year	**Grade I** • Historical of earthquakes • Existing information on geological and geomorphological maps	**Grade II** • Microtremor studies • Simplified geotechnical studies	**Grade III** • Complete geotechnical investigations • Ground response analysis
Palu, 2015 [13]	Deterministic and probabilistic approach, and	Microtremor measurement	-
Jakarta, 2015 [14]	Deterministic, probabilistic approaches, and seismotectonic map	Microtremor array measurements, output parameters: PSA contour, bedrock depth, site classification	Detailed seismic risk map
Semarang, 2015 [15], [16]	Deterministic approach, and geological map	Single station feedback seismometer measurement, output parameters: Spectral Acceleration	-
Bojonegoro, 2017 [2], [17]	Deterministic, probabilistic approach, and geological map	Microtremor measurement, output parameters: f_0, Kg	-
Padang, 2017 [18]	Geological map	Microtremor measurements, output parameter: type of soil	-
Singaraja, 2017 [19]	Geological map	Microtremor measurement, output parameters: Am, type of soil	-
Yogyakarta, 2017 [20]	Deterministic and probabilistic approach, soil condition, and geological map	Boreholes and microtremor measurement, output parameter: Empirical prediction of PGA	-
Bandung, 2018 [21], [22]	Geological map	Microtremor/ambient noise data, output parameters: f_0, Am, Kg	-

The methods for micro-zonation in Indonesia used the microtremor for the geotechincal enginering survey. And the parameter for micro-zonation produced the type of soil.

3. Methodology

The micro-zonation methods based on the level zonation had been done in some areas in Indonesia. The framework plan of micro-zonation can illustrate in Figure 2. The first step of the micro-zonation by using a geotechnical engineering survey by using MASW, SPAC and the HVSR for dominant period. Figure. 2 show the pre-processing of the geotechnical engineering survey produces the shear wave Velocity, Density function, and Lithological information. The second step by analyzing site response, soil properties and spectra acceleration to get the seismic micro-zonation map. The new framework of micro-zonation in Medan, North Sumatera will be done in 2020.

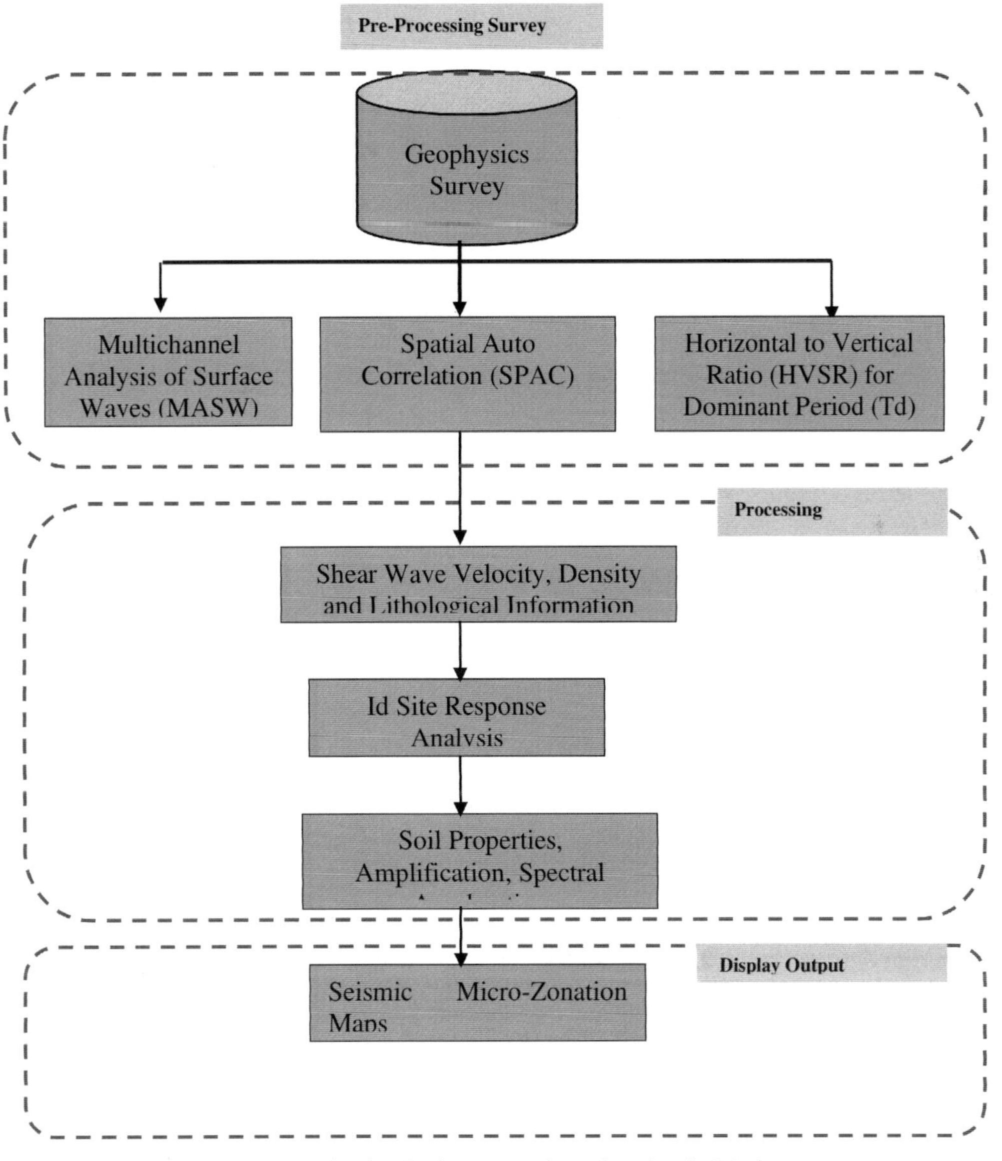

Figure. 2. The methods of micro-zonation planning in Medan

The micro-zonation methods in this study, based on the geotechnical engineering survey. The first method is MASW. The MASW is a seismic exploration. The technique first introduced in geophysics by [23], [24].The MASW evaluates ground stiffness by measurement shear wave velocity (Vs). MASW is useful the micro-zonation methods for exploring shallow geological structure and in particular, the relative shear strength of subsurface materials.

The second methods in geotechnical engineering survey is SPAC method. The SPAC was first proposed by [25] for understanding horizontally propagating waves or the relations between the temporal and spatial spectra of micro-tremors and their phase velocity characteristics [1], [13], [21], [24], [26]–[29]. The last methods for geotechnical engineering survey in this study is HVSR for dominant period (Td). The natural periods of vibration of the soil profile are most significant dynamic parameters involved in response analysis [27], [30], [31].

4. Result and Discussions

The framework of seismic micro-zonation result for the analysis of seismicity of North Sumatera as shown in Figure. 2 indicates the micro-zonation in Medan City as the capital of Prov. North Sumatera is a good investigation to evaluate and identify the seismology and geology structure in the field. There are three approaches for micro-zonation will be computed, the first approach by using geotechnical engineering survey by using MSAW, seconds, geotechnical engineering survey by using the SPAC and the last is using HVSR for dominant period in geotechnical engineering survey. The HVSR is a method for getting subsurface material from single station measurement by comparing the spectrum of horizontal component to the vertical components. HVSR method applied to ambient noise (Nakamura]. The goal of HVSR methods to feature the resonance frequency or fundamental mode of the soil. The deterministic approach is based on grade and level characteristics.

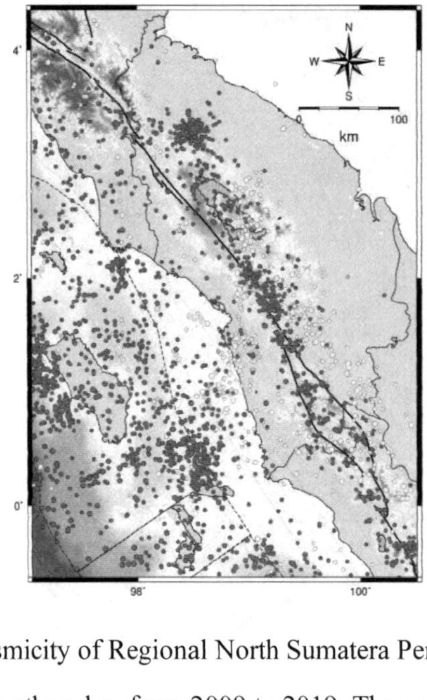

Figure. 3 Seismicity of Regional North Sumatera Period 2009-2019

Figure. 3 is the distribution of earthquakes from 2009 to 2019. The map seismicity had been classified based on the depth and the Magnitude. The depth and classification are divided into three-part, shallow (0-60 km), middle (60-100 km) and deep (> 100 km). The classification of Magnitude in the seismicity maps shows between small Magnitude (< 4 SR), Magnitude (4-5 SR) and Magnitude biggest 5 SR.

The location of Medan City, North Sumatera, Indonesia is close to Sumatera Fault, the earthquake activity in Medan is dominated of horizontal fault.

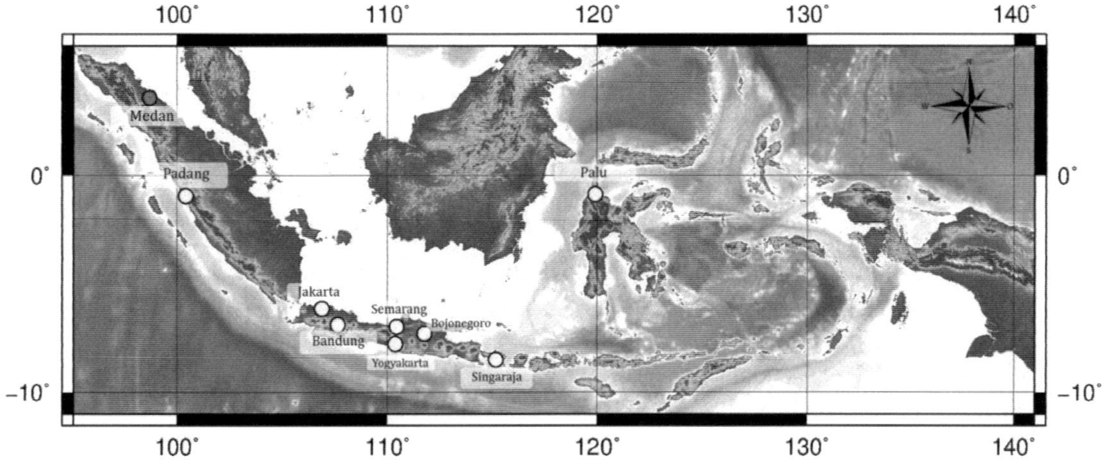

Figure. 4 The frame work map of micro-zonation in Indonesia

Figure.4 is a combination of the micro-zonation (yellow circle) that has been done and being done (red circle) in Medan, North Sumatera, Indonesia. The determining of Medan city as the next micro-zonation is crucial, seeing the location and condition of Medan City are disaster-prone areas and impact and risk of earthquake potential. Based on the [32] try to identify the earthquake activity in Deli Serdang on 2017 by analyzed the waveform from broadband station using machine learning approach. Based on Figure. 2, the seismicity of Medan City as the capital of North Sumatra or near to the Medan area like Deli Serdang indicated the distribution of earthquake in the field can affect the living condition and these building areas.

5. Conclusions

In this study, we present new micro-zonation in Medan, North Sumatera, Indonesia. The micro-zonation is urgent to identify the structure of geology and the environment of the Medan field. Based on the review methods of micro-zonation in Indonesia, we suggest using the approach from framework micro-zonation in the geotechnical engineering survey based on MASW, SPAC, and HVSR methods. The micro-zonation in Medan City as a potential disaster impact is recommended and a good way to identify the structure and geology in the Medan field based on the history of Medan as a potential impact and risk of earthquake activity. The new framework of micro-zonation in Medan, North Sumatra will be offer urban planners and city to mitigate earthquake risk.

Acknowledgment

We are grateful to the Department of Environmental and Natural Resource Management, Universitas Sumatera Utara and BMKG for making the geophysics survey available in 2020.

References

[1] Fernández J, 2004 *Geodetic and Geophysical Effects Associated with Seismic and Volcanic Hazards*. Birkhäuser Basel.
[2] Cummins P R, Meilano I,2017 and Geological Society of London, *Geohazards in Indonesia : earth science for disaster risk reduction*. Geological Society of London.
[3] Witton P, 2003 *Indonesia*. Melbourne: Lonely Planet.

[4] Hall R and Derek J B, 1996 *Tectonic evolution of Southeast Asia*. Geological Society, 1996.
[5] Anbazhagan P, 2013 Method for seismic microzonation with geotechnical aspects, *Disaster Adv.*, **6**(4), pp. 66–85.
[6] Ansal *et al.* A, 2004 Seismic microzonation for earthquake risk mitigation in Turkey, *13th World Conf. Earthq. Eng.*, no. 1428.
[7] Nath S K, 2007 Seismic Microzonation Framework – Principles & Applications, *Proc. Work. Microzonat.*, pp. 9–35.
[8] Albarello D, Socco L V, Picozzi M, and Foti S, 2015 in Italy : the role of seismic investigation, *First Break*, **33**, no. August, pp. 79–85.
[9] Mohanty W K and Walling M Y, 2008 First order seismic microzonation of haldia, Bengal Basin (India) using a GIS platform, *Pure Appl. Geophys.*, **165**(7), pp. 1325–1350.
[10] Nath S K, 2005 An initial model of seismic microzonation of Sikkim Himalaya through thematic mapping and GIS integration of geological and strong motion features, *J. Asian Earth Sci.*, **25**(2), pp. 329–343.
[11] Shiuly A and Narayan J P, 2012 Deterministic seismic microzonation of Kolkata city, *Nat. Hazards*, vol. 60, no. 2, pp. 223–240.
[12] Sari AM and Fakhrurrozi A, 2018 Earthquake Hazard Analysis Methods: A Review, *IOP Conf. Ser. Earth Environ. Sci.*, **118**(1).
[13] Thein P S *et al.*,2015 Estimation of S-wave Velocity Structure for Sedimentary Layered Media Using Microtremor Array Measurements in Palu City, Indonesia, *Procedia Environ. Sci.*, vol. 28, no. SustaiN 2014, pp. 595–605.
[14] Irsyam M *et al.* ,2015 Development of seismic risk microzonation maps of Jakarta city, *Geotech. Catastrophic Flooding Events - Proc. 4th Int. Conf. Geotech. Eng. Disaster Mitig. Rehabilation, GEDMAR 2014*, no. December, pp. 35–47.
[15] Partono W, Wardani S P R, Irsyam M, and Maarif S, 2015 Development of seismic microzonation maps of Semarang, Indonesia, *J. Teknol.*, vol. 77, no. 11, pp. 99–107, 2015.
[16] Partono W, Wardani S P R, Irsyam M, and Maarif S, 2016 Seismic microzonation of Semarang, Indonesia based on site response analysis using 30 M soil deposit model, *J. Teknol.*, vol. 78, no. 8–5, pp. 31–38.
[17] Hidayat S, Warnana D D, Koesuma S, and Cari C, 2017 Local Site Effects Evaluation Using Microtremor Measurements at North Side of Pandan Mountain, *J. Phys. Theor. Appl.*, vol. 1, no. 2, p. 89.
[18] Putra, *Seismic Microzonation For Urban Planning and Vulnerability Assessment OF non-Engineered Struture in Earthquake Prone Area, Padang, Indonesia*. 1384.
[19] Sabrianto L and Awali A, 2017 Seismic Microzonation Using Microtremor Measurement for Natural Disasters Mitigation of Earthquake at Regions Singaraja City the Province of Bali Indonesia, *ARPN J. Earth Sci.*, vol. 6, no. 1, pp. 1–6.
[20] Fathani T F and Wilopo W, 2017 Seismic microzonation studies considering local site effects for Yogyakarta City, Indonesia, *Int. J. GEOMATE*, vol. 12, no. 32, pp. 152–160, 2017.
[21] Pranata B *et al.*, 2018 Seismic microzonation of Bandung basin from microtremor horizontal-to-vertical spectral ratios (HVSR), *AIP Conf. Proc.*, vol. 1987.
[22] Srigutomo W, Handayani G, Pribadi A R, and Nugraha A M, 2018 Investigation of 1D shear wave velocity profile using the spatial autocorrelation (SPAC) method: Case study at West Park ITB Campus, Bandung, Indonesia, *ARPN J. Eng. Appl. Sci.*, vol. 13, no. 6, pp. 2096–2107.
[23] S. A. to Q. M. of HMA, MASW-Park Seismic. [Online]. Available: http://www.parkseismic.com/. [Accessed: 09-Mar-2020].
[24] C. H. B. P. Ark, R. I. D. M. Iller, J. I. X. Ia, J. U. I. Vanov, and K. G. Survey, The conventional seismic, 2007.
[25] Bin Huang A and Mayne P W, 2008 Geotechnical and Geophysical Site Characterization - Proceedings of the 3rd International Conference on Site Characterization, ISC'3: Introduction, *Geotech. Geophys. Site Charact. - Proc. 3rd Int. Conf. Site Charact. ISC'3*, no. January.

[26] Cooperation G T and Series D, 2014 Microzonation Studies using Microtremor - Manual - Microzonation Studies using Microtremor - Manual -, no. October.

[27] Gosar A, 2017 Study on the applicability of the microtremor HVSR method to support seismic microzonation in the town of Idrija (W Slovenia), *Nat. Hazards Earth Syst. Sci.*, vol. 17, no. 6, pp. 925–937.

[28] Chávez-García F Jand Aguirre J, 2012 Exploration of subsoil structure in México city using correlation of micrometremors, *Geofis. Int.*, vol. 51, no. 3, pp. 251–270, 2012.

[29] Wathelet M, Jongmans D, and Ohrnberger M, 2005 Direct inversion of spatial autocorrelation curves with the neighborhood algorithm, *Bull. Seismol. Soc. Am.*, vol. 95, no. 5, pp. 1787–1800, 2005.

[30] Isburhan R W P, Nuraeni G, Verdhora R, Yudistira T, Cipta A, and Cummins P, 2019 Horizontal-to-Vertical Spectral Ratio (HVSR) Method for Earthquake Risk Determination of Jakarta City with Microtremor Data, *IOP Conf. Ser. Earth Environ. Sci.*, **318**(1)

[31] Grippa A *et al.*, 2011 Use of the HVSR method to detect buried paleomorphologies (filled incised-valleys) below a coastal plain: The case of the metaponto plain (Basilicata, Southern Italy), *Boll. di Geofis. Teor. ed Appl.*, **52**(2), pp. 1–16.

[32] Darnila E, Ula M, Tarigan K, Limbong T, and Sinambela M, 2018 Waveform analysis of broadband seismic station using machine learning Python based on Morlet wavelet, *IOP Conf. Ser. Mater. Sci. Eng.*, **420**(1)

IOP Conf. Series: Materials Science and Engineering **851** (2020) 012044 doi:10.1088/1757-899X/851/1/012044

Classification of elderly based on attitude for health promotive in Deli Serdang using machine learning approach

Masdalifa Pasaribu[1,2,*1], Albiner Siagian[1*], Rahayu Lubis[1], Dewi Ramadani[2], Rika Subarniati Triyoga[3], Etti Sudaryati[1]

[1]Faculty of Public Health, Universitas Sumatera Utara, Medan, North Sumatera-20222, Indonesia
[2]Sekolah Tinggi Ilmu Kesehatan Rumah Sakit Haji, Deli Serdang, Indonesia
[3]Faculty of Public Health, Universitas Airlangga, Surbaya, Indonesia

Corresponding author: * masdalifa.hajisumut@gmail.com

Abstract. Health promotion is the process of enabling people to increase control over and to improve their health. Health promotion is a global issue. In the most city in Indonesia, health is improving. In this case, we collected the elderly data from Deli Serdang, North Sumatera, Indonesia. The aim of this study to classify the elderly dataset using Support Vector Machine and Logistic Regression as a machine learning approach. The result of this study indicated the SVM and Logistic Regression is available to train and test the dataset. The dominant sample in this study is female. The evaluation of the elderly dataset using two models of machine learning indicated there is a positive attitude with the age for health promotion.

1. Introduction

The situation and the main problem faced in Indonesia show undergoing a major process of social, and economical change. The global issued about health promotion makes a different situation in Indonesia. Indonesia has a large population in the world. The demographics composition of Indonesia shown great economic potential and a large population with more islands. The public health problem in a big population is more related to health promotion. The health promotion based on the [1] need not only to be built into all the policies but also if utilized efficiently will lead to positive health outcomes. Indonesia with many islands and the city is a priority for health promotion. The health promotion for the elderly is very important. In Indonesia, health promotion for many older adults is motivated to avoid processing negative information. Health promotion is meant for the entire population [2], [3], if a specific group within a population is singled out as the recipient of health promotion interventions, it's because of valid reason, such us epidemiological concern or preferences in social policy [4]. The elderly has long been deserted as the addressee of health promotion activities. In this study, we focused on collected and observed the dataset from Deli Serdang, North Sumatera, Indonesia. The demographic of the Deli Serdang showed that the productive age (15-64 years) was greater than the non-productive age (children and elderly). The local government of Deli Serdang had been given information about the demographic bonus for a teenager in 2016 [5]. The location Deli Serdang unclose to Medan shows the population increase per year. Based on the histories of the population in Deli Serdang, we try to collect and observe the data in the field, for recoded the dataset of the elderly from more than 54 years. The elderly data will be pre-processing and processing first by classify using some model in statistic approach or machine learning. The main aim of this study to identification and classification the elderly to compare the attitude observed in Deli Serdang, North Sumatera, Indonesia. Classification of the elderly dataset by training the data and testing the data using a machine learning approach. We run the data by using the Support Vector Machine and Logistic Regression.

Machine learning is a perspective on modeling data with focused model skill, and close to the statistical learning as a perspective on modeling data. To train and test the data, in this case, we use the SVM as a supervised learning model and logistic regression that analyze data. The classification of SVM is a supervised machine learning algorithm that can be used for the regression problem. And the classification of logistic regression based on the concept of probability. The classification of the elderly for health promotion is very urgent for training and testing the data, it will be better to evaluate the data by using some model like SVM and logistic regression.

2. Data and Methods

2.1 Data

The dataset of the elderly had been collect in Deli Serdang, North Sumatera from 2018 years. There is more than 200 sample in this study was recorded. From the dataset of the elderly, the female is dominant than male. There are 242 females and 40 males in these samples. The percentage of the dataset can show in Figure. 1.

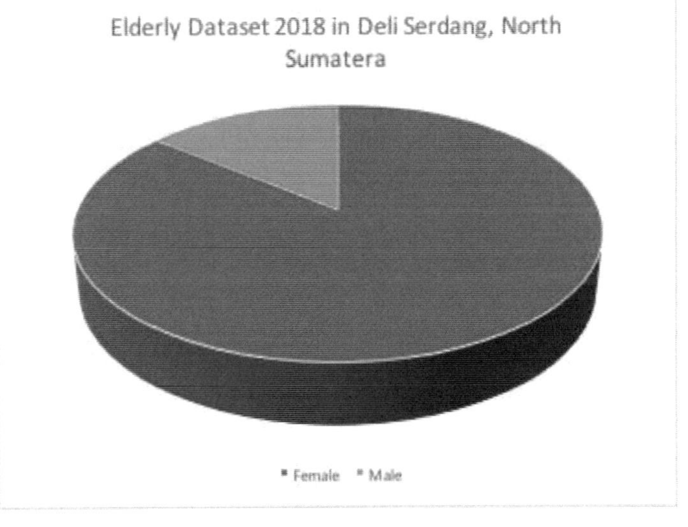

Figure. 1 Elderly dataset 2018 in Deli Serdang, North Sumatera, Indonesia

The health promotion in Deli Serdang had been studied [6]. There is a positive attitude with the sample with the predisposing factor associated with clean and healthy life behavior. The map of this study show in Figure.2. Deli Serdang is located in North Sumatera and the capital of the district is Lubuk Pakam. The research area is located approximately 30 km east of Medan.

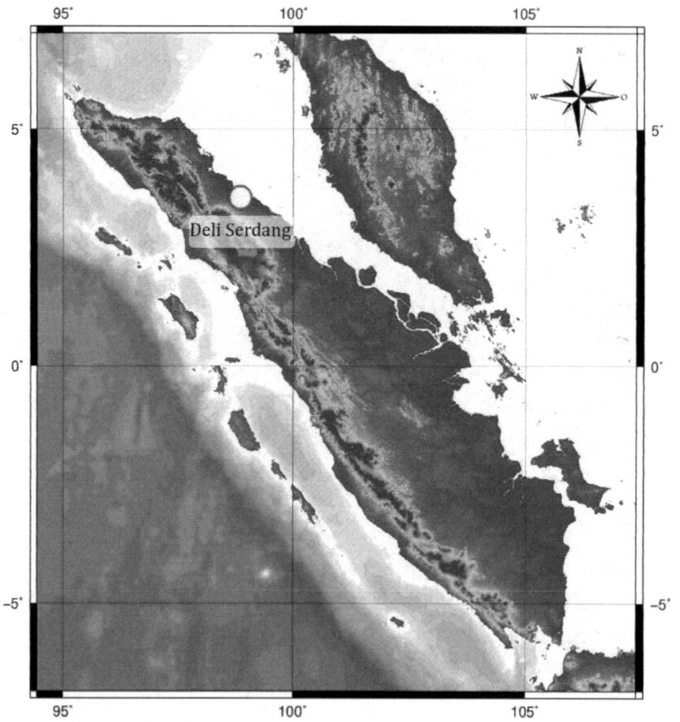

Figure. 2 The research area for collected the elderly dataset in Deli Serdang, North Sumatera

2.2 Methods

The model to classify the elderly dataset by using SVM and Logistic Regression. SVM is one of the machine learning approaches. Other methods such as Learning Vector Quantization also can be applied[20]. Logistic regression is a supervised classification algorithm, in a classification problem, the target variable or output y can take only discrete values for a given set of features X [7], [8]. The model of logistic regression using the sigmoid function as shown in Eq. 1

$$g(z) = \frac{1}{1+e^{-z}} \quad (1)$$

The logistic regression means binary which having binary target variables, but there can be two more categories of target variables that can be predicted by it [9]. SVM is a discriminative classifier formally representation of different classes in hyperplane in multidimensional space. SVM is a supervised machine learning algorithm that can be used for both regression challenges and classification [10]–[13]. The SVM mostly used in classification problems, in this case, we plot each data item as a point in n-dimensional space with the value of each feature being the value of a particular coordinate [14]–[17]. SVM classification had been studied for multidiscipline field [18], [19].

3. Result and Discussions

The result of the analysis of the classification of the elderly dataset for health promotion with 282 sample data in Deli Serdang can show in Figure. 3 and Figure. 4. The visualizing of SVM result in Figure. 3 based on selected the age (elderly) with the score of attitude remark in this research in Deli Serdang, North Sumatera. In the SVM model, we train the dataset to one or the other of two categories, and assigns new examples to one category or other, making it a non-probabilistic binary linear classifier.

IOP Conf. Series: Materials Science and Engineering **851** (2020) 012044 doi:10.1088/1757-899X/851/1/012044

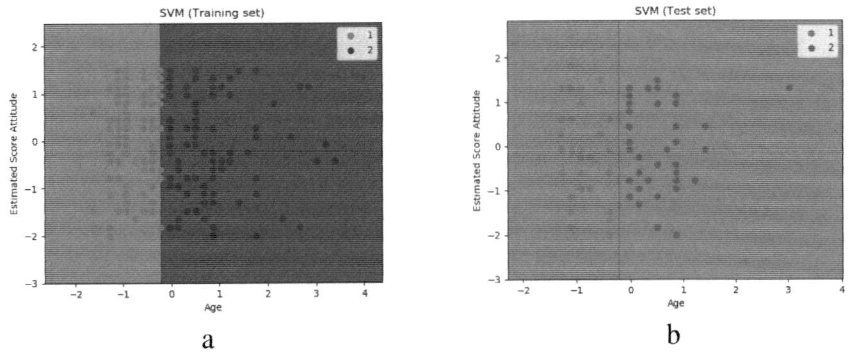

Figure. 3 Visualizing SVM result for elderly dataset: a. Training set, b. Testing set

The visualizing of SVM in Figure. 3 shows the result of the elderly data set since in the result a hyper-plane has been found in the training set result and verified to be the best one in the set result. The SVM result in Figure. 3 (a) shows the separation boundary of the two classifiers. This creates a way to classify the vector or the features of the elderly dataset. In this study, there are observable blue dots in the red region and red dots in the green region. The SVM result in Figure. 3 (b) show the results of the testing elderly dataset, which could minimize the correct prediction. The green dots in the green region and red dots in the red region could be separated unless anon-linear boundary was used since the hyperplane is linear.

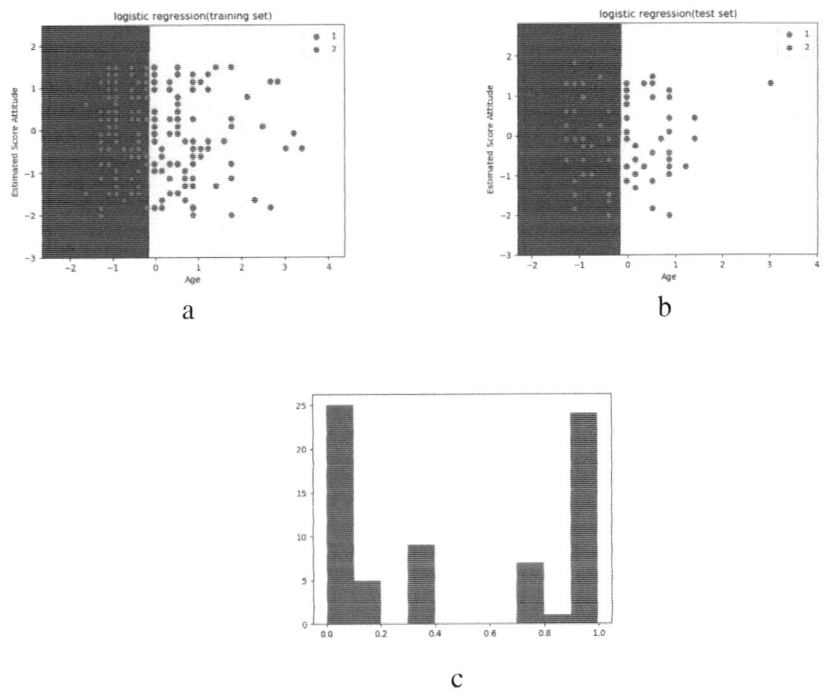

Figure. 4 Visualizing Logistic Regression result: a. Training elderly dataset, b. Testing elderly dataset, c. Probability values

Figure. 4 (a) is a logistic regression result of the elderly dataset, the graph shows the clear correlation between the dependent and independent variables. It is obvious that as Age (elderly dataset) and Attitude score increase, each individual has a higher likelihood of being green dots. There will almost always be some degree of error - or at least there should be, otherwise our model is probably guilty of overfitting. Figure 4. (b) shows the testing result using logistic regression. The graph looks very similar to the last, but with fewer observations. We can classify the age (elderly dataset) and attitude score increase will lead to a higher probability of clicking the advertisement. Figure. 4 (c) is probability value for classify the elderly dataset and create a prediction based on specified cut of value.

4. Conclusions

In this study, we present the elderly dataset from collected in Deli Serdang, North Sumatra, Indonesia. Based on the result, indicated the training and testing dataset with age and attitude scores in pre-processing indicated there are positive attitudes. The classification using two models in this study is better to identify the elderly dataset. The SVM model shows 90 percent accuracy and the logistics regression show 93 percent accuracy. Both of the SVM and logistics regression models show good performance for train and testing the dataset of the elderly in health-promoting.

Acknowledgment

We are grateful to STIKES RS. Haji and Universitas Sumatera Utara. We would like thank to Kabupaten Deli Serdang.

References

[1] Kumar S and Preetha G S, 2012 Health promotion: An effective tool for global health, *Indian J. Community Med.*, **37**(1), pp. 5–12.

[2] Sujudi, 1997 Health promotion towards the 21st century - Indonesian policy for the future, *Health Promot. Int.*, **12**(4), pp. 279–282.

[3] Delhi, N 2001 Health promotion in indonesia,

[4] Golinowska S, Groot W, Baji P, and Pavlova M, 2016 Health promotion targeting older people, *BMC Health Serv. Res.*, **16**, no. Suppl 5, pp. 4–6..

[5] Santosa H, Mutiara E, and Juanita J, 2017 An Understanding Demographic Bonus and Its Implication among Teenagers in Deli Serdang District, vol. 1, no. PHICo 2016, pp. 404–408,

[6] Simbolon P and Sianipar C M, 2018 Predisposing Factors Associated with Health Behavior in Deli Serdang, North Sumatera, *J. Heal. Promot. Behav.*, vol. 3, no. 2, pp. 130–138, 2018.

[7] Introduction to Logistic Regression - Towards Data Science. [Online]. Available: https://towardsdatascience.com/introduction-to-logistic-regression-66248243c148. [Accessed: 09-Mar-2020].

[8] Understanding Logistic Regression - Tutorialspoint.dev. [Online]. Available: https://tutorialspoint.dev/language/python/understanding-logistic-regression. [Accessed: 09-Mar-2020].

[9] Hutabarat L T, H. A. S, M. Sinambela M, and Limbong T, 2019 Classification of Student ' S Air Traffic Control Skill Using, **3**(2), pp. 166–169.

[10] Hsu C W and Lin C J, 2002 A comparison of methods for multiclass support vector machines, *IEEE Trans. Neural Networks*, **13**(2), pp. 415–425.

[11] Smola A J and Schölkopf B, 2004 A tutorial on support vector regression, *Statistics and Computing*, vol. 14, no. 3. pp. 199–222.

[12] Meyer D, Leisch F, and Hornik K, 2003 The support vector machine under test, *Neurocomputing*, **55**, no. 1–2, pp. 169–186.

[13] Support-vector machine - Wikipedia. [Online]. Available: https://en.wikipedia.org/wiki/Support-vector_machine. [Accessed: 09-Mar-2020].

[14] Decoste D and Schölkopf B, 2002 Training invariant support vector machines, *Mach. Learn.*, vol. 46, no. 1–3, pp. 161–190.

[15] Boser B E, Guyon I M, and Vapnik V N, 1992 Training algorithm for optimal margin classifiers,

in *Proceedings of the Fifth Annual ACM Workshop on Computational Learning Theory*, pp. 144–152.
[16] Wenzel F, Galy-Fajou T, Deutsch M, and Kloft M, 2017 Bayesian Nonlinear Support Vector Machines for Big Data, in *Lecture Notes in Computer Science (including subseries Lecture Notes in Artificial Intelligence and Lecture Notes in Bioinformatics)*, vol. 10534 LNAI, pp. 307–322.
[17] Cortes C and Vapnik V, 1995 Support-vector networks, *Mach. Learn.*, **20**(3), pp. 273–297.
[18] Ritonga A P, Adithya A R, Agustina I, and Limbong T, 2019 Classification of Labor Using Support Vector Machine in North Sumatera, vol. 04, pp. 240–243.
[19] Anwar D, Sembiring K, Tarigan K, Sinambela M, Abubar A R, and Hasibuan D, 2020 Machine learning approach for turbulence forecasting using support vector machine, *IOP Conf. Ser. Mater. Sci. Eng.*, vol. **725**, no. 1
[20] Rahmat R F, Pulungan A F, Faza S, Budiarto R 2017 Image Classification on Ribbed Smoked Sheet using Learning Vector Quantization *Journal of Physics: Conference Series* **801**(1).

IOP Conf. Series: Materials Science and Engineering **851** (2020) 012045 doi:10.1088/1757-899X/851/1/012045

Time Frequency Signal Classification Using Continuous Wavelet Transformation

Helbert Sirait[1,*1], Kerista Sebayang[1*], Syahrul Humaidi[1], Timbangen Sembiring[1] Kerista Tarigan[1], Kurnia Sembiring[1], Teguh Rahayu[2], Ayun Ria Ainun[2], Marzuki Sinambela[1,2]

[1]Department of Physics, Faculties of Mathematics and Natural Science, Universitas Sumatera Utara, Medan, North Sumatera-20155, Indonesia

[2]Badan Meteorologi, Klimatologi dan Geofisika, Indonesia

Corresponding author: *keristasebayang@usu.ac.id, nikygavinarsene@gmail.com

Abstract. Time-frequency analysis can provide useful information in digital signal seismic data processing and interpretation. The energy concentration of the spectrum depends on the consistency of function of the time-frequency analysis and instantaneous frequency variation digital signal. In this case, we used the digital signal seismic from selected seismometer broadband which deployed in Sumatera Island. The aim of this study to classify the waveform based on the time-frequency analysis using continuous wavelet transform (CWT). The sample data used the earthquake of 20 February 2018 in North Sumatera. The result indicated the classification between the horizontal and vertical components from the seismometer broadband is different. The classification of vertical is affected by seismic source and horizontal component affected the site effect.

1. Introduction

Time-frequency analysis can reveal useful information in the digital seismic data. The high resolution of the time-frequency representation is of great importance to characteristic the signals. Digital signal seismic or earthquake wave characterization is essential for better understanding wave and local site effect subject of the earthquake [1], [2]. Earthquake activity in Indonesia is potential and will be interested in the study. The location of Indonesia in the ring of fire-making earthquake activity and disaster will be happening every time. Based on these situations, Indonesia deployed more seismometer broadband to monitoring and analysis of the earthquake. Earthquake waves can record from seismometer broadband in digital signals in SEED data [3]–[5]. In seismic design, the waveform of seismometer broadband consists of horizontal (East and North) and vertical (Z) component [6]. The vertical component of input ground motion has been taken into account and an important role to be considered, regarding the seismic response of the critical structure. In this case, we used the waveform which recorded from selected seismometer broadband in the Sumatra earthquake in 2018, 20 February. The waveform had been recorded form any sensor not only close to the event but also sensor from other

IOP Conf. Series: Materials Science and Engineering **851** (2020) 012045 doi:10.1088/1757-899X/851/1/012045

fields. The characteristic of the digital signal is very important to evaluate the performance of the sensor. To evaluate each sensor which recoded in this case, we can use time-frequency analysis to classify the vertical and horizontal component. The time-frequency is an important technology in digital signal seismic processing. Time-frequency analysis computed using CWT. The CWT has played a key role in time-frequency information in many fields of digital signal processing.

The main problem of the digital signal seismic is how to classify the waveform based on the vertical and horizontal component as affected by the seismic source and site effect.

This study aim to classification and identification of the time-frequency based on the vertical and horizontal component.

2. Data and Methods

2.1 Data

The digital signal seismic data were employed from the BMKG-IA network. We use the Sumatra earthquake in origin time 2018-02-20 T 23:40:49, Mag. 4.8 in lat. 1.03, long. 98.69 and depth 80.0 Km in Northern Sumatra, Indonesia. We select the station of seismometer broadband in Table 1.

Table 1. The Station of Seismometer Broadband

No	Code	Network	Lat (N)	Long (E)	Stream
1	BKNI	IA	0.33	101.04	BHE, BHN, BHZ
2	GSI	IA	1.30	97.58	BHE, BHN, BHZ
3	KCSI	IA	3.52	97.77	BHE, BHN, BHZ
4	MNSI	IA	0.80	99.58	BHE, BHN, BHZ
5	PBSI	IA	-0.05	98.28	BHE, BHN, BHZ
6	SNSI	IA	2.41	96.33	BHE, BHN, BHZ
7	TDNI	IA	0.57	97.82	BHE, BHN, BHZ

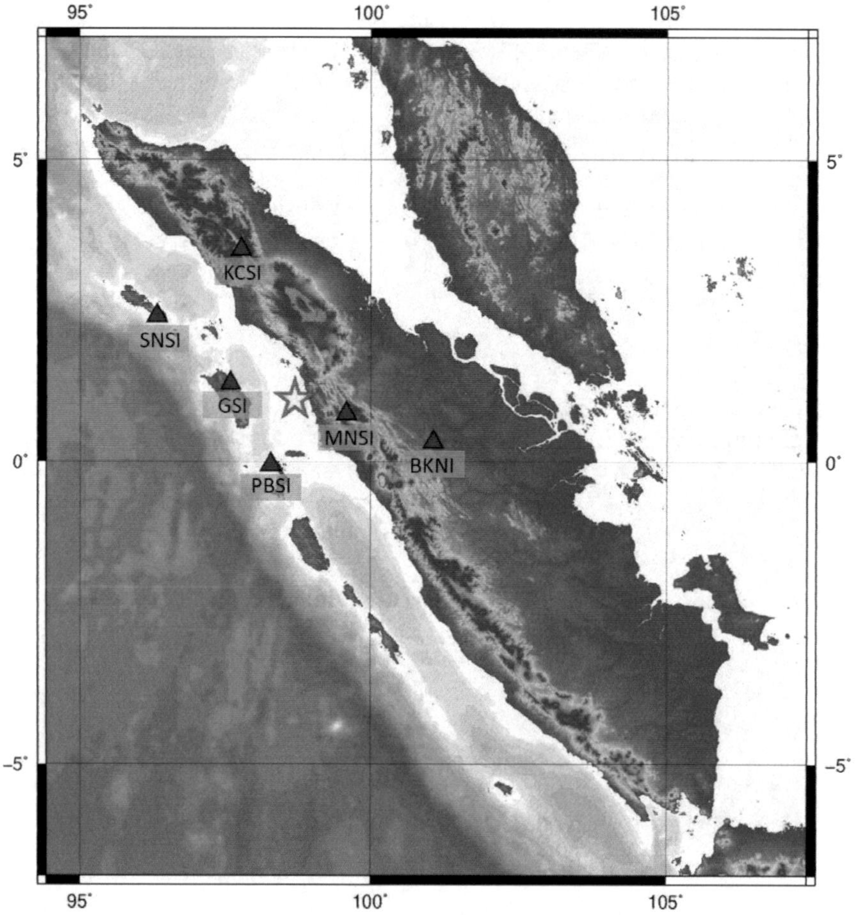

Figure. 1 Earthquake Location and Distribution of Selected Seismometer Broadband

The distribution of selected seismometer broadband and earthquake location shown in Figure. 1. In this study, we selected the sensor which deployed in Sumatera Island, we selected sensor based on the location of the earthquake to the seismometer broadband.

3. Methods

The methods of this study using Continuous Wavelet Transform, CWT is descending cross-correlation between a signal s(t) and family of wavelets as shown in Eq.1

$$W_s(a,b) = \frac{1}{\sqrt{a}} \int s(t)\Psi * \left(\frac{t-b}{a}\right) dt \tag{1}$$

where a and b are scales and time shifts of a indication wavelet Ψ, respectively, $\Psi *$ is the complex conjugate of the reference wavelet, t is time and $W_s(a,b)$ is the time-scale representation of the signal [7]–[11].

The temporal length of the wavelets used in the cross-correlation is varied differing on the frequency component under investigation. The effective and operative at identifying a feature in digital seismic data as combining the machine learning wavelet-based. In this study, we use the mlpy as phyton open-source machine learning library built on NumPy/SciPy and GNU Scientific Libraries. Mlpy provides a wide range of state of the art machine learning methods [12], [13] such as Multi layer Perceptron or Deep Learning [14].

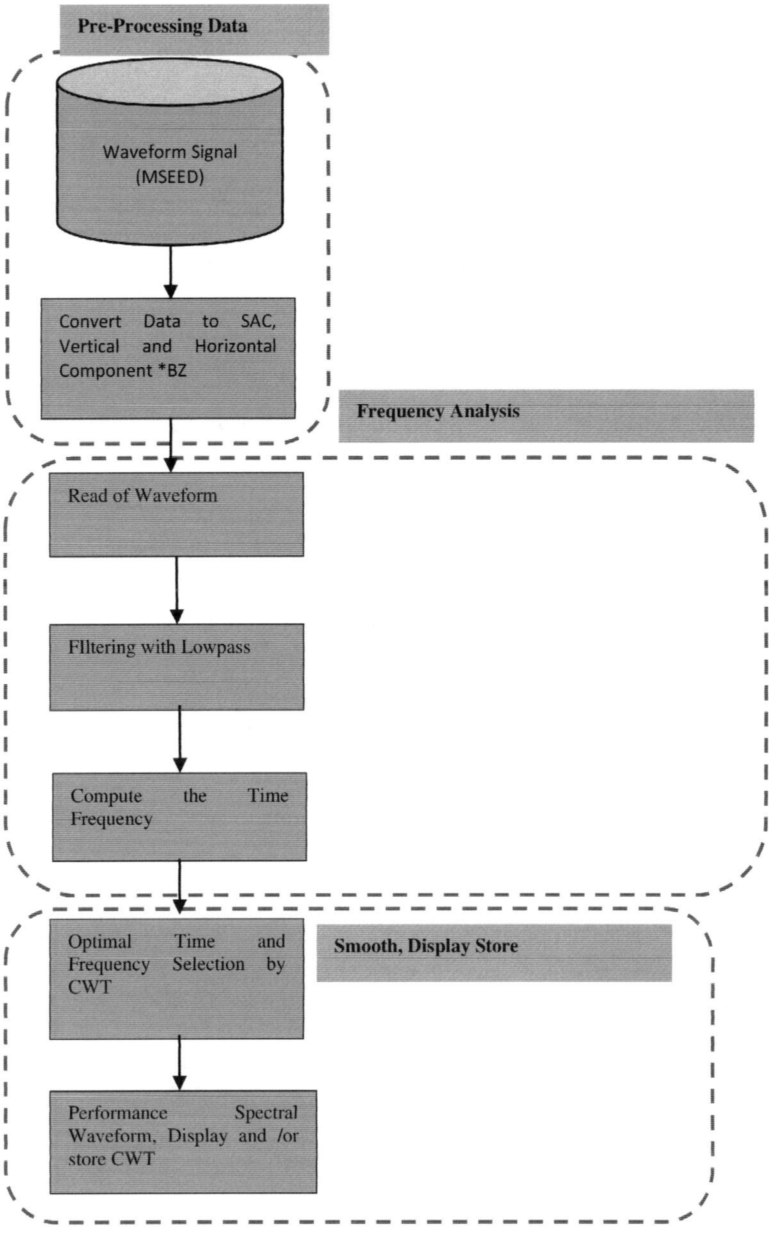

Figure. 2. The methods of Time Frequency Classification

The general methods of this study show in Figure. 2, the methods of Time-Frequency classification starting from converting the row data MSEED to the vertical and horizontal component. Time-Frequency analysis computed by CWT and using package machine learning python (mlpy). The CWT for time-frequency analysis produced a high resolution in time and frequency in spectral.

IOP Conf. Series: Materials Science and Engineering **851** (2020) 012045 doi:10.1088/1757-899X/851/1/012045

4. Result and Discussion

The result of the analysis of the time-frequency of the selected sensor in this study can show in Figure 3. The classification of time-frequency based on CWT had been present, the computation of CWT is an accurate efficacious and efficient method to improve the quality of the digital earthquake signal. Classification of a digital signal earthquake in this study had been a matter of extensive research.

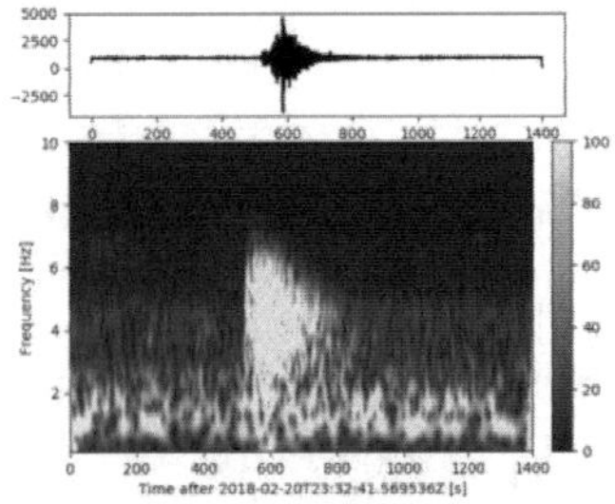

a. BKNI –BHZ

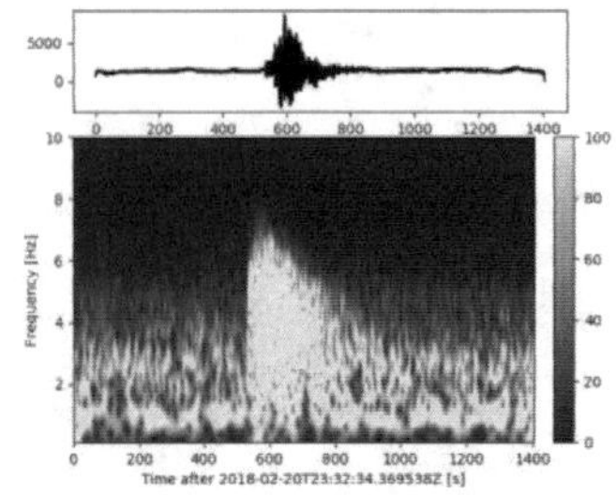

b. BKNI –BHE

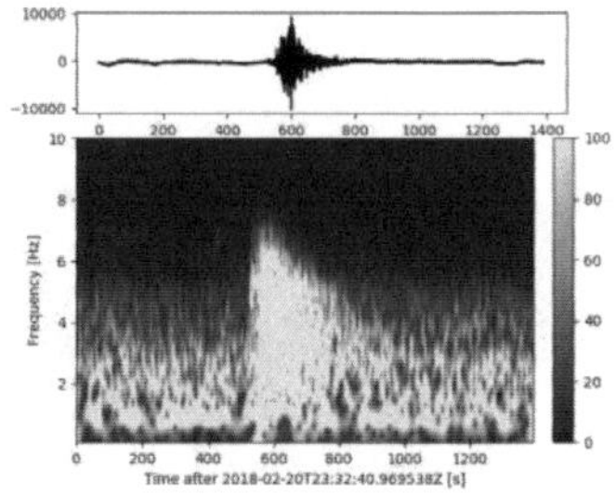

c. BKNI -BHN

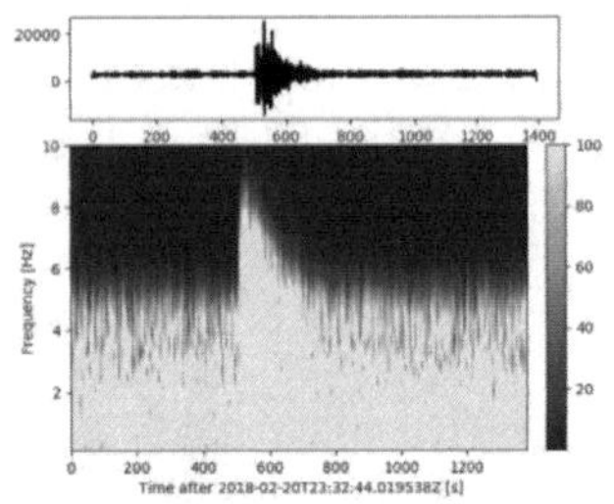

d. GSI- BHZ

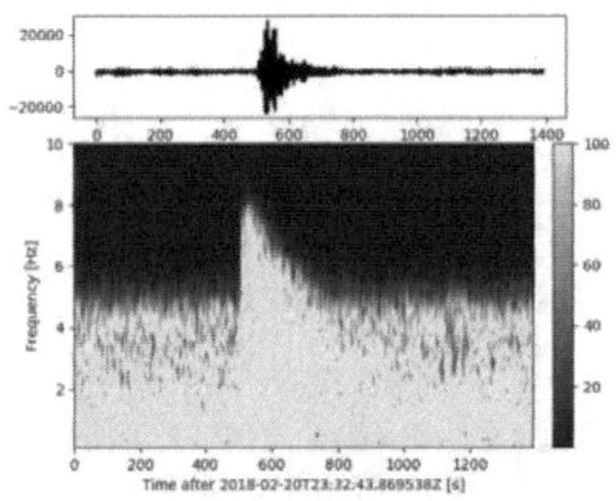

e. GSI-BHE

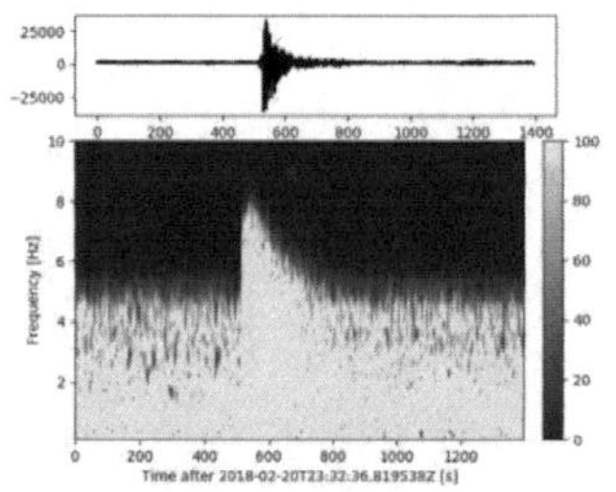

f. GSI-BHN

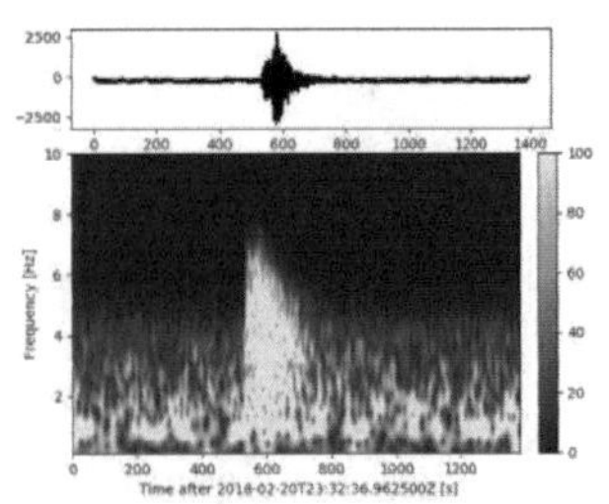

g. KCSI – BHZ

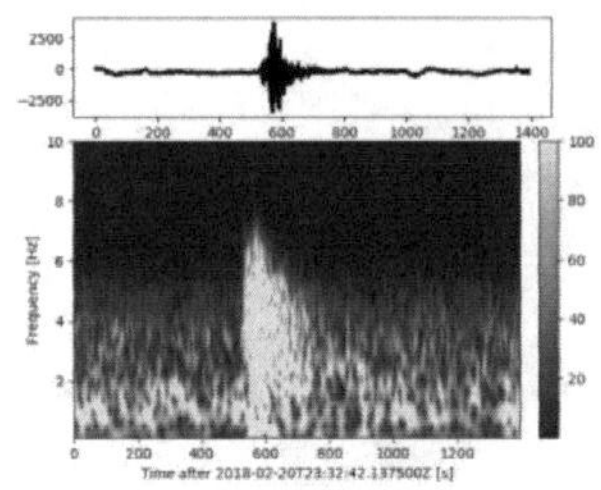

h. KCSI-BHE

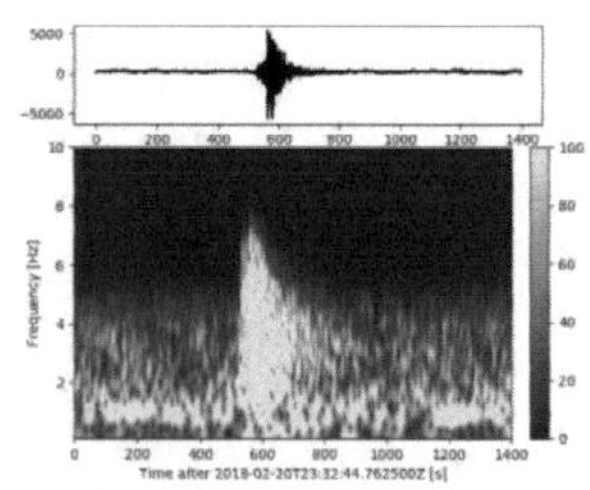

i. KCSI-BHN

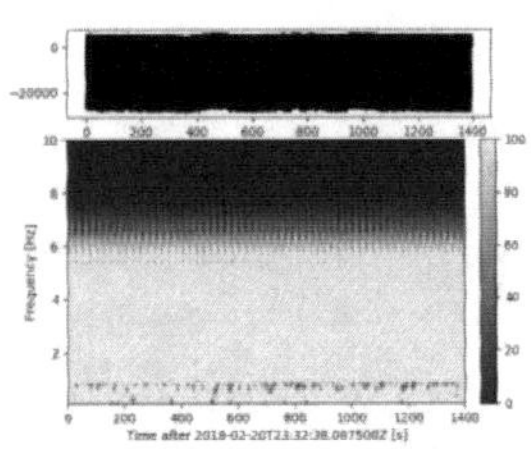

j. MNSI-BHZ

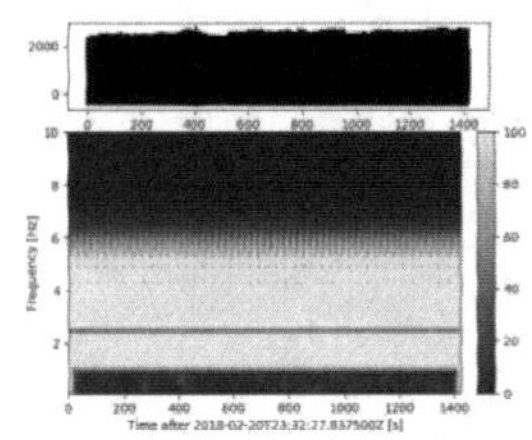

k. MNSI- BHE

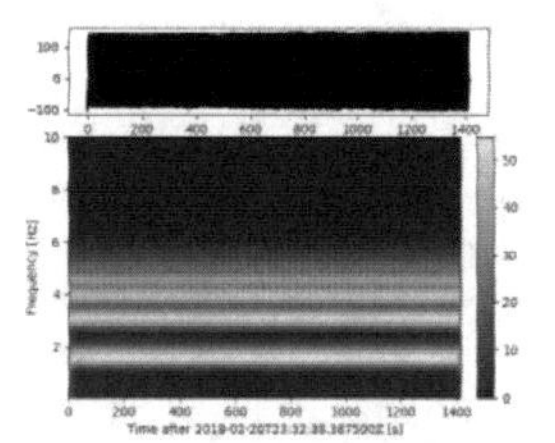

l. MNSI-BHN

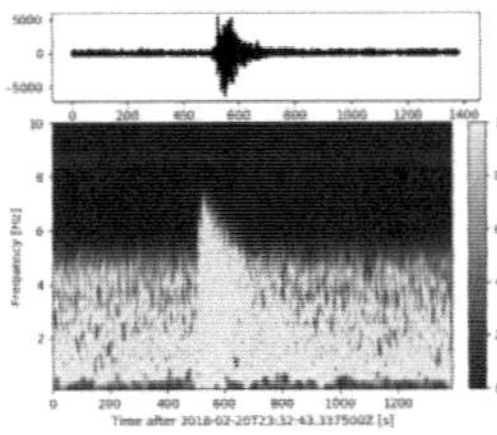
m. PBSI –BHZ

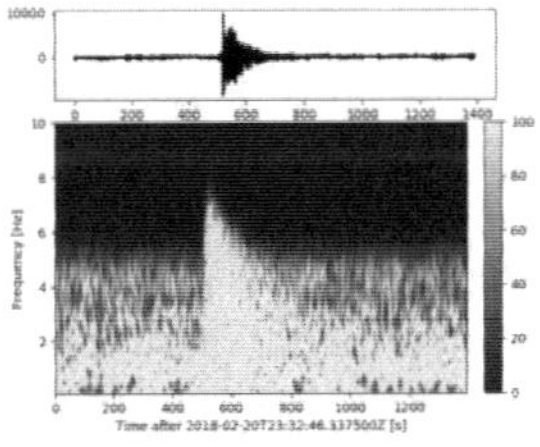
n. PBSI – BHE

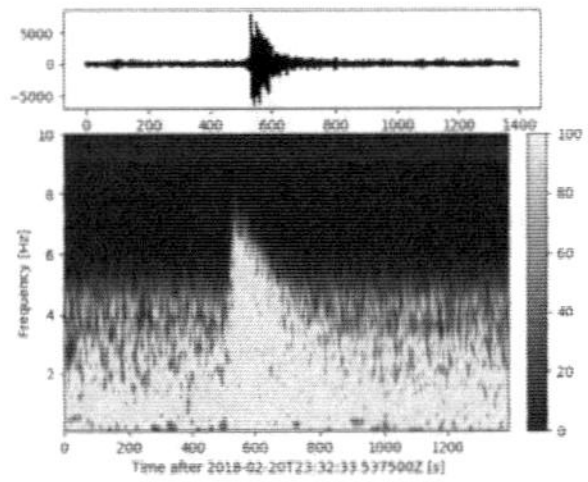
o. PBSI=BHN

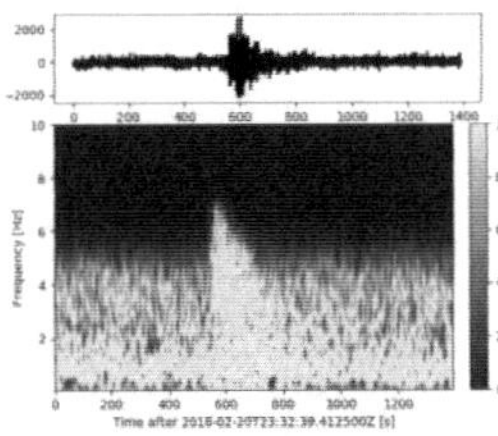
p. SNSI-BHZ

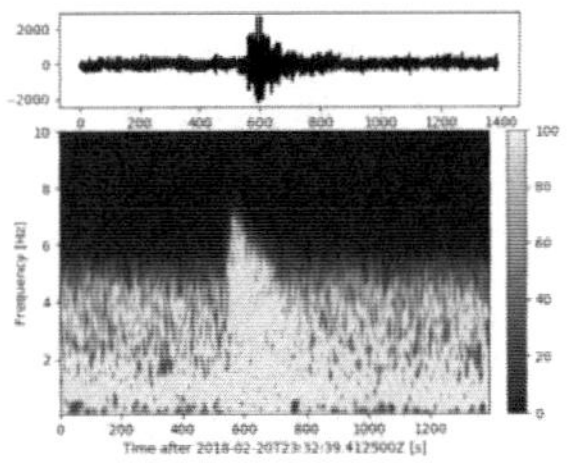
q. SNSI-BHE

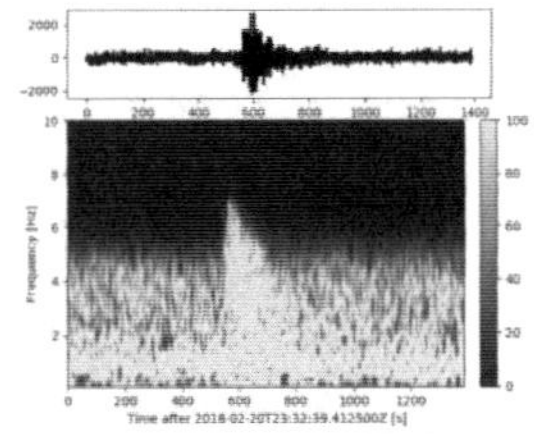
r. SNSI-BHN

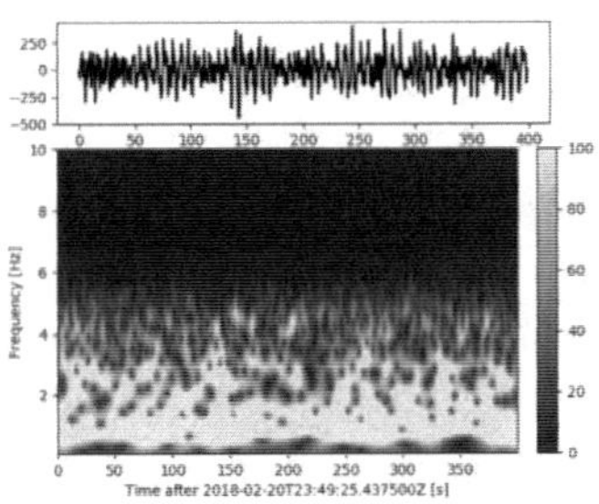
x. TDNI-BHZ

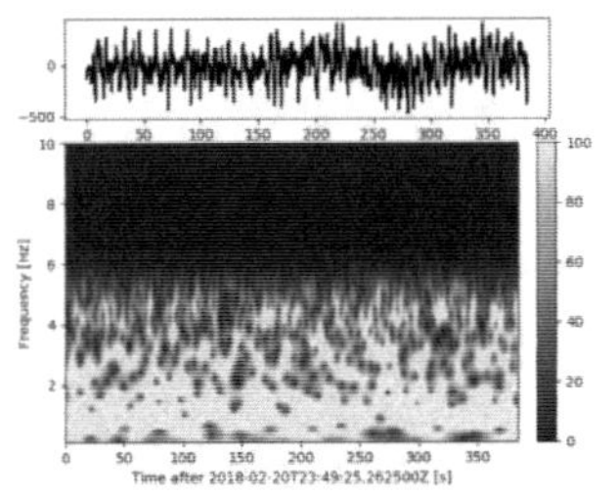
y. TDNI-BHE

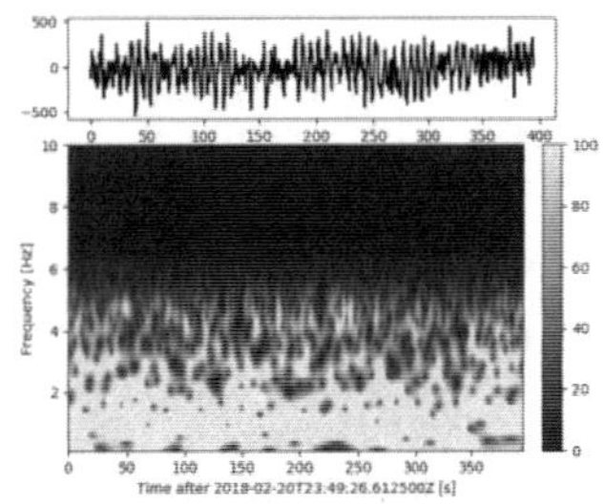
z. TDNI-BHN

Figure. 3 Time-Frequency Classification based on CWT

The Time-Frequency classification based on CWT had been plotted in Figure. 3. The classification of digital signals computed for the vertical and horizontal components. The different time-frequency representation is presented in Figure. 3. The CWT representation is calculated using a morlet wavelet. We select seven sensor stations close to the earthquake location. Figure. 3 (a-c) is spectral of the BKNI sensor in the time-frequency. The waveform of the signal on the top indicates the observation seismic record of an earthquake. The computation of BKNI for vertical (BHZ) and horizontal (BHE, BHN) component is clear to detect the event in the P phase and S phase. The result of the BKNI sensor shows the frequency energy. Figure. 3 (d-f) is the time-frequency spectral of the GSI sensor form record event location. Both of the GSI station in Figure. 3 (d-f), KCSI station in Figure. 3 (g-i), PBSI station in Figure. 3 (m-o) and SNSI station in Figure. 3 (p-r) is clear to detect the earthquake waveform in this study. The classification of time-frequency shows the P phase, S phase, and energy. Different time-frequency spectrogram like a representation of MNSI in Figure. 3 (j-l) and TDNI in Figure. 3 (x-z). The result of the MNSI station shows the sensor not recorded the event and had seismic noise. The TDNI station shows not detected the waveform of earthquake activity, the time-frequency classify can be not described P phase and S phase. The classification of vertical component is affected by the seismic source and the horizontal component is affected by the site effect.

5. Conclusions

In this study, we present the time-frequency classification using CWT. Based on the result, the variable time-frequency resolution of the CWT enables analysis of a truly vertical and horizontal component of broadband digital signals. The characteristic of vertical and horizontal components shows the phase difference is likely to show a simple relationship when it plotted in frequency. The phase difference becomes constant in terms of the horizontal and vertical components in the lower frequency range. The classification of time-frequency using CWT has designed an information confidence measure of time-frequency and multi-scale.

Acknowledgment

We are grateful to the Department of Physics, Universitas Sumatera Utara and BMKG for collected the data waveform.

References

[1] Bard P Y and Riepl-Thomas J, 2016 Wave propagation in complex geological structures and their effects on strong ground motion Interpacific Project View project Dam-Seismic Hazard View project Wave propagation in complex geological structures and their effects on strong ground motion

[2] Huerta-Lopez C and Shin Y 2000 Time-frequency analysis of earthquake records, ... *Earthq. ...*, pp. 1–8, 2000.

[3] Bormann P, Klinge K, and Wendt S 2018 Data Analysis and Seismogram Interpretation, *Network.*

[4] Doody C D *et al.*, 2018 Effects of thermal variability on broadband seismometers: Controlled experiments, observations, and implications, *Bull. Seismol. Soc. Am.*, **108**(1), pp. 493–502, 2018.

[5] Alsagaby S A 2019 Understanding the fundamentals of proteomics *Curr. Top. Pept. Protein Res.*, **20**(3), p. 25

[6] Jaiswal O R and Jain S K, 2004 13 th World Conference on Earthquake Engineering, *Indian Concr. J.*, **78**(11), pp. 74–76.

[7] Darnila E, Ula M, Tarigan K, Limbong T, and Sinambela M, 2018 Waveform analysis of broadband seismic station using machine learning Python based on Morlet wavelet, *IOP Conf. Ser. Mater. Sci. Eng.*, **420**(1).

[8] Ula M, Darnila E, Siagian P, Siagian L, Peristiwanto, and Sinambela M, 2018 Machine learning on waveform spectral analysis of nuclear explosion from broadband seismic station in Indonesia, *IOP Conf. Ser. Mater. Sci. Eng.*, **420**(1).

[9] Koutsoukos ET and Melis NS, 2005 A horizontal component broadband seismic sensor based on an inverted pendulum, *Bull. Seismol. Soc. Am.*,**95**(6), pp. 2462–2471.

[10] Kristeková M, Kristek J, and Moczo P, 2009 Time-frequency misfit and goodness-of-fit criteria for quantitative comparison of time signals, *Geophys. J. Int.*, **178**(2), pp. 813–825

[11] Shan X, Tian F, Cheng, Yang C, and Xin W, 2019 Spectral decomposition and a waveform cluster to characterize strongly heterogeneous paleokarst reservoirs in the Tarim Basin, China, *Water (Switzerland)*, **11**(2), pp. 1–21

[12] mlpy - Machine Learning Python. [Online]. Available: http://mlpy.sourceforge.net/. [Accessed: 11-Mar-2020].

[13] Albanese D, Visintainer R, Merler S, Riccadonna S, Jurman G, and Furlanello C, 2012 mlpy: Machine Learning Python.

[14] Rahmat R F, Dennis, Sitompul O S, Purnamawati S, Budiarto R, 2019 Advertisement billboard detection and geotagging system with inductive transfer learning in deep convolutional neural network *Telkomnika* **17**(5), pp 2659-2666.

IOP Conf. Series: Materials Science and Engineering **851** (2020) 012046 doi:10.1088/1757-899X/851/1/012046

Crude Palm Oil Product Quality Control Using Seven Tools (case study: XYZ Company)

Rosnani Ginting[1], Wanli[2] and Alfin Fauzi Malik[3]
[1,2,3]Department of Industrial Engineering, Universitas Sumatera Utara, Medan, Indonesia 20155

E-mail: rosnani_usu@yahoo.co.id

Abstract. Competition in the palm oil processing industry in Indonesia is very competitive. One of the factors that influence the good name of the company is the quality of the product produced. PT. XYZ is a palm oil processing factory with export and domestic market share. In order for the company to be able to compete with other companies engaged in similar industries, the company is required to need to carry out strict checks on product quality. PT. XYZ faces several obstacles to meet the quality standards of its main products namely Crude Palm Oil (CPO). The problem that is often found related to quality in CPO is the quality of CPO that does not meet consumer standards even though the company has implemented quality control in CPO production. This can reduce consumer confidence in the products sold by the company. CPO product quality standards set are free fatty acid levels (FFA) (<5%), water content (<0.15%) and impurities content (<0.02%). This study aims to determine what types of defect affect the quality of CPO and what causes the defect. This research uses seven tools method consisting of stratification, check sheet, histogram, pareto diagram, scatter diagram, control chart and cause and effect diagram. From the results of the Pareto diagram analysis it appears that the dominant defect that occurs in CPO is high free fatty acid levels with a percentage of 39.3%, then followed by defect with high impurities content of 35.4% and high water content of 25.3%. By using the Pareto 80/20 principle, the defect factor which is further analyzed is the defect of high free fatty acids and high levels of impurities. Through scatter diagrams and correlation analysis, the value of r = 0.5, where the value indicates that high free fatty acid defects and high impurities have a positive effect on the number of defects. By using a causal diagram analysis (fishbone diagram) it was found that the factors causing the high level of damage to this palm oil product are influenced by 3 main factors, namely raw materials, humans, and machinery.

1. Introduction

In the midst of globalization and free competition at this time it is very necessary for the ability of companies to compete in the midst of uncertain economic conditions. Companies must determine the right direction and policies in order to maintain economic stability in the company. The increasing competition between companies, encourage each company to increase efficiency and quality standards precisely in all fields. Only companies that have high competitiveness can survive one of them is to prioritize quality improvement.

Quality control is a system of verification and maintenance of a level or degree of quality of the product or process desired by careful planning, use of appropriate equipment, continuous inspection and corrective actions when needed, so quality control is not just an inspection activity or determine whether the product is good (accept) or bad (reject).

XYZ company is a palm oil processing factory with export and domestic market share. Companies are required to produce high quality products so that the company is able to compete with other companies engaged in similar industries. In meeting the needs of consumers, companies still find obstacles from within the company, namely regarding product quality, where the products produced still have a high level of defects. Where the quality standards of the factory consist of free fatty acid levels (<5%), impurities levels (<0.15%) and water content (<0.02%) where the factory production capacity is ± 60 tons / hour. Based on the above conditions, it is necessary to analyze the company using seven tools consisting of stratification, check sheets, histograms, pareto diagrams, scatter diagrams, control charts and cause and effect diagrams (fishbone), where the purpose of this study is to determine what types of defect affect the quality of CPO and what causes the defect using fishbone diagrams.[1].

2. Theoretical Background

The study was conducted at PT. XYZ, where the data taken are free fatty acid levels (FFA), impurities and water content contained in CPO. The study uses seven tools in Statistical Process Control (SPC). The function of the seven control devices is to improve process improvement, so that it will be obtained[2]:

a. Increased competitiveness.
b. Decreasing cost of quality and increasing price flexibility.
c. Increase resource productivity.

Data can be obtained from two main sources namely primary sources and secondary sources. data obtained from primary sources is called primary data, that is data obtained by searching / digging directly from the source by the relevant researcher. Secondary data is data that has been collected and processed by other parties so that it does not need to be explored / sought by the relevant researcher but only to quote or retrieve [3].

The data processing steps taken are as follows [4]:

a. Stratification, namely the Division and Grouping of Data into smaller categories and have the same characteristics.
b. Check Sheets is a chart for collecting data. Clearly and regularly designed check sheets will help to collect accurate and related data and allow data to be read and used easily
c. A histogram is a portrait of product variations or the results of a process. The histogram displays the frequency distribution of various measurements. The measured variables are along the horizontal x-axis and are grouped into various measurements. The frequency at which each measurement occurs is arranged along the vertical y axis
d. Pareto diagrams can be used to display problem categories graphically so that they can be prioritized correctly. Pareto charts or diagrams show which problems must be dealt with first by showing the proportion of the total problems, each of which consists of smaller problems.
e. Scatter plot diagrams are used to uncover possible causal relationships. it is constructed by plotting two variables against each other on a pair of axes. scatter plots cannot prove that one variable cause another, but it shows how a pair of variables are related and the strength of the relationship.
f. Control charts usually display limits that can be explained by statistical variability as usual. if your process is performing within these limits, it is said to be in control, if not, it's out of control.
g. Cause and effect diagrams (fishbone diagrams) organize and display relationships between different causes for the effect being examined. This chart helps organize the brainstorming process. the main cause categories are placed on the main branches which are connected back to the backbone, and various sub-causes are attached to the branches.

3. Research Methodology

This study was conducted at XYZ company on CPO products. The data used in this study is secondary data while the data is obtained indirectly which can form file documents, archives or company records. XYZ Company itself collects defect data by taking samples 12 times a day by using an erlenmeyer bottle, after which it is tested for free fatty acid levels, air levels and impurities in the laboratory.

This study uses seven tools in data processing, of which seven tools consist of

a. Stratification
b. Check Sheets
c. Histogram
d. Pareto
e. Scatter Plot Diagram
f. Control Charts
g. Cause and Effect Diagram.

4. Result

After all the necessary information has been collected, an analysis is carried out using seven tools. The initial data collected can be seen in table 1.

Table 1. Total Defects in Palm Oil

No	Number of Samples examined (n)	Number of High Free Fatty Acids	Number of High Impurities	Number of High Water Content	Total Defects/ Day
1	12	3	8	6	17
2	12	11	7	1	19
3	12	9	8	1	18
4	12	6	8	2	16
5	12	9	5	5	19
6	12	4	7	4	15
7	12	7	5	3	15
8	12	6	5	3	14
9	12	5	4	3	12
10	12	7	9	9	25
11	12	12	9	2	23
12	12	12	7	3	22
13	12	11	2	8	21
14	12	7	8	1	16
15	12	9	6	5	20
16	12	8	5	9	22
17	12	5	7	8	20
18	12	5	3	4	12
19	12	3	5	7	15
20	12	9	5	5	19
21	12	3	6	5	14
22	12	3	10	7	20
23	12	7	9	3	19
24	12	3	3	6	12
25	12	12	8	4	24
26	12	6	5	3	14
Total	**312**	**182**	**164**	**117**	**463**

Table 1. shows that the dominant defect is high FFA levels with 182 defects in 1 month, followed by impurities and water content with 164 and 117 defects respectively. The steps to identify defect using the seven tools are as follows.

a. Stratification

Stratification of types of defects in CPO production include:

IOP Conf. Series: Materials Science and Engineering **851** (2020) 012046 doi:10.1088/1757-899X/851/1/012046

- High levels of Free Fatty Acid (FFA)
- High levels of impurities
- High levels of water content

b. Check Sheet
CPO defects check sheet can be seen in table 2.

Table 2. Average CPO Defects Check Sheet

No.	Type of Defects	Average defects of CPO
1	High levels of Free Fatty Acid (FFA)	182
2	High levels of impurities	164
3	High water content	117
Total		**463**

c. Histogram
The histogram is used to describe the spread of defect. The histogram can be seen in Figure 1.

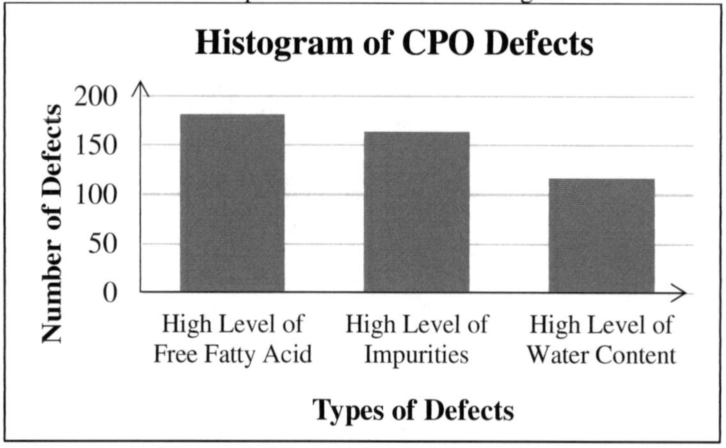

Figure 1. Histogram of CPO Defects

Based on histogram checking, obtained defect factors that mostly occur in CPO are high levels of free fatty acids, followed by high levels of impurities and high levels of water content.

d. Pareto Diagram
Pareto diagram shows the percentage of damage, where the percentage of damage can be seen in table 3.

Tabel 3. CPO Defects Percentage

No.	Types of Defects	Number of Defect Products	Percentage (%)	Cumulative Percentage (%)
1	High levels of Free Fatty Acid	182	39,3	39,3
2	High levels of impurities	164	35,4	74,7
3	High levels of water content	117	25,3	100

The concept of the pareto diagram used is 80/20, where 80% of disabilities are caused by 20% of the causes of defect. From the pareto diagram the highest percentage of defects is High levels of Free Fatty Acid with a percentage of 39.3%, followed by high levels of impurities (35.4%) and high water content levels (25.3%)[5].
Pareto diagram can be seen in Figure 2.

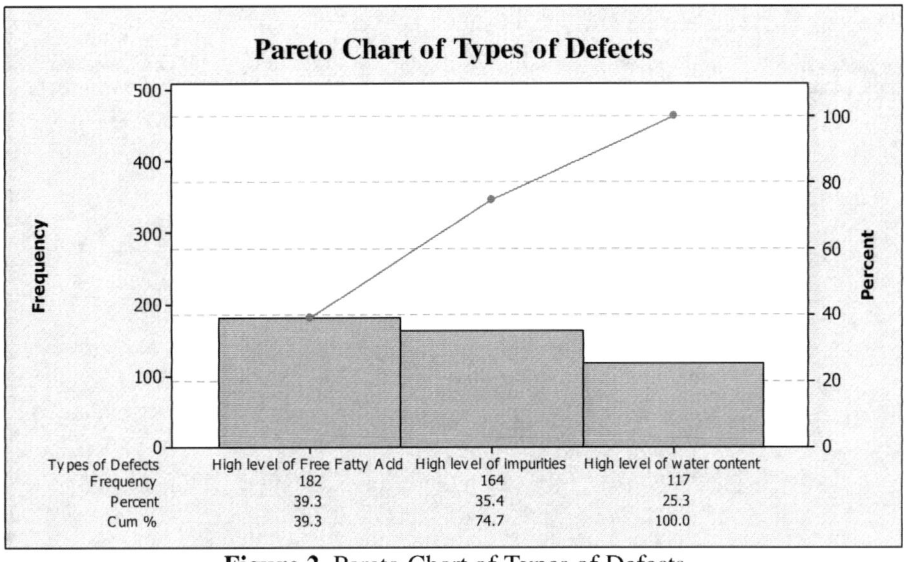

Figure 2. Pareto Chart of Types of Defects

e. Scatter Diagram

The first analysis was performed with two variables between the amount of high FFA levels as an independent variable and the number of disabilities as the dependent variable. The regression equation obtained after calculating the parameter values a and b is:

$$Y = 11{,}97 + 0{,}83x$$

After that, scatter diagrams are made with the equation that has been obtained using SPSS software. Furthermore, the correlation values of the two variables were calculated and the value of r = 0.665 was obtained, where the value of 0.665 indicates that high FFA has a strong positive influence on the number of defects [6].

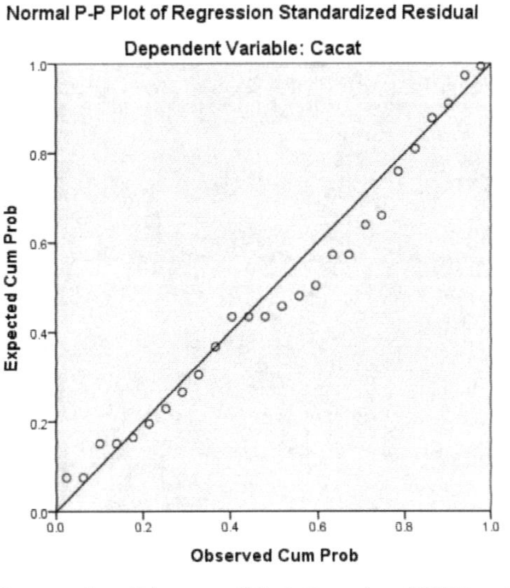

Figure 3. Scatter Regression Diagram High Levels of FFA with Total Defects

The second analysis is carried out between the variable amount of high impurities and the number of defects, where the regression equation obtained can be seen as follows.

$$Y = 11{,}97 + 0{,}83x$$

After that, scatter diagrams are made with the equation that has been obtained using SPSS software.

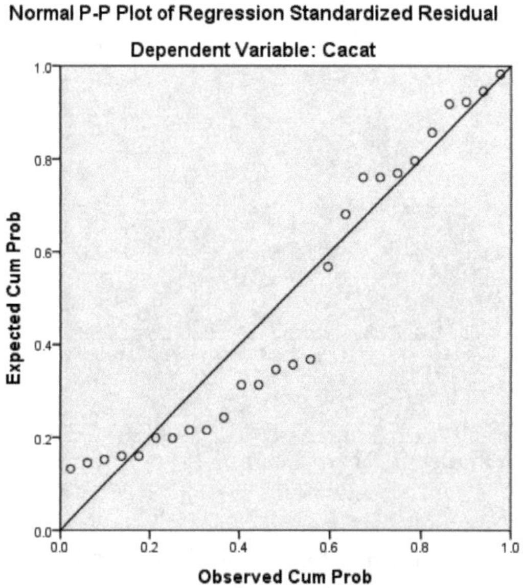

Figure 4. Scatter Regression Diagram of High Impurities Level with Total Defects

Furthermore, correlation analysis is performed to see the relationship between the 2 variables. The correlation values of the two variables were calculated and the value of r = 0.5, where the value of 0.5 indicates that high FFA and high impurities gives a positive enough influence on the number of defects [7].

f. Control Chart

The control chart aims to ensure that a process is in control and continuously monitors variations of the process. The type of control chart that will be used is a map c. The first test is carried out on high FFA data levels, where the test results show that all data are within control limits. Mapping data into a control chart can be seen in Figure 5.

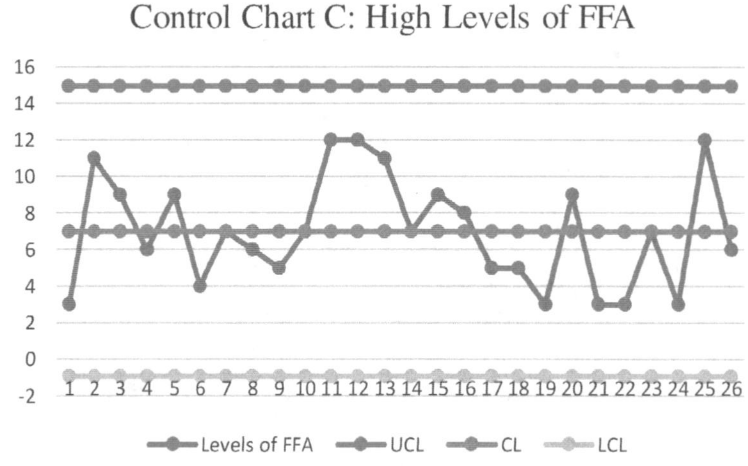

Figure 5. Control Chart C High Fatty Acid Levels of CPO Products

The second test was carried out on high impurities data. The results of tests conducted on high dirt content data indicate the data are within control limits so there is no need for revisions.

Control chart chart of high impurities data on CPO can be seen in Figure 6.

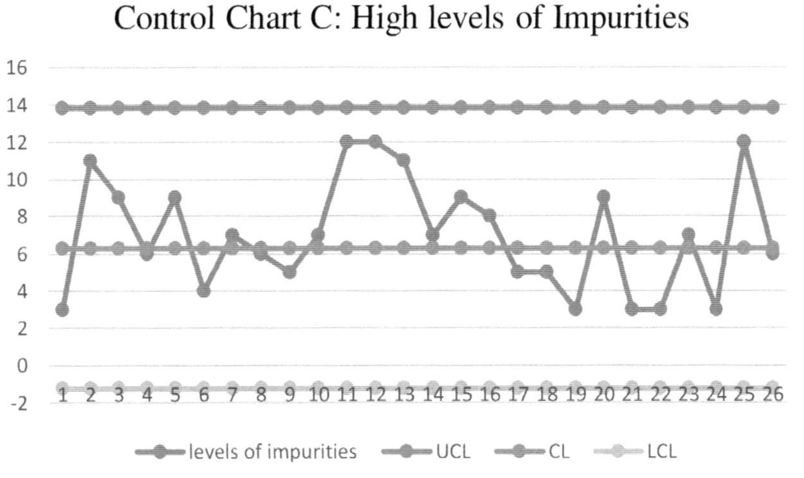

Figure 6. Control Chart C High Levels of Impurities

g. Cause and Effect Diagram

The settlement method using this diagram is used to analyze and determine the factors that cause high FFA levels and high impurities in CPO oil. A causal diagram of high FFA levels in CPO can be seen in Figure 7.

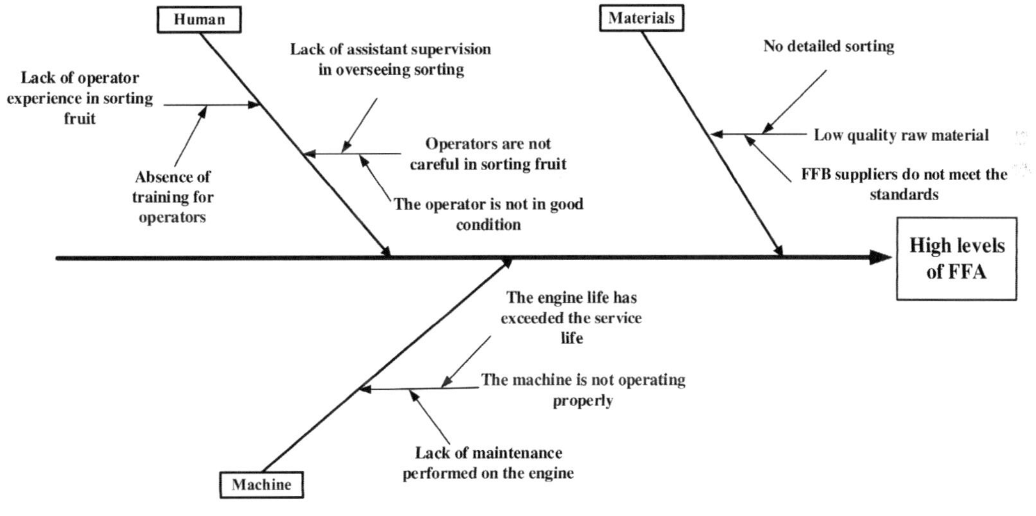

Figure 7. Cause and Effect Diagram (Fishbone) Causes of High FFA Levels in CPO Oil

Aspects that affect high levels of impurities in CPO can be seen in Figure 8.

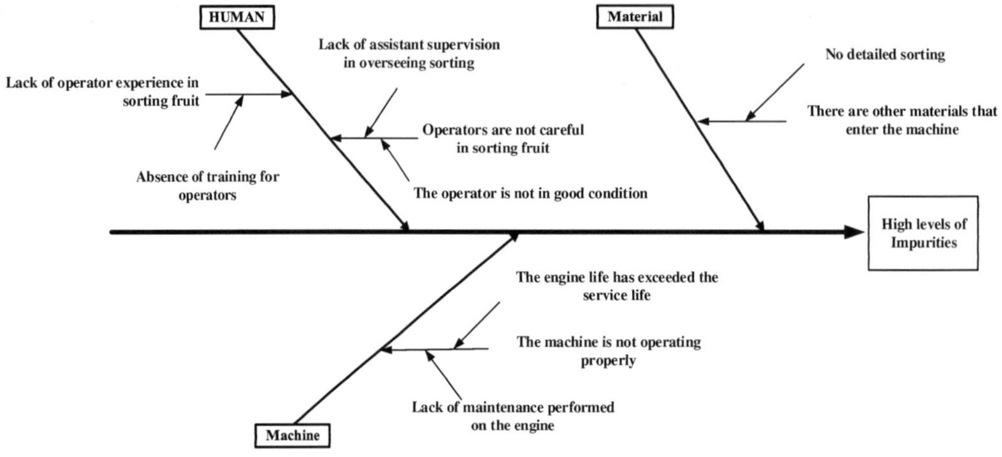

Figure 8. Cause and Effect Diagram (Fishbone) Causes of High Impurities in CPO Oil

5. Discussion

The results of the analysis show that high FFA levels and high impurities are the dominant defects that occur in CPO obtained using the Pareto 80/20 principle, where the two defect factors have a strong positive influence on the number of defects. By using a fishbone diagram it is found that the main causes of defect consist of 3 main factors, namely humans, machines and materials.

6. Conclusion

Based on the analysis of the discussion that has been done, it can be concluded that the defects contained in CPO are high FFA levels, high impurities and high water content, where the dominant disabilities occur in CPO are high FFA levels and high impurities identified by using the principle Pareto 80/20, the cumulative percentage of both defect reached 74.7%. The results of scatter diagram analysis showed high FFA levels and high impurities have a strong positive influence on the number of defects. Thus an analysis of the causes of defect is carried out using a causal diagram and it is found that the dominant cause of defect is high levels of FFA in terms of material (such as low quality raw materials), humans (for example: operators who are not careful in sorting fruit) and machines (like the age of an old machine). The dominant cause of defect for high levels of impurities also consists of 3 factors, namely humans, machines and materials.

Acknowledgements

The author would like to thank those who have helped in completing the research.

References

[1] Lubis R P, Poerwanto and Anizar. 2013. Usulan Perbaikan Kualitas Produk CPO dengan Menggunakan Konsep Kaizen di PT XYZ. E-jurnal Teknik Industri FT USU, 2, 1.
[2] Ginting, R 2007. Sistem Produksi. Graha Ilmu: Yogyakarta
[3] Sinulingga, S 2018. Metode Penelitian Edisi ke 3. USU Press: Medan
[4] Hart M K dan Robert F H 2007. Introduction to Statistical Process Control Technique. Statit Software, Inc: Oregon.
[5] Gaspersz V. 2001. Metode Analisis untuk Peningkatan Kualitas. Gramedia Pustaka Utama: Jakarta.
[6] Sugiyono, 2007. Statistik Untuk Penelitian. CV Afabeta. Bandung.
[7] Hartono. 2004. Statistik untuk Penelitian. LSFK$_2$P: Yogyakarta

Analysis of Drivers and Barriers The Implementation of Sustainability Supply Chain Management (SSCM) in PT. ABC

Nurhayati Sembiring*[1], Riky Yurisditira[1], Jeanica Devany[1]

[1]Department of Industrial Engineering, Faculty of Engineering, Universitas Sumatera Utara, Jl. Almamater Campus USU, Medan 20155

*nurhayatipandia68@usu.ac.id

Abstract. PT. ABC is one of government industry. Although the company has implemented SSCM, but in its implementation many problems still occurred. The problems that this company faced such as the unavailability of sufficient information to achieve sustainability goals to top management, budget allocation for the sustainability of the operation that not provide enough, the technology to adopt sustainability has not been efficient, etc. Some studied by another researcher had been done in this company but none studied about this topic. This study aims to analyze the factors that driving and inhibiting factors of supply chain implementation of sustainability management (SSCM) in PT. ABC. Based on TOPSIS method, it was found that the integrating factor sustained in a proactive plan with a value v = 0.8765, adequate budget v = 0.6204, support infrastructure and facilities for sustainable operations v = 0.5032, access to sophisticated technology for sustainable operations v = 0 , 3122, to enforce the government draft legislation effectively v = 0.3122, raising public awareness of the value of sustainable products v = 0.2780, and implementing sustainable waste management with the value v = 0.2423.

Keywords : SSCM, TOPSIS, drivers, barriers

1. Introduction

Environmental issues today become one of the worlds industry discussion. This is evidenced by the level of public awareness of environmentally products With the level of public awareness of environmentally products and then encourage the industries to apply the concept of environmental awareness into their businesses. Components, process and flow of information needed to build as a supply chain management system simultaneously. Hence there is a government regulation to raise awareness in terms of protecting the environment. So the company currently can not ignore environmental issues if they want to survive in the global market [1,2]. Integration with SCM sustainability increasingly emphasized in almost all industries through the cooperation of all parties, including the government, suppliers and customers. Responding to rapid changes as a result of environmental uncertainty and strong demands from governments, communities and markets to implement sustainable principles (sustainability), as well as the development of technology and increasingly global competition caused many companies to implement the strategy of Sustainability Supply Chain Management (SSCM [3.4]. Sustainable supply chains account environmental aspects in all activities, ranging from raw materials to the final disposal of the goods, but environmental issues in the supply chain decision-making process does not stand alone, the aspect of financial gain either [5,6].

The ability of PT. ABC as a company that has implemented a quality management system certified ISO 9001, management of the effectiveness of environmental activities certified ISO 14000, the management of occupational safety and health certified HSE (SMK3), supply chain management of sustainable certified RSPO and ISPO would have supported the production process of a planned sustainable. In order to produce a product that can meet the needs of consumers, enterprises should carry out activities from the beginning of the raw material in the get up of these products to the end users [7,8]. Focus on Supply Chain Sustainability is doing widespread adoption and development of sustainability issues from upstream to downstream. The issue of sustainability has to start from the beginning of raw materials to delivery to the end consumer. The issue of sustainability also have to integrate the issues and currents that go beyond the core of Supply Chain Management [9]. As described previously, sustainable SCM is the integration of sustainable development and management of the supply chain that contains three dimensions, which integrate environmental, social and economic issues that give effect to the company's strategy [10].

Although PT. ABC has implemented a system of SSCM but in the application still occurred many problems. The problems experienced by PT. ABC among others, is not an adequate information is available to achieve sustainability goals to top management, do not provide enough budget allocation for the operation of sustainability, technological tools to adopt sustainable yet efficient, yet effective rules to ensure government regulations for sustainability enforced, and the perception of stakeholders factory management is not the same with the top leadership.

2. Research methods

The research was conducted at PT. ABC as a palm oil processing plant with a capacity of 75 tons / hour which operates 22 hours / day and 6 working days / week. The data used in this study are primary data obtained directly from the production process at the plant and information of staff authorized in PT. ABC.

Data processing is performed in this study include the presentation the consensus of the experts on the factors driving and inhibiting the implementation of supply chain management sustainability by using Technique for Order Preference by Similarity to Ideal Solutions (TOPSIS) [11,12].

Data collection techniques used in this study are as follows:

- Interview
  The collection of data by conducting research directly by visiting and interview the company's management to obtain the necessary data.
- Questionnaires
  Questionnaire is a collection of data by spreading or provide a list of questions to respondents with expectations of respondents gave the answer according to the statement. Data collected by giving questionnaires to employees of various levels of experts such as the staff and the manager at the company of palm oil who have understood deeply about the sustainability of the company's supply chain management in charge and give them the measured structural modeling interpretative method.
- Data processing
  Data processing by using the method Technique for Order Preference by Similarity to Ideal Solutions (TOPSIS).

3. Result and discussion

3.1. TOPSIS (Technique for Order Preference by Similarity to Ideal Solution)

IOP Conf. Series: Materials Science and Engineering **851** (2020) 012047 doi:10.1088/1757-899X/851/1/012047

Topsis method could be applied in solving problem and it started with choose respondent that would give information by answering questionnaire that arranged for this research.

Questionnaire sustainability factors that supporting supply chain management, as follows.

Table 1. The driving factor of SSCM.

No.	Factors that drivers of sustainable performance
1	An adequate budget
2	Developing the infrastructure and facilities to support ongoing operations
3	Access to advanced technologies for sustainable operations
4	Raising public awareness of sustainable products
5	Enforcing regulations and laws effective
6	Applying sustainable waste management
7	Integrating sustainable operations in a proactive plan

Source: Expert opinion

Table 1 shows the factors that drivers of sustainable performancee for company. These factors came from interview and discussion with the experts at the company. Company's budget, infrastructure and facilities, advanced technologies, public awareness, regulations and laws, sustainable waste management and sustainable operations were the factors that influences the sustainable performance. Besides the questionnaire about the drivers of sustainability factors, here also a questionnaire that inhibiting supply chain management (barriers), as follows:

Table 2. Barriers of SSCM.

No.	Inhibitors for organizational change for sustainability
1	Commitment of leader is not optimal
2	Financial constraints
3	Inefficient technology
4	Inefficient legal framework
5	Supporting infrastructure is not optimal
6	Perception of stakeholders is not the same
7	Lack of sustainable waste management

Source: Expert opinion

In Table 2 could be seen that there are some factors that inhibit sustainability such as lack of commitment, financial, technology and legal framework that inefficient, infrastructure, different perception and waste management. After the questionnaire answered by respondent, and processing the data then we found obstacle factor with normalization of decision matrix score as could be seen in Table 3.

Table 3. Decision Matrix Score Normalization Influence To All Expert

Factors	Obstacle factor		
	K1	K2	K3
An adequate budget	3.5869	2.9771	3.8731
Developing the infrastructure and facilities to support	2.7448	2.9638	3.5869

ongoing operations			
Access to advanced technologies for sustainable operations	2.7448	3.3071	2.5420
Enforcing regulations and effective law	2.2781	2.9771	2.3362
Raising public awareness of sustainable products	3.3219	2.3362	2.2781
Applying sustainable waste management	3.3219	2.4599	1.6093
Integrating sustainable operations in a proactive plan	4,2009	3.8731	3.3219

From Table 3 we know that all of the factor had different value for each of the obstacle factor that would be used to calculated the preferences result. In Table 4 the result of that calculation is shown.

Table 4. Decision Matrix Score Normalization Influence To All Expert

Factors	D_i^+	D_i^-	$D_i^+ + D_i^-$	V	Rank
An adequate budget	0.4416	0.7218	1.1634	0.6204	2
Developing the infrastructure and facilities to support ongoing operations	0.5733	0.5807	1.1539	0.5032	3
Access to advanced technologies for sustainable operations	0.5708	0.5159	1.0868	0.4748	4
Enforcing regulations and effective law	0.7582	0.3441	1.1023	0.3122	5
Raising public awareness of sustainable products	0.8399	0.3234	1.1633	0.2780	6
Applying sustainable waste management	0.8871	0.2836	1.1707	0.2423	7
Integrating sustainable operations in a proactive plan	0.1363	0.9678	1.1041	0.8765	1

From the value of v in Table 4. it can be seen that the most good motivating factor for minimizing the inhibiting factor in PT. ABC is sustainable operations integration in a proactive plan factor with a value of v = 0.8765, then factor an adequate budget to the value of 0.6204, support infrastructure and facilities for the operation continuing with the value of v = 0.5032, access to advanced technology for continuous operation with a value of v = 0.4847, enforcing regulations and laws effective with the value of v = 0.3122, raising public awareness about sustainable products with the value of v = 0.2780, and the last is applying sustainable waste management with the value v = 0.2423. The all process for the drivers factors to get the rank was applied to the barriers factor and the value for the most highest factors are as follow, the technology is inefficient, facilities and supporting infrastructure is not optimal, and the perception of stakeholders is not the same.

4. Conclusions

Based on the research results and analysis of driving and inhibiting factor in the application of SSCM in PT. ABC, it can be concluded as follows:

- The driving factors for the result is that, develop the infrastructure and facilities for sustainable operations, access to advanced technologies for sustainability, enforce regulations and laws reactive, implementing waste management is sustainable, as the four drivers of the most influential in the implementation of SSCM in PT. ABC
- Factors inhibiting the result is that the technology is inefficient, facilities and supporting infrastructure is not optimal, and the perception of stakeholders is not the same is a key barriers factor to implementation SSCM

References

[1] Manavalan, E. and Jayakrishna, K 2019 An Analysis on Sustainable Supply Chain for Circular Economy , *Procedia Manufacturing*, 33, pp. 477-484. doi: 10.1016 / j.promfg.2019.04.059.

[2] Waskito J. et.al. 2011 Green Consumer: Description of the level of public awareness and concern about environmental sustainability Joglosemar, pp 29-39 *JDM* Vol 3 No 1.

[3] Abdul R 2017 Modeling the interrelationships among barriers to sustainable supply chain management in leather industry.S0959-6526 (18) 302 762 DOI: 10.1016 / j.jclepro.2018.01.245

[4] Balkan et al 2011 Sustainable Supply Chain. Springer-Verlag Berlin Heidlberg. Doi 10.1007 / 978-3-642-12023-7

[5] BaNaite D and Tamošiūnienė R 2016 Sustainable Development: the Circular Economy Indicators" Model Selection ", *Journal of Security and Sustainability Issues*, 6 (2), pp. 315-323. doi: 10.9770 / jssi.2016.6.2 (10).

[6] Cetinkaya, the Balkans, Richard Cuthbertson, Graham Ewer, and et al. 2011. *Sustainable Supply Chain Management*, Heidelberg, Germany: Springer Heidelberg Dordrecht.

[7] Hong J, Zhang Y and Ding M 2018 Sustainable supply chain management practices, supply chain dynamic capabilities, and enterprise performance, *Journal of Cleaner Production*. Elsevier Ltd, 172, pp. 3508–3519. doi: 10.1016/j.jclepro.2017.06.093.

[8] Juliet O. 2018. Examining barriers to organizational change for sustainability and drivers of sustainable performance in the metal manufacturing industry. *Resources, Conservation and Recycling* 140 (2019) 102-114

[9] Lozano O 2012. Organizational Changes for Corporate Sustainability. *Greener Management International Journal*

[10] Mohammad Hussain, et al. 2015. ISM-ANP framework for evaluating alternatives integrated supply chain management forsustainable TiwariPII: S0307-904X (15) 00552-1DOI: 10.1016 / j.apm.2015.09.018

[11] Azimifard et al., 2018. Selecting sustainable supplier countries for Iran's steel industry at three levels by using AHP and TOPSIS methods. doi.org/10.1016/j.resourpol.2018.01.002,

[12] Sembiring N et.al. 2019. Supplier selection in rubber industry using analytic network process (ANP) and technique for order preference methods by similarity to ideal solution, *IOP Conf. Ser.: Mater. Sci. Eng.* **508** 012091

IOP Conf. Series: Materials Science and Engineering **851** (2020) 012048 doi:10.1088/1757-899X/851/1/012048

Comparison of Crude Oil Supply Cost Between Ship Rental and Single Point Mooring

I Rizkya*, K Syahputri, RM Sari, OC Syardhi
Departement of Industrial Enginnering, Faculty of Engineering, Universitas Sumatera Utara

*e-mail: indahrizkya@usu.ac.id

Abstract. Today, the need for fuel oil and gas are increasing. Fuel oil and gas is the energy needed by the public to support daily activities. To meet the people needs for fuel oil and gas, the company is not only focused on the Indonesian people needs. Production principle carried out is make to stock, so the company makes direct purchase of imported and domestic crude oil and through government entitlement. Currently this fuel oil company has a problem that is quite urgent because the currency value is related to the crude oil procurement. This causes the profit obtained by the company is not optimal. So it needs to be evaluated relates to crude oil procurement policies in the company. As a result, the total costs spent by GP ship rental policy is $ 53.572.938, but by using the SPM (Single Point Mooring) method will save around $ 114.300.205 for 4 years scale or $ 28.575.051 in a year.

1. Introduction

Inventory as the products and materials quantity stored. Inventories consist of all products and materials used in the production or distribution process [1]. Inventory is related to capital, storage space usage, maintenance needs, product value decrease, taxes expenditure, insurance needs, theft risk, and loss [2]. Inventories are company assets that can be sold or goods still in the production process or raw material inventories are waiting to be used in the production process [3].

Inventory management is a business process that is responsible for developing and managing inventory levels, which are raw materials inventory, work in process or finished goods, so that right inventory quantity must always available and the company must ensure the cost of excess or shortage stock is always low [4]. Inventory control is very important for a company because it is made specifically to reduce cost or increase profit while meeting customer demands by ensuring that stock balanced items are maintained at quality, quantity, and it can be obtained at the right time and at the right place [5]. Thus, inventory control is very important to achieve the company goals, namely to create a production balance and the ability to distribute these products optimally with minimum costs to achieve maximum profits.

The company is engaged in the energy sector including oil, natural gas and other energy. The need for fuel oil and gas is increasing along with the increasing needs of households and industry. This industry produces 3 groups of products, namely fuel oil (BBM), Non-BBM, and Petrochemical to distribute all over Indonesia.

Receiving raw materials can be through pipelines planted underground and through the shipping system by Musi river and the barging system. Oil receipts still originate from the South Sumatra regions are using pipelines. While the reception of crude oil from outside South Sumatra is done via

MR tankers or barging through the Musi river. The obstacle faced by the acceptance of the waterway is the Musi river depth, which is approximately of 6 meters, which does not allow large tankers to enter the dock. Therefore, before entering the Musi river, large tankers will be unloaded and transferred to the smaller sized of GP ships type with a maximum load about 70 MB.

Crude oil procurement using GP ships type takes 2 days to load and unload until it reaches the dock. As the result, the procurement cost of crude oil has increased. In addition, weather factors also gives the great affect the supply crude oil to the dock because the river tidal cause the ship cannot lean on the dock. This research will try to find an alternative supply crude oil so that the procurement cost by renting GP GP ships type can be minimized.

2. Methodology

The objects studied by the Petroleum Industry in the South Sumatra region are mainly the Prabumulih and Jambi Upstream Operations Area. Historical data form on Crude Oil demands received via ship. The study begins by making observations directly to the Marine Region to observe the conditions occur in the Marine. After that, make a description of the problem occurred, namely the high cost of crude oil procurement due to ship rental with a fishbone diagram. Data collection was carried out to analyze the demand for crude oil. Data collected in the form of crude oil demand data, type of crude oil and ship rental cost. The next step is forecasting.

Forecasting is the value estimation at a certain time in the future, while predictions are used for more general estimation [6]. Forecasting use formal statistical methods with time series, cross-sectional or longitudinal data, or as an alternative to less formal assessment methods. Quantitative forecasting methods with time series models are used in this study to predict future data as a function of past data [7]. Models for horizontal data patterns are Single Moving Average, Single Exponential Smoothing and Weighted Moving Average. Moving Average is one of the widely known technical indicators used to predict future data in the time series analysis [8]. In general, Moving Average is used as a refinement instrument in time series analysis. In beginning, moving averages of various dimensions (n = 5, 10, 20, 50 and etc) are plotted against a value to get a picture of the trend. Shorter moving averages (n = 5, 10) have a higher degree of leveling [9]. Crude Oil demand estimation begins by the demand determination for the next 5 months period, the ship type rented, the total cost of ship rental. Furthermore, the costs spent by the company are calculated compared to the Single Point Mooring Method (SPM). Single Point Mooring (SPM) is a type of offshore facility used for loading and unloading oil and gas tankers at sea [10].

3. Result and Discussion

The problem of the high total cost of crude oil procurement occurs in the Petroleum Industry in Indonesia is described of the fishbone diagram. Fishbone diagram or also called Ishikawa diagram or cause-effect diagram) is a technical graph to show some causes of certain events or phenomena. Specifically, fish bone diagrams (shaped like fish skeletons) are commonly tools used for a purpose and effects analysis to identify causes complex interactions for a particular problem or event. Fishbone diagrams can be a comprehensive theoretical framework for representing and analyzing sources of innovation [11].

Based on the fishbone diagram, the cause of the high total cost of crude oil procurement due to ship rentals happen because:

1. Man, the need for additional labor served as inspector and security in order to maintain the quality of crude oil in the mother ship which is will increase the labor costs.
2. Material, transport ships used sometimes arrive earlier than the specified date and even later than the specified date which causes an increase in inventory costs.
3. Method, ship used to transport oil are rented ships and cause increase rental costs for transportation.

4. Environment, there are competitors who require companies to maintain performance to meet consumer needs. From the diagram it was found that the high total cost spent for crude oil procurement due to ship rental.

Crude Oil demand by type from January to August 2019 can be seen in Table 1.

Table 1. Actual Data of Crude Oil (January-August 2019)

Period	Crude Type			
	Arjuna (MB)	Banyu Urip (MB)	Grissik (MB)	Geragai Crude (MB)
January	280	600	180	205.5
February	200	600	180	210
March	210	600	165	0
April	0	600	100	0
May	0	600	0	0
June	0	400	140	100
July	200	600	230	155
August	180	600	0.00	250
Total	1070	4600	995	920.5

Based on the table above, it can be seen that the crude types are often ordered of Banyu Urip, Arjuna, Grissik and Geragai Crude.

3.1. Forecasting of Crude Oil

The selection of the best method is used to determine the forecasting results that can be used to calculate the total costs spent in the next period. The following is a table shows the comparison of error values with Mean Absolute Deviation, Mean Square Error, Mean Absolute Percentage Error[14] from the Single Moving Average method, Single Exponential Smoothing, Weighted Moving Average.

Table 2. Selection of Crude Oil Forecasting Methods

The Best Method	MAD	MSE	MAPE
Single Moving Average (n=3)	308.5	139861.09	31,71%
Single Exponential Smoothing (α=0,3)	249.3	99236.55	25,85%
Weighted Moving Average	254.93	75719.2	26.8 %

Based on the calculation results of errors with the Mean Absolute Deviation (MAD), Measure Square Error (MSE), and Mean Absolute Percentage Error (MAPE) methods in Table 2, shows the single moving average method with $\alpha = 0.3$ has the smallest error rate. So this method is used to predict crude oil demand in the next 5 periods. The disaggregate demand for Crude Oil can be seen in the following table:

Table 3. The Disaggregate Demand of Crude Oil

Period	Forecasting Result	Arjuna (MB)	Banyu Urip (MB)	Gressik (MB)	Geragai Crude (MB)
September 2019	1143,52	161,24	693,43	149,91	138,59
October 2019	800,47	112,87	485,40	104,94	97,02
November 2019	560,33	79,00	339,78	73,46	67,92

December 2019	392,23	46,42	273,84	51,42	47,53
January 2020	274,56	38,71	116,50	35,10	33,28

Based on the table above shows that the demand for crude oil requires the rental of MR and GP ship for the oil transportation to the dock so that the total cost has increases.

3.2. Total Cost of Crude Oil Procurement by Ship Rental

The procurement cost of crude oil types is estimated with the rental cost of MPO bunker and MDO bunkers for the MR Tanker, Mother Ship and GP Ship. Calculation the total cost of various types of crude oil procurement can be seen in the following table:

Table 4. Procurement Cost Based on Crude Oil Arjuna Type (in dollar)

Type of Crude Oil	September 2019	October 2019	November 2019	December 2019	January 2020	Total
Arjuna	3.311.094	3.311.094	1.711.398	1.711.398	1.711.398	11.756.382
Banyu Urip	5.915.094	4.427.097	4.427.097	3.311.094	2.939.097	21.019.479
Gressik	2.939.097	2.939.097	1.711.398	1.711.398	1.711.398	11.012.388.
Geragai Crude	2.939.097	1.711.398	1.711.398	1.711.398	1.711.398	9.784.689
TOTAL						**53.572.938**

Table 4 shows the estimated cost of procuring crude oil with a ship rental policy. The current policy using mother ships by rent the GP type ship to send crude oil to the dock, so that the total cost of supplying crude oil becomes expensive. Crude oil arrives at the dock takes 2 days by sending crude oil in a smaller capacity using a GP type vessel. As a result, the cost of spending has increased, in addition to weather factors and tides affect the length of time the ship can lean towards the dock.

3.3. Design of Single Point Mooring (SPM)

Single Point Mooring (SPM), also known as Single Buoy Mooring or SBM is an offshore port buoy, which is function as a mooring point and interconnection for gas loading or unloading tankers or liquid products functions as tethering and interconnection for tanker loads or unloading gas products or fluid. For the SPM manufacture requires into the sea with a range of 200 ft - 600 ft. There are several types of SPM. In its SPM development varies in several types. Some types of SPM are turret, Catenary Anchor Leg Mooring (CALM), Single Anchor Leg Mooring (SALM), Vertical Anchor Leg Mooring (VALM), and Single Point Mooring Towers (SPMT) [12]. For the solution given by the author, that is using SPM type Catenary Anchor Leg Mooring (CALM) bouy which is a type where the anchor chain buoy extends to the anchor point at a certain distance from the buoy. This type of SPM has several important parts including anchors (anchors or gravity anchors connecting mooring chains to the sea floor), anchor chains (systems with six or eight chain anchors are the most common), and stopper chains (connecting chains to buoys) [13] Costs spent to make the SPB Bouy Catenary Anchor Leg Mooring (CALM) "Costs Estimation for making SPM include engineering costs, and construction costs and construct a 92-kilometer pipeline from the edge of the South Sumatra Sea to dock 2 and dock 10 with a costassumption of around ± **$ 400.000.000** ". (source of Pertamina Plant Directur Chrisna Damayanto.2013)

3.4. Comparison Between SPM and GP Ship Rental

Like a design, there must be advantages and disadvantages to the design created. These strengths and weaknesses are important to know, so that management understand the advantages and risks faced from making this design system. The advantage is if it is assumed that the total cost of the oil company in a year is ± $ **128.575.051** and for four years it reaches ± $ **514.300.205** and making SPM only around ± $ **400.000.000**, this succeeded in reducing the total oil company cost by $ **114.300.205**. Then

IOP Conf. Series: Materials Science and Engineering **851** (2020) 012048 doi:10.1088/1757-899X/851/1/012048

the design minimize idle time, prevent fraud, and the design minimize the sell loss. Some disadvantages to the design are the long manufacturing processes and understand the right selection to minimize leakage and work accidents. If this recommendation is not implemented then the total cost will increase in a certain period and then when the weather is bad, the sender of crude oil to the dock will be late. This condition is very disadvantage, because there is may be a sell loss eliminates the customer or the inventory quantity so that the storage cost increases, this is happened when the crude oil is late arriving at the dock.

4. Conclusion

Based on the previous discussion in oil companies that there are some disadvantages in the Crude Oil procurement system cause the total costs spent by the company. The total costs spent by the company with GP ship rental policy is **$ 53.572.938**. To minimize the total costs of oil company spent, using the SPM (Single Point Mooring) method with a total cost of $ 400.000.000 to save costs of **$ 114.300.205** for 4 years scale or **$ 28.575.051** in a year. With this SPM method, it can improve the efficiency, effectiveness and productivity of oil companies in South Sumatra.

Acknowledgement

Authors would to thank to Universitas Sumatera Utara for their financial support to publish this reasearch. Authors also thanks to the other participants who helped this research.

References

[1] Oláh J et al 2017 Scientific Journal of Logistics 13 (4), pp 439-454
[2] Afolabi O J, Onifade M K and Olumide O 2017 LOGI – Scientific Journal on Transport and Logistics 8 (2), pp 1-11
[3] Shenoy D and Rosas R 2007 Problems & Solutions in Inventory Management (Mexico: Springer)
[4] Gokhale P P and Kaloji M B 2018 International Journal of Latest Technology in Engineering, Management & Applied Science 7 (9), pp 29-31
[5] Atnafu D and Balda A 2018 Cogent Business & Management, 5:1, 1503219
[6] Ezelori D C, Umeh N M, Mbeledeogu N N and Okoye P U 2014 International Journal of Science, Engineering and Technology Research 3 (2), pp 184-202
[7] Ivanovski Z, Milenkovski A and Narasanov Z 2018 UTMS Journal of Economics 9 (2), pp 121–132.
[8] Du R and Yang H 2014 The Open Automation and Control Systems Journal 6, pp 1886-1890
[9] Adrian Z I 2015 Procedia Economics and Finance 32, pp 1353 – 1361
[10] B Eghbali, M Daghigh, Y Daghigh and F Azarsina 2018, International Journal of Maritime Technology 9, pp 41-49
[11] Coccia M 2017 Journal of Social and Administrative Sciences 4 (4), pp 291-303
[12] Rutkowski G 2019 The International Journal on Marine Navigation and Safety of Sea Transportation 13 (1), pp 187-195
[13] Lee S W and Kim Y D 2019 Journal of the Korean Society of Marine Environment & Safety 25 (3), pp 287-297
[14] Rahmat R F, Nurmawan S, Sembiring S, Syahputra M F, Fadli 2018 Adaptive neuro-fuzzy inference system for forecasting rubber milk production *IOP Conference Series: Materials Science and Engineering* **308**(1),012014

Product Placement Based on Throughput at PT XYZ Warehouse

Mangara M Tambunan[1], Regina Tambunan[2]

[1,2]Industrial Engineering Department, Engineering Faculty, Universitas Sumatera Utara, Medan, Indonesia

E-mail : reginaprisella97@gmail.com araapul_tambunan@yahoo.com

Abstract. PT. XYZ is a company engaged in the manufacture of beverages with various types of drinks both carbonated soft drinks and non-carbonated soft drinks. The company's products are coca cola, fanta, sprite, minute maid, ades, nutriboost, etc. The company has a warehouse to store products that are produced before being distributed to consumers. There are 79 types of products stored in the warehouse. Problems faced by PT. XYZ is a product that does not have a fixed storage block yet and the company does not have a clear product grouping where the product placement is not based on the similarity of product types and the level of product activity so it takes time to look for products stored in the warehouse when needed. Therefore, it is necessary to design warehouse layout PT. XYZ by using the dedicated storage method to place the product in a fixed location and based on the level of activity. The results of product placement in the dedicated storage method show that the highest T / S value is product code 975341, with a value of 5,200 so that the product is placed in the storage block closest to the staging bays (I / O point).

Keywords : Dedicated Storage, Space Requirement, Throughput

1. Introduction

Layout is the arrangement of production facilities to obtain efficiency in a production. Included in production services are material / goods receiving facilities, storage / warehouse facilities, and shipping facilities. Warehouse or storage is a place to store goods both raw materials that will be carried out manufacturing processes, and finished goods that are ready to be marketed. [1].

A good warehousing system is a warehousing system that is able to utilize space for storage effectively in order to increase space utilization and minimize material handling costs [2].

PT. XYZ is a company engaged in the manufacture of beverages with various types of drinks both carbonated soft drinks and non-carbonated soft drinks. The company's products are coca cola, fanta, sprite, frestea, minute maid, ades, nutriboost, etc. The company has a warehouse to store products that are produced before being distributed to consumers.

Problems faced by PT. XYZ is a product that does not have a fixed storage block yet and the company does not have a clear product grouping where the placement of products is not based on the similarity of product types and the level of product activity so it takes travel time for products stored in the warehouse when needed. Therefore, it is necessary to design the layout of the warehouse of PT. XYZ by using the dedicated storage method to place the product in a fixed location and based on the level of activity.

Hadiguna (2008) states that there are several methods that can be used to store products in the warehouse, namely Random Storage, Dedicated Storage, Classed Based Storage and Shared Storage. Random Storage is storage of items that come in every available location, where each item has the same probability at each location. Dedicated Storage, which is an item stored in a certain location depending on the type of item. This policy is designed with the storage area of each item equal to the maximum level of inventory. Classed Based Storage is a pareto effect application where 80% S / R activity by 20% items, 15% S / R by 30%, and 5% S / R by 50%. Shared Storage is a policy that is at the extreme point of random and dedicated storage policy [3].

The method used to solve this warehouse layout problem is the Dedicated Storage Method. Dedicated storage is also called a fixed slot storage that places a specific storage location or address for each item stored. This is because a storage location is given to a specific product [4]. With the design of the preparation using the dedicated storage method, it is expected that the products to be stored can occupy a fixed location to facilitate operators in storing and retrieving products.

Research on the design of warehouse layout using the dedicated storage method has been conducted by Reinny Patrisina and Indrawati (2010) in the warehouse of commercial lightings product distributors from national lighting manufacturers who place products with the largest T / S ratio values in the slots with the smallest spacing, products with values the second largest T / S ratio will be placed in the slot with the second smallest distance and so on so as to obtain total operator distance minimization with an expected total mileage of 4137 m per day [5].

Research using this method has also been carried out by Ayunda Prasetyaningtyas, et al (2013) who did the redesign of the layout of the steel plate products in the warehouse using the dedicated storage method, where the laying process is based on the product ranking of T / S starting from the largest to the highest. The smallest is placed at the shortest distance to the furthest from the I / O point. The results of the design using the dedicated storage method obtained a distance of material handling by 2030,642 m so that it decreased by 1138,391 m or about 35,924% from the previous distance, which is 3168,853 m [6].

According to the research of A Vildha and S Setio (2016) in an electric shop, the warehouse of goods looks irregular which results in inefficient collection and storage of goods in the warehouse. To overcome this problem, 3 alternative layouts are proposed, namely frequency, volume and dedicated storage. The alternative layout for dedicated storage was chosen because goods that are close to the entrance and exit are the types of goods that have the largest exit frequency followed by goods that have the same type and then the type of goods with the second largest frequency etc so that it provides better capabilities with a clear grouping system and provides a faster search, retrieval and storage process time [7].

2. Method

This research was conducted at PT. XYZ where the time of the research takes place during May 2019 to August 2019. This type of research is descriptive research. The object of research was warehouse layout at PT. XYZ. This research consists of several stages as follows:

1. Preliminary studies are conducted to identify problems found in the company, observations and other information.
2. Studying the literature and supporting theories to get solutions for problem solving.
3. Formulate a problem in the company that is the product does not have a fixed storage block yet and the company does not have a clear product grouping where the laying of the product is not based on the similarity of types of products and the throughput so that it takes time to look for products stored in the warehouse when needed.
4. Data collection, are data of product type and throughput value.
5. Processing data using a dedicated storage method, the step taken is to place the product based on the throughput value
6. Analysis and discussion.

3. Result and Discussion

There are 79 types of products stored in PT. XYZ. Warehouse PT. XYZ has an area of around ± 8200m2. The warehouse is divided into several areas, namely loading bays, 4 staging bays (I / O points), and product storage zones. The product storage area is divided into 2 areas namely racking area and bulk area.

Racking area is an area that holds the finished product that is placed on the rack. Racking area has 8 shelves. 1 shelf has 4 levels, 1 level can hold 2 pallets, which means that the capacity of 1 block is 8 pallets. Products stored on the shelves are products with packaging named CBA, TWA, and BIB, which are types of products with packaging characteristics that cannot be stacked by other pallet.

Figure 1 shows the actual layout of PT.XYZ's warehouse.

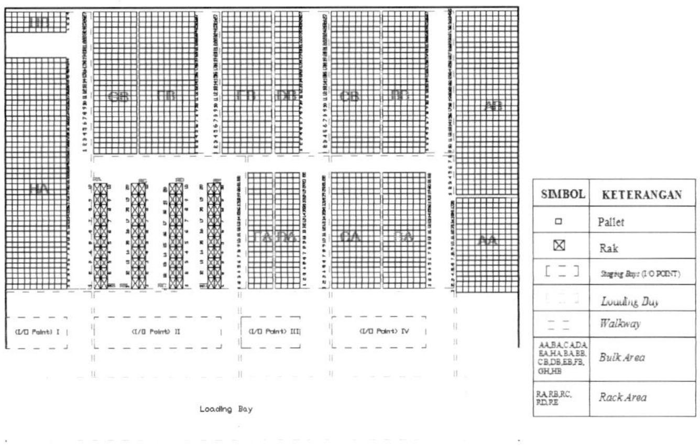

Figure 1. Actual Layout of PT. XYZ's Warehouse

After obtaining the value of T / S of each product ordered from the largest to the smallest, the product can be provided in the warehouse. Product placement is based on the largest T / S value untill the smalest value that is allocated to the block close to the staging bay (I / O point) to the furthest away from the staging bay (I / O point). In a special storage method, each type of product must have a fixed place and only one block for the same type of product.

In the actual warehouse layout, there are inter-storage zones that do not have a material handling aisle / road yet, so its difficult to make product retrieves in FEFO. So in the proposed warehouse layout, the addition of material handling lines is based on the length of the diagonal forklift. The size of the forklift used is 220 cm x 110 cm x 250 cm. The calculation is as follows:

$$\text{length of diagonal forklift} = \sqrt{(\text{width of forklift})^2 + (\text{lenght of forklift})^2}$$

From the calculation of the length of diagonal forklift, it is obtained that the width of the aisle needed for material handling is 2.5 m.

The proposed warehouse layout of PT. XYZ can be seen in Figure 2.

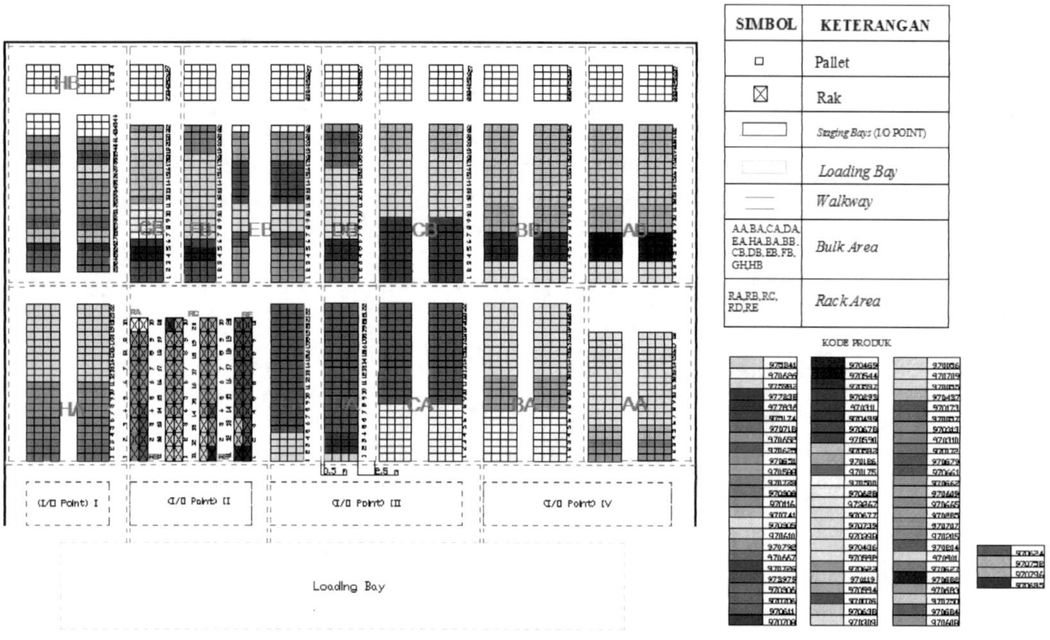

Figure 2. Proposed Warehouse Layout of PT Coca Cola Amatil Indonesia Unit Medan

4. Conclusion

Based on the results and discussion conducted above, it can be concluded that the layout of the proposed product placement is based on the largest T / S sequence in the storage block closest to the staging bays (I / O point). Based on throughput data, it is obtained that the highest T / S value is product code 975341, with a value of 5,200 so that the product is placed in the storage block closest to the staging bays (I / O point).

Acknowledgement

This research supported by Industrial Engineering Department, Faculty of Engineering, Universitas Sumatera Utara

Reference

[1] Purnomo H. 2004. Perencanaan dan Perancangan Fasilitas (Edisi Pertama). Graha Ilmu. Yogyakarta.
[2] Heragu S. 2006. Facilities Design (Second Edition). Lincoln: iUniverse Inc.
[3] Hadiguna R, Setiawan H. 2008. Tata Letak Pabrik. ANDI. Yogyakarta.
[4] Francis R L, McGinnis Jr, Leon F, dan White J. 1992. *Facility Layout and Location: An Analitycal Approach* (Second Edition). Prentice Hall. New Jersey.
[5] Patrisna R, dan Indawati. 2010. Perancangan Tata Letak Gudang dengan Metode Dedicated Storage Location Policy (Studi Kasus: PT X). *Jurnal Optimasi Sistem Industri*, 9: 37-44.
[6] Praseningtyas, A, dkk. 2013. Usulan Tata Letak Gudang untuk Meminimasi Jarak Material Handling Menggunakan Metode Dedicated Storage. *Jurnal Teknik Industri*, 1:29-34
[7] Efrataditama A V, Wigati S S. 2016. Perancangan Tata Letak Gudang Dengan Metode Dedicated Storage di Toko Listrik Anugrah Jaya. *Seminar Nasional IENACO*.
[8] Siregar I, Muchtar M A, Rahmat R F, Andayani U, Nasution T H, and Sari R M 2018 Method of calculation overall equipment effectiveness in fertilizer factory. *IOP Conference Series: Materials Science and Engineering* **308**(1), p. 012053)

IOP Conf. Series: Materials Science and Engineering **851** (2020) 012050 doi:10.1088/1757-899X/851/1/012050

Improvement of endoclip design using 3D CAD approach

I Siregar[1], M I C Izham[2], A C Ariffin[3] and M S Adenan[2], J B Saedon[2]

[1]Universitas Sumatera Utara, Kampus USU, Medan 20155, Medan Indonesia
[2]Smart Manufacturing Research Institute, Universiti Teknologi MARA, 40450, Shah Alam, Selangor, Malaysia
[3]Faculty of Medicine and Health Sciences, Universiti Sains Islam Malaysia, Bandar Baru Nilai, 71800, Nilai, Negeri Sembilan, Malaysia

ikhsan.siregar@usu.ac.id

Abstract. Endoscopic clip (endoclip) is a metallic mechanical device used in gastrointestinal endoscopy practices with the intention of close mucosal surfaces deprived of the requirement for surgery and sew up. Nowadays, endoclip is accessible in the market predominantly in single firing mode, where the clip can only be used once in a period of process. Due to that, the clip is loaded to its applicator repeatedly during the operation which cause delay for the surgery that can lead to fatality. Therefore, this project aims to develop a multi-firing endoclip cartridge that can reduce the difficulty of reloading the endoclip system to its applicator by improving current design. Three design of multi-firing endoclip cartridges were produced by using different mechanism to apply the endoclip. The casing of the cartridge was designed in cylindrical shape with 40 mm diameter and 68 mm length with CATIA V5 software. These three designs formerly assessed by using decision matrix method for assortment of the best concept design. The result from the decision matrix table shows that design with revolve internal clip holder got the highest decision matrix selection amongst the three designs, advantaging in holding more clips and can reduce operation time compared to the other two designs

1. Introduction

Metallic clipping such as endoclip were first introduced for the primary purpose of achieving haemostasis which is to stop the flow of blood of focal gastrointestinal bleeding. Endoscopic request of clips has been exposed to be effective in management of bleeding from peptic ulcer, Mallory-Weiss tear, Dieulafoy's lesion, duodenum and colonic diverticula [1]. It is also playing an important role in non-haemostatic application such as markers to direct fluoroscopic stent placement and target surgical, binders to attach enteral tube and stent devices to the luminal wall and others.

Endoclip is used mainly alongside with the flexible gastrointestinal endoscope devices which are flexible instruments that combine fibre-optics and charged-coupled devices to facilitate illumination and visualization. The endoscope is made up of three basic parts: the tip, the insertion tube and the controller unit. The supplement tube is the part that enters the body of the endoscope and it contains a variable diameter working channel, which allows endoscopic accessories such as endoclip device to pass through [2].

There are several endoscopic clipping devices that are available on the market and others are being developed. Although they are different from each other but they share two similarities which are the metallic double or triple pronged clips and a delivery catheter-handle assembly. The most used types of endoclips are the Rotating Clip (Olympus Corporation), the QuickClip2 (Olympus Corporation), the Triclip (Wilson-Cook Medical Incorporated) and the Resolution Clip (Boston Scientific Corporation).

The endoclip applicator expedient contains of handle, slider, sheath and the clip. The procedural steps of endoclip are straightforward. The Teflon sheath functions as a cover for the clip and the catheter to prevent any damage and breakaway during the process of clip insertion inside the body. The orientation of the clip's prongs can be accustomed by revolving the handle clockwise [3]. The clip must be press in contradiction of the lesions, put on suction and the slider necessity be rapidly dragged back to close the clip and then deploy it. A number of limitations had been identified by researchers when operating the endoclip devices and some improvement is essential to overcome the issue in the future. The development process is a essential stage in order to advance the device's possible, efficiency and the usability in the gastrointestinal endoscopy practice. The limitations of the current endoclip device includes size of the lesion, access to the lesion, Fibrotic lesion and the single use submission.

This project aims to tackle the single use application of the existing endoclips. It is described that an average of 2.9 clips were required per patient to achieve haemostasis in a single procedure. Multi-firing clipping device such as the Clipmaster3 and the Multi-Clip are being developed and tested to overcome the same problem. Therefore, the need for design optimization of the endoclip device is a must to solve the issue at hand.

Using 3D computer aided design software (CATIA V5), a cartridge that covers multiple clip is planned to enable the endoclip to be multi-firing. The cartridge is designed to substitute the serious part of the endoclip, which is its catheter's tip. Therefore, the scope of work of this project is to make an upgrading for the endoclip by designing a cartridge to replace the tip of its catheter and to create a prototype of the cartridge design by using 3D printing for the proof of concept of the design. The design of the cartridge is based on the Olympus endoclip (HX110-LR).

2. Concept generation

There are a few elements that necessity be measured when designing the improvement for the cartridge of the endoclip that lets the device to be multi-firing such as mechanism with attached endoclip, critical component identification and arrangement of clips inside cartridge.

The improvement idea for this project is to design a mechanism for the multi-firing endoclip cartridge is based on "Cartridge in Cartridge" design. The concept project has two cartridges attached together where the smaller cartridge which is for the settlement of the clips and the other bigger cartridge will house the smaller cartridge and it is design to replace the catheter of the clip application device.

The idea of the deployment mechanism is to hold the clip with a static part that is integrated with the larger cartridge and with the help of the spring to provide potential energy for the clips to move. Furthermore, other design must consider the difficulty of assembling the cartridges

3. Concept selection

Different cartridge deployment mechanism concepts are associated to control which concept is the best among the concept design. The selection of the concept design contains key components that must be considered and compared in order to accomplish the impartial of the project.

These components include design complexity, functionality, size of cartridge, ease of assembly and number of clips. So as to conduct the assortment process, all the conceptual design of endoclip cartridges were evaluated using the decision matrix method.

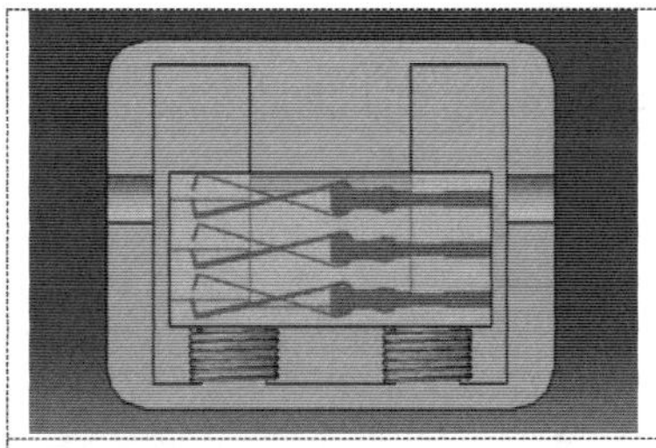

Figure 1. Design 1, side view

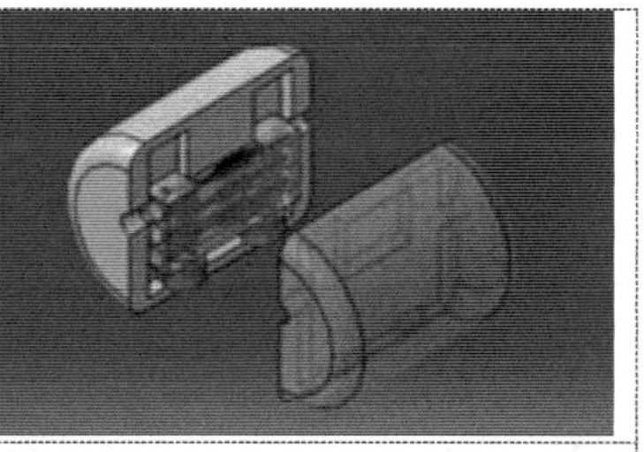

Figure 2. Design 1, isometric view

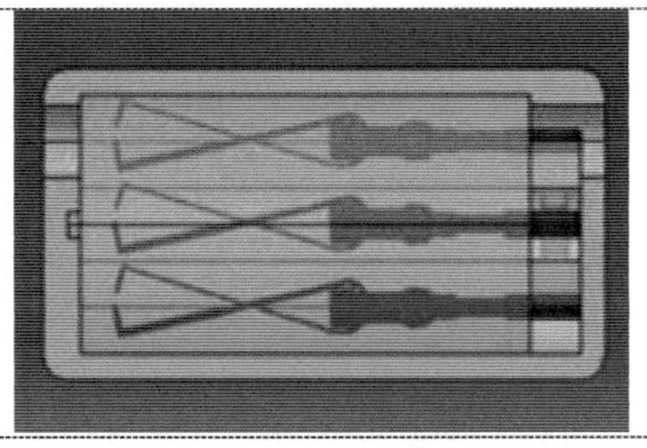

Figure 3. Design 2, side view

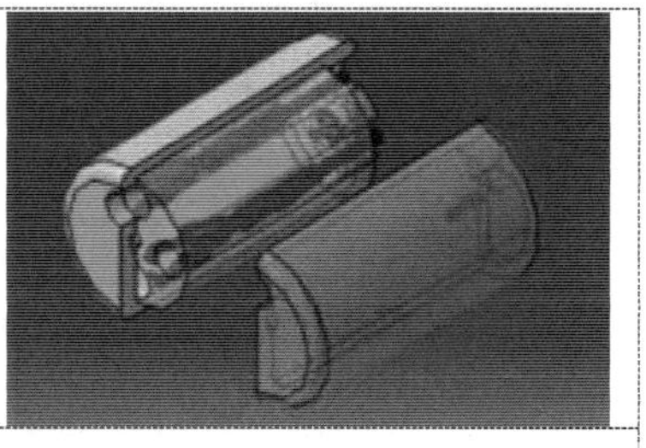

Figure 4. Design 2, isometric view

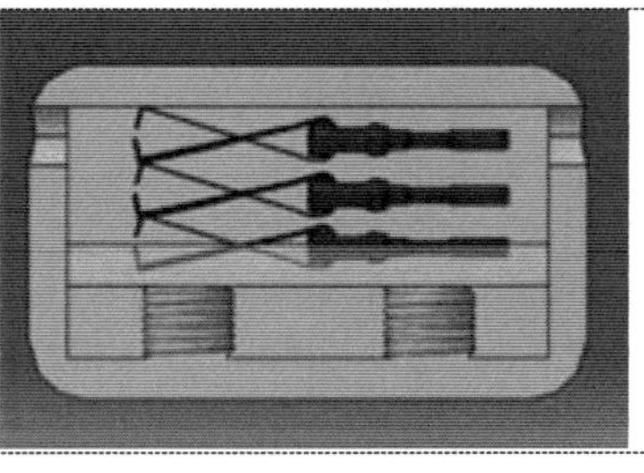

Figure 5. Design 3, side view.

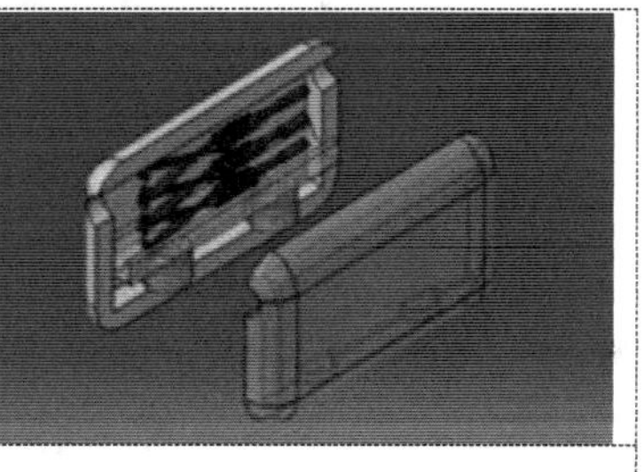

Figure 6. Design 3, isometric view.

Table 1. Decision matrix for concept selection

Criteria	Design 1	Design 2	Design 3
Design complexity	4	6	2
Functionality	4	4	4
Size of cartridge	2	4	2
Ease of assembly	4	6	2
Number of clips	4	4	2
Total Average	**18**	**24**	**12**
Qualitative Score Assignments			
Good		6	
Fair		4	
Poor		2	

4. Improvement design
Based on the evaluation from the decision matrix, the best designated was Design 2. Further details design regarding Design 2 was generated in the process, and all components of the design were

sketched using CATIA V5 software. Design 2 cartridge is composed with several important parts such as rotating cartridge, housing cartridge, spiral spring and clips.

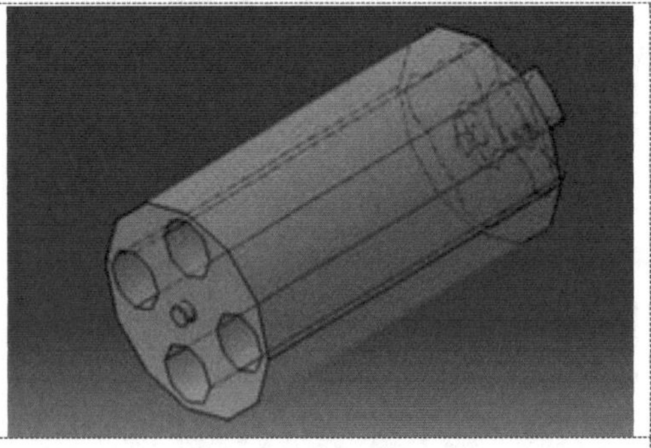

Figure 7. Isometric view of rotating clip

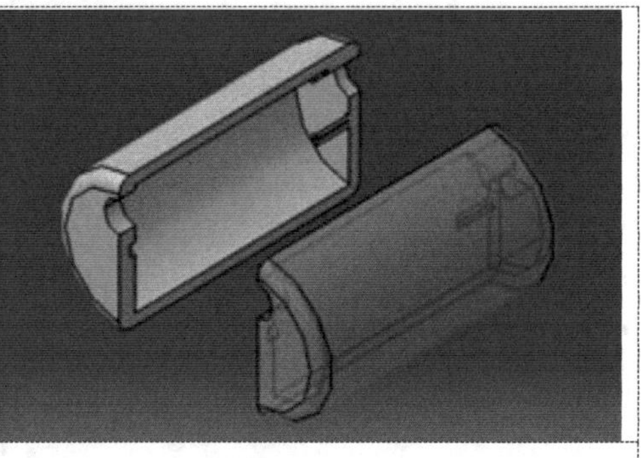

Figure 8. Isometric view of housing cartridge

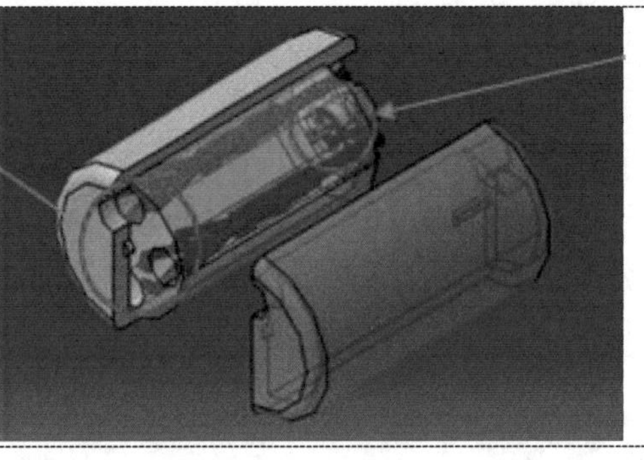

Figure 9. Isometric view of the final design assembly

Each component was sketched based on the appropriate measurement and requirement, before being accumulated.

5. Endoclip digital mock up

The mechanism of the multi-firing cartridge was simulated by using Digital Mockup (DMU) Kinematics workbench in CATIA V5. The simulation shows the measure of the clips privileged the cartridge and the movement of the rotating clip which is the most critical part of this design. The spiral torsion spring stores the potential energy and can be converted as kinetic energy once the clip is deployed.

IOP Conf. Series: Materials Science and Engineering **851** (2020) 012050 doi:10.1088/1757-899X/851/1/012050

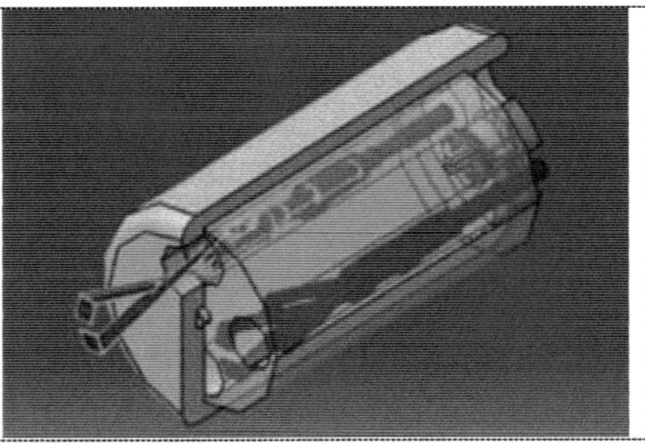

Figure 10. Deployment of first clip

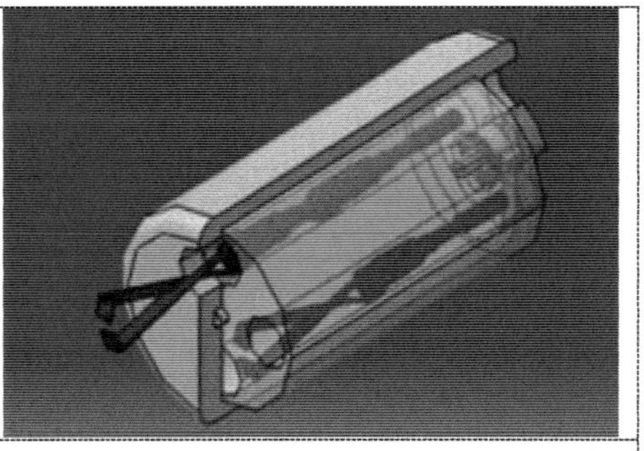

Figure 11. Deployment of second clip

Fig. 10 shows that the first clip is being deployed to the affected area with a deflection of θ = 0°, this can be considered as the original orientation of the rotating cartridge. Fig. 111 shows that the first chamber is empty, and the rotating cartridge rotates anti-clockwise with a deflection of θ = 90° to the next chamber that contains the second clip. The chamber is aligned with the hole at the housing cartridge because the static part at the rear of the housing cartridge holds the clip in place.

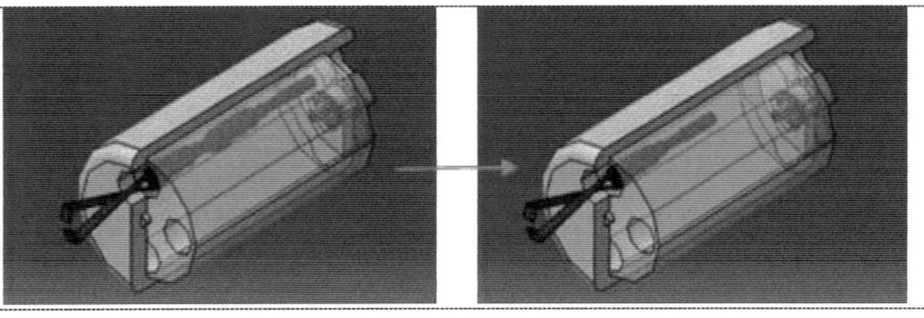

Figure 12. Deployment of third and last clip

Fig. 12 shows that the third and final clip is being deployed to the affected area and rotates anti-clockwise with a deflection of θ = 180°, 270° respectively. It uses the same mechanism concept in Fig. 16 where the spiral torsion spring rotates the rotating cartridge until the chamber and the hole at the housing cartridge is properly aligned.

6. Conclusions

In this project, the upgrading of endoclip design by designing a cartridge that covers numerous clips had met the objectives which is to design the mechanism for multi-firing endoclip cartridge and also to fabricate the optimize design using rapid prototyping. Three design concept of multi-firing endoclip cartridge were developed and each of the concept designs uses different deployment mechanism for the endoclip. The best design was selected by evaluating the critical component to achieve the objective by using decision matrix method.

The result of the decision matrix shows that design 2 got the highest mark in the table because it can hold more clip than the other two design and also it also can reduce operation time. The cartridge is designed to substitute the tip of the catheter of the endoclip so its design can be improved. Nonetheless, this project only gets the mechanism pattern of the multi-firing cartridge where it allows multiple clips to be deploy to the affected area without the need of reloading after single use.

For further research and design, the materials and properties for the endoclip cartridge and also the spring must be identified to continuingly improve the endoclip design. It is important also to develop on the attachment of the cartridge to the clip applicator so it can operate the endoclip.

References

[1] Ronald W Y, Tonya K, Roy S, *Endoclips, Techniques in Gastrointestinal Endoscopy*, Volume 8, Issue 1, 2006, Pages 2-11

[2] Divyanshoo R. K, John B, 3 - *How Endoscopes Work, Clinical Gastrointestinal Endoscopy (Third Edition)*, 2019, Pages 24-31.e2,

[3] Christopher E. Devereaux, Kenneth F. Binmoeller, *Endoclip: closing the surgical gap, Gastrointestinal Endoscopy*, Volume 50, Issue 3, 1999, Pages 440-442,

[4] Shuchen G, Chengli S, Shiju Y, Liaoyuan A, Jingjing X, Mingyang L, Bing H and Alfred C 2016 Novel endoscopic multi-firing-clip applicator for endoscopic closure of large colonic perforations, *Minimally Invasive Therapy & Allied Technologies*,**25**(4), pages 188-195

[5] Romagnuolo J 2009 Endoscopic clips: past, present and future. *Canadian journal of gastroenterology = Journal canadien de gastroenterologie*, **23**(1), pages 158–160.

Analysis of Raw Mill Machines Maintenance in Cement Industry

R M Sari*, K Syahputri, I Rizkya, N Nadhirah

Department of Industrial Engineering, Faculty of Engineering, Universitas Sumatera Utara, Indonesia, 20155.
*Email: rahmi.m.sari@usu.ac.id

Abstract. Cement processing plan is a state-owned enterprise engaged in the cement industry. The problem that this factory has is such as frequent damage to the engine. The machines used in cement production at the factory include raw mill machines, kiln machines, and cement mill machines. The more often a machine is damaged, the average time between failures of a machine is lower. So that it can cause the production process is hampered. In this study, the engine studied was the Raw Mill machine at plant. The raw mill machine functions as a main material grinder in the initial milling process such as limestone, silica stone, clay, and iron sand to become raw mix. To find out the frequency of damage to the machine or the time interval between damage to the tool (MTBF) and the length of time used to repair the tool (MTTR), the Total Productive Maintenance (TPM) method is used to perform maintenance by optimizing the effectiveness of the equipment and reducing / eliminating sudden damage (breakdown) by first identifying and then the cause of the problem is identified using fish bone analysis and Failure Mode and Effect Analysis (FMEA).

Keywords: Maintenance, Mean Time Between Failure (MTBF), Mean Time To Repair (MTTR), Fault Mode and Effect Analysis (FMEA).

1. Introduction

In an effort to increase competitiveness with the other companies, the cement processing company must compete to apply management practices that lead to achieve efficiency and effectiveness of the company. One of the company's objective is to improve the effectiveness of machinery/equipment through equipment maintenance.

Maintenance is all activities related to maintaining a machine/equipment so that it is in a ready condition to operate, and if there is damage it is endeavored that the machine/equipment can be returned to a good condition. The role of new maintenance felt when the system starts to experience disruption or cannot be operated again. [1]

Maintenance activities are usually carried out according to a schedule determined by the company. Maintenance activities have been carried out at the cement processing company. However, at sometime there are still engine failures that occur. This condition certainly disrupt the production process. Maintenance activities that disrupt the production process will increase downtime. Downtime in a production system means loss of production time. Loss of production time can be caused by equipment breakdown due to damage and the time required to set up and adjust the damaged equipment. Maintenance activities will also affect the Mean Time Between Failure (MTBF) and the length of Mean Time To Repair (MTTR). The effectiveness of MTBF and MTTR in optimal conditions can be seen from the level of availability of the tool / machine. To optimize the performance of tools/machines, a good concept is needed, namely Total Productive Maintenance (TPM). Based on this, the author conclude the title of Analysis of Maintenance Raw Mill Machines in Cement Processing Company.

IOP Conf. Series: Materials Science and Engineering **851** (2020) 012051 doi:10.1088/1757-899X/851/1/012051

2. Research Methodology

2.1. Total Productive Maintenance (TPM)

Total Productive Maintenance (TPM) is a maintenance program that includes the latest concept definitions to maintain equipment and one of them can also be used by calculating performance maintenance. Performance consists of 3 parts, namely: [3]

1. *Reliability*
   Reliability is the possibility that the equipment can operate under normal conditions properly. [7] MTBF (*Mean Time Between Failures*) which is the time estimation between an improvement and subsequent failure (damage) of a component, machine, process or product. [6] MTBF this is formulated as the quotient of the total operating time of the machine divided by the number/ frequency of failures. Operation of the machine due to breakdown. The following is a calculation from MTBF :
   $$MTBF = \frac{Total\ Operation\ Time}{Breakdown\ Frequency} \quad (1)$$
2. *Maintainability*
   Maintainability is a business and costs to carry out maintenance (maintenance). A measure of maintainability is *Mean Time To Repair* (MTTR), the increase of MTTR identifies the decrease of *maintainbility*.[8] Dimana *Mean Time To Repair* (MTTR) is the average time the repaired machine when a failure occurs. [9] The following is an calculation from MTTR:
   $$MTTR = \frac{Breakdown\ Time}{Breakdown\ Frequency} \quad (2)$$
3. *Availability*
   Availability is the time percentage of the machine is used. Availability (A) is calculated using the formula:[10]
   $$A = \frac{Total\ Operation\ Time}{Loading\ Time} \text{ x } 100\% \quad (3)$$

2.2. *Overall Equipment Effectiveness (OEE)*

Overall Equipment Effectiveness (OEE) the overall effectiveness of the facility obtained by calculating the *Avaibility, Performance, Efficiency*, dan *Rate of Quality Product.* [2]

OEE = *Availability* x *Performance* x *Quality*

The following is an explanation of three ratios to measure machine effectiveness.

1. *Availability*
   Availability is the probability of a system operating at a time or in a certain interval. [11] The following is the calculation of Availability:
   $$Availability = \frac{standby\ time + operational\ time}{Calendar\ time} \text{ x } 100\ \% \quad (4)$$
2. Performance
   Performance in OEE calculation is the number of machine products produced in the time available. The following is a calculation of performance.
   $$Performance = \frac{Production\ Total / Operational\ Time}{Design\ Production\ Capacity} \text{ x } 100\ \% \quad (5)$$
   Design Production Capacity Raw mill (Indarung VI) = 18000 tpd (ton per *day*)
3. *Quality*
   Quality n this OEE is the number of machines of good products successfully produced compared to the total number of machines of products produced. Quality of cement production is calculated of 100%

2.3. Failure Mode and Effect Analysis (FMEA)

FMEA is a systematic method of identifying and preventing problems that occur in products and processes. [5]

IOP Conf. Series: Materials Science and Engineering **851** (2020) 012051 doi:10.1088/1757-899X/851/1/012051

3. Result and Discussion

3.1. Data Collection

Data collection of MTBF, MTRR, and OEE of Raw Mill Machine is the following below.

1. Operational Data

Table 1. Operational Data of Raw Mill Machine

Period	Operational Time (Days)	Total Operational Time (Hours)
1	16,8	403,3
2	13,5	324,6
3	24,5	587,0
4	21,6	519,4
5	23,6	567,5
6	20,8	498,8
7	24,9	597,9
8	21,4	513,8
9	16,8	402,7
10	24,2	581,8
11	21,0	504,4
12	20,8	499,0
Total	250	6000,20
Average	20,83	500,02

2. Data of *breakdown time*

Table 2. Breakdown Time Data of Raw Mill Machine

Period	Operational Time (Days)	Total Operational Time (Hours)
1	1.20	29.87
2	2.90	69.32
3	2.10	50.22
4	6,5	155,67
5	4,0	95,45
6	4,6	110,72
7	-	0,00
8	4,9	116,57
9	2,5	60,37
10	3,7	87,70
11	5,1	122,67
12	2,5	59,67
Total	33,80	808,82
Average	3,64	79,85

3. Data of *breakdown* frequency

Table 3. *Breakdown Frequency Data of* Raw Mill Machine

Period	Frequency of breakdown stop
1	42,0
2	25,0

3	45,0
4	103,0
5	88,0
6	48,0
7	-
8	55,0
9	47,0
10	70,0
11	88,0
12	42,0
Total	653
Average	59,36

4. Data *Loading Time*

Table 4. *Loading Time Data of* Raw Mill Machine

Period	Standby Time (Days)	Total of Standby Time (Hours)
1	11,9	286,6
2	11,2	268,1
3	1,4	34,4
4	0,4	9,8
5	1,1	25,7
6	0,5	11,3
7	6,0	144,5
8	0,4	10,7
9	7,9	190,0
10	0,8	18,3
11	1,8	42,5
12	4,0	96,5
Total	47,40	1138,40
Average	3,95	94,87

5. Data of Cement Production

Table 5. Cement Production Data of *Raw Mill* Machine

Period	Production (Ton)
1	245.718
2	212.485
3	405.304
4	396.998
5	416.437
6	376.047
7	328.490
8	382.792

9	298.093
10	413.719
11	354.231
12	357.761
Total	4.188.075
Average	349.006,25

3.2. Data Processing

The following is a recapitulation of *Mean Time Between Failure* (MTBF), *Mean Time To Repair* (MTTR), and *Overall Equipment Effectiveness* (OEE).

Table 6. Recapitulation of MTBF, MTTR and OEE

MTBF	8,42 Hours
MTTR	1,35 Hours
Availability	81%
Performance	83%
Quality (set by company)	100%
OEE (overall equipment effectiveness)	76%

The following is the cause of the raw mill machine experiencing a breakdown.

Table 7. Cause of the Raw Mill Machine Experiencing a Breakdown

No.	Problems	Description	Improvement	Frequency	Duration (hours)
1	*Max Vibration*	Vibration crosses the threshold		190	179:53
2	*Chain reclaimer* broke	The reclaimer chain broke	Replacement of a broken reclaimer chain	24	43:08
3	*Belt* 6R1J03Z1 *runtime*	Appear *alarm runtime* in *belt*	Check and reset again	19	18:00
4	6R1J04 *trip high current*	Appear of *alarm high current*	Check and reset again	15	4:44

3.3 Fault Mode and Effect Analysis (FMEA)

Table 8. Problem Resolution Priority Value in Raw Mill Machines.

Part/Process Function& specification	*Potential failure mode*	*Potential effect of failure*	*Sev*	*Potential Causes/ Mechanism failure*	*Occ*	*Current Controls Design*		*Det*	*RPN*	*Recommended Action*
						Prevention	*Detection*			
Man	Lack of concentration when machine controlling	Machine stop due to maximum vibration	7	Operators do not focus when there are parameters change	4	Workers focus when operating the machine	Do control when the machine is operated	4	112	Workers must be more focus when operating the machine
Machine	Gate used to sort strange material does not run	Machine stop due to maximum vibration	8	Material enters and gets stuck between the gates	7	Do a gate check	Do checking machine parts	1	56	More often to check the machine parts
Material	There is a strange object in raw materal object such as metal in the material above the table	Machine stop due to maximum vibration	8	Loose Raw Mill body plates or metal detection sensors do not detect strange objects	6	Do routine inspection of machine parts	Do checking machine parts	5	240	More often to do a machine inspection

	Hard material (dark/black)	Machine stop due to maximum vibration	6	The hardness of mining material is not homogeneous	3	Take material samples for testing in the laboratory	Test the material hardness	4	72	More improvement in material testing
	Hanging clay	Machine stop due to maximum vibration	7	Sticky and wet materials	7	Control feeding clay in the field	Rpm speed monitor apron clay and monitor the amount of clay being fed actually	2	98	More improvement in controlling the trend speed rpm apron feeder clay
Method of Operational Control	*Bed* material under ideal (operational ideal 50 mm)	Machine stop due to maximum vibration	8	Water injection is given to the material which causes dry material so the roller pressure must be harder	6	Adjust the injection of water injection	Do control of *water injection*	1	48	More improvement in controlling water injection

4. Conclusion

From the data processing and analysis that has been done, the following conclusions can be conclude:

1. The value of the MTBF unit of Raw Mill is 8,42 hours. With the achievement of a target about 4% of the target set by the company of 250 hours, the MTTR value of the Raw Mill unit in 2018 was 1,35 hours, and the value of the Overall Equipment Effectiveness (OEE) unit of the Raw Mill was 76%. This achievement has reached the target set by the company by 75%.
2. The biggest cause of damage to the raw mill machine is the maximum vibration problem with a frequency of 190 times with a duration of 179:53 hours.
3. Based on the results of the FMEA, it was found that the biggest factors causing maximum vibrations in Raw Mill machines were material, human, and machine factors that were part of the process.

Acknowledgment
Authors special thanks to Universitas Sumatera Utara for funding this work from the beginning until its done.

References

[1] Assauri S 2004. *Manajemen Produksi dan Operasi.*Penerbit Fkultas Ekonomi Universitas Indonesia. Jakarta

[2] Ade A 2015. *Analisis Reliabilitas dan Availabilitas pada Mesin Produksi dengan Sistem Seri MenggunakanPendekatan Analisis Markov di PT. "X".* Surabaya: Institut Teknologi Sepuluh November.

[3] Bentley J 1999. *Introduction to Reliability and Quality Engineering*, 2nd edition, Addison-Wesley, England.

[4] Deni J, 2009,*The Implementation of Maintenance Quality Function Deployment (MQFD) for Improving Maintenance Quality at Mining Industry*. Jakarta: FT UI.

[5] Dervitsiotis K N, 1981. *Operational Management.* New York: Mc Graw Hill Book Company

[6] Ebeling CE, 1997.*An Introduction to Reliability and Maintanability for Engineering*.First edition.New York: McGraw Hill

[7] Hansen R C, 2001, *Overall Equipment Effectiveness: A Powerful Production / Maintenance Tool For Increased Profits,* Industrial Press Inc, United Stated of America

[8] Hegde, Harsha G, Mahesh N S, Doss K 2009, *Overall Equipment Effectiveness Improvement by TPM and 5S Techiniques in a CNC Machine Shop*.

[9] Peter W and Dennis M. 2001. *TPM –A Route to World-Class Performance, A division of Reed*

Educational and Professional Publishing Ltd. Oxford

[10] Pujotomo D 2011. *Analisis Total Productive Maintenance pada Line 8/Carbonated Soft Drinkpt Coca-Cola Bottling Indonesia Central Java*. Semarang: Universitas Diponegoro.

[11] Siregar I, Muchtar M A, Rahmat R F, Andayani U, Nasution T H, and Sari R M 2018 Method of calculation overall equipment effectiveness in fertilizer factory. *IOP Conference Series: Materials Science and Engineering* **308**(1), p. 012053) IOP Publishing.

Proximate, ultimate and calorific value analyses of paper industry sludge at different moisture content

Sugimin Yoh[1], T. Sitepu[1], H. Ambarita[1,2]

[1]Mechanical Engineering Department, Faculty of Engineering, Universitas Sumatera Utara, Jl. Almamater Kampus USU, Medan 20155 Indonesia
[2]Sustainable Energy and Biomaterial Centre of Excellent, Universitas Sumatera Utara, Jl. Almamater Kampus USU, Medan 20155 Indonesia

[*]Email: himsar@usu.ac.id

Abstract. Paper industry has a big potency of energy that can be reused in its sludge. However, the water content of the sludge is still high and need to be dried. In this study, the characteristics of the sludge at different water content are explored. The sludge is collected from paper industry in Sumatera Utara province of Indonesia mainly at 70% water content. The experiments are carried out at lower water content that is from 10% to 60%. The proximate and ultimate analyses are carried out and the calorific values are discussed. The objective is to explore the potency of the sludge if it is employed as solid fuel. The results show that reducing water content of the sludge affects the characteristics of the sludge. The lower water content increases fixed carbon value, calorific value and volatile matter.

1. Introduction

The global warming, if is not controlled, can be a catastrophe in the future. In order to save the human being from catastrophe, emission of Greenhouse Gases (GHGs) must be reduced. Reducing fossil fuel consumption is a significant mitigation action to reduce the GHGs emission. This can be done by enhancing the utilization renewable energy and increase the energy efficiency [1]. Increasing the energy efficiency include improvement the process and also recovery system [2]. Here the strategy include reuse the waste [3]. There many industries can implement the reuse strategy, one of those is paper industry. The industry has a big potency of energy that can be reused in its sludge. However, the water content of the sludge is still high and need to be reduced. Many researchers have proposed strategy and innovation to reuse the sludge of paper industry. Chen et al. [4] reported a study on recycling of paper sludge powder for achieving sustainable and energy-saving building materials. The effects of paper sludge powder on properties of Calcium silicate board materials (CSBM) were reported. The experimental results indicate that single addition of 30% paper sludge ash of 20% alkali recovery paper sludge can improve the flexural strength of CSBM effectively. The conclusion of the study is that the utilization of paper sludge powder presents a great potential to develop sustainable and energy-saving CSBM with enhanced flexural strength and reduced water absorption. Azevedo et al. [5] reported a study on the evaluation the effect of partial substitution of paper sludge, up to 20 wt%, for lime in cementitious mortars. The results reveals that for use in wall and ceiling mortal coatings, the level of incorporation should not exceed 10%. The reason is that higher levels yield lower values of mechanical strength resistance and will incompatible with market requirement. An et al. [6] reported the study on the performance and energy recovery of a single and two stage biogas production from paper sludge. The study suggests that the potential application of the established two-stage process in valorization of low-value waste biomass. Azevedo et al. [7] asses the potential of sludge generated by the pulp and paper in assembling locking blocks. The objective is to analyze the feasibility of incorporating sludge from the pulp and paper industry in soil-cement block using 10% cement and 90% compensated soil. The energy saving in the process of bioethanol production from paper mill sludge examined by Salameh et al. [8]. In specific the objective is to reuse the wastewater produced during the de-ashing step as a substitute for freshwater addition during the conversion of paper mill sludge into ethanol. The results showed that wastewater recycling minimized the energy demands in the distillation and evaporation units by 1206 kJ/kg. The energy reduction due to the increase of metals and total soluble solids in both stream after fermentation. Glinska et al. [9] studied the recovery and characterization of cellulose from industrial paper mill sludge using tetrakris and imidazolium based ionic liquids. It was shown that the recovery cellulose from paper industry sludge is feasible with ionic liquid. However, improvement of the performance is needed to allow scale-up of the technology. Recently, Azevedo et al. [10] make comparison of the processing of primary sludge waste from paper industry for mortar from technological and environmental aspects. The conclusion is that the beneficiation process, which produces the thin liquid is the most efficient for the incorporation of waste into mortars when compared to the dry method.

Those literature show that mainly the sludge of the paper industry is recycled to produce building materials. A few study uses the strategy converting the waste into energy. The used method is converting the waste into ethanol production. Our work focuses on how to convert the sludge into energy. The used method is by reducing its water content and becoming biomass. In this work the characteristics of the sludge at lower water content will be explored. The results are expected to support the necessary information to convert sludge into energy.

2. Experimental method

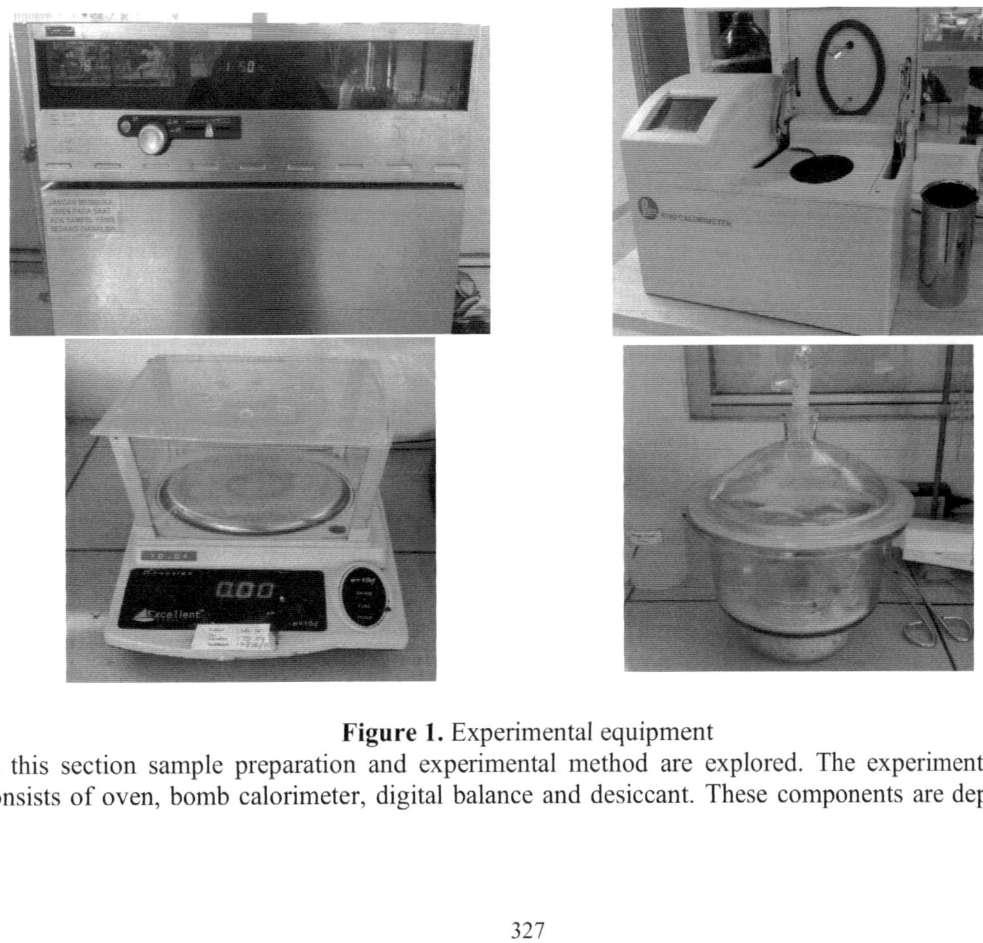

Figure 1. Experimental equipment

In this section sample preparation and experimental method are explored. The experimental setup consists of oven, bomb calorimeter, digital balance and desiccant. These components are depicted in

Figure 1. The tested sludge collected from paper indutry in Sumatera Utara province of Indonesia. The sample of the sludge depicted in Figure 2.

Figure 2 Sludge used in this study

3. Results and Discussions

In the experiment the slude is then divided into six groups. Each group is dried to a certain number of water content. The water content is dry based. The target of the water contents are 10%, 20%, 30%, 40%, 50% and 60%. To each component the proxymate analysis, ultimate analysis and calorific value analyses are carried out. The results are presented in the below section.

3.1. Proxymate analysis

The proxymate analysis has been carried out for all samples and the results are presented in Table 1. In the top line it is mentioned that there are six samples have been tested. The water content varied from 11.72% to 61.5%. The tested parameters are Ash, Volatile matter and Fixed Carbon. The ash content in the sludge is not affected by water content. On the other hand, volatile matter increase with decreasing water conttent in the sludge. Furthermore, the fixed carbon in the sludge inrease with decreasing water content.

Table 1 Proxymate Analysis

Parameter	Sample					
	A	B	C	D	E	F
Water content	61.5%	49.4%	39.64%	28.96%	19.28%	11.72%
Ash	11.18%	11.47%	11.52%	10.94%	9.10%	8.20%
Volatile matter	22.37%	28.48%	31.48%	35.64%	34.95%	33.29%
Fixed Carbon	5.30%	10.71%	17.36%	24.46%	36.67%	46.79%

3.2. Ultimate analysis

The results of the ultimate analysis is shown in Table 2. The tested parameters are Carbon, Hydrogen, Nitrogen and Sulphur. It can be seen that there is no effect of water content in the sludge to all those parameters.

Table 2 Ultimate Analysis

Parameter	Sample					
	A	B	C	D	E	F
Water content	61.5%	49.4%	39.64%	28.96%	19.28%	11.72%
Carbon	14.28%	16.12%	18.51%	21.57%	25.98%	29.42%
Hydrogen	7.70%	7.62%	7.54%	7.46%	7.37%	7.29%
Nitrogen	0.42%	0.45%	0.49%	0.53%	0.57%	0.61%
Sulphur	0.11%	0.11%	0.12%	0.12%	0.12%	0.13%

3.3. *Calorific value*

The calorific value for all samples has been tested and shown in Table 3. In order to make comparison the theoretical Low Heating Value (LHV) and High Heating Value (HHV) of the samples are also presented in Table 3. It can be seen that those parameters show the same trend. The calorific value increase as water content decrease. In addition, the present calorific value drawn from experimental is within the range of LHV and HHV. Same for all samples.

Table 3 Calorific value

Parameter	Sample					
	A	B	C	D	E	F
Water content	61.5%	49.4%	39.64%	28.96%	19.28%	11.72%
LHV (Theoretical) [kcal/kg]	1175.43	1288.62	1522.94	1786.65	2208.49	2478.46
HHV (Theoretical) [kcal/kg]	1215.43	1321.22	1549.13	1805.78	2221.23	2470.72
Experimental [kcal/kg]	1216.8	1304.3	1544.4	1782.6	2226.2	2438.10

4. Conclusions

In the present work the characteritics of sludge from the paper industry has been tested experimentally. There are six samples at different water content have been analzed. The proxymate analysis, ultimate analysis and calorific value analysis have been carried out. The main conlclusion are as follows. The ash content in the sludge is not affected by water content. On the other hand, volatile matter increase with decreasing water conttent in the sludge. Furthermore, the fixed carbon in the sludge inrease with decreasing water content. The calorific value increase as water content decrease. In addition, the present calorific value drawn from experimental is within the range of LHV and HHV. Same for all samples.

References

[1] Ambarita H 2018 *IOP Conference Series: Earth and Environmental Science* **126** 012021
[2] Ambarita H 2017 IOP Conference Series: Materials Science and Engineering **237** 012013
[3] Ambarita H, Soeharwinto S, Ginting N, Basyuni M and Zen Z 2018 *IOP Conference Series: Earth and Environmental Science* **126** 012114
[4] Chen M, Zheng Y, Zhou X, Li L, Wang S, Zhao P, Lu L and Cheng X 2019 *Construction and Building Materials* **229** 116874
[5] Azevedo ARG, Alexadre J, Xavier GC and Pedroti LG 2018 *Journal of Cleaner Production* **192** 335-346
[6] An Q, Cheng JR, Wang YT and Zhu MJ 2020 *Renewable Energy* **148** 214 – 222
[7] Azevedo ARG, Marvilla TM, Fernandes WJ, Alexadre J, Xavier GC, Zanelato EB, Cerquira NA, Pedroti LG and Mendes BC 2019 *Journal of Building Engineering* **23** 334-340

IOP Conf. Series: Materials Science and Engineering **851** (2020) 012052 doi:10.1088/1757-899X/851/1/012052

[8] Salameh T, Tawalbeh M, Al-Shannag M, Saidan M, Melhem KB and Alkasrawi M **2020** *Energy* **196** 117085
[9] Glinska K, Bin Ismal MS, Goma-Camps J, Valencia P, Stuber F, Giralt J, Fabregat A, Torrens E, Olkiewicz M and Bengoa C 2019 *Industrial Crops & Products* **139** 111556
[10] Azevedo ARG, Alexadre J, Marvila MT, Xavier GC, Montiero SN and Pedroti LG 2020 *Journal of Cleaner Production* **249** 119336

IOP Conf. Series: Materials Science and Engineering **851** (2020) 012053 doi:10.1088/1757-899X/851/1/012053

Analysis of the use of SEN0161 pH sensor for water in goldfish ponds

T H Nasution[1*], S Dika[1], E P Sinulingga[1], K Tanjung[1], and L A Harahap[2]

[1]Department of Electrical Engineering, Universitas Sumatera Utara, Medan Indonesia
[2]Department of Mathematics and Computer Science, Faculty of Science, Chulalongkorn University, Bangkok, Thailand

Correspondent Author: tigor.nasution@usu.ac.id

Abstract. Goldfish are considered economically important because they are popular with the community. Goldfish is an important type of freshwater aquaculture that is quite developed and can be cultivated in Indonesia. The pH value will affect the growth of fish. The acidity of water that is good for goldfish life is 7 to. The lowest limit that causes fish death is pH 4 and the highest at pH 11. Therefore, in this study, we will analyze the use of a SEN0161 pH sensor to measure the pH level of water in goldfish ponds. The pH measurement of fish pond water is compared with measurement data using a standard pH meter. Data collection and testing is done using two sensor nodes, where the first sensor node is connected to the Lattepanda which also functions as a base station. LattePanda communicates simultaneously with a data speed of 9600 bps with a second node sensor that uses Arduino as a microcontroller. In communicating, 2 nodes use the Xbee radio module. From the results of testing and implementation that have been done p, it can be concluded that the percent accuracy of the pH sensor reading SEN0161 at each node is ± 98.39% and ± 99.07%.

1. Introduction

Goldfish are considered economically important because they are popular with the community. Goldfish is one type of important freshwater aquaculture that is quite developed and can be cultivated in Indonesia[1]. Goldfish are freshwater fish that are usually sold alive. Goldfish production which is targeted to be 500,000 tons in 2013 has not been able to reach the target (68.17% of the target) as well as the production value which only reached 90.89% of the target[2]. Several things cause the achievement of these targets, including the emergence of disease and fish death. Live fish transportation techniques that can guarantee that the fish will come to consumers in a state of survival are urgently needed. These efforts have not been followed by efforts to increase fish survival and physiological studies of fish so there are still many problems faced. one of the physical factors that influence the physiological process of fish is temperature and is one source of stress that can affect the physiological changes in the fish body[3].

Another factor is the degree of acidity or pH, dissolved oxygen, and ammonia levels are some of the chemical parameters of water that has a great influence on the organisms that live in them[4]. Discrepancies in temperature, pH, DO and ammonia levels where goldfish live (the environment) will result in slow fish growth and can result in fish mortality. The pH value will affect the growth of fish. The acidity of water that is good for goldfish life is 7 to 8[1]. The lowest limit that causes fish death is

IOP Conf. Series: Materials Science and Engineering **851** (2020) 012053 doi:10.1088/1757-899X/851/1/012053

pH 4 and the highest at pH 11. Waters with a pH range of 4-6 result in slow growth for cultured fish[5,6]. Therefore, in this study, we will analyze the use of the SEN0161 pH sensor to measure the pH level of water in a goldfish pond.

2. Material and method

2.1. SEN0161 pH sensor

This sensor is one of the components in this system because it is used to measure the pH value of water[7]. These sensors are installed at each node to monitor water pH and calculate ammonia levels. The output from this sensor is in the millivolt (mV). So, to get the output in pH value, convert the output from pH millivolts to this formula:

$$Value\ of\ pH = \frac{(7-(ouput))}{((59.16\,mV)(2.2))}$$

The output of the sensor in millivolts will be automatically converted into a pH value after this formula is applied to programming[8]. The pH sensor display can be seen in Figure 1.

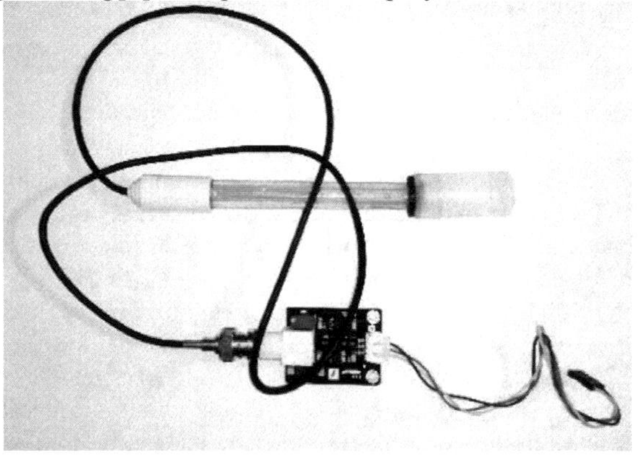

Figure 1. SEN0161 pH sensor

The specifications of this sensor are as follows:

Measuring range	: 0 ~ 14 pH
Operating temperature	: 0 ~ 60 ºC
Accuracy	: ± 0.1pH (25 ℃)
Response time	: ≤ 1 m

The SEN0161pH sensor is specifically designed for Arduino controllers and has simple connections and features ... As for the schematic circuit, see Figure 2. This sensor can be used to measure pH with a range of 0 ~ 14, with an accuracy of ± 0.1pH (25 ℃).

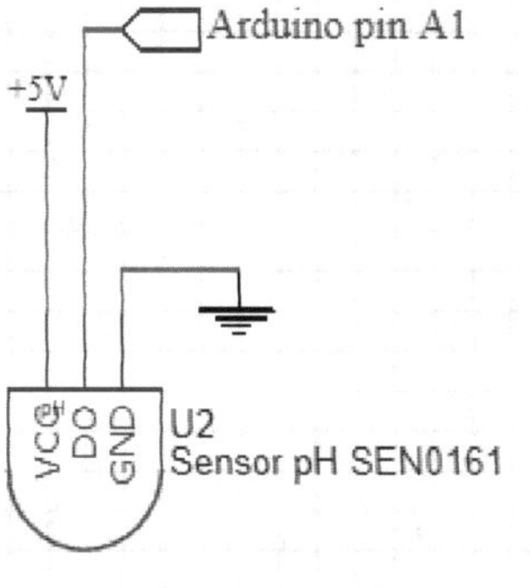

Figure 2. SEN0161 pH Sensor Circuit with Arduino

2.2. Data collection and testing

Data collection and testing are done using two sensor nodes, where the first sensor node is connected to Lattepanda which also functions as a base station[9,10]. LattePanda communicates simultaneously with a data speed of 9600 bps with a second node sensor that uses Arduino as a microcontroller[11,12]. In communicating, 2 nodes use the Xbee radio module[13–15].

The test location is done on Jl. Bunga Cempaka, Tj.Sari, Medan Selayang, Medan. Testing is done by placing the sensor node on the edge of the pool, because of limited access to the middle of the pool and the power supply for LattePanda is on the edge of the pool. The location of the test can be seen in Figure 3. and the location of the nodes can be seen in Figure 4.

Figure 3. Goldfish pond location

Figure 4. Node 1 and node 2 at the test site

Testing is done by uploading a pH reading program on LattePanda for node 1 and Arduino UNO for node 2. Testing the pH using a pH meter at the location of node 1 and node 2 can be seen in Figure 5. below this.

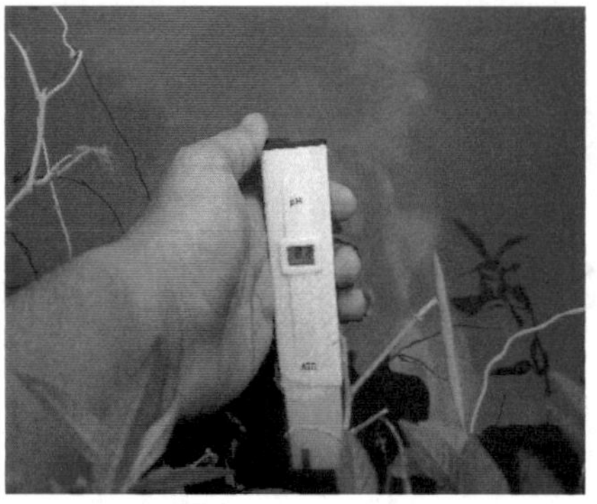

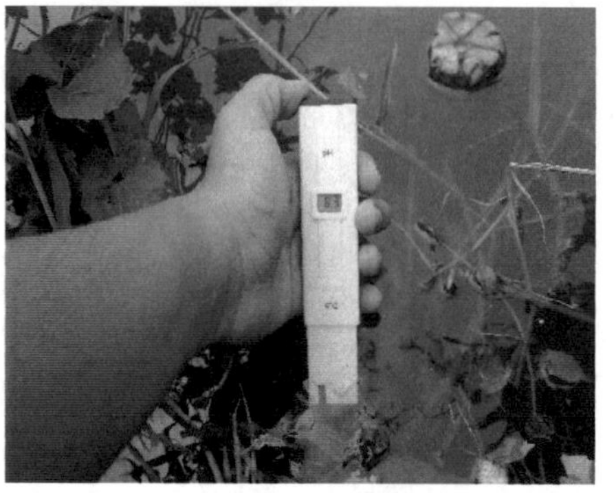

Figure 5. Testing the pH at node 1 and node 2 uses a pH meter

3. Results and discussion

The pH measurement of fishpond water is compared with measurement data using a standard pH meter. The results of the measurement of pH at the node designed and the measurement of pH using a pH meter at node 1 can be seen in Table 1. The results of pH measurements at the nodes designed and pH measurements using a pH meter on node 2 can be seen in Table 2.

Table 1. Measurement of pH on Node 1

No.	pH measurements using the SEN0161 pH sensor in Node 1 (mg/L)	pH measurements using a pH meter on node 1 (mg /L)
1	8.17	8.2
2	7.98	8.2
3	8.22	8.2
4	8.46	8.2

Table 2. Measurement of pH on Node 1

No.	pH measurements using the SEN0161 pH sensor in Node 2 (mg/L)	pH measurements using a pH meter on node 1 (mg /L))
1	8.19	8.3
2	8.19	8.3
3	8.33	8.3
4	8.36	8.3

From Table 1 and Table 2 the accuracy values obtained from the measurement of the pH sensor SEN0161 for node 1 are ± 98.39% and ± 99.07% for node 2. Besides, there are also insignificant differences between the readings of the two sensor nodes, with a reading difference of 0.1 back digits comma, and the percent difference of accuracy is 0.68%. That is because the difference in the location of the sensor node reading on the sensor is not fixed.

4. Conclusion

From the results of testing and implementation that have been done in the previous section, it can be concluded that the percent accuracy of the reading of the SEN0161 pH sensor at each node is ± 98.39% and ± 99.07%. There was also an insignificant difference between the readings of the two sensor nodes, with a reading difference of 0.1 digits behind the comma, and a difference in percent accuracy of 0.68%. That is because the difference in the location of the sensor node reading on the sensor is not fixed

Acknowledgment
We thank the Ministry of Education and Culture and Universitas Sumatera Utara for supporting the implementation of this research.

References

[1] Sabrina S, Ndobe S, Tis'i M and Tobigo D T 2018 Pertumbuhan Benih Ikan Mas (Cyprinus carpio) Pada Media Biofilter Berbeda *J. Penyul. Perikan. dan Kelaut.*
[2] Soetrisno C K 2014 Laporan Tahunan Direktorat Produksi Tahun 2013, Direktorat Jenderal Perikanan Budidaya, KKP. *Program*
[3] Rachmawati F N Respon fisiologi ikan nila Oreochromis niloticus, yang distimulasi dengan daur pemuasaan dan pemberian pakan kembali
[4] Kordi M G H 2010 Panduan lengkap memelihara ikan air tawar di kolam terpal *Jakarta Lily Publ.*
[5] Boyd C E 1998 Water Quality for Pond Aquaculture *Int. Cent. Exp. Stn.*
[6] Rahmat R F, Syahputra M F and Lydia M S 2016 Real Time Monitoring System for Water Pollution in Lake Toba *2016 International Conference on Informatics and Computing (ICIC)* pp 383–8
[7] Quintero-Polanco J D, Betancourt F R and Molina-Mosquera J 2018 Control of dissolved oxygen in water for intensive tilapia culture using IoT *ARPN J. Eng. Appl. Sci.*
[8] Shahrulakram M A I and Johari J 2017 Water storage monitoring system with pH sensor for pharmaceutical plants *Proceedings of the 2016 6th International Conference on System Engineering and Technology, ICSET 2016*
[9] Anon LattePanda Docs
[10] Nasution T H, Muchtar M A, Seniman S and Siregar I 2019 Monitoring temperature and humidity of server room using Lattepanda and ThingSpeak *J. Phys. Conf. Ser.* **1235** 012068
[11] Nasution T H, Muchtar M A, Siregar I, Andayani U, Christian E and Sinulingga E P 2017 Electrical appliances control prototype by using GSM module and Arduino *2017 4th International Conference on Industrial Engineering and Applications, ICIEA 2017*
[12] Nasution T H, Siagian E C, Tanjung K and Soeharwinto 2018 Design of river height and speed monitoring system by using Arduino *IOP Conference Series: Materials Science and Engineering* vol 308
[13] Yumang A N, Paglinawan C C, Sejera M M, Lazam A S, Pagtakhan J C and Santos J S B 2016 ZigBee Based Monitoring of Temperature and Humidity of Server Rooms using Thermal Imaging *2016 6th IEEE International Conference on Control System, Computing and Engineering (ICCSCE)* (IEEE) pp 452–4
[14] Çetin G, Özdemir M Ş, Karadaş S and Yılmaz Y 2017 IEEE 802.15.4 / ZigBee based wireless sensor network design for monitoring server rooms *2017 International Conference on Computer Science and Engineering (UBMK)* pp 565–70
[15] Nasution T H, Siregar I and Yasir M 2017 UAV telemetry communications using ZigBee protocol *Journal of Physics: Conference Series* vol 914

IOP Conf. Series: Materials Science and Engineering **851** (2020) 012054 doi:10.1088/1757-899X/851/1/012054

Optimization of Sei Wampu Dam Project Schedule Using Critical Path Method

K Syahputri, P Mukhlidani , RM Sari, I Rizkya

Department of Industrial Engineering, Faculty of Engineering, Universitas Sumatera Utara

syahputri.khalida@gmail.com

Abstract: There are several companies engaged in construction, EPC (Engineering, Procurement, Construction), Infrastructure Investment, Property, and Real Estate experiencing project delays. One project is experiencing delays is Sei Wampu Dam Construction Project, Langkat Regency, North Sumatera Province starting from December 4, 2015 spending cost of IDR 309.093.870.437,59. Based on the data obtained, there was a delay in the project work by 13,575% in the cumulative 10th month results in material losses. Therefore, the project schedule optimization of Sei Wampu Dam is performed. The method used is Critical Path Method (CPM). Based on the calculation, the duration of the lead time is 1136 days with a total cost of IDR 362.200.345.57,-.

1. Introduction

Project management is a project activity applies knowledge, skills, tools and techniques to meet project requirements [1,2,3,4]. Project Management is the application of knowledge, expertise and skills, the best technical and with limited resources, to achieve the goals and objectives that have been determined in order to obtain optimal results in cost performance, quality and time and work safety aspects [5].

A construction project can be said to be successful if it is capable to meet its lead time objectives at the specified time, according to the costs allocated and meet the required quality. Project management is tasked by planning, organizing, leading and controlling existing resources in order to achieve project objectives. There are two tools or techniques for analyzing a development project optimally, namely Gantt Chart and Network Planning [6,7].

The case study discussed in this research is the Sei Wampu Dam Construction Project, Langkat Regency, North Sumatra Province. The project is calculated from 4 December 2015 with an expense of IDR 309,093,870,437.59. Based on data obtained from PT. Adhi Karya Persero Tbk, comparative data between planning and actualization of the Sei Wampu Dam project can be seen in Table 1.

Table 1. Comparison of Planning and Actualization Data

Time Application (Month)	Progress Planning (%)	Progress Realization (%)
1	0,024	0,008
2	0,297	0,003

Table 1. Comparison of Planning and Actualization Data (Continued)

Time Application (Month)	Progress Planning (%)	Progress Realization (%)
3	0,940	0,006
4	0,935	0,120
5	2,173	1,125
6	6,058	0,787
7	2,798	2,679
8	1,867	1,462
9	2,013	3,021
10	4,271	4,364
Total	**21,555**	**13,575**

Based on Table 1, the process plan with the process realization has a cumulative difference in the 10th month of 13,575%. This is shows that the delays occurrence in project work due to several problems appear during the project work, including delay in material delivery, labor delay, equipment procurement delay, subcontractors delay, incorrect planning, weak project time control, weak coordination, lack of project work supervision, lack of technical personnel, weak communication, and others.

To overcome the project schedule delay, an optimization of the project schedule is carried out using the Critical Path Method (CPM).

2. Methodology

The study was conducted on the Sei Wampu Dam construction project in Langkat Regency. The study was conducted by direct observation in the field and conducting interviews with employees in the field. From the immediate observation, the problem formulation is determined based on conditions so that the research objectives can be applied. The purpose of the research specified is the solution to the problem that occurs. Furthermore, data collection is required as an input to resolve the project schedule issue. Based on this data, problem analysis occurs. The analysis is carried out with several stages, namely the making of a SIPOC (Supplier-Input-Process-Output-Customer) diagram and the critical path calculation using the Critical Path Method (CPM). The SIPOC diagram is useful to know the processes involved, the sequence of processes and interactions between processes, and what materials are involved in the process. While the critical path calculation is done to determine the minimum lead time project. The critical path calculation is done by forward calculation, backward calculation, and total float calculation at each activity on the project. By calculating the total float, a critical path and project duration work will be obtained.

2.1. *Critical Path Method* (CPM)

CPM is a deterministic technique uses a dependencies network between tasks and deterministic values given for the task duration, the critical path calculation which is the minimum time of lead time [8]. CPM provides parameters considered on each activity including the quickest start and end time, last start and end time, and the delay [1,9,10]. The most important advantage of CPM is it allows facilitating the optimization models usage for time estimation and costs and for sequentially [11,12]. To list the activities needed to finish the project with the duration at each activity and dependence at each activity, scheduling critical paths for network activity on nodes can be formulated, mathematically. ES (i) is the earliest activity start time for activity i, EF (i) is the earliest activity finish time for activity i, LS (i) is the latest start for activity i and LF (i) is the latest finish for activity i. Calculations relevant to nodes ie forward calculations (running from the first to the last), and countdowns (running from the last back to the first node in the network) is presented as follows:
Stage 1: Initial activities numbered by 0.

IOP Conf. Series: Materials Science and Engineering **851** (2020) 012054 doi:10.1088/1757-899X/851/1/012054

Stage 2: The next number assigned to the numbered activity which is the previous activity has been numbered. Stage 2 repeated by until all activities are numbered [10].

3. Result and Discussion

3.1. SIPOC Diagram (*Supplier-Input-Process-Output-Customer*)

SIPOC Diagram is a tool used by the team to identify all relevant elements of the project upgrade process before the work starts. This helps define complex projects that may not have good coverage, and are typically used in the Measure phase of the Six Sigma DMAIC methodology (Define, Measure, Analyze, Improve, Improve, Control). The tool name asks the team to consider the supplier ('s ' in SIPOC) of your process, input (' I ') into the process, the process (' P ') your team improves, the output (' O ') of the process, and the customer (' C ') that receives the process output. In some cases, customer requirements may be added to the end of SIPOC for further details. The SIPOC diagram of Sei Wampu Dam project can be seen in Figure 1.

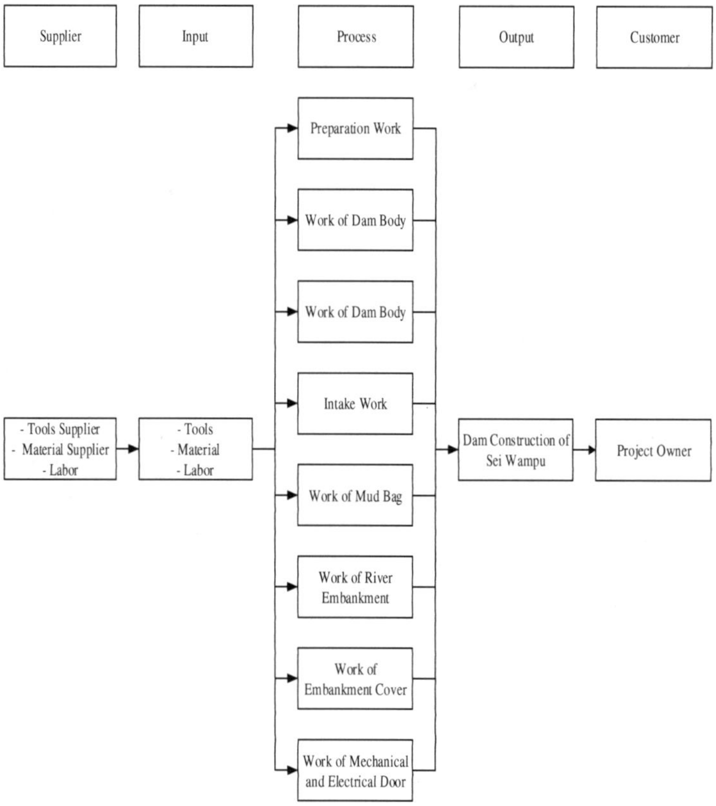

Figure 1. SIPOC Diagram

Based on the diagram above, it can be seen the components in the project and the relationship between existing components.

3.2. Calculation of Critical Path

Critical activity is the minimum lead time project. If there is a delay in activities of the critical path it will result in an increase all of project duration. Project network analysis is carried out to obtain critical path on the project network.

3.2.1. Forward Calculation

Activity A is not preceeded by any activity, then:

Early Start $(ES)_{(1,2)} = 0$, so that *Early Finish* $(EF)_{(1,2)}$:

$$EF_{(1,2)} = ES_{(1,2)} + D_{(1)}$$
$$EF_{(1,2)} = 0 + 180$$
$$EF_{(1,2)} = 180$$

The calculation continues for all activities.

3.2.2. Backward Calculation

BW activities do not precede any activities, then:

Lastest Start $(LS)_{(104,105)} = 1116$, so that *Lastest Finish* $(LF)_{(104,105)}$:

$$LF_{(104,105)} = LS_{(104,105)} + D_{(105)}$$
$$LF_{(104,105)} = 1116 + 20$$
$$LF_{(104,105)} = 1136$$

The calculation continues for all activities.

3.2.3. Calculation of Total Float

In activity A, the total float calculation is performed:

$$TF_{(A)} = LF_{(A)} - ES_{(A)} - D_{(A)}$$
$$TF_{(A)} = 180 - 0 - 180$$
$$TF_{(A)} = 0$$

The calculation continues for all activities.

Based on the results of forward calculations, backwards calculations, and total float, a project critical path is carried out: A-C-D-E-I-J-AA-d8-AC-d9-AD-AE-AF-AG-AI-AJ-d11-AL-AM-BO-BP-BQ-BR-BW dan A-C-D-E-I-J-AA-d8-AC-d9-AD-AE-AF-AG-AH-d10-AJ-d11-AL-AM-BO-BS-BT-BU-BW with a total lead time of 1136 days. The project critical path can be seen in Figure 2.

Figure 2. Critical Path of Sei Wampu Dam Project

3.3. *Calculation of Project Cost*

The activities description of Sei Wampu Dam construction has a critical path = A-C-D-E-I-J-AA-d8-AC-d9-AD-AE-AF-AG-AH-AI-d10-AJ-d11-AL-AM-d19-BO-BP-BQ-BR-BS-BT-BU-d27-d28-BW.

Lead Time	= 1136 days
Total Cost	= Direct cost + Indirect cost
	= IDR 309.093.870.438 + (1136 × 17.763.355)

	= IDR 329.273.041.415
PPN (10%)	= 10 % × IDR 329.273.041.415
	= IDR 32.927.304.142
Total Cost (include PPN)	= IDR 329.273.041.415 + IDR 32.927.304.142
	= IDR 362.200.345.557

4. Conclusion

The problem of project delays in Sei Wampu Dam Construction can be overcome by using Critical Path Method (CPM). Based on the calculations result using CPM obtained that lead time of Sei Wampu Dam project is 1136 working days from December 4, 2015 with a total cost of IDR 362.200.345.557.

References

[1] P H Soe and T M Htike 2018 *MATEC Web of Conferences* **192**, 01027 (2018)

[2] Henry A. Homstein *International Journal of Project Management* **33** (2015), pp 291-298

[3] Rismawan A and Muhardi 2018 *Seminar Penelitian Sivitas Akademika Unisba* (Bandung, Februari) **4** (1), pp 474-479

[4] Foulds LR and Wilson J M 2006 *Journal of Food Engineering* **70** , pp 281-292

[5] Ramadhanti N dan Muhardi 2019 *Seminar Penelitian Sivitas Akademika Unisba* (Bandung, Februari) **5** (1), pp 373-377

[6] Maryati S, Muhardi, and Koesdiningsih N 2017 *Seminar Penelitian Sivitas Akademika Unisba* (Bandung) **3** (2), pp 885-893

[7] Yang J B 2007. *How the Critical Chain Scheduling Method is Working for Construction*, *Cost Eng* **49** (4), pp 25–32

[8] Fahimifard S M and Kehkha A A 2009 Application of Project Scheduling in Agriculture (Case Study: Grape Garden Stabilization), *American-Eurasian J. Agric. & Environ. Sci.,* **5** (3), pp 313-321

[9] Smith P. 1984 Agricultural Project Management: Monitoring and Control of Implementation, *Elsevier Applied Science Publisher* , pp 190

[10] Zareei S 2017 *Renewable and Sustainable Energy Reviews* **81** (2018), pp 756-759

[11] Kelley J E 1961 *Critical-Path Planning and Scheduling: Mathematical Basis, Operations Research* **9** (3), pp 296–320

[12] Kelley JE, 1963 *The Critical Path Method: Resources Planning and Scheduling, Chapter 21*, Muth, J., Thompson, G.L. (Eds.), *Industrial Scheduling*, Prentice-Hall, (1963), pp. 347–365

IOP Conf. Series: Materials Science and Engineering **851** (2020) 012055 doi:10.1088/1757-899X/851/1/012055

Maternal mortality classification for health promotive in Dairi using machine learning approach

Henry Manik[1,2,*1], M.Fidel G.Siregar[1*], R.Kintoko Rochadi[1], Etti Sudaryati[1] , Ida Yustina[1] Rika Subarniati Triyoga[3]

[1]Faculty of Public Health, Universitas Sumatera Utara, Medan, North Sumatera-20222, Indonesia
[2]Rumah Sakit Umum Daerah Sidikalang, Dairi, Indonesia
[3]Faculty of Public Health, Universitas Airlangga, Surbaya, Indonesia

Corresponding author: * **henry_m2368@yahoo.com**

Abstract. Reducing maternal mortality rate is a key concern of health promotion in developing countries or city face. The investigated and survey for maternal mortality had been done in Dairy City. There are 149 samples got from the survey directly in this area for 2017. In this study, we use a machine learning approach to train and test the data of maternal mortality. The aim of this study to classification maternal mortality in health promotion for reducing the maternal mortality rate in Dairi. The result of this study indicated the decision tree and Naïve Bayes are available to train and test the dataset. The accuracy of the decision tree of maternal mortality is 100 % and the Naïve Bayes model indicates 97.37 % of maternal mortality.

1. Introduction

Maternal mortality is unacceptably high in the word. Based on the word, health organization report in 2019, about 295 000 women died during and following pregnancy and childbirth in 2017 [1], [2]. The situation and the main problem faced in Indonesia show the maternal mortality rate is higher than those of neighboring countries in ASEAN. [3] report the risk of maternal death per five million is 1 in 65 mothers, compared to 1 in 1100 mothers in Thailand, or they estimate that 20000 women die from complications of pregnancy per five million birth each year. Based on this situation, Indonesia tries to solve this issue by reducing maternal mortality using education and health promotion action. To solve this issued need to mapping and survey of maternal in each city in Indonesia. The represent of maternal mortality show the risk associated with each pregnancy. Maternal mortality has become important in this case, a measure of human and social development in Dairi, North Sumatera, Indonesia. It is particularly revealing of women's overall status, access to health care, and the responsiveness of the health care system to their needs. In this study, we collect the maternal mortality dataset from Dairi which 149 samples. Dairi regency is the area on the northwest shore of Lake Toba in North Sumatera [4], [5].

Machine learning is a perspective on modeling data with focused maternal mortality on modeling data and evolving research fields that are gaining popularity in many areas including health care, health promotion, and education [7]–[9]. One major health issue is maternal mortality, that needs to be worried about. Womens are the pillars of any family and society but maternal death during pregnancy is a great loss to baby, family, society, and country in Dairi, Indonesia. The goal of this study to classify the maternal mortality observed in Dairi, North Sumatera, Indonesia. Classification of the maternal mortality dataset by training the data and testing the data using a machine learning approach. We run the data by using the Discussion Tree and Naïve Bayes model.

2. Data and Methods

2.1 Data

The dataset of maternal mortality had been collected in Dairi, North Sumatera from 2017 years. There is more than 149 sample in this study was recorded. From the dataset of maternal mortality is female as age productive. The health promotion in Dairi had been studied [8]. There is a positive attitude with the sample with the predisposing factor correlated with clean and healthy life behavior. Dairi is located in North Sumatera and the capital of the district is Sidikalang. The research area is located approximately 148.3 km of Medan [5].

2.2 Methods

The model to classify the maternal mortality dataset by using Decision Tree and Naïve Bayes. A decision tree is the most influential and popular tool for prediction and classification. A decision tree is a flowchart like a tree structure, where each internal node denotes a test on an attribute, each branch signifies outcomes of the test and each leaf node holds a class label. A tree can be *"learned"* by splitting the source set into subsets based on an attribute value test. This process is repeated on each derived subset in a recursive manner called *recursive partitioning*. The recursion is completed when the subset at a node all has the same value of the target variable, or when splitting no longer adds value to the predictions [10], [11].

Naive Bayes algorithms are naturally used in sentiment analysis, spam filtering, recommendation systems. They are fast and easy to use but their biggest drawback is that the constraint of predictors to be independent. In the reals of life case, the predictors are dependent, this hinders the performance of the classifier [12]–[14].

A Naive Bayes classifier is a probabilistic machine learning model that's used for the classification task. The crux of the classifier is based on the Bayes theorem.

The Bayes theorem can show in eq. 1.

$$P(A|B) = \frac{P(B|A)P(A)}{P(B)} \quad (1)$$

From Bayes theorem, we can discovery the probability of A happening, given that B has ensued. Here, B is the evidence and A is the hypothesis. The supposition made here is that the predictors/features are independent. That is the presence of one particular feature does not affect the other. Hence it is called naïve.

3. Result and Discussions

The result of the analysis of the classification of the maternal mortality dataset for health promotion with 149 sample data in Dairi can show in Figure. 1 and Figure. 2. The visualizing of the Decision Tree result in Figure. 1 based on selected the age (maternal mortality) with the score of attitude remark in this research in Dairi, North Sumatera. In the Decision Tree model, we train the dataset to one or the other of two categories, and assigns new examples to one category or other, making it a non-probabilistic binary linear classifier. The confusion matrix result use Decision Tree show the classification rate or accuracy is 100 %, where the true positive from maternal mortality dataset is nine, the true negative is twenty-nine, false negative and false positive is nol.

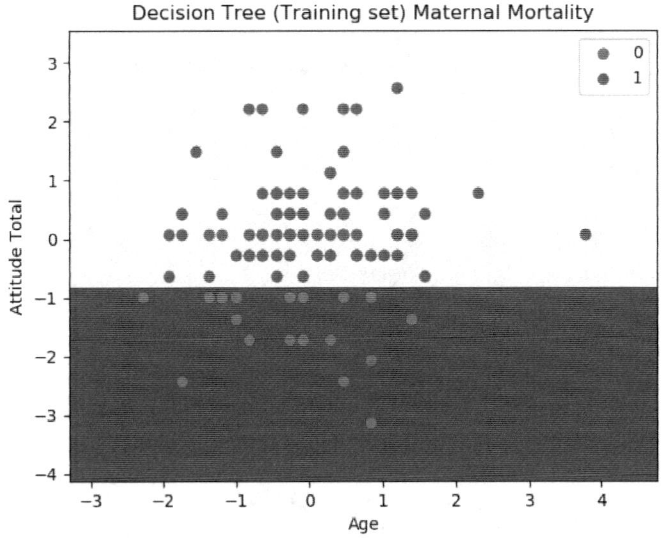

Figure. 1 Visual of decision Tree in Training Set of Maternal

The visualizing of the Decision Tree in Figure. 1 shows the result of the maternal mortality data set since in the result a hyper-plane has been found in the training set result and verified to be the best one in the set result. The decision tree result in Figure. 1 shows the separation boundary of the two classifiers. This creates a way to classify the vector or the features of the maternal mortality dataset. In this study, there are observable green dots in the yellow region and red dots in the blue region.

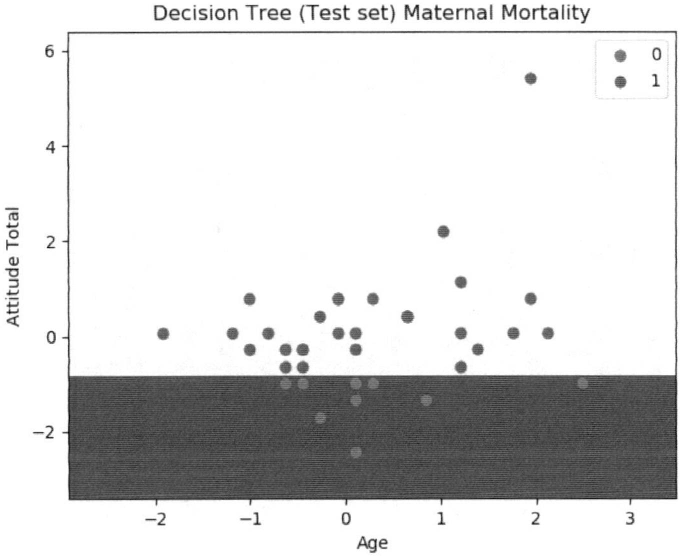

Figure. 2 Visualizing of decision Tree result for testing maternal dataset

The decision result in Figure. 2 shows the results of the testing of the maternal mortality dataset, which could minimize the correct prediction. The green dots in the yellow region and red dots in the blue region could be separated unless anon-linear boundary was used since the hyperplane is linear.

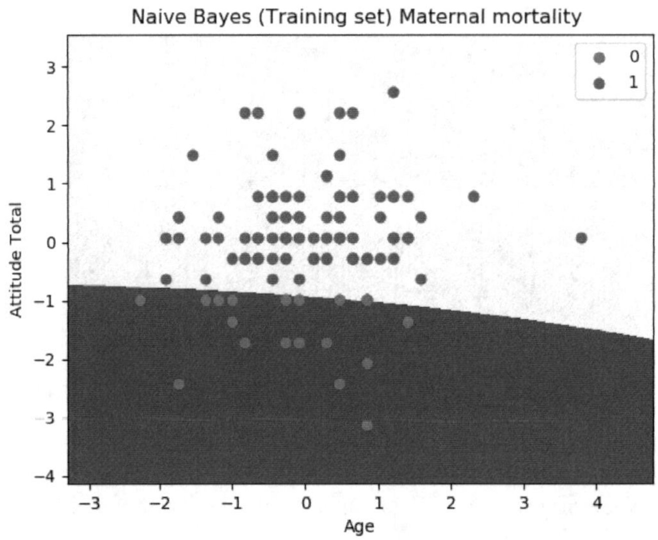

Figure. 3 Visualizing of Naïve Bayes result for training maternal dataset

Figure. 3 is a Naïve Bayes result of the training maternal mortality dataset, the graph shows the clear correlation between the dependent and independent variables. It is obvious that as Age (maternal mortality dataset) and Attitude score increase, each individual has a higher likelihood of being green dots. There will almost always be some degree of error - or at least there should be, otherwise, our model is probably guilty of overfitting.

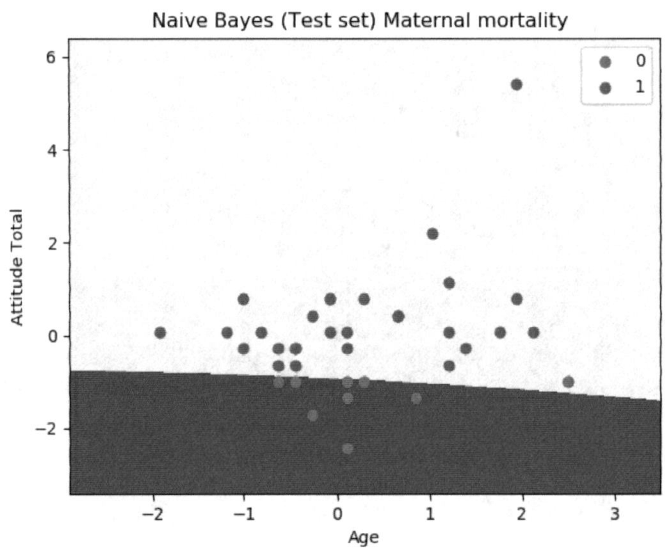

Figure. 4 Visualizing of Naïve Bayes result for testing maternal dataset

Figure 4. shows the testing result using Naïve Bayes. The graph looks very similar to the last, but with fewer observations. We can classify the age (maternal mortality dataset) and attitude score increase will lead to a higher probability of clicking the advertisement. The accuracy value of Naïve Bayes is 97.36 % when the confusion matrix indicated the true positive in maternal mortality is 8, the true negative is 29, false positive is 0 and false negative is 1.

4. Conclusions

In this study, we present the maternal mortality dataset from collected in Dairi, North Sumatra, Indonesia. Based on the result, indicated the training and testing dataset with age and attitude scores in pre-processing indicated there are positive attitudes. The classification using two models in this study is better to identify the maternal mortality dataset. The classification accuracy of the decision tree is 100 % and the Naive Bayes Model is 97.37 %. The Decision Tree and Naïve Bayes models show good performance for train and testing the dataset of the maternal mortality in health-promoting.

References

[1] "WHO | Maternal mortality." [Online]. Available: https://www.who.int/gho/maternal_health/mortality/maternal_mortality_text/en/. [Accessed: 21-Mar-2020].
[2] "WHO | Maternal mortality," *WHO*, 2019.
[3] "UNFPA Indonesia | Disparity of Access & Quality Review of Maternal Mortality in Five Region in Indonesia." [Online]. Available: https://indonesia.unfpa.org/id/node/11920. [Accessed: 21-Mar-2020].
[4] *Demografi*. Government of Dairi Regency.
[5] "Dairi Regency - Wikipedia." [Online]. Available: https://en.wikipedia.org/wiki/Dairi_Regency. [Accessed: 21-Mar-2020].
[8] Purba J, Bahri S, Khair H, Darnila E, and Sinambela M, 2020 Performance of TB health center service officers based on compensation effect, supervision and competence using machine learning technique *IOP Conf. Ser. Mater. Sci. Eng.*, vol. **725**(1)
[9] Hutabarat L T, Sunardy, Sysvia T, Dwiyanto, and Sinambela M, 2020 Machine learning approach to air traffic control skill based on mastery theory of aerodrome control procedures, self-concept and practice drills *IOP Conf. Ser. Mater. Sci. Eng* **725**(1)
[10] Hutabarat L T, H. A. S, Sinambela M, and Limbong T, 2019 Classification of Student's Air Traffic Control Skill Using **3**(2), pp. 166–169.
[11] Tahmassebi A, Gandomi A H, Schulte M H J, Goudriaan A E, Foo S Y, and Meyer-Baese A, 2018 Optimized Naive-Bayes and Decision Tree Approaches for fMRI Smoking Cessation Classification *Complexity*.
[12] Rahmat R F, Nasution F R, Seniman, Syahputra M F, Sitompul O S 2018 Implementation of Bayesian model averaging on the weather data forecasting applications utilizing open weather map *IOP Conf. Ser. Mater. Sci. Eng.*, **309**(1), 012054.
[13] Safri Y F, Arifudin R, and Muslim M A, 2018 "K-Nearest Neighbor and Naive Bayes Classifier Algorithm in Determining The Classification of Healthy Card Indonesia Giving to The Poor," *Sci. J. Informatics*, **5**(1)
[14] Aravinthan K and Vanitha D M, 2016 A Novel Method For Prediction Of Heart Disease Using Naive Bayes," *Int. J. Adv. Res. Trends Eng. Technol.*, **3**(2), pp. 56–60.
[15] Kour P *et al* 2019 Classification of Maternal Healthcare Data using Nave Bayes *Int. J. Comput. Sci. Eng.*, **7**(3), pp. 388–394

IOP Conf. Series: Materials Science and Engineering **851** (2020) 012056 doi:10.1088/1757-899X/851/1/012056

Investigation of Ti-6Al-4V surface integrity from milling process

I Siregar[1], F N M Dzulkifli[2], J B Saedon[2] and M S Adenan[2]

[1]Universitas Sumatera Utara, Kampus USU, Medan 20155, Medan Indonesia
[2]Smart Manufacturing Research Institute, Universiti Teknologi MARA, 40450, Shah Alam, Selangor, Malaysia

ikhsan.siregar@usu.ac.id

Abstract. The applications of titanium alloys vary from aerospace industry, medical device and military, however having many challenges in the machining process due to difficult-to-machine characteristic of the material. This study focused on surface integrity of titanium alloy grade Ti-6Al-4V after milling process with different set of parameters. This research investigates the effect of surface integrity of titanium alloys (Ti-6Al-4V) after milling operation with difference cutting parameters such as feed rate, depth of cut and cutting speed. Characterization and measurements were taken on the surface morphology, residual stress, surface roughness and surface hardness. It was found out that from the experiment, the highest cutting speed at 8000 rpm produced lowest surface roughness value of 2.066 μm, while lowest cutting speed at 4500 rpm produced highest surface roughness of 2.55μm. The residual stress analysis shown that all the surfaces produced compressive stresses of more than 500 MPa. It is also found out no evidence of transformation at the morphology of the machined surface. Microhardness measurement at the depth 10 μm to 500 μm from the machined surface only shows slight increase of hardness, however, at the range from machined surface to 10 μm indicates low hardness due to softening effect from the machining process. This demonstrated that surface integrity of titanium alloys does effect by the machining parameters.

1. Introduction

Titanium alloy machining process is considered as a challenging process, with regards to its material property which is acknowledge as difficult-to-machine material. Besides, the machining process may also influence by huge chemical reactivity, causing the modification at its surface integrity and may cause the tool failure [1-2]. These challenges make the research focus on the machining of titanium alloys become limited for only specific application in aerospace, medical and military industry. Other influence that affect the machining process of titanium alloys are including residual stress, high heat stress and inconsistency of chip thickness. The impact of these factors contributes to overall production costs and final quality machined surface due to poor machinability of titanium alloys [3].

The machining performance of titanium alloys can be defined by the machining output of its surface integrity, for correlation with machining parameters used. Amongst the basic characteristics that can be evaluated for the surface integrity is such as morphology, residual stress, surface roughness and surface hardness. For topological parameters, the surface finish of the material is one of the examples of the surface integrity used. If essential objective of the product development is to reach

IOP Conf. Series: Materials Science and Engineering **851** (2020) 012056 doi:10.1088/1757-899X/851/1/012056

extra reliability levels such as in aerospace industry, the most significant parameters used to determine the quality of the product is surface integrity [4].

The desired machining outputs from the machining of titanium alloys may be hindered by surface defects. Mostly, the defects happen due to the high heat generated during the machining process, thus defects such as plastic deformation and microcracking may occurred during the process [5].

In order to obtain some insight about the correlation, this study had focus on initial data for the investigation. The attention was to characterize and measure the morphology, residual stress, surface roughness and microhardness for depth profile. The machining parameters used were depth of cut, feed rate and cutting speed.

2. Research methodology

In this experiment, titanium alloy of grade Ti-6Al-4V was used with chemical elements properties as in Table 1. As the aim for this study is to obtain primary results, the titanium alloy block was machine with three different set of parameters as in Table 2, with identification name of Sample A, Sample B and Sample C as shown in Figure 1.

Table 1. Chemical composition of Ti-6Al-4V

Element	Ti	Al	V	Fe	O	C	N	N
Wt. %	Base	5.8	3.8	0.8 max.	<0.2	<0.05	<0.05	<0.02

Table 2. Machining parameters

	Cutting Speed (rpm)	Feed Rate (mm/min)	Depth of Cut (mm)
Sample A	4500	240	0.1
Sample B	4500	320	0.3
Sample C	8000	240	0.1

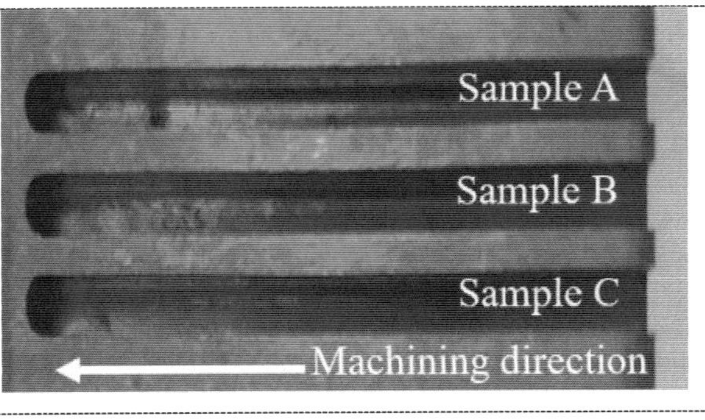

Figure 1. Machined specimens with different machining parameters

Upon completion of the machining process, the machined areas were cut to various size with wire EDM in order to characterize the microstructure, residual stress and microhardness. For microstructure characterization, the samples were ground and polished until mirror finish, then etched with Nital solution for 30 second. The microstructure then observed with Hitachi SU3500 Scanning Electron Microscope (SEM) was used.

For surface roughness measurement, the samples were measured by using Alicona Infinite Focus not contact measurement, several data such as smoothening depth (Rp), roughness (Ra), maximum peak of the profile (Rt) and root mean square (Rq) were obtained, however on Ra value was considered in this study. roughness value [16]. The process then repeated with sample A, B and C.

The residual stress measurement was taken by using Rigaku Ultima IV, X-Ray Diffractometer (XRD) with inclination scan range between 112° - 116° and step width of 0.02°. Other setting for the XRD machine was 1 second for the fixed time, and of 10 diverse tilt angles (ψ) at the range of -5° to 45° at each measurement point.

Microhardness measurement was taken by using Mitutoyo MVK-HI Hardness Testing Machine. Measurement was taken at the cross section of samples taken in order to obtain the microhardness depth profile, from the machined surface towards downward to the base of titanium alloy. The setting applied was 200 g load with indention dwell time of 20 s.

3. Results and discussion

Morphological characterization was observed at the machined area towards downward area. It can be clearly seen that no white layer formation and phase transformation happen as shown in Figure 2. It also shown that sample B and C did not show changes at the microstructure and grain size as compared to Sample A. This happen due to relatively low cutting force applied and the behaviour of titanium alloy material itself.

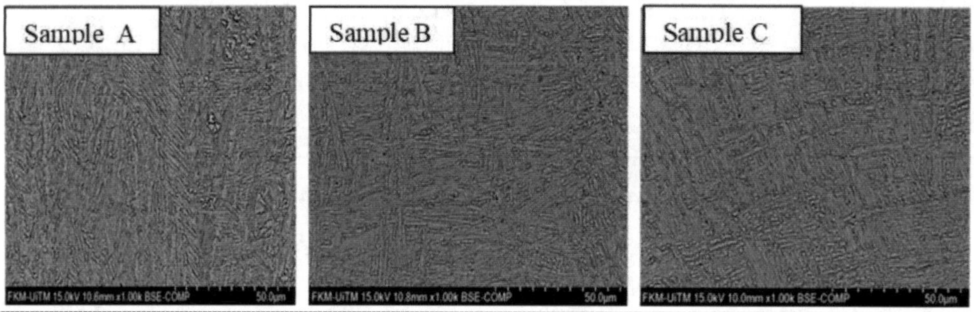

Figure 2. SEM morphology of machined Ti-6Al-4V

For the surface roughness, it can be seen in Table 3 that there is improvement of surface roughness for the three samples after machining process. For Sample A was found to obtain highest average roughness value with 2.507 µm, while Sample C exhibited the lowest Ra value with 2.066 µm. Despite utilising highest feed rate and depth of cut on Sample B, the surface roughness value obtained 2.144 µm, slightly above Sample C. Therefore, it can be shown that both feed rate and depth of cut is not a determining factor for surface roughness on machined titanium alloys. It can be observed that among three different milling parameters, Sample C produced the most excellent surface roughness value.

Table 3. Average roughness value

	Average Roughness, Ra (µm)
Sample A	2.507
Sample B	2.144
Sample C	2.066

Table 4 shows the residual stress result for Sample A, B and C. Highest recorded compressive state happens to be at Sample C with magnitude of 970.90 MPa, followed by Sample A with 878.18 MPa and Sample B recorded the lowest reading with 579.00 MPa.

Table 4. Residual stress measurement

	Residual stress, (Mpa)
Sample A	878.18

Sample B	579.99
Sample C	970.90

Microhardness value across machine depth was shown in Figure 3. At the position of 10 µm below the machined surface, it indicates the microhardness value for all samples is below initial hardness value Ti-6Al-4V specimen. This is due to the softening effect after the machining process take place. Then, hardness values were increase at 100 µm, the trend shows increase of hardness and slightly increase accordingly. The significant increase happens due to work hardening effect.

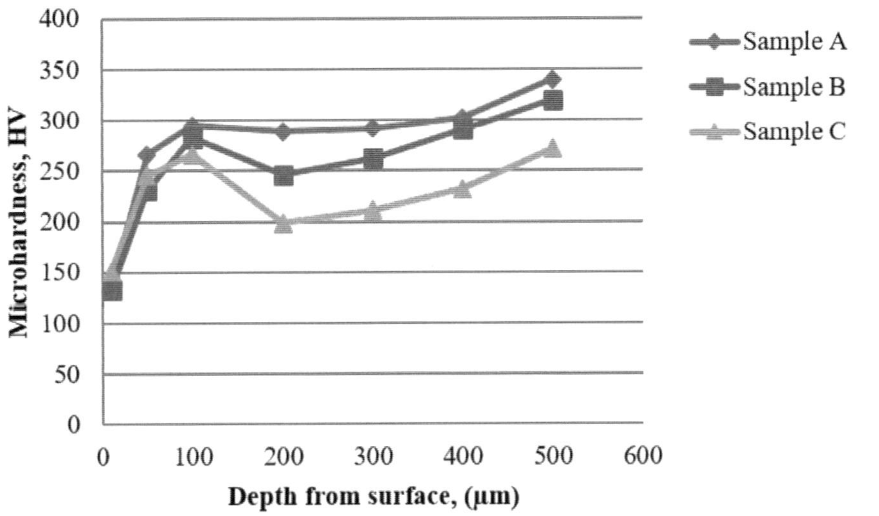

Figure 3. Microhardness measurement along the direction of milling process

The reason on the correlation between parameters of cutting speed, the surface hardness value may be related to the excessive temperature generate at the titanium alloys surface during machining process which will induce the thermal softening of the surface. Similar behaviour also occurred at 100 µm - 500 µm depth machined surface that shows Sample C recorded lower hardness value compare to Sample A and Sample B, mostly due to higher value of cutting speed.

4. Conclusions

From this study, it can be concluded that, the machined area does not show any occurrence of phase transformation, microcracks and white layer formation on the titanium alloys. No changes also found not happen at the microstructure and grain. This is due to the cutting parameters used in this study which did not generate enough cutting force that will affect the microstructure and phase transformation on the surface and subsurface of the titanium alloys.

Residual stress shows that all the surfaces contained compressive stresses > 500 MPa. It was found out that with the increase of cutting speed, the residual stress value will decrease. Moreover, the residual stress generated becomes less compressive when the feed increases. Surface roughness was found to be very influenced by cutting speed during the machining process. Sample C with highest cutting speed of 8000 rpm produced the lowest surface roughness value of 2.066µm, while Sample A with lowest cutting speed 4500 rpm produced highest surface roughness 2.507µm. Hardness value at the cross section was influenced by softening effect, occurred until the depth around 100 µm. Overall conclusion from this initial study shows that cutting speed is the most influential parameter in the machining process of Ti-6Al-4V.

IOP Conf. Series: Materials Science and Engineering **851** (2020) 012056 doi:10.1088/1757-899X/851/1/012056

References

[1] Nouari M and Makich H 2013 Experimental investigation on the effect of the material microstructure on tool wear when machining hard titanium alloys: Ti-6Al-4V and Ti-555 Int. J. Refract. Met. Hard Mater. 41 259–69

[2] Hasçalik A and Çaydaş U 2008 Optimization of turning parameters for surface roughness and tool life based on the Taguchi method Int. J. Adv. Manuf. Technol. 38 896–903

[3] Saedon JB 2011 Micromilling of Hardened (62 HRC) AISI D2 Cold Work Tool Steel. Master's thesis, The University of Birmingham, UK 94-120.

[4] Oosthuizen G A , Nunco K , Conradie P J T, Dimitrov D M S 2016 The Effect Of Cutting Parameters On Surface Integrity In Milling Ti6al4v Africa S, Science E and Africa . ARTICLE INFO 27 115–23

[5] Filip R, Kubiak K, Ziaja W and Sieniawski J 2003 The effect of microstructure on the mechanical properties of two-phase titanium alloys J. Mater. Process. Technol. 133 84–89

[6] Siregar I, Muchtar M A, Rahmat R F, Andayani U, Nasution T H, and Sari R M 2018 Method of calculation overall equipment effectiveness in fertilizer factory. In IOP Conference Series: Materials Science and Engineering **308**(1), p. 012053) IOP Publishing.

IOP Conf. Series: Materials Science and Engineering **851** (2020) 012057 doi:10.1088/1757-899X/851/1/012057

Effect of shape of test specimen on the shear strength of bamboo from eastern region of sumatera utara

M A P Handana*[1], B Surbakti[1], D D Harisdani[2], R Karolina[1] and D P Abdi[1]
[1]Departemen Teknik Sipil, Universitas Sumatera Utara, Perpustakaan Street, USU Campus, Medan, Indonesia
[2] Departemen Arsitektur, Universitas Sumatera Utara, Perpustakaan Street, USU Campus, Medan, Indonesia

*E-mail: agung.putra@usu.ac.id

Abstract. Wood is the most common forest-based product used in the construction industry, but due to reducing area of forest in the world, substitution materials for wood are needed. Bamboo has a huge prospect for the future as a wood substitute, and extensive and continuous research of bamboo, especially in mechanical properties of bamboo must be carried out to support bamboo as a material substitution for woods. This research focused on the shear strength of bamboo by comparing two types of standard tests used to determine the shear strength of bamboo. The standards are ISO 22517 and ASTM D 143. The compression test to determine the compressive strength is also conducted in this research. The bamboo species that used in this research are the species that grow in various district in the eastern region of Sumatera Utara. The result of the test shows that the highest shear strength for the specimens tested by ISO 22157 standard is Bambusa vulgaris with 6.32 MPa from Asahan and Tanjung Balai, and for the specimens tested by ASTM D 143 standard is Bambusa vulgaris with 12.31 MPa from Batubara. From the result it can be seen that all samples tested with ASTM D 143 standard have higher shear strength than those who tested with ISO 22157, with conversion ratio of 0,63. For compression test, the highest compressive strength is *Gigantochloa robusta* with 65.41 MPa from Asahan and Tanjung Balai.

1. Introduction

For forest-based products in Indonesia, wood is one of the most commonly used materials almost in every aspect, especially in the construction industry. The price of wood material is increasing each day as the forest itself has been reduced, so that the substitution material is needed. Bamboo has a shorter cycle of growth than wood, can be harvested every 3-5 years while softwoods have 20 years at most to be harvested [1], and it's mechanical properties comparable to some wood species [2]. Therefore, it makes bamboo more suitable than wood. In Indonesia, bamboo has been used in rural life. Bamboo planted around the countryside or even gardens to be used as goods for daily use [3].

Bamboo is an orthotropic material with high strength along and low strength transversal to its fibers. The structure of bamboo is a composite material, consisting of long and aligned cellulose fibers immersed in the ligneous matrix [4]. Like other materials, the mechanical properties of bamboo are very dependent on the correlation of proportions and fiber distribution in the cross-section. Mechanical property is strongly influenced by the density which depends on the fiber content, fiber diameter, and thickness of the wall on the bamboo [5]. For the durability, untreated bamboo has a design life reach 10-15 years if it keeps appropriately [6]. For having bamboo in the construction, it is lack of design guidance and standardization which means that the engineering design of bamboo has not yet been fully addressed. Further application of bamboo relevant to construction includes its use as scaffolding, farmhouse, and bridges.

This research aims to determine the mechanical properties of the bamboo species, especially the shear strength which is obtained by the shear test. Shear is important in bamboo because its the weakest point. Bamboo is as strong as wood in tension but much weaker in shear. In the bamboo construction of joints, holes are needed, and these give the effect of shear in the bamboo [7]. Several studies have tested bamboo's mechanical properties [8] [9] [10], but most only test the compressive and tensile strength of bamboo, rarely test the shear strength.

Test specimens and shear tests are conducted by following ISO 22157 standards as a test standard for bamboo, and also be compared with the test by following the ASTM D 143 standard as a test standard for small clear sample for woods. As an additional test, the compression test for bamboo also conducted along with the shear test because it is using the same bamboo specimen size. The specimen for the test was taken from 7 species of bamboo grow in Simalungun, Batubata, Asahan and Tanjung Balai. The species were selected for test specimen because they are suitable as the material the construction [11].

2. Method

2.1. Preparation Of Bamboo

There are 7 bamboo species involved in this research, which are sourced from Simalungun, Batubata, Tanjung Balai and Asahan district in Sumatera Utara province. Those 7 bamboo species include *Bambusa blumeana, Bambusa vulgaris ,Dendrocalamus asper, Gigantochloa atter, Gigantochloa robusta, Schizostachyum bracycladum, Schizostachyum zolingeri.* For the availability in each section (Simalungun, Batubara, Tanjung Balai and Asahan) is shown in table 1.

Table 1. The availability of bamboo species in east region of Sumatera Utara.

Species	availability		
	Simalungun	Batubara	Asahan And Tanjung Balai
Bambusa blumeana	X		X
Bambusa vulgaris	X	X	X
Dendrocalamus asper	X	X	X
Gigantochloa atter			X
Gigantochloa robusta			X
Schizostachyum bracycladum		X	X
Schizostachyum zolingeri		X	X

These bamboo culms from every species were selected and dried for about three weeks. The moisture content more or less than 15 % to have good mechanical properties and is less vulnerable to fungus attacks [12]. The moisture content to conduct mechanical properties testing is 12 % more or less regarded as the standard and references[13][14]. The selected bamboo culm must be free from fungus and has no defect for the mechanical properties test. Figure 1 shows the bamboo culm being pile in the area to get it dried.

Figure 1. 7 species bamboo from different area and left to dry for 3 weeks

2.2. Mechanical testing of bamboo

Each of bamboo culm was cut into small specimens. The small specimens were taken from the internode and node of the bamboo culm. There are 2 tests in this research which are, shear, and compression test following ISO 22147-1:2004[13], ISO 22157-2:2004 [15], and ASTM D 143 [16]. The specimen was taken from the top, middle, and bottom sections of the culm and each specimen will be mark respectively. For bamboo specimens to be tested with ASTM D 143 standard, modifications must be made to the specimen. In ASTM D 143 shear parallel to grain test, the specimen size is specific, where the specimen width is equal to the height of the shear plane to be compressed to get the shear strength. But in bamboo, the thickness of the specimen has to modified equal to the thickness of the bamboo culm. Additional area also added to the bottom of the specimens where the area is to be attached to the test eqipment.

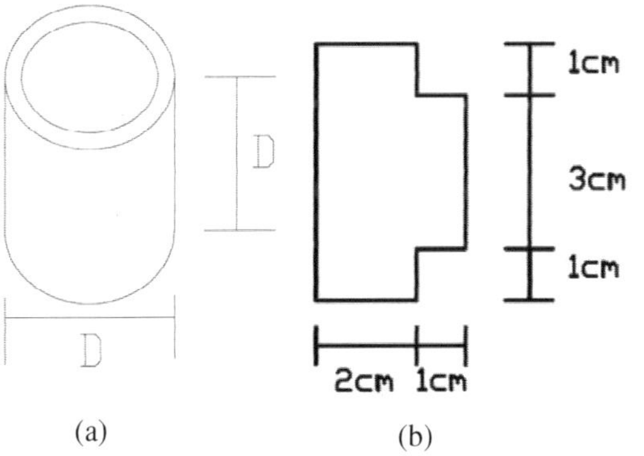

Figure 2. (a) compressive and shear test specimen according to ISO 22157 (b) Modified Shear Test Specimen according to ASTM D 143.

Figure 3 shows the conditions when the specimens test conducted. Figure 3a shows when the compression test is conducted and the maximum load will occur along with the failure on the specimen which is shown with the crack and the deformation on the specimen. Figure 3b shows failure starts to occur when the constant load is given at the Universal Testing Machine and the specimen will be split. The maximum load for each test carried out is recorded and then the average value of each type of test is calculated to determine the compressive and shear strength from each of bamboo species.

(a) (b)

Figure 3. (a) Compressive test setup (b) Shear test set up according to ISO 22157 (c) Shear test set up according to modified ASTM D 143.

3. Result and Discussion

3.1. Shear strength

Failure pattern caused by the shear test comes along with the visual crack of the specimen [19]. Shear test failure crack at the specimen tested by the modified ASTM D 143 standard is shown in figure 4a, and the failure tested by ISO 22157 standard is shown in figure 4b. Table 2 shows that the highest average value of the shear strength tested by ISO 22157 standard is Bambusa vulgaris, with 6.32 MPa from Asahan and Tanjung Balai district, and for the modified ASTM D 143 is Bambusa vulgaris with 12.31 MPa from Batubara. From this result, it can be seen that the average shear strength value obtained from the modified ASTM D 143 testing is higher than the ISO 22157 results. The bottom section has the highest value followed by middle and top respectively. For the ISO 22157 test, the difference from node to internode average shear strength values are 1.006 MPa, 0.96 MPa, and 1.01 MPa for Asahan and Tanjung Balai, Batubara, and Simalungun specimens respectively. For the modified ASTM D 143 test, the difference from node to internode average shear strength values are 0.98 MPa, 0.99 MPa, and 0.84 MPa for Asahan and Tanjung Balai, Batubara, and Simalungun specimens respectively.

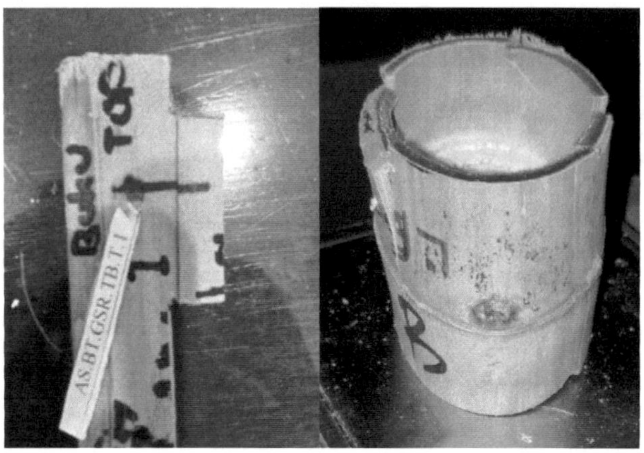

(a) (b)

Figure 4. Specimen failure after the shear test (a) Modified ASTM D 143 (b) ISO 22157

Table 2. Shear strength according to ISO 22157 and modified ASTM D 143.

Origin	Species	Test Condition	Shear Strength (MPa) (ISO 22157)			Shear Strength (MPa) (modified ASTM D 143)		
			Top	Middle	Bottom	Top	Middle	Bottom
Asahan and Tanjung Balai	*Bambusa blumeana*	Node	3.81	5.01	5.15	5.57	6.39	6.75
		Internode	4.67	6.71	7.01	6.47	8.14	8.23
	Bambusa vulgaris	Node	5.20	5.73	7.63	7.11	9.06	8.91
		Internode	6.22	6.39	6.77	7.40	8.14	9.24
	Dendrocalamus asper	Node	5.05	5.71	6.74	7.72	8.98	10.14
		Internode	5.54	6.41	6.46	9.36	7.92	7.31
	Gigantochloa atter	Node	5.00	5.15	6.19	5.44	5.85	9.56
		Internode	4.84	5.39	5.63	9.15	10.78	12.37
	Gigantochloa robusta	Node	5.04	5.63	7.11	12.88	8.16	9.98
		Internode	5.41	6.50	6.61	9.58	10.15	10.52
	Schizostachyum bracycladum	Node	4.12	4.46	4.77	9.26	9.76	15.08
		Internode	4.19	4.28	4.64	10.24	10.46	11.18
	Schizostachyum zolingeri	Node	5.74	6.07	6.30	6.82	9.16	9.75
		Internode	4.33	4.38	4.52	6.41	6.78	8.93
Batubara	*Bambusa vulgaris*	Node	5.81	5.77	5.93	11.55	13.69	14.44
		Internode	4.68	4.95	5.57	9.29	11.13	13.78
	Dendrocalamus asper	Node	4.15	4.50	6.53	4.41	7.09	7.69
		Internode	5.01	6.49	6.67	8.01	8.33	8.57
	Schizostachyum bracycladum	Node	3.61	3.59	4.99	9.46	9.53	9.74
		Internode	3.83	3.86	4.83	8.28	8.29	9.70
	Schizostachyum zolingeri	Node	4.61	5.41	6.13	7.56	7.83	8.09
		Internode	5.16	6.22	6.57	7.95	8.14	8.62
Simalungun	*Bambusa blumeana*	Node	5.59	6.16	6.98	7.28	8.20	9.08
		Internode	4.71	6.62	7.51	12.26	12.41	13.37
	Bambusa vulgaris	Node	5.87	6.27	6.84	9.07	10.04	11.08
		Internode	5.52	5.91	6.22	9.07	10.92	11.81
	Dendrocalamus asper	Node	4.67	5.83	7.64	6.69	8.01	8.10
		Internode	5.51	5.67	7.79	9.04	10.21	12.86

3.2. Conversion Ratio modified ASTM D 143 standard to ISO 22157 standard

The average shear strength value obtained from testing following the modified ASTM D 143 is significantly different from the value following ISO 22157, with the difference ratio varies from 0.41 to 0.94. Table 3 presents the ratio between shear strength tested with ISO 22157 and the modified ASTM D 143, for nodes and internodes, from the top, middle, and bottom section. The average value of the overall ratio calculated in Table 3 is 0.63. This average value can be recommended for use when researchers use the modified ASTM D 143 standard which is intended for wood in testing the bamboo specimens and wants to convert the result obtained to ISO 22157 standard as the reference in testing bamboo.

Table 3. The conversion ratio of ASTM D 143 to ISO 22157.

Origin	Species	Test Condition	Ratio		
			Top	Middle	Bottom
Asahan and Tanjung Balai	*Bambusa blumeana*	Node	0.68	0.78	0.76
		Internode	0.72	0.82	0.85
	Bambusa vulgaris	Node	0.73	0.63	0.86
		Internode	0.84	0.79	0.73
	Dendrocalamus asper	Node	0.65	0.64	0.66
		Internode	0.59	0.81	0.88
	Gigantochloa atter	Node	0.92	0.88	0.65
		Internode	0.53	0.50	0.46
	Gigantochloa robusta	Node	0.39	0.69	0.71
		Internode	0.56	0.64	0.63
	Schizostachyum bracycladum	Node	0.44	0.46	0.32
		Internode	0.41	0.41	0.42
	Schizostachyum zolingeri	Node	0.84	0.66	0.65
		Internode	0.68	0.65	0.51
Batubara	*Bambusa vulgaris*	Node	0.50	0.42	0.41
		Internode	0.50	0.44	0.40
	Dendrocalamus asper	Node	0.94	0.63	0.85
		Internode	0.63	0.78	0.78
	Schizostachyum bracycladum	Node	0.38	0.38	0.51
		Internode	0.46	0.47	0.50
	Schizostachyum zolingeri	Node	0.61	0.69	0.76
		Internode	0.65	0.76	0.76
Simalungun	*Bambusa blumeana*	Node	0.77	0.75	0.77
		Internode	0.38	0.53	0.56
	Bambusa vulgaris	Node	0.65	0.62	0.62
		Internode	0.61	0.54	0.53
	Dendrocalamus asper	Node	0.70	0.73	0.94
		Internode	0.61	0.56	0.61
Average				0.63	

3.3. Compressive strength

Table 4 shows that the highest average compressive strength came from the bottom section, followed by the middle section and then the lowest came from the top section. The node and internode sections compressive strength is not significantly different, with the difference of the average values are 0.88 MPa, 0.98 MPa, and 0.95 MPa obtained from Asahan and Tanjung Balai, Batubara, and Simalungun respectively. The latest research in the mechanical properties of bamboo showed that nodes did not give a significant effect on tensile strength, flexural strength and compressive strength [17] but it only restricted further deformation of the specimen [18]. The highest average compressive strength is from *Gigantochloa robusta* with 65.41 MPa for Asahan and Tanjung Balai, *Dendrocalamus asper* with 53.12 MPa for Batubara, *Bambusa vulgaris* with 56.83 MPa for Simalungun. It can be seen that the highest compressive strength for *Bambusa bleumana* and *Bambusa vulgaris* are from Simalungun, for *Dendrocalamus Asper* is from Batubara, and for *Schizostachyum bracycladum* and *Schizostachyum zolingeri* are from Asahan and Tanjung Balai. Figure 5 shows the specimen condition after the compression test conducted.

IOP Conf. Series: Materials Science and Engineering **851** (2020) 012057 doi:10.1088/1757-899X/851/1/012057

Table 4. Compressive strength according to ISO 22157.

Origin	Species	Test Condition	Compressive Strength (MPa)		
			Top	Middle	Bottom
Asahan and Tanjung Balai	*Bambusa blumeana*	Node	39.97	51.75	60.97
		Internode	51.94	60.48	61.57
	Bambusa vulgaris	Node	37.69	40.07	49.50
		Internode	43.56	49.44	58.92
	Dendrocalamus asper	Node	43.99	44.19	46.58
		Internode	49.85	50.48	51.90
	Gigantochloa atter	Node	32.17	39.55	41.39
		Internode	34.68	48.86	48.93
	Gigantochloa robusta	Node	58.31	62.16	72.64
		Internode	60.99	66.52	71.85
	Schizostachyum bracycladum	Node	37.85	38.35	46.21
		Internode	46.30	46.36	49.46
	Schizostachyum zolingeri	Node	30.41	31.50	32.88
		Internode	30.24	36.99	38.63
Batubara	*Bambusa vulgaris*	Node	43.80	51.06	53.98
		Internode	44.54	53.93	56.56
	Dendrocalamus asper	Node	55.04	57.94	59.22
		Internode	46.55	47.69	52.27
	Schizostachyum bracycladum	Node	31.96	36.29	38.43
		Internode	35.53	44.75	47.83
	Schizostachyum zolingeri	Node	23.12	24.43	30.24
		Internode	26.86	28.01	28.55
Simalungun	*Bambusa blumeana*	Node	44.97	51.06	57.77
		Internode	51.79	63.68	67.27
	Bambusa vulgaris	Node	53.44	61.88	62.61
		Internode	47.55	56.74	58.75
	Dendrocalamus asper	Node	46.77	47.42	52.80
		Internode	51.79	53.84	56.81

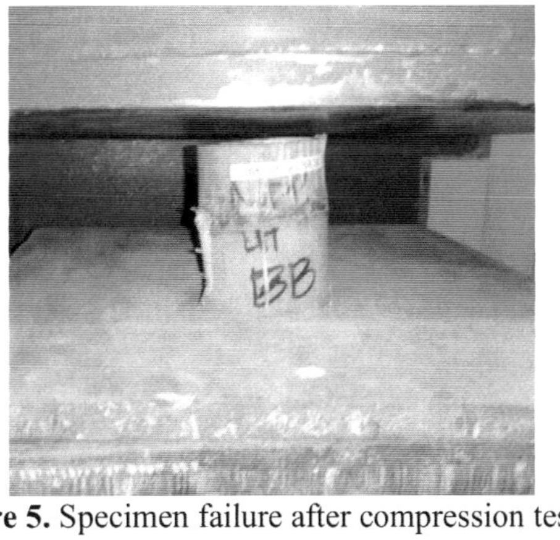

Figure 5. Specimen failure after compression test

4. Conclusion

Shear strength is an important element in the mechanical properties of bamboo but its testing is often ignored. In this study shear test was carries out to obtain bamboo shear strength according to ISO 22157 standard and also modification of the ASTM D 143 standard. This was done to see the comparison of the result of the test between these two methods and also to open up the possibility of carrying out tests other than using ISO 22157 standard as the reference for bamboo shear testing. From the test result, it can be seen that shear test with modified ASTM D 143 standard gives higher average shear strength compared to the shear test result with ISO 22157 standard, with the highest average value of the shear strength tested by ISO 22157 standard is Bambusa vulgaris, with 6.32 MPa from Asahan and Tanjung Balai district, and for the modified ASTM D 143 is Bambusa vulgaris with 12.31 MPa from Batubara The average value of the overall ratio from comparison between ISO 22157 and modified ASTM D 143 is 0.63. For compressive strength, the highest average compressive strength is from *Gigantochloa robusta* with 65.41 MPa for Asahan and Tanjung Balai.

5. Acknowledgment

This research is funded by The Ministry of Research, Technology, and Higher Education, Republic of Indonesia.

References

[1] Environmental Bamboo Foundation, *Uses of Bamboo*. Bali, Indonesia: Environmental Bamboo Foundation, 1999.

[2] Leitner P 2015 Structural Bamboo - Mechanical Properties and Potential Fields of Application, Graz University of Technology.

[3] Widjaya E A, 2000 Bamboo Diversity And Its Future prospect in Indonesia, *Proc. Third Int. Wood Sci. Symp.*, no. November,.

[4] Liese W, 1992 The Structure of Bamboo in Relation to its Properties an Utilization, in *Proceedings of the International Symposium on Industrial Use of Bamboo*, , pp. 99–100.

[5] Janssen J, 2000 Designing and Building with Bamboo, in *International Network for Bamboo and Rattan*,

[6] Ghavami K, 2008 Bamboo: Low cost and energy saving construction materials, *Int. Conf. Mod. Bamboo Struct.*

[7] Janssen J, 1980 The Mechanical Properties of Bamboo Used in Construction, in *International Development Research Centre*.

[8] Gutu T, 2013 A Study on the Mechanical Strength Properties of Bamboo to Enhance Its Diversification on Its Utilization, *Int. J. Innov. Technol. Explor. Enginnering*, no. 5, pp. 314–319.

[9] Appiah-kubi E, Owusu F W, Tekpetey S., and Essien C., 2014 Investigating the mechanical properties of some bamboo species for efficient utilization in Ghana, *J. Bamboo Ratt.*, **13**, pp. 81–89.

[10] Adedipe O, Abdullahi A A,. Ogunwole O. A, and Ogu S. I., 2013 Investigation of some mechanical properties of 'Agaraba' - a native Nigerian bamboo., *Aust. J. Technol.*,**16**(3), pp. 181–186.

[11] Saputri A, 2012 Biodiversitas Bambu Di Sumatera Utara Bagian Timur, Universitas Sumatera Utara

[12] Wakchaure M R and Kute S Y, 2012 Effect of moisture content on physical and mechanical properties of bamboo, *Asian J. Civ. Eng. (Building Housing)*, **13**(6) , pp. 753–763.

[13] I. 22157-1, 2004 Bamboo —Determination of physical and mechanical properties - Part 1: Requirements, Sitzerland.

[14] Chinese Standard Agency, Testing methods for physical and mechanical properties of bamboo used in building, 2007

[15] I. 22157-2, Bamboo - Determination of physical and mechanical properties - Part 2: Laboratory Manual, Switzerland, 2004.

[16] A. D143, Standard Test Methods for Small Clear Specimens of Timber 1, 2000.

[17] Shao Z P, Zhao L, Liu Y M, 2010 Differences in structure and strength between internode and

IOP Conf. Series: Materials Science and Engineering **851** (2020) 012057 doi:10.1088/1757-899X/851/1/012057

node sections of mosobambo, *J. Trop. For. Sci.*, vol. 22, pp. 133–138.
[18] Fabiani M, 2015 Physical and Mechanical Properties of Italian Bamboo Culm, in *10th World Bamboo Congress*.
[19] Awalluddin D *et al.*,2017 Mechanical properties of different bamboo species, *MATEC Web Conf.*, vol. 138, p. 01024

IOP Conf. Series: Materials Science and Engineering **851** (2020) 012058 doi:10.1088/1757-899X/851/1/012058

Lean Manufacturing: Waste Analysis in Crude Palm Oil Process

I Rizkya*, K Syahputri, R M Sari, D S Situmorang
Departement of Industrial Engineering, Faculty of Engineering, Universitas Sumatera Utara

*E-mail: indahrizkya@usu.ac.id,

Abstract. Lean Manufacturing is an approach to system efficiency by reducing waste. Companies need to identify value added activities on products, non-value added activities (waste) must be eliminated to decrease the production lead time. Palm Oil Industry produces 2 main products namely Crude Palm Oil and Kernel Palm Oil. This paper describes the waste analysis occur in the palm oil production process. Waste management activities are identified along the production process flow. Then investigates the waste root causes occur the most. The result obtained is value added activities of 47%, non-value added activities of 13%, and necessary non value added activities of 40%.

1. Introduction

The production process is a stage for converting inputs into desired outputs. This production process involves all the resources are owned by a company. Increasing business competition and high demands from consumers requires companies capable to manage production processes more efficient and effective and produce high quality products. Lean manufacturing is a method to increase responsiveness through efforts to reduce waste, continuous improvement and cost reduction [1].

A company with a lean system is the goal that the company wants to achieve. Because lean system have many advantages from efforts to reduce and eliminate waste [2]. Lean Manufacturing is a production activity focuses on reducing waste in all aspects of the company's production activities [3,4]. Lean manufacturing is a technique allows work carried out without a problem. This method will eliminate wasteful activity by balancing the number of work steps, allowing the product to be consumed directly to the next step, step by step until it's done [5,6].

Waste is any work activity does not provide value added throughout the process flow in the process of converting inputs into outputs. Waste are divided into two types, namely type 1 and type 2. Type 1 is a waste does not provide value added along the production flow but this activity cannot be avoided for various reasons. Type 2 is a waste that does not provide value added and must be eliminate immediately [7].

Lean Manufacturing is a systematic approach to eliminate waste and processes converting. This is done by identifying and reducing waste by continuous improvement [8]. Lean Manufacturing encourages the flexibility creation in production systems reduce unnecessary inventory, increase knowledge about the production process, save cost, reduce defects so that quality increases, reduce production lead times and reduce waste [1]. Lean manufacturing aims to increase productivity, improve product quality and manufacturing cycle times, reduce inventory, reduce lead times and eliminate manufacturing waste [9]. Many studies have applied lean principles in various industries. In the manufacturing industry especially applied the implementation of lean manufacturing and concurrent engineering has been carried out on the assembly production floor in the tractor industry [10].

Another implementation carried out in the plastic injection molded auto-parts industry showed that lean manufacturing increase productivity by up to 25%. Lean manufacturing has also been used in industries produce Fast Moving Consumer Goods (FMCG) products. The lean approach has also been carried out in the service industry namely Indian Software Services Firm [11], government agencies to improve efficiency and provide better service quality to their customers [12], banking industry [13], and local public service provider councils [14]. This paper aims to identify and analyze the waste occurs throughout the palm oil production process.

2. Methodology

There are five Lean Principles, namely 1) Value Determination. Value only defined by key customers. Values are distorted by preexisting organizations, especially engineers and experts. They add to the complexity which is not attractive to customers. 2) Identification of Value Streams. Value Stream is all actions needed to bring a product to the customer. 3) Flow. Eliminate the department that runs out a large number of single task processes. 4) Attraction. Let customers attract or request products from you. 5) Pursue Perfection. There is no end to the process of time, space, costs, and errors reduction [15].

To implement Lean Manufacturing can be done with several stages [16] namely (a) problem identification, (b) observation, (c) waste analysis to see possible causes and find the root causes of waste problems with cause and effect diagrams, and (d) action. The problem faced by the factory is that there is still waste causes production is not optimal. Then it is necessary to observe waste with value stream mapping to identify the waste. The application of lean manufacturing with value stream mapping has been successful in reducing waste and total lead time [17]. Fishbone diagram to analyze the causes of waste.

3. Result and Discussion

3.1. Problem Identification

The palm oil industry produces 2 main products, namely crude palm oil and kernel palm oil. The palm oil industry is still experiencing problems related to the lack of supply of fresh fruit bunches, and high manufacturing lead time due to activities that do not provide value added. At the sterilizer/boiling station where the loading ramp capacity (each door has a capacity of 16,5 tons) is not proportional to the truck capacity (each truck capacity is 3,5 tons). Transfer of lorries at the sterilizer station must use a capstand and transfer carriage to enter the sterilizer to be boiled and do the same thing when lorries containing boiled FFB must be pulled using a capstand and transfer carriage move to the tippler machine. Each machine operator often leaves the machine during the production process for other purposes. There are still many wastes occur along the production flow causing high manufacturing lead time. Lean manufacturing is used to identify and analyze the waste occurs along the value of the palm oil production process flow.

3.2. Observation

Observations were made on the palm oil industry with the determination of 10 observations for each activity in a 5 minutes cycle or less and 5 times of observation for each activity in the 5 minutes cycle of more than 5 minutes. The results of observations will be tested for uniformity and normality. The processing time of each activity is the normal time obtained from the calculation stages of cycle time and standard time.

Manufacturing Lead Time is the time required to carry out the production process from the beginning to the end. Calculation of Manufacturing Lead Time is done by adding up the entire work process time consisting of 20 work processes. Value Stream Mapping which is equipped with total cycle time information which is categorized in the non-value added activity (NVA), necessary non-value added activity (NNVA), and value added activity category (VA).

Table 1. Classification of Palm Oil Production Process Activities

No.	Activities	Time (sec)	Category
1	FFB (Fresh Fruit Bunches) weighed in Weighbridge	44	NNVA
2	FFB transfer to sortation station	72	NNVA
3	FFB collection to disassembly and sorting	2.753	NNVA
4	FFB sortation inserted into the Loading Ramp	1.674	NVA
5	FFB inserted into the truck	1.133	NNVA
6	Truck is pulled into sterilizer	559	NNVA
7	Sterilizer process to produce ripe fruit	5.747	VA
8	Truck is pull out	1.069	NVA
9	Truck of capacity fulfilment	4.196	NNVA
10	Ripe fruit is inserted into the thresher machine to separate the palm seeds with empty bunches	3.531	VA
11	Palm seeds digest using a digester machine	1.040	VA
12	Palm seeds is pressed into the presser machine to produce dirty CPO and core	4.178	VA
13	Dirty CPO sedimented into Sand Trap	2.153	NNVA
14	Dirty CPO is filtered using the vibrating screen	1.055	VA
15	Dirty CPO is temporarily collected in Crude Oil Tank	1.680	NVA
16	Dirty CPO is separated between oil, water and sludge using Continuous Storage Tank	965	VA
17	Oil and water are sedimented in the Pure Oil Tank	2.090	NNVA
18	The oil is inserted into a Vacuum Dryer Machine to separate the oil from the water content to obtain pure CPO	1.205	VA
19	Pure CPO is streamed into Storage Tank for storage	413	NVA
20	CPO fulfillment in trucks	2.245	NNVA
	TOTAL	37.802	

Table 1 shows the manufacturing lead time of the palm oil production process which consists of 20 activities of 37.802 seconds. These activities include of 7 value added activities (VA), 4 non value added activities (NVA), and 9 necessary non value added activities (NNVA). The proportion of each VA, NVA, and NNVA activity towards manufacturing lead time can be seen in Figure 1.

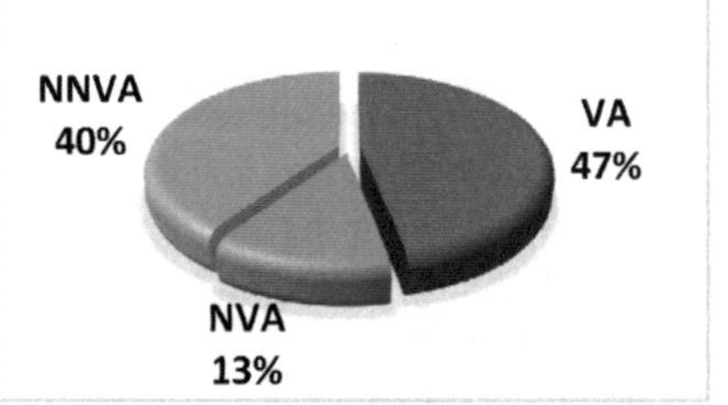

Figure 1. Comparison of VA, NVA, and NNVA Activities

Figure 1 shows the percentage of each activity towards the manufacturing lead time of the palm oil production process. About 53% of activities carried out are non-value added activities and only 47% are included in value added activities. However, 40% of non-value added activities is difficult because it is needed and support the production process to achieve customer value. But these activities can be reduced. While the other 13% is non value added activities that can be eliminated because it includes waste. Value stream mapping includes current state mapping and future state mapping. Current State Map is a description of the production process flow in the company which includes material flow and information flow. Value stream mapping process of palm oil production can be seen in Figure 1.

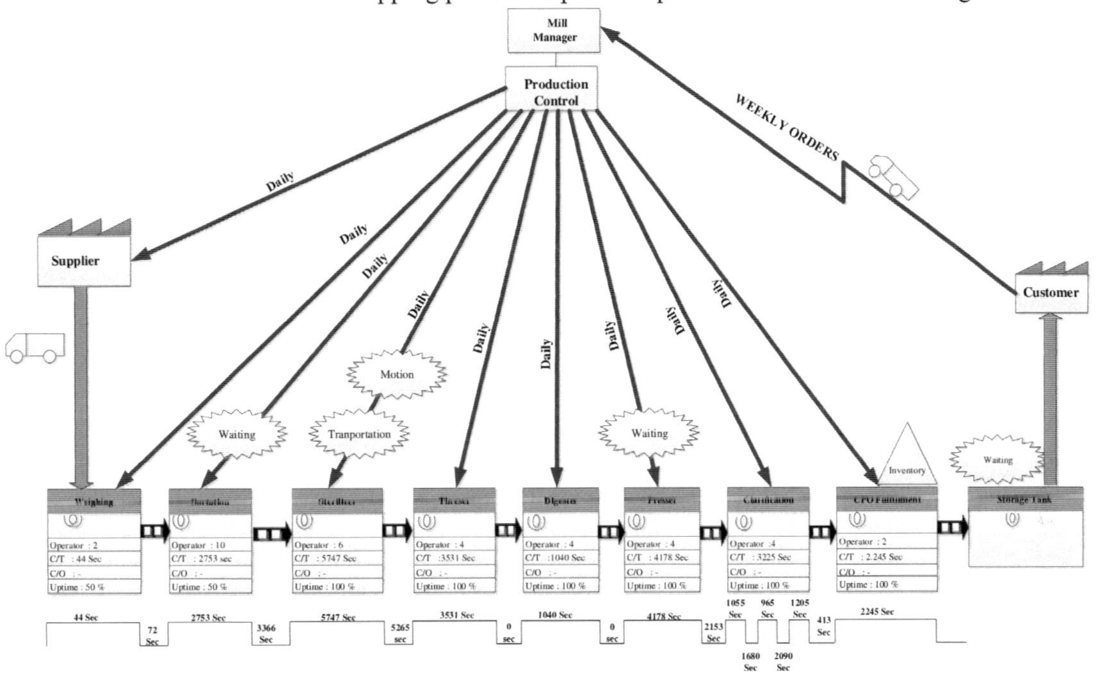

Figure 2. Value Stream Mapping Production Process in the Palm Oil Industry

All stages/processes of actual palm oil or Crude Palm Oil (CPO) are described in Value Stream Mapping (VSM). This aims to facilitate the process of non-value added activities identification throughout the production process. So, based on the observation results on the production floor, do seven identification. The identification results are as follows.

3.2.1. Transportation Waste

Transportation waste can be identified by looking at there is material movement or not, work in process, or finished products on the same flow. In general, the factory layout has been set well, but there is still a work in process flow that goes through the same flow. Such as the sterilizer station there are still waste activities. Where at the FFB inserted into a truck to be boiled in a boiling machine moment. The truck flow surrounds the entire area of the sterilizer station. We can see that in the figure below.

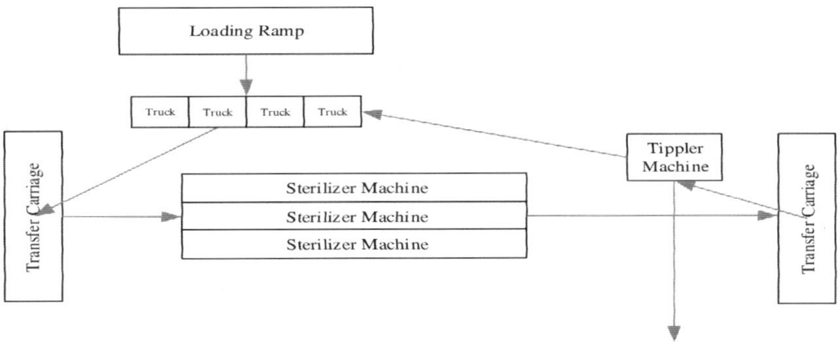

Figure 3. Boiling Process Flow

3.2.2. Inventory Waste
Inventory Waste in this CPO industry is the CPO bottleneck occurrence or storage of CPO production materials in the factory for a long time. At this time the factory has 2 storage tanks which are used as a CPO storage. As a result, CPO has experienced thickening or sedimentation in a storage tank with a capacity of 2000 tons. At the time of the CPO pickup by the truck transportation certainly takes a very long time. Because to overcome the thickening of CPO required heat from a very long Boiler. You can say there is still a storage waste in the factory.

3.2.3. Motion Waste
In general the movement activities on the production floor at the factory has been run smoothly and systematically. This is because of Standard Operating Procedures (SOP) usage are made perfectly to help every step of the work on the production floor. All operators on the production floor also have their respective roles to carry out the same work in a sustainable method. Each operator works according to their expertise, with facilities and supporting equipment have been schematically arranged on the production floor. But there are also operators who ignore the Standard Operating Procedures (SOP) and leave the machine without a replacement operator observation on the machine.

3.2.4. Waiting Waste
Waiting waste is a significant problem in the factory supply chain. At the sorting station, the press and CPO fulfilment stations are experienced waiting waste. The raw materials of Fresh Fruit Bunches do not arrive on time and in a different quantities. Sorting stations with collect FFB raw materials collection for unloading and sorting first. The press station had experienced damage so that production activities experienced a temporary delay. When CPO fulfilment in the truck transportation experienced waiting waste because the CPO has sedimentation/thickening so that it requires heat from the boiler for the CPO fulfilment process.

3.2.5. Over processing Waste
Overproduction process activity in factory is rarely happen. There is no unnecessary process in the process of producing FFB into CPO products and no repetition of processes is found. Every activity in the production process has been arranged and arranged properly with the existence of Standard Operating Procedures (SOP) guide employees when working on the production floor.

3.2.6. Overproduction Waste
Overproduction waste in general it can be interpreted as something related to the overproduction resulting in the finished products bottleneck or semi-finished products exceed the factory capacity which causes storage for a long period of time. Production results produced will be taken by a Group subsidiary to be managed into oil and launched to the market, so it can be said that the factory does not have overproduction waste.

3.2.7. Defect Waste
Defect waste is insignificant, because the raw materials used such as dura and tenera palm oil must be in the specified quality standards and employees at the sorting station must join the training. Of course this is affect the quality standards of CPO products. In the production process the quality level of CPO products is inspected by laboratory employees and factory and also maintaining and guaranteeing the quality level of CPO production.

Table 2. Quality of CPO Production

CPO Production	Actual	Standard
Moist CPO	0.176	0.2
Dirt CPO	0.019	0.02

FFA	5.979	< 3
DOBI	2.7	> 2.5

Table 2 shows the quality standards applied by the company and the actual achievements of the production process. Crude Palm Oil production still has a FFA value greater than the standards set by the company, but for the moist level, CPO and DOBI dirt is in the standard limits. High levels of FFA in CPO are influenced by the length of FFB waiting time to be produced and the quality of FFB.

3.3. Waste Analysis

Fishbone Diagrams are tools used to identify and place the root causes of waste problems systematically. A causal diagram (fishbone) for obstacles occur in the production process of palm oil can be seen in Table 3.

Table 3. Analysis of Fishbone Diagram

Problem	Causes	Roots of Problem
Transportation Waste	Man	Operators are slow at work
		Operators are not responsive
		Fatigue operators
	Machine	Low capacities
		Only one truck of pouring machine
	Method	SOP misunderstanding
		Attention lack of work instructions
		SOP is not implemented properly
Inventory Waste	Man	Operators do not supervise intensively
		Fatigue operators
	Machine	Long distance of storage tank
		The pump is not in ideal condition
	Method	CPO collection schedule is missing
		SOP ignorance
Motion Waste	Man	Non intensive supervision
		Fatigue operators
	Machine	Temperature of boiling machine
		Body of boiling machine close to the operator's station
	Method	SOP ignorance when production process
		Operator do not understand of SOP
	Environment	Hot temperature of boiling work area
Waiting Waste	Man	The sorted FFB is not separated immediately
	Machine	Damage trucks are not removed
		Loading ramp capacities over the truck capacity
	Method	Do FFB bottleneck
	Material	The quality of FFB inserted is not standard
		FFB delay
		CPO experiencing sedimenting/thickening

Table 3 shows the problem analysis for each waste found by describing the problem roots. After it is known that the cause of the waste occur so that the non-value added does not change, then next actions will be made to reduce the nonvalue added time based on the analysis of the root cause discovered. Based on the root causes described above, corrective action plans can be made to minimize waste.

4. Conclusion

At the palm oil factory, activities occur include value added activities by 47%, non-value-added activities by 13% and necessary non value added activities by 40%. In running its production, the palm oil industry is still experiencing waste, namely Transportation Waste, Inventory Waste, Motion Waste, and Waiting Waste. Every waste is searched for the root cause of the problem by using a fishbone diagram.

Acknowledgement

Authors would to thank to Universitas Sumatera Utara for its financial support to publish this paper. Authors also thanked to one of the palm oil industries in Aceh, Indonesia for helping to carry out this research.

References

[1] Melton T 2005 The Benefits Of Lean Manufacturing, Chemical Engineering Research and Design **83** (6), pp 662-673
[2] Revelle B J 2002 Manufacturing handbook of best practices an innovation, productivity, and quality focus (Florida: CRC Press LLC)
[3] Sun S 2011 International Journal of Business and Management, **6** (2), pp 1833-3850
[4] Charron R, Harrington H J, Voehl F and Wiggin H 2015 The Lean Management System Handbook (Boca Raton: CRC Press)
[5] Behrouzi F and Wong K Y 2011 Procedia Computer Science **3**, 388395
[6] Bon A T and Rahman N A 2009 International Conference on Instrumentation, Communication, Information Technology, and Biomedical Engineering Bandung, pp 1-7
[7] Gasperz V and Fontana A 2011 Lean Six Sigma for Manufacturing and Service Industries (Bogor, Indonesia: Vinchristo Publication)
[8] Alaca H and Ceylan C 2011 International Conference on Industrial Engineering and Operations Management Kuala Lumpur, pp 6
[9] Russell R S and Taylor B W 1999 Operations Management, 2nd edition (Uppre Saddle River, NJ: Prentice Hall)
[10] Singh G and Belokar R M 2012 International Journal of Innovative Technology and Exploring Engineering (IJITEE) 1 (2), pp 71-74
[11] Siregar I, Nasution A A, Prasetio A and Fadillah K 2017 IOP Conf. Series: Materials Science and Engineering **237**, 012002
[12] Staats B R, Brunner D J and Upton D M 2011 Journal of Operations Management **29** (5)
[13] Asnan R, Nordin N and Othman S N 2015 Proceeding Social and Behavioral Sciences **211**
[14] Rizkya I, Syahputri K, Sari R M and Siregar I 2018 E3S Web of Conferences **73**, 07004
[15] Barraza M F, Smith T and Dahlgaard-Park S M 2009 The TQM Journal **21** (2)
[16] Aulakh S S and Gill J S 2008 Lean Manufacturing a Practitioner"s Pespective (India: Department of Mechanical Engineering, RIMTInstitute of Engineering & Technology)
[17] Salunke S S dan Hebbar S 2015 International Journal of Current Engineering and Technology 5 (2), pp 931-934

IOP Conf. Series: Materials Science and Engineering **851** (2020) 012059 doi:10.1088/1757-899X/851/1/012059

Characteristics of Process Parameters in Fused Deposition Modelling Process

Ikhsan Siregar[1], J.B. Saedon[2]*, Mohd Shahriman Adenan[2], Mohammad Firdaus Mohammed Azmi[3]

[1]Faculty of Engineering, Universitas Sumatera Utara, Kampus USU, Medan
[2]Faculty of Mech. Eng, Universiti Teknologi MARA, Shah Alam, Selangor, Malaysia.
[3]Politeknik Sultan Azlan Shah, Behrang Stesen, Behrang, Perak

E-mail: ikhsan.siregar@usu.ac.id

Abstract: Additive manufacturing has received incredible attention as a crucial emerging technology as three-dimensional printing offers wide range of industrial applications. This article outlines empirical analyses on the influence of key process parameters such as print speed, infill speed and layer resolution on roughness of the surface and time taken of Fused Deposition Modelling (FDM) processed of Polylactic Acid (PLA) part. Influence of each parameter couple with analysis of variance were utilize using full factorial design. The optimum parameters setting for best surface roughness and lower machining time also being determined. Experimental results indicate that optimal factor settings for each performance characteristic are different.

Keywords: Additive Manufacturing, 3D Printing, Fused Deposition Modelling, Polylactic Acid, Taguchi Method

1. Introduction

Additive manufacturing is recognized as a crucial element in technological advancement and has gained considerable interest from many industrial applications [1]. Additive manufacturing technique integrates the use of computer aided design (CAD) and computer aided manufacturing (CAM) in the manufacture of the specific geometry component using a three-dimensional printer. Through AM technique, the first step is to transform the 3D model which was modeled using CAD application into STL format which is an object's triangular mesh. The computer program will divide the model piece into 2D profile layers in the STL format. Next, the 3D component is developed by each layer that is sliced into 2D model layers and bonded to the previous layer on the build platform until it is completed as required in the original CAD software as in the 3D model [2].

According to the standard of ISO/ASTM 17296 standard on Additive Manufacturing (AM) Technologies [3], it stated that there are seven types of AM processes and can be distinguished into material jetting, material extrusion, direct energy deposition, sheet lamination, binder jetting, powder bed fusion and vat photo polymerization. 3D printing can be classified as laminated object manufacturing, selective laser sintering, fused deposition modeling (FDM), and PolyJet technology. Between those techniques, the most prominent 3D printing technologies is FDM as it offers basic AM concept, robust technology, ease of use, relatively small scale and cost effective. Polymeric materials, metallic materials, inorganic nonmetallic materials, and composite materials are commonly used material in 3D printing. However, in FDM, thermoplastic polymer materials including acrylonitrile-butadiene-styrene (ABS) resin, polylactic acid (PLA), nylon, and polycarbonate are mainly used.

Researchers have been actively investigating ways to attain high-quality FDM finished products in recent years by investigating the FDM process parameter in order to optimize the 3D printing process in accordance to industrial demands. Rational processing parameter is the essential in producing high quality FDM products. However, the effect of each processing parameter on 3D printed products has not yet been widely understood due to the complexity of processing parameters and the uncertain relation between processing parameters and product performance [2]. Thus, to prepare suitable FDM products, the law of parameter selection must be explored.

This study investigates the characteristics of the FDM process parameter includes print speed, infill speed and layer resolution in relation to the fabrication performance namely surface roughness and machining time using full factorial design.

2. Details experimental

2.1 Processing Parameters

In order to investigate the impact of various processing parameters on PLA specimen's 3D printing products, this work scrutinizes the effects of print speed, infill speed and layer resolution on FDM PLA parts through a full factorial design paired with variance analysis (ANOVA) for printing process optimization. The processing parameter defined as follows:

a. Printing speed: The moving speed of the nozzle.
b. Infill speed: Percentage count for compact models to be filled with the material when printed
c. Layer resolution (thickness): The layer slice thickness by the software.

For the FDM product performance responses:

a. Surface roughness (Ra): FDM final products surface finish and can being used directly.
b. Total machining time: Total time taken to complete the product by FDM.

To assess the influence of process variables, impact on printed products, a same model see Figure 1, was built under various input parameters conditions. The process parameters involved are (1) print speed, (2) infill speed and (3) layer resolution. The printed products were manufactured using RPMAKER by MYCRO 3D printer machine and using PLA plastic as model build material. The selected values of process parameters are tabulated in Table 1. To design the experiments run, three factors with full factorial design (L_8) is applied by using the process parameters as shown in Table 2.

Figure 2 shows, the 3D printer machine, RPMAKER manufactured by MYCRO while in Table 3 shows some of the technical specifications. The surface roughness measurement was attained by using ALICONA Infinite Focus Microscope IFM-Infinite Focus Surface Metrology System as shown in Figure 3. The measurement recorded follows the guideline from ISO 4288 standards. For each sample test, the data were taken at three different positions on the sample surface and averages of those values were taken as final experimental data of surface roughness.

Analysis of variants (ANOVA) is being used to assess the effect and variance of each factor and interaction compared to the total variance observed in the result. The linear trends of the results is a reasonable assumption to articulate the performance in ideal condition. Experiment analysis is made using Minitab R14 software. For the analysis of data, results were evaluated using analysis of variance (ANOVA) techniques at 5% significant level. The main plot effect is used to predict the degree of the optimum factor. ANOVA establishes the relative effect of each factor for each interaction.

Figure 1: Part model

Table 1: Factors and Levels

Machining Parameters	Low	High
Print speed (mm/s)	100	200
Infill speed (%)	10	30
Layer resolution (μm)	100	300

Table 2: Experimental Layout

Test Run	Print speed mm/s	Infill speed (%)	Layer resolution (μm)

Test 1	100	10	100
Test 2	200	10	100
Test 3	100	30	100
Test 4	200	30	100
Test 5	100	10	300
Test 6	200	10	300
Test 7	100	30	300
Test 8	200	30	300

Figure 2: RPMAKER 3D printer

Table 3: RPMAKER Specification

Items	Specification
Build volume	150 x 150 x 150mm
Layer resolution	60 -300 μm
Max printing speed	80mm/s
Max travel speed	200mm/s
Nozzle diameter	0.4mm
Filament diameter	1.75mm
Electrical power	12 Watt

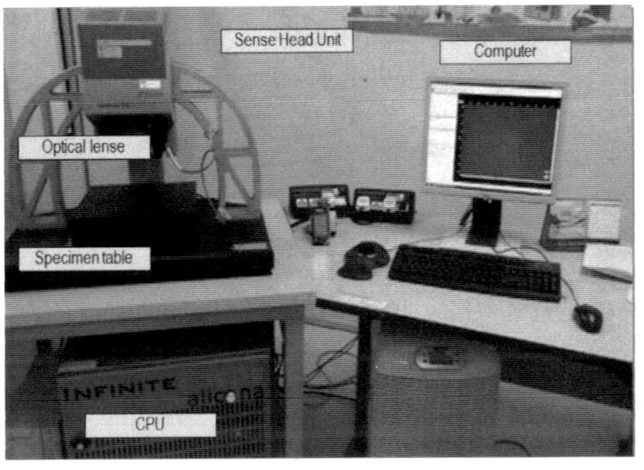

Figure 3: Alicona Infinite Focus 3D Metrology System

3. Results and discussion

3.1 Surface roughness

The results of mean surface roughness of the FDM products are presented in Figure 4. The value of the surface roughness, Ra for each experimental was the average value of three roughness readings. Based from the result in Figure 4, the resulted mean surface roughness, Ra obtained range from 1.9164 to 2.5812μm. The smoothest Ra is achieved when cutting condition of Test 3 (print speed 100mm/s, infill speed 30% and layer resolution 100μm) is employed whereas the roughest Ra is detected with cutting condition Test 8 (print speed 200mm/s, infill speed 30% and layer resolution 300μm).

Analysis of variance techniques (ANOVA) is used to evaluate the adequacy of the developed model. Using this technique, the measured F- ratio was found to be larger than the tabulated value at a confidence level of 95 per cent; thus, the model is considered adequate. For the models developed, the adequacy of a fitted regression models' coefficient of calculated R^2 and adjusted R^2 values were above 99.9% and 99.97%, respectively. These values indicate that the regression models are quite adequate.

Table 4 shows the details of the ANOVA results with respect to surface roughness. All three variable factors were identified to be statistically significant with the print speed shows the highest effect on surface roughness with a percentage contribution ratio (PCR) of 54%. This was followed by layer resolution and infill speed at reduced PCR's of 21% and 3% respectively. Two level interactions between print speed vs layer resolution and infill speed vs layer resolution also found to have 5% significant level with contributions of 13% and 8% respectively. The associated main effects plot for mean surface roughness shown in Figure 5. The interaction plot of parameters for surface roughness is shown in Figure 6. It shows print speed vs layer resolution and infill speed vs layer resolution have significant interaction between factors. Furthermore, it clearly shows print speed vs layer resolution have more strong interaction (PCR 13%) compare others. Figure 7 shows, 3D surface plot of Ra vs print speed, infill speed of interaction graph for surface roughness performance measure.

It observed at A1B2C1 (Test 3- inclusive with interaction) are the optimal levels of the design parameters for improve surface roughness which implies print speed at low level, infill speed high level and lower layer resolution. These combinations give the best surface finished (1.9164 μm) within the specified range.

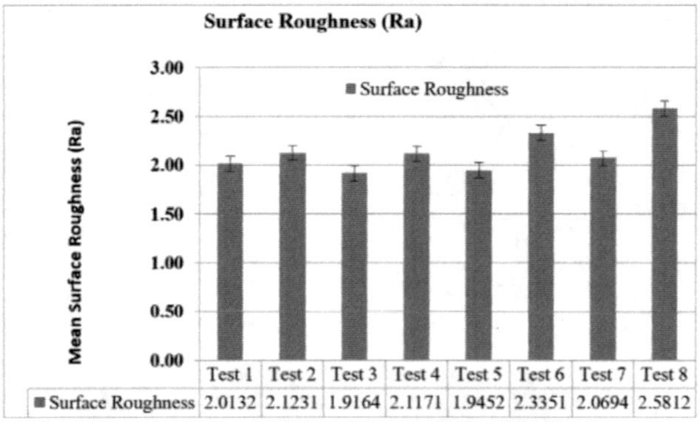

Figure 4: Mean surface roughness

Table 4: ANOVA result for surface roughness

Factor	Sum of Squares	DF	Mean of Squares	F Value	Prob> F	PCR
Model	0.342371	6	0.057062	471.97	0.0352	Significant
Print Speed (A)	0.183709	1	0.183709	1519.496	0.0163	54%
Infill Speed (B)	0.008945	1	0.008945	73.98212	0.0737	3%
Layer Resolution (C)	0.072409	1	0.072409	598.9115	0.0260	21%
AB	0.005655	1	0.005655	46.77504	0.0924	2%
AC	0.043675	1	0.043675	361.2444	0.0335	13%
BC	0.027978	1	0.027978	231.4116	0.0418	8%
Residual	0.000121	1	0.000121			
Total	0.342492	7	*Significant at 5% level			

Std. Dev	0.010996	R-Squared	0.999647
Mean	2.137588	Adjusted R-Squared	0.997529
C.V	0.514389	Predicted R-Squared	0.977408

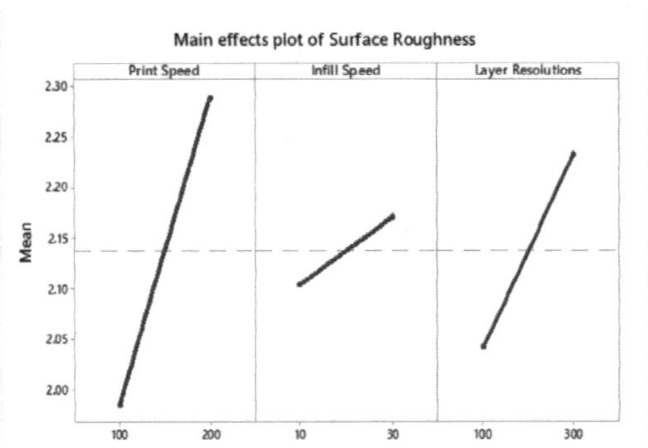

Figure 5: Main effect plots of surface roughness

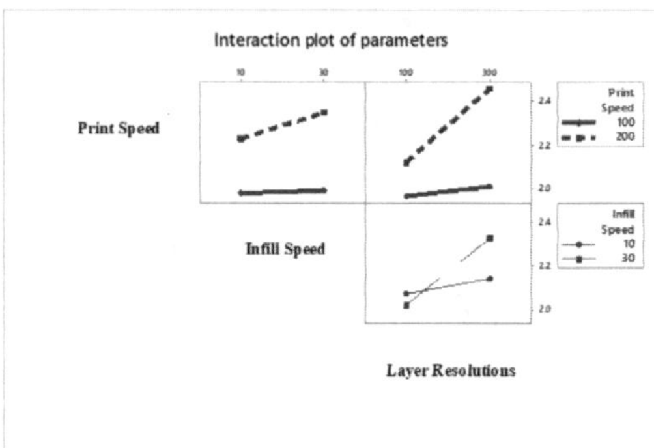

Figure 6: Interaction plots of parameters

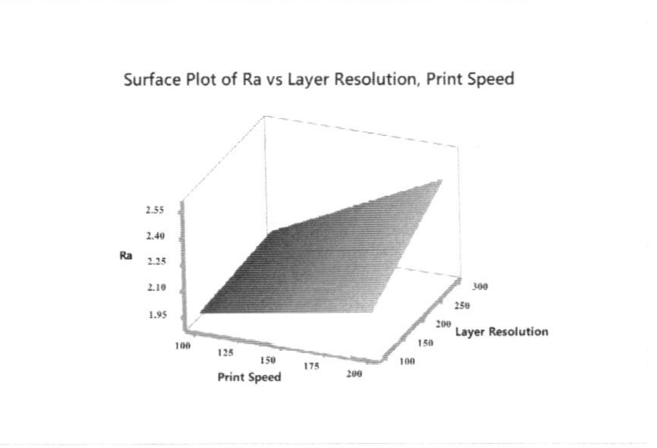

Figure 7: Surface plot of Ra vs print speed, Infill speed

3.2 Total Machining Time

The results of total machining time of the FDM products are presented in Figure 8. The total machining time obtained range from 25 to 53 minute. The shortest time is achieved when cutting condition of Test 1 (print speed 100 mm/s, infill speed 10% and layer resolution 300 μm) is employed whereas the longest time taken is detected with condition Test 8 (print speed 200mm/s, infill speed 30% and layer resolution 300μm).

For the models developed, the adequacy of a fitted regression model coefficient of calculated R^2 and adjusted R^2 values were above 99.9% and 99.8%, respectively. These values indicate that the regression models are quite adequate. Table 5 details the ANOVA results with respect to machining time. All 3 variable factors were found to be statistically significant with layer resolution having the highest influence on machining time with a percentage contribution ratio (PCR) of 45%. This was followed by infill speed and layer resolution at reduced PCR's of 21% and 11% respectively. However, two level interactions only infill speed vs layer resolution found to have 5% significant level with contributions of 21%. The associated main effects plot for machining time shown in Figure 9. The interaction plot of parameters for machining time is shown in Figure 10. It clearly shows, only infill speed vs layer resolution has significant interaction between factors. Figure 10 shows, 3D surface plot of machining time vs layer resolution, infill speed of interaction graph for machining time.

It observed at A1B1C1 are the optimal levels of the design parameters for improve machining time which implies print speed at low level, infill speed and layer resolution at low level. These combinations give the best machining time (25 minute) within the specified range.

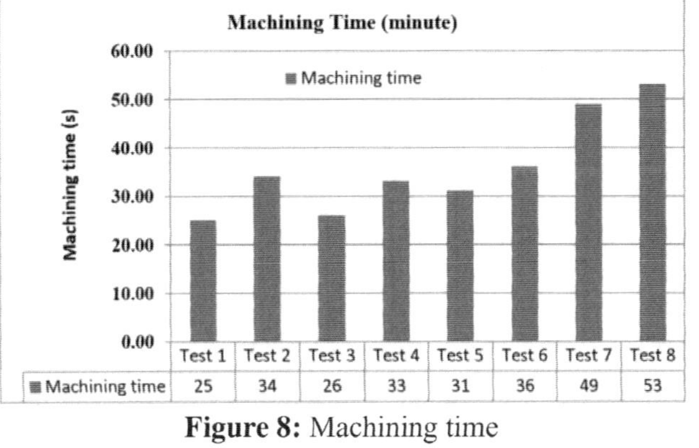

Figure 8: Machining time

Table 5: ANOVA result for machining time

Factor	Sum of Squares	DF	Mean of Squares	F Value	Prob> F	PCR
Model	716.75	6	119.4583	955.6667	0.0248	Significant
Print Speed (A)	78.125	1	78.125	625	0.0255	11%
Infill Speed (B)	153.125	1	153.125	1225	0.0182	21%

IOP Conf. Series: Materials Science and Engineering **851** (2020) 012059 doi:10.1088/1757-899X/851/1/012059

Layer Resolution (C)	325.125	1	325.125	2601	0.0125	45%
AB	1.125	1	1.125	9	0.2048	0%
AC	6.125	1	6.125	49	0.0903	1%
BC	153.125	1	153.125	1225	0.0182	21%
Residual	0.125	1	0.125			
Total	716.875	7	*Significant at 5% level			

Std. Dev	0.353553	R-Squared	0.999826
Mean	35.875	Adjusted R-Squared	0.998779
C.V	0.985515	Predicted R-Squared	0.98884

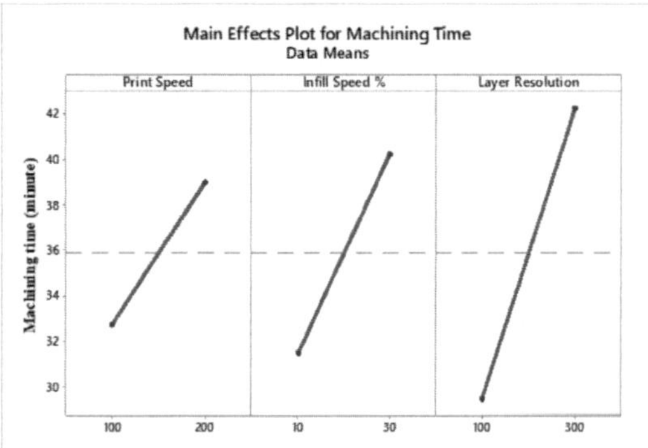

Figure 9: Main effect plots of machining time

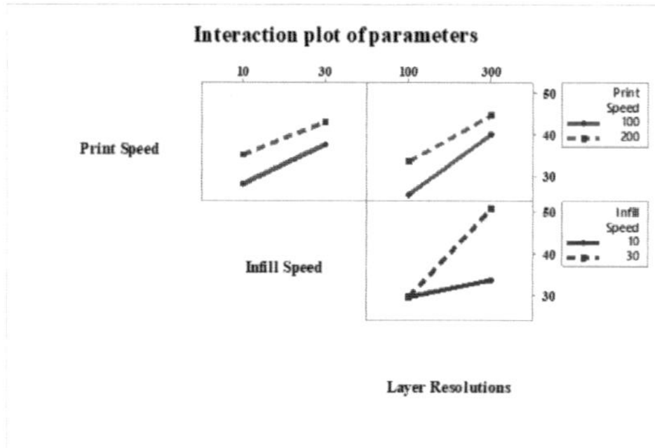

Figure 10: Interaction plots of parameters

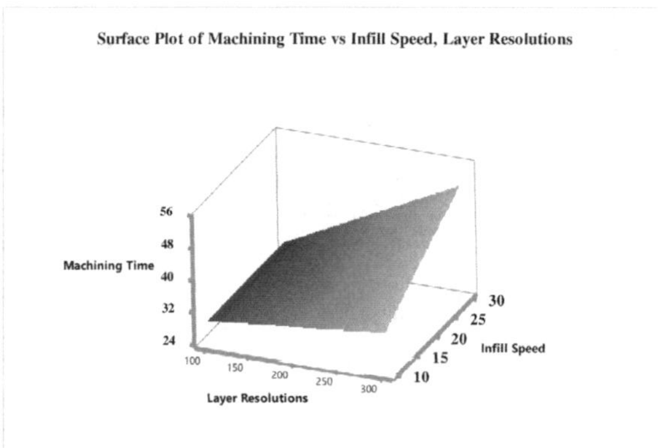

Figure 11: Surface plot of machining time vs infill speed, layer resolution

4. Conclusions

Experimental investigations on the influence of main process parameters such as print speed, infill speed and layer resolution on surface roughness and time taken from the Polylactic Acid (PLA) processed part of Fused Deposition Modelling (FDM) were studied. Based on the analytical results obtained, the following conclusions can be drawn.

1. Print speed and layer resolution are verified to be critical factors affecting surface roughness of the product. Surface finish obtained at high layer resolution was obtained the best quality product. The optimum surface roughness was at parameters; print speed at low level, infill speed high level and lower layer resolution.
2. The thickness of the layer resolution was found to be an influencing parameter which affects total machining time. The build time for higher layer resolution thickness values was lower.

By using the results derived using this study, this analysis could be applied in the direction of prototype development by functionally graded materials.

Acknowledgments

The authors would like to thank the Faculty of Mechanical Engineering (AMTEX), Universiti Teknologi MARA, Universitas Sumatera Utara, Politeknik Sultan Azlan Shah and Ministry of Education, Malaysia for the research funding

References

1. Bai H ,Shuna M, Hui H, Yunchao J, Yingbin X, Huankun H 2018 Study of Processing Parameters in Fused Deposition Modeling Based on Mechanical Properties of Acrylonitrile-Butadiene-Styrene Filament *Polymer Engineering And Science*, pp 1-9,
2. Anoop K S, Ohdar R K, Mahapatra S S, 2009 Improving dimensional accuracy of Fused Deposition Modelling processed part using grey Taguchi method *Materials and Design* **30** (2009) 4243–4252
3. ISO 17296-2:2015(en) Additive manufacturing — General principles — Part 2: Overview of process categories and feedstock

IOP Conf. Series: Materials Science and Engineering **851** (2020) 012060 doi:10.1088/1757-899X/851/1/012060

The Feasibility Study of Palm Oil Replanting Using Chipping Technique

Candra E. Ritonga[1], Rulianda P. Wibowo[1]*, Satia N. Lubis[1]
[1]Universitas Sumatera Utara, Medan, Indonesia

*Email :rulianda_wibowo@usu.ac.id

Abstract This study aims to determine the feasibility of palm oil replanting using the chipping technique. The data were obtained through interview, observation and secondary data collection. The study is using NPV, Net B / C, IRR, Payback Period and sensitivity analysis for data analysis. The results indicate that the palm oil replanting using chipping technique is financially feasible to run. The results of the feasibility analysis show that the chipping technique generate financial benefits of Rp. 24,241,138 / ha per planting period with Net B/C ratio at 1.5. The payback period is 10 years and 7 months. This study also indicates that loan interest greater than 12.98% will not be feasible for palm oil plantation. The sensitivity analysis shows that the change in productivity has greater impact on profitability compared to the change in cost. Thus, palm oil plantation should focus on maintaining high level of crop productivity.

1. Introduction

The productivity of palm oil plant varies based on age and variety. The replanting program for palm oil plantation are mandatory to maintain plantation profitability as the crop productivity will decline rapidly when the age is being matured until a certain stage of palm oil growth. Furthermore, the palm oil replanting program will not produce any revenue until the crop reaches 3-4 years of age so that replanting program should be planning considering the company cash flow.

The replanting program should be implemented if the palm oil crop had across its economic productive ages. The productivity standard for implementing replanting program is if an average of crop production are below 14 tons per hectare per year. Another consideration for replanting program other than crop age is crop height and density. Crop height more than 12 meter and crop density less than 80 plant per hectare are considered not economically efficient so that the replanting program should be commence immediately. This study examines the replanting program using chipping technique. The benefit of chipping technique is the soil preparation are much intensive which provide better media for crop growth. Other benefit of chipping technique are the increase in the rate of plant weathering, maintaining the soil moisture, slower the risk of *O.rhinocerus* pest infection spread and decrease the risk of plant disease, Ganoderma sp. This study aims to see the financial feasibility of chipping technique for replanting program. The sensitivity analysis on profitability will be analyzed with the scenario of changing in production and cost.

2. Methods

This study was conducted at Afdeling II PT X, Padang Lawas Regency, Sosa District in September – December 2019. The data used for the analysis comprises primary and secondary data. The primary data is obtained through interview with top company management. The secondary data are collected from various resources likewise Dinas Perkebunan Provinsi Sumatera Utara and Badan Pusat Statistik (BPS).

This study uses financial analysis includes Net Present Value (NPV), Internal Rate of Return (IRR), Net Benefit Cost Ratio (Net B/C), and sensitivity analysis (Gittinger, 1986) as follows:

2.1 *Net Present Value (NPV)*

NPV is the difference between benefit and cost valuing the time value of money. The NPV analysis are using discount rate from Bank of Indonesia, 11%, for calculating the present value over time. The NPV analysis is estimated with 24 years' time frame. There are three criteria for NPV analysis as follows:

- NPV > 0, means that the business is feasible to run
- NPV = 0, means that the benefit are equal with social opportunity cost
- NPV < 0, means that the business is not feasible to run

The formula of NPV is:

$$\mathrm{NPV} = \sum_{t=1}^{n} \frac{\mathrm{Bt} - \mathrm{Ct}}{(1 - \mathrm{i})^{\mathrm{t}}} \quad (1)$$

Where:
$B_t = Benefit\ at\ time\ t$
$C_t = Cost\ at\ time\ t$
$i = discount\ rate$ (11%)
$t = time\ with\ n\ at\ 24\ years$

2.2 *Internal rate of Return (IRR)*

IRR is a discount rate which makes the NPV of all cash flows from a business project equal to zero. IRR estimation rely on the same formula as NPV does. There are three criteria for IRR analysis:

- IRR > discount rate, means that the business is feasible to run
- IRR = discount rate, means that the benefit are equal with social opportunity cost
- IRR < discount rate, means that the business is not feasible to run

The formula of IRR is :

$$IRR = i_t + \frac{NPV1}{NPV1 - NPV2} x(i_2 - i_1) \quad (2)$$

2.3 *Net Benefit Cost Ratio (Net B/C)*

Net B/C is a ratio used in a cost-benefit analysis to summarize the overall relationship between the relative benefits and cost of business project. Net B/C can be exhibited in qualitative terms. If a business project has a Net B/C greater than 1.0, the business project is expected to generate a positive net present value. If a Net B/C equal to 1.0, the business project only generates net benefit which equal with social opportunity cost. Net B/C less than 1.0 means the business project is not feasible to run.
The formula of Net B/C is:

$$NetB/C = \frac{\sum_{t=1}^{n} \frac{Bt - Ct}{(1-i)^t}}{\sum_{t=1}^{n} \frac{Ct - Bt}{(1-i)^t}} \quad (3)$$

2.4 *Payback Period*

The payback period is the time duration in which an investment reaches a breakeven point. Shorter payback period means that the investment has higher desirability. The formula of payback period relies on the initial investment and annual cash flow. The annual cashflow are the difference between benefit and cost for each period. The payback period is not including the time value of money on the estimation.

3. Result and Discussion

PT X, company which used as a study case, is located at Siborna Bunut Village, Sosa District, Padang Lawas Regency, North Sumatera Province. PT X is a modern palm oil plantation enterprise that impose ISO 9001:2008 and ISO 14001:2004 on management. PT X has plantation area of 1,495.68 Hectares that splits into 3 different areas comprises Afdeling I (399.65 Ha) , Afdeling II (629.02 Ha) and Afdeling III (536.01 Ha). PT X were implementing replantation program when the age of the tree are above 27 years and the height of the tree are higher than 12 meters.

The revenue generated for PT X are from the selling of palm oil fresh fruit bunch. The projection of the 24 years fresh fruit bunch revenue is from the historical data of the company. The price assumption is Rp 1,100 per kg, which generated from average historical palm oil fresh fruit bunch price. The company is assumed to borrow Rp 20.200.000.000 from Bank Rakyat Indonesia (BRI) with interest rate of 11% per year and payback period of 8 years.

The projection of operational cost per hectare for 24 years at table 1 is generated from historical operational cost of Afdeling II at PT X. The plantation area of Afdeling II is 629.02 Ha, which is the biggest among plantation area that PT X owned. The biggest operational cost is fertilizer as shown at table 1, followed by harvest and transportation cost. The loan and interest cost are relatively high as it constituted 11.44% of total operational cost. The high loan and interest cost because of the high debt that company borrow from BRI, in which the debt settlement need to be finalized in 8 years. Another big spending is on tax cost that accounts for 18.60% of total operational cost. The total operational cost of PT X per year is Rp 680,141,272.

Table 1. Operational Cost of Afdeling II at PT X per hectare for 24 years

No	Item	Cost	
		Rupiah	Percentage
1	Harvest and Transportation	Rp 151,189,500	22,20%
2	Replanting	Rp 4,040,000	0,59%
3	Cultivation	Rp 7,335,188	1,08%
4	Legumes maintenance	Rp 5,640,000	0,83%
5	Tree maintenance	Rp 911,330	0,13%
6	Sodding	Rp 32,997,662	4,85%
7	Pest and disease maintenance	Rp 5,493,436	0,81%
8	fertilizer	Rp 244,607,552	35,96%
9	Sprout maintenance	Rp 7,155,000	1,05%
10	Machine and building	Rp 16,457,335	2,42%
11	Loan dan interest	Rp 77,831,182	11,44%
12	Tax	Rp 126,483,088	18,60%
	TOTAL	Rp 680,141,271	100,00%

The income statement is a report that shows the amount of revenue earned and expenses, costs incurred in each period. The income statement can describe the performance of companies in each period. Components of income statement derived from the sale (income), cost of goods sold, operating expenses, interest and tax expense. The projection of income statement for Afdeling II PT. X. is shown in table 2. Table 2 shows that the plantation gain positive net income after 7 years of production. Total net income for 24 years, without including time value of money, is Rp 330,032,929.

Table 2. Income statement projection of Afdeling II PT X per hectare

Period	Year	Net Income (Rp)	Period	Year	Net Income (Rp)

TBM 1	2017	Rp, (24,422,599)	TM 12	2031	Rp 18,409,293
TBM 2	2018	Rp, (7,870,126)	TM 13	2032	Rp 19,685,976
TBM 3	2019	Rp, (7,944,358)	TM 14	2033	Rp 20,653,965
TM 1	2020	Rp, (9,461,111)	TM 15	2034	Rp 21,591,878
TM 2	2021	Rp, (3,193,626)	TM 16	2035	Rp 22,127,029
TM 3	2022	Rp, (1,484,552)	TM 17	2036	Rp 26,074,023
TM 4	2023	Rp, 1,330,754	TM 18	2037	Rp 26,383,273
TM 5	2024	Rp, 1,608,840	TM 19	2038	Rp 26,788,074
TM 6	2025	Rp 3,363,461	TM 20	2039	Rp, 27,907,920
TM 7	2026	Rp 4,488,608	TM 21	2040	Rp, 25,893,822
TM 8	2027	Rp 6,123,771	TM 22	2041	Rp, 26,913,510
TM 9	2028	Rp 17,174,533	TM 23	2042	Rp, 32,114,486
TM 10	2029	Rp, 18,426,400	TM24	2043	Rp, 20,144,952
TM 11	2030	Rp, 17,204,729	TOTAL		Rp, 330,032,929

The results of the feasibility analysis shows that the replantation program using chipping technique generate NPV of Rp. 24,241,138 / ha per planting period, 24 years, with Net B/C ratio at 1.5. Net B/C ratio is 1.5 which implies for every Rp 1,000 in cost will generate a benefit of Rp 1,500. This study also indicates that loan interest greater than 12.98% will not be feasible for palm oil plantation to borrow. The payback period is 10 years and 7 months, which point out a loan from BRI with payback period of 8 years are not financially feasible to borrow. So that PT X should find a debt that offer longer payback period.

Table 3. Financial Feasibility Analysis

Investment criteria	Value	Decent / Not Decent
Net Present Value	Rp, 24,241,138	Decent
Net B / C	1,5	Decent
IRR	12,98%	Decent
Payback Period	10,2	Not Decent

The sensitivity analysis is conducted by simply changing the size of important variables and determine how the results of calculations are changing (Kadariah, 1988). The sensitivity analysis are conducted by applying a scenario of change in crop production up to ±10% and change in production cost up to ±30%.

Table 4. The sensitivity analysis with the scenario of changing the crop production

	Investment criteria			
Change in production	NPV	Net B / C	IRR	PP
Down 10%	Rp, (5,940,385)	0,9	9,54%	12,8
Down 7,5%	Rp, 1,618,343	1	10,43%	12
Down 5%	Rp, 9,177,071	1,2	11,30%	11,3
UP 5%	Rp, 39,411,985	1,8	14,60%	9,3
UP 7,5%	Rp, 46,970,713	2	15,40%	8,9
UP 10%	Rp, 54,529,442	2,1	16,17%	8,6

The sensitivity analysis results in table 4 and 5 shows that the impact of change in crop production is higher than the impact of change in production cost to the profitability. The decrease in crop production at 10% has generated a negative NPV and Net B/C lower than 1.0. The palm oil plantation is no longer financially feasible to run if the decrease in crop production has reached 10%. Based on net B/C criterion, palm oil plantation is not financially feasible to run if the crop production is decreasing by 7.5%.

Table 5. The sensitivity analysis with the scenario of changing the production cost

	Investment criteria			
Change in cost	NPV	Net B / C	IRR	PP
Down 30%	Rp, 39,674,418	1,8	14,69%	9,2
Down 20%	Rp, 34,547,788	1,69	14,12%	9,5
Down 10%	Rp, 29,421,158	1,57	13,55%	9,8
UP 10%	Rp, 19,167,898	1,36	12,41%	10,6
UP 20%	Rp, 14,041,268	1,26	11,84%	11
UP 30%	Rp, 8,914,638	1,16	10,32%	11,5

Palm oil plantation profitability is less sensitive to the change in production cost. The palm oil plantation is still feasible to run even though the production cost is increase by 30%. The decrease in production cost also has moderate effect on the profitability, which shown by the decrease in production cost by 30% will only increase the Net B/C ratio by 0.3 point. Results in table 4 and table 5 shows that PT X should maintain the crop production to have high productivity by increase the optimality of plant maintenance program. PT X should not rely its profitability strategy on production cost efficiency.

4. Conclusion

The financial feasibility results indicate that the palm oil replanting program using chipping technique is financially feasible to run. The results of the feasibility analysis shows that the chipping technique generate NPV of Rp. 24,241,138 / ha per planting period with Net B/C ratio at 1.5. The debt from the bank that has payback period less than 10 years and 7 months is not financially feasible to borrow. This study also indicates that loan interest greater than 12.98% will not be feasible for palm oil plantation to borrow. The sensitivity analysis shows that the change in crop productivity has greater impact on profitability compared to the change in cost. Palm oil plantation profitability is less sensitive to the change in production cost. Thus, palm oil plantation should focus on maintaining high level of crop productivity and not rely its profitability strategy on production cost efficiency.

References

[1] JP Gittinger. 1986. Economic Analysis of Agricultural Projects. Translation of: Economic Analysis of Agriculture. Sutomo S and K. Mangiri UI Press, Jakarta.

[2] Kadariah. 1988. Evaluation Project (Economic Analysis). Jakarta: Faculty of Economics UI

[3] Rahmat R F, Nurmawan S, Sembiring S, Syahputra M F, Fadli 2018 Adaptive neuro-fuzzy inference system for forecasting rubber milk production *IOP Conference Series: Materials Science and Engineering* **308**(1),012014

[4] Siregar I, Muchtar M A, Rahmat R F, Andayani U, Nasution T H, and Sari R M 2018 Method of calculation overall equipment effectiveness in fertilizer factory. *IOP Conference Series: Materials Science and Engineering* **308**(1), p. 012053) IOP Publishing.

IOP Conf. Series: Materials Science and Engineering **851** (2020) 012061 doi:10.1088/1757-899X/851/1/012061

Design of Hydrogen Fuel Supply Chain Distribution Network from The Potency of Fresh Fruit Empty Bunches

R M Sari[1], I Rizkya[1], T B Nur[1], K Syahputri[1], W A Syahputra[1], R Nouvan[1]

[1] Industrial Engineering Department, Faculty of Engineering, Universitas Sumatera Utara, Medan, Indonesia

rahmi.m.sari@usu.ac.id

Abstract. In Indonesia, the production of biomass is wasted because it is considered as "waste" that has no use anymore such as bagasse, rice grain skin, corn cob, and oil palm waste. Oil palm empty bunches have C and H chemical bonds, where most of the materials with chains C and H store heat energy so that hydrogen can be obtained as fuel. Process to convert a biomass into hydrogen is by gasification and pyrolysis. The purpose in this research is to design the distribution network from the potential fresh fruit empty bunches. The hydrogen produced is only 0.469% of the available potential and is similar to the existing of the used empty bunches potential. Selection of factory location is based on 4 criteria. From the criteria, obtained Labuhan Batu District as the best regencies in North Sumatra. Potential plantations can supply as much as 6,390,423,000 kgs with the ability of oil palm factory distribution of 1,469,797,290 kgs.

1. Introduction

As the depletion of fossil fuel and natural gas reserves make the world look for replacement energy. Biomass is an alternative fuel source that is still abundant and has not been fully utilized [1].

Exploration for renewable alternative energy sources for the long term will takes an environmentally friendly source of energy, one of which is hydrogen. In the long term, especially in Indonesia, the production of hydrogen through coal gasification and through biomass gasification will provide economic benefits. Consider the potential of biomass in Indonesia today and in the future, hydrogen production will have a positive effect on the national economy.

Globally the need for hydrogen as an alternative energy has been increasing in recent years hydrogen as one of the hydrocarbon fuels that has the advantage of being eco-friendly and having great energy efficiency especially if used on vehicle fuel [2].

Biomass that has the potential to be processed is empty bunch palm oil. Oil palm empty bunches have considerable potential to be utilized. However, empty bunch utilized as organic fertilizer presently. The remaining empty bunches will make problem for the place and its disposal transport resulting in additional production costs for the processor [3][4].

Hydrogen in the industry is generally produced using the Steam Methane Reforming (SMR) method from natural gas resulting from pyrolysis and gasification methods [5][6]. Steam Methane Reforming is more efficient with low cost through pyrolysis and gasification process [7].

Supply chain management is now important for oil palm companies and other companies as the competition grows and the position of palm oil and mill suppliers and millers as partners. The process of fulfilling the customer needs in the supply chain is a value-adding chain that not only stops at one company but includes all the companies that are members of the supply chain [8].

Pyrolysis is decomposed with the aid of heat in the absence of oxygen or with a limited amount of oxygen. Usually there are three products in the process of pyrolysis such as charcoal, gas, and liquid. Pyrolysis process is influenced by several factors such as time, temperature, particle number, particle size, particle weight [9].

Strategic business location can maximize the business profit. Location is where to serve consumers, defined as a place to introduce goods. Thus the definition of a location is a place to run an activity that serves consumers, production activities, storage activities, or to control the activities of the company as a whole part of the company. The main factors in choosing the location are (1) Availability of Raw Materials, (2) The location of the intended market, (3) Availability of energy source, water, and communication facilities, (4) Availability of manpower and (5) Availability of transportation [10].

Supply chain is a network of agencies or companies that work together to create and deliver a product into the hands of end users. These companies usually include suppliers, agencies, distributors, stores or retailers, as well as support agencies such as logistics services agencies [11]. In this definition, there are a number of key characteristics that have been used to portrait a supply chain. First, a supply chain is formed and can only be formed if there are more than one participating companies. Second, the participating companies within a supply chain normally do not belong to the same business ownership, and hence there is a legal independence in between. Third, those companies are inter-connected on the common commitment to add value to the steam of material flow that run through the supply chain. This material flow, to each company, comes in as the transformed inputs and goes out as the value-added outputs [12].

2. Methodology

Research object is hydrogen fuel that produced from palm oil empty bunch with pyrolysis method. The data gathered from the result of empty bunch, plantation location and factory locations in Sumatera Utara. This data then processed to gain the information of distribution network palm oil empty bunch.

First step is to figure out the needs of raw material by define the capacity per day based on the steam methane reformer machine. Second step is to define the needs of raw material needed based on the percentage of hydrogen by gas and the percentage of gas from empty bunch from the result of pyrolysis process and also percentage data from empty bunch and fresh fruit bunch. In this step, also resulting the hydrogen potential that can be get from empty bunch in Sumatera Utara.

Next, designing the distribution network is done by defining the location of hydrogen plant in Sumatera Utara with qualitative method. The assessment based on raw material supply and added material, utility, transportation and surroundings. The result is the best location to build the hydrogen plant.

The last step is figuring the distribution network after resulting the ideal plat location and factory data in Sumatera Utara. The result is shown by the distribution figure from farm to hydrogen plant.

3. Result and Discussion

3.1. Production Capacity

The determination of production capacity is based on rated power capacity (MW) using Steam Methane Reformer. The results of a study by the National Renewable Energy Laboratory for the production of hydrogen using pyrolysis and Steam Methane Reformer methods obtained a production capacity of 1,000 kgs / day

3.2. Raw Materials Needs

The needs of raw material calculated by the following steps :

1. Hydrogen Calculation Result
   Based on the calculation, Percentage of empty bunch turns in to gas is 52.02%. Then the percentage of hydrogen in gas is 8.69%.
2. Potential of Hydrogen Produced in North Sumatera
   The processing of fresh fruit bunches (FFB) will produce 23% palm empty fruit bunches, 6.5% shell waste, 4% wet decanter, 13% fiber and liquid waste as much as 50%.
   The following is the calculation of the amount of hydrogen produced in North Sumatra [13].
   a. The number of FFB produced in 2015 is 6,390,423,000 kg.
   b. The percentage of oil palm empty fruit bunches from fresh fruit bunches is 23% so that there are empty bunch of 1,469,797,290 kg.
   c. The percentage of oil palm empty bunches into gas as much as 52.02% to obtain gas of 764,588,550 kg by pyrolysis method.
   d. The percentage of hydrogen in the gas as much as 4.52% to obtain hydrogen of 690.265.368 kg per year.
3. Needs of Empty Bunch per hour Calculation
   The hourly empty bunch requirement is calculated using the hydrogen percentage in TKKS. The results can be seen in Table 1.

Table 1. The Needs of Palm Oil Empty Bunch

	Percentage (%)	Weight (Kg)	Type	Weight (Kg)	Type
Percentage Hydrogen in Gas	4,52	22,123.89	Gas	1,000	Hydrogen
Percentage Gas in Empty Bunch	52,02	42,529.58	Empty Bunch	22,123.89	Gas
Percentage Empty Bunch from Fresh Fruit Bunch	23	184,911.24	Fresh Fruit Bunch	42,529.58	Empty Bunch

Table 1. shows that the need of empty bunch fruit to gain 1000 kg hydrogen per day is 42,529.58 kg.

3.3. Plant Site Selection

The ideal plant site selection is the location that will provide the minimum total cost of production and transportation and maximum profit for the company. The selected regions are districts located in the province of North Sumatra. In the selection process of several alternative locations will be used qualitative methods to determine the alternative locations should be selected.

Then the determination of the weight of the criteria that affect the selection of factory location. Giving weight of each factor or criteria identified based on degree of importance. The highest market share is Labuhan Regency with 37.09% market share followed by Asahan District with 13.13% market share and Langkat Regency with market share of 12.05% as location 1, location 2 and location 3. [15]

Table 2. Weight and Value Matrices from Each Location Factor and Alternative

Criteria	Weight	Alternatives		
		Alternative 1	Alternative 2	Alternative 3
Easy supply of raw materials and auxiliary materials	35%	9	7	7
Utility	30%	8	9	7
Easy Transportation	20%	8	8	8

IOP Conf. Series: Materials Science and Engineering **851** (2020) 012061 doi:10.1088/1757-899X/851/1/012061

Environmental and social circumstances	15%	8	8	8

The value comes from research considerations based on the actual conditions in which location. Then get the values of each alternative location that can be seen in Table 3.

Table 3. Assessment of Location Alternatives

Criteria	Alternatives		
	Alternative 1	Alternative 2	Alternative 3
Easy supply of raw materials and auxiliary materials	3.15	2.45	2.45
Utility	2.8	3.15	2.45
Easy Transportation	2.8	2.8	2.8
Environmental and social circumstances	2.8	2.8	2.8
Total	11.55	11.2	10.5

From the results of the calculation above, then the selection of alternative locations that are considered good alternative location that has the greatest value of z / total. Thus location 1 has the largest total value of 11.55, so the best location according to this analysis is alternative 1.

3.4. Distribution Network

Supply chain problem can be done by using a facility location capacity approach. The development of supply chain networks is carried out at three levels of supply chain distribution (suppliers, distribution centres, and consumers) in consider of multi-period, multi-product issues, and adding inventory issues to anticipate uncertainty in supply. Hydrogen fuel supply chain distribution network can be seen in figure 1. below

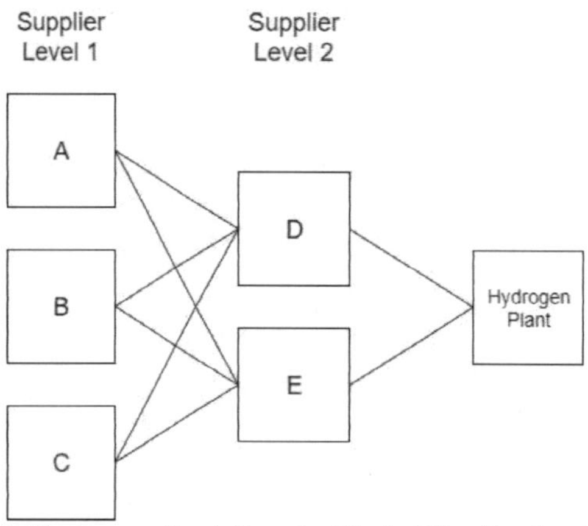

Figure 1. Hydrogen Fuel Supply Chain Distribution Network

The supply chain is divided into three groups: (1) Suppliers level 1, that is oil palm plantations that produce FFB for processing in Palm Oil Mill, (2) Suppliers level 2, which is leaving empty bunch to become the raw material for the process of making hydrogen fuel, and the last destination which is Hydrogen Plant that produces hydrogen fuel.

4. Conclusions

Sumatera Utara has a great potential of empty bunch for making a hydrogen fuel. To make it come true, it needed a distribution network start from fresh fruit bunch from farm until hydrogen fuel from factory. Supply chain consist of three part such as supplier level 1, supplier level 2, and hydrogen plant as consumer. Supplier level 1 is farm that send a fresh fruit bunch to supplier level 2. Supplier level 2 is a palm oil factory that sent palm oil empty bunch to hydrogen plant. The calculation shows that the palm oil farm in Sumatera Utara can supply empty bunch as the production capacity that makes this supply chain can be applied.

Acknowledgement
This study was conducted by the research team with collaboration of industrial engineering and mechanical engineering of engineering faculty and fully supported by Universitas Sumatera Utara.

References

[1] M Muresan, *et.al* 2013 *Chemical Engineering Research and Design* **91**, pp.1527-1541
[2] M Matzen, *et.al* 2015 *Advanced Chemical Engineering* **5**(3)
[3] C L Jin, *et.al* 2017 *Journal Energy Sources: Economics, Planning, and Policy* **12**
[4] Y Zhang, *et.al* 2014 *Bioresources Technology* **157**
[5] J Eichman, *et.al* 2016 "Economic Assessment of Hydrogen Technologies Perticipacating in California Electricity Markets" *National Renewable Energy Laboratory*
[6] M Melaina and M Penev 2013 "Hydrogen Station Cost Estimates" *National Renewable Energy Laboratory*WIB
[7] G Delgado, *et.al* 2016 *Chemical Engineering Transactions* **52**(2016)
[8] M Jahirul 2012 *Biofuels Production through Biomass PyrolysisA Technological Review*. (Queensland University of Technology, Brisbane, Queensland *4000*. Australia)
[9] A Fathoni 2016 *Jurnal Penelitian Ilmu Manajemen* **1**(1)
[10] S Chopra, *et.al* 2013 *Supply Chain Management. Strategy, Planning, and Operation*. Fifth Edition. England: Penerbit Pearson.
[11] Lu, Dawei. *Fundamentals of Supply Chain Management*. London: Ventus Publishing. 2011.
[12] Direktorat Jendaral Perkebunan.2017.Statistik Perkebunan Indonesia 2015-2017 Kelapa Sawit
[13] Pusat Data dan Sistem Informasi Pertanian Sekretariat Jendral-Kementrian Pertanian.2016. Outlook Kelapa Sawit Komoditas Pertanian Subsektor Perkebunan
[14] Siregar I, Muchtar M A, Rahmat R F, Andayani U, Nasution T H, and Sari R M 2018 Method of calculation overall equipment effectiveness in fertilizer factory. In IOP Conference Series: Materials Science and Engineering **308**(1), p. 012053) IOP Publishing.

Node-based versus Server-based Data Validations in Internet of Things

Naemah Mubarakah[1*], Poltak Sihombing[2], Suherman Suherman[3], Soeharwinto[4]

[1,3,4]Electrical Engineering Department, Universitas Sumatera Utara, Medan-Indonesia
[2]Faculty of Computer Science, Universitas Sumatera Utara, Medan-Indonesia

*naemah@usu.ac.id

Abstract. Internet of Things (IoT) collects and sends data over TCP/IP network for the intended application. Some applications designed sensors to be lost after battery exhausted. Therefore, battery lifetime is important. On the other hand, data quality can be disrupted by the harsh environment and interfering noises, so that the transmitted data may be invalid. Transmitting invalid data may exhaust battery sooner, while implementing validation algorithm may absorb more power. This paper proposed a simple peak detection method in IoT sensor to avoid transmitting invalid data. The experimental evaluation was set up and the method was compared to server-based support vector machine classifier. The results show that the proposed method is able to cancel 86% error data transmission with 1.25% power savings.

1. Introduction

The progress on some technologies such as sensors, actuator, and embedded computing enable distributed small devices controlling many applications. Those devices can be controlled and connected remotely through TCP/IP network. The internet of things is emerged to satisfy networking requirements to interconnect those devices, including sensor and interconnection framework [1].

Wireless World Research Forum projected that there are 7 trillion wireless devices around the world serving about 7 million people [2]. IoT enables collaborations and cooperation among those wireless devices, allows self-organizing and adaptive operation as well as intelligent applications [3]. Wireless sensors are implemented mainly for wireless sensor networks (WSNs), radio identification network (RFID)[12], machine-to-machine (M2M), supervisory control and data acquisition (SCADA) and other applications. IoT is expected to cover anything any device, anytime any context, anyone anybody, any service any business, any path any network, and any place anywhere [4].

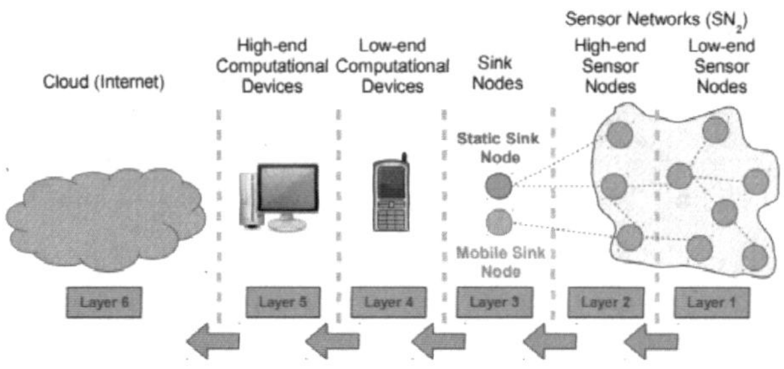

Figure 1. IoT components [4]

Figure 1 shows the essential component in IoT [4]. The architecture contains sensors at the network edge (layer 1). These sensors route data to the edge sensor (layer 2) which forward all data to sensor static sensor referred to as sink node (layer 3). Sink node sends data to receiver with low processing capabilities such as mobile node or a single personal computer (layer 4). In order to maximize the work, data may forwarded to a high processing unit such as servers (layer 5). At the end, data may be stored, shared and processed at the cloud (layer 6). Middleware is required to integrate those systems in order to accommodate various applications and requirements. Survey performed by Razzaque et al [5] showed that some middleware architecture have been proposed to support multi-propose IoT networks. However, mostly is WSN-centric. The research also said that there no middleware framework thas is available to cover all IoT applications.

Sensor node has limitation mainly related to processing capabilities and electrical supply. Those sensors are mainly using wireless link to connect, so they are mostly unreachable for recharging purpose [6]. Even dough many energy-saving methods proposed, including the use of energy harvesting, steps on energy efficient are still pressing and to increase battery and network lifetime for IoT application [7]. As the consequence, techniques and methods used in IoT should consider energy limitation and efficiency.

Other emerging problem is the irregularity of radio link that potentially causes disturbances and interferences to the transmitted data. The quality of the transmitted data surely influences the applications [8]. Even its importance for application performances, study on this matter is still limited. On the other hand, data quality implementations are mainly performed in database level. Some issues related to data quality covers accuracy, currency, and completeness [9]. Quality guarantee is generally performed in database which is more stable and well defined. There is few work found on real-time data validation, implemented during data generations. Meanwhile, pervasive system performances depend on data acquisition and energy consumption. So this topic is still challenging.

Data quality can be measured as data validity as well as data integrity. Validity means data represent the unit that is measured. While integrity guarantees that data is free of separation, privileges, virus free, and intrusion free [10]. This paper focuses on real-time data validation on IoT, by comparing validation process in wireless sensor node and validation process in server. Trade-off between microcontroller capabilities and energy limitation were assessed trough implementations. A lightweight validation system is proposed.

Fongen [10] collected sensor data in the IoT/ WSN applications. It showed that during data collection, some collected values are not correct. This event leads to incorrect data received by server which decreases the accuracy and reliability. Fongen [10] stated that the statistical observation for data outliers exerts lower complexity rather than supervisor and cluster method. However, the statistical methods cannot be performed real-time as it is performed in a fixed collected data. Work on [2] implemented one class quarter sphere support vector machine (OCSVM) for detecting outliers. However, the method is to complex for real-time applications. Xiong et al [3] used one-class principal component classifier. It has lower computational cost than OCSVM but requires retraining. Lower layer such as transport layer [11] may also be used for validation. This paper proposes a lightweight data validation based on peak detection.

2. Drone design and measurement method

In order to evaluate the proposed method, a simple IoT device implemented by using ESP8266 module is set up. This wireless sensor collects temperature data and sends it to a server by using a 802.11 network. Figure 2 shows the designed circuit.

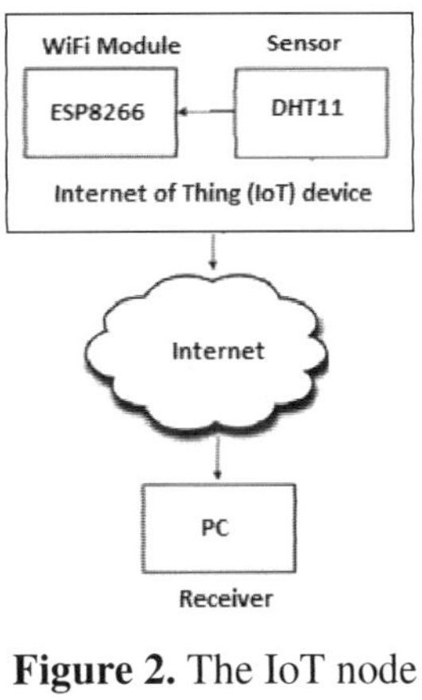

Figure 2. The IoT node

In order to simulate errors, an interfering circuit was designed as in Figure 3. This circuit periodically turns the light-bulb on so that the measured temperature increases as an error. The effectiveness of the method is also measured in term of energy consumption. The current and voltage measurement device is arranged by using arduino as in Figure 4.

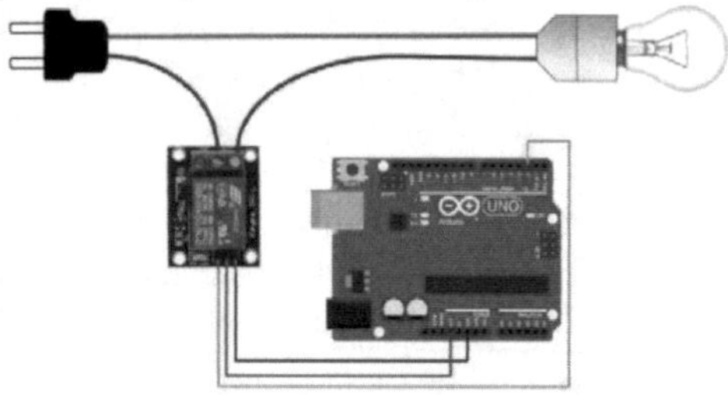

Figure 3. Interfering circuit

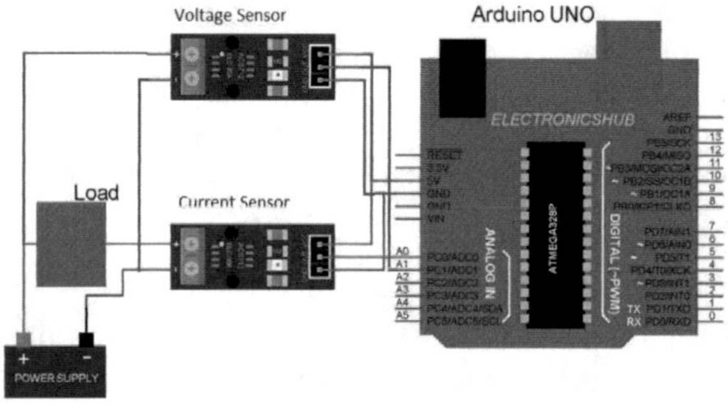

Figure 4. Power consumption device

The interfering unit is set to turn light-bulb in some periods. Light-bulb to sensor distance is also varied to represent degree of disturbances. The proposed validation is by using peak detection. Peak point (Xn) is obtained when it is larger than previous (Xn-1) and next (Xn+1) values as denoted by Equation 1. The outliers are determined by difference, ∂, so that outliers are fulfilling the Equation 2.

$$X_{n-1}<X_n>X_{n+1} \quad [1]$$

$$|X_n - X_{n-1}|>\partial \text{ and } |X_{n+1}-X_n|>\partial \quad [2]$$

3. Assessment results

Fig. 5 shows the non-validated temperature data generated by the IoT sensor with 5% interferences. Data contains outliers as well as errors which are presented by sparking high measurement results and zero values. In server-based validation, this data is validated by removing the outliers and errors. In Server side, the presented data is as depicted in Figure 5 depending upon the validation technique used. This paper used support vector machine classifier in server-based validation.

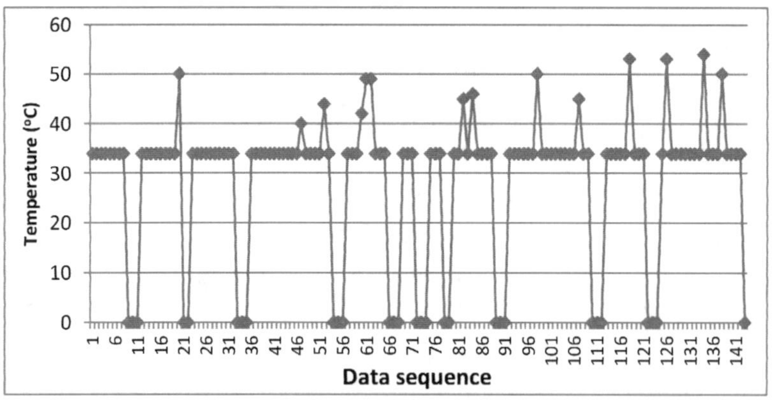

Figure 5. IoT data with 5% interferences

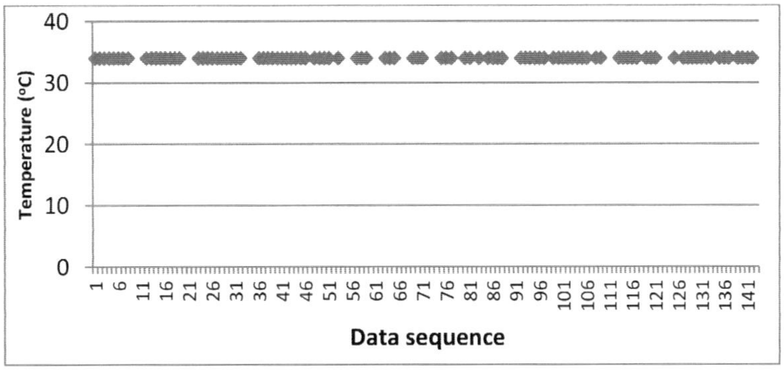

Figure 6. Validated data in server

The validation for the first evaluated data sequence results 100% cleaned data with 43 errors out of 142 data. In term of power consumption, this paper considers IoT sensor only as server is not critical for low power application. Figure 7 shows the IoT sensor power consumption. Power consumption changes overtime with average power consumption per data retrieval is 0.650544 mW and total power consumption for the period of observed data is 93.028 mW.

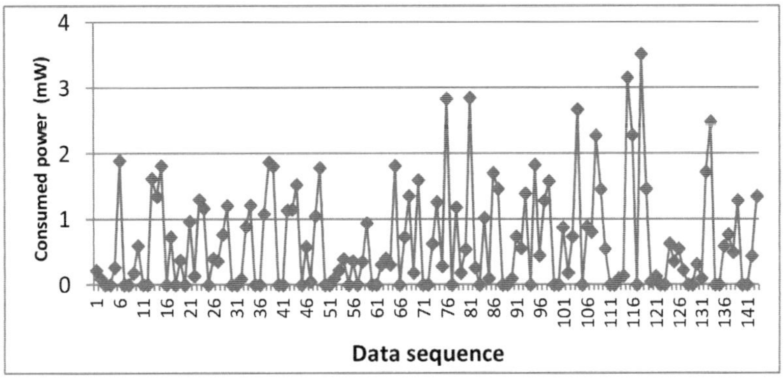

Figure 7. IoT power consumption of server-based validation

Figure 8 shows the transmitted data by IoT sensor when data validation as in Equation 1 applied in IoT sensor. Since at time of assessment, the amplitude of outliers and errors are distinguishable, value δ is set constant 8°C. As results some outliers and errors are not transmitted which expectedly saving more energy on IoT sensor. The advantage of the peak detection validation technique is that the process is simple and requires low power. The proposed method is unable to detect the sequential outliers or errors as sensor will consider them as normal data. Validation on IoT sensor results 86% unsent error. Tthe proposed sensor node validation is able to save power consumption; with average power consumption per data retrieval is 0.642 mW and total power consumption for the observed period is 91.865 mW (Figure 9). This means that IoT sensor with proposed peak detection validation is able to conserve about 1.25% power.

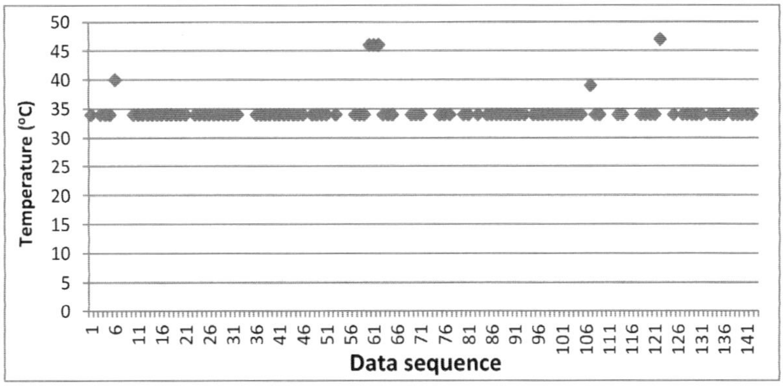

Figure 8. IoT data with node-based validation

IOP Conf. Series: Materials Science and Engineering **851** (2020) 012062 doi:10.1088/1757-899X/851/1/012062

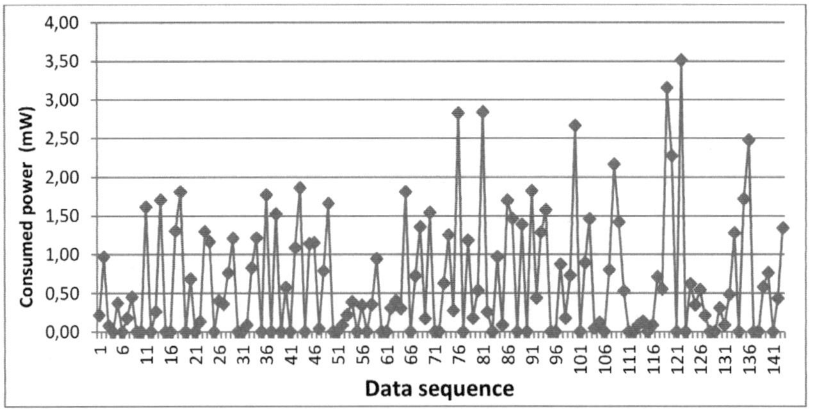

Figure 9. IoT power consumption of node-based validation

In order to enhance validation performance, the combination of node- and server-based validations is suggested, so both low power consumption in IoT sensor and better validation results are achieved.

4. Conclusions

This paper has proposed IoT sensor node data validation by using peak detection method. The assessment was performed in real environment, but harsh circumstances were simulated by using interfering circuit. The node-based validation method was compared to server-based validation that made use of support vector machine. The results showed that node-based validation unsent 86% of error data, which makes 24% lower valid data than server-based validation. However, the proposed method is able to reduce power consumption about 1.25%, from 93.028mW to 91.865mW. Further improvement was gained when both node- and server-based validations were employed.

Acknowledgement

This research has been supported by TALENTA USU research grant 2020.

References

[1] Palattella M R, Accettura N, Vilajosana X, Watteyne T, Grieco L A, Boggia G, and Dohler M/ 2013. Standardized protocol stack for the internet of (important) things, *IEEE Communications Surveys & Tutorials*, 15(3), 1389-1406.

[2] Sheng Z, Mahapatra C, Zhu C, and Leung V C. 2015. Recent advances in industrial wireless sensor networks toward efficient management in IoT, *IEEE Access*, 3, 622-637.

[3] Xiong S, Ni Q, Wang X, and Su Y. 2017. A connectivity enhancement scheme based on link transformation in IoT sensing networks, *IEEE Internet Things J.*, 4(6), 2297–2308

[4] Perera C, Zaslavsky A B, Christen P, and Georgakopoulos D,. 2013. Context aware computing for the Internet of Things: A survey, *IEEE Commun. Surveys Tuts.*, 16(1), 414–454.

[5] Razzaque M A, Milojevic-Jevric M, Palade A, and Clarke S. 2016. Middleware for Internet of Things: A Survey. *IEEE Internet of Things Journal*, 3(1), 70–95. doi:10.1109/jiot.2015.2498900

[6] Diao J, Zhao D, Wang, J, Nguyen H M, Tang, J, and Zhou Z. 2019. Energy-Efficient Boundary Detection of Continuous Objects in IoT Sensing Networks. *IEEE Sensors Journal*, 1–1. doi:10.1109/jsen.2019.2919580

[7] Olatinwo S O, and Joubert T H. 2019. Energy ef?cient solutions in wireless sensor systems for water quality monitoring: A review, *IEEE Sensors J.*, 19(5), 1596–1625

[8] Suherman S, Ulfa Hasnita, Rambe A H, Mubarakah N, Sembiring R. 2019. Data transmission period consideration in the internet of thing node, *International Conf. on Computing and Applied Informatics*.

[9] Li F, Nastic S, and Dustdar S. 2012. Data Quality Observation in Pervasive Environments. *IEEE 15th International Conference on Computational Science and Engineering*. doi:10.1109/iccse.2012.88

[10] Fongen A, and Mancini F. 2015. Integrity attestation in military IoT. *IEEE 2nd World Forum on Internet of Things (WF-IoT)*. doi:10.1109/wf-iot.2015.7389102.
[11] Abdurrahman, H., Sitompul, O. S., and Mubarakah, N. 2017. UDP-Lite Enhancement Through Checksum Protection. In *IOP Conference Series: Materials Science and Engineering* 180(1), 012146). IOP Publishing.
[12] Rahmat R F, Herly E T, Siregar B, Syahputra M F, Sitompul O S 2018 RFID presence monitoring system as an input to measure the workload of employee *4th International Conference on Computer Applications and Information Processing Technology*, pp. 1-6

IOP Conf. Series: Materials Science and Engineering **851** (2020) 012063 doi:10.1088/1757-899X/851/1/012063

Bearing Damage Detection using Support Vector Machine

Sukendi, Ikhwansyah Isranuri, Suherman*

Faculty of Engineering, Universitas Sumatera Utara, Medan, Indonesia

*suherman@usu.ac.id

Abstract. Vibration can be used for preventive condition monitoring in industrial equipment. Vibration measurement may avoid unexpected damages. The measurement data can be analyzed either by using signal processing or machine learning. This work assessed bearing damage detection as bearing is intensively used in industrial mechanical equipment. The assessment was performed experimentally by developing system prototype and measurement tools, including the use of laser vibrometer and a labjack as well as support vector machine to predict some types of damages. The work has been able to predict the BPFI, BPFO, BSF, and FTF damages by 93.125% accuracy. However, accuracy plunged to 61.25% when bearing damages were of three types: BPFO, BSF and FTF.

1. Introduction

An ideal engine should not exert vibration as the mechanical energy should be converted into work. Vibrations show the system imperfection that wasted the energy. Vibration is emerged as a cyclic force of mismatch elements that potentially deforming the surface and breaking the engine. One of ways to detect potential damaging sources is by detecting vibration as part of predictive maintenance [1].

Fast Fourier Transform (FFT) is generally used for extracting vibration signals. Ameid et al [2] used FFT to detect rotor imbalance that produces vibration. The generated spectrum enables imbalance detection as the source of vibration. FFT has also been used for detecting bearing damage [3]. FFT signal analysis was combined with Wigner-Ville distribution and modified poincare mapping to locate bearing failure.

The use of sophisticated sensors such as ultrasonic, eddy current, piezoelectric and strain gage sensors enable live monitoring on the engine vibrations. Some signal processing methods are available to extract signal features to define what happens to the mechanical devices. Further, practical classifier of machine learning can be used for differentiating between damages and goods. Damages were able detected by comparing features of extracted FFT spectrum.

Since failure diagnostic is an important part of condition monitoring to prevent un-predictive failure, the accurate method is still challenging to explore. The progress on machine learning is helpful to be implemented in this matter. Support vector machine (SVM) is the excellent pattern recognition system that is able to analyze features from bulk data in either time or frequency domain. The SVM performances are mainly depending on the feature extraction [4].

Several techniques have been proposed to enhance SVM performance in classifying vibration information. Intrinsic mode function (IMF) for spectrum enveloping has been proposed to find failure spectral line of decomposed modulated signal by using empirical mode decomposition (EMD). The amplitude and frequency of the failure characteristics were then used for SVM input. SVM classifier decided the engine status [5]. Other researcher used IMF envelope sample entropy (SampEn) to diagnose bearing problems [6]. The EMD extract vibration data adaptively into some IMFs. IMFs were sorted and envelops were used to find failure information. The envelop sample entropy (SampEn) prepares features for SVM input.

IOP Conf. Series: Materials Science and Engineering **851** (2020) 012063 doi:10.1088/1757-899X/851/1/012063

Bearing is an important device in industrial equipment. The broken bearing may result the damages on equipment. Local damage in bearing results scratch or holes in inner race (ball pass frequency inner race, BPFI), outer race (ball pass frequency outer race, BPFO), damage in rolling element (ball spin frequency, BSF) or cage damage (fundamental train frequency, FTF) [7]. Those damages should be detected before affecting the overall equipment.

Considering the aforementioned methods, there are two important components for vibration detection to reveal damages: vibration detection and spectrum extraction. Spectrum extractions either by using FFT, signal processing or machine learning are depending to the vibration detection. Since many sensors have been available to detect vibration, this paper proposes laser sensor and machine learning combination to detect vibration and to extract vibration data to determine bearing damages. The proposed method was examined by developing a mechanical prototype, setting up sensors and applying support vector machine.

2. Experiment design

The bearing evaluated in this paper is NTN UCP 204 D1, with specifications plotted in Table 1. Bearings were tested under conditions of new and damage. The bearings were examined for rotation speed of 400 to 1200 rpm, outer race damage, inner race damage, rolling element damage, and cage damage.

Table 1. Bearing specifications

No	Item	Size
1	Number of balls (Nb)	8
2	Diameter (Bd)	7.94 mm
3	Pitch Diameter (Pd)	33.48 mm
4	Contact angle (α)	0
5	Relative frequency (Fr)	23 Hz

The observed system was a prototype as shown in Figure 1. Table was made from metal with size of 500x500x500 mm, two poles with diameter of 20 mm and length of 300 mm and 500 mm were placed on table top. Four units of pillow block NTN UCP 204 were employed with mild steel holders; two spur gears of 33 and 2.5 were attached to the poles. Two 60 mm type-A V-pulleys and a V-belt A39 connects the mechanics to a 3-phase, 1380 rpm, 0.75 kW LM-Motor, which was supplied by 1 HP 3 phase inverter.

Figure 1. The mechanical prototype

The bearing assessment set up is shown in Figure 2. Bearings were assessed with speeds and poles axis varied. Vibration was read by using a laser vibration photo sensor, interfaced to a computer by using a labjack that converts analog measurement signals into digital form readable for computer. The vibrometer was located 376 mm from bearing. The laser was pointed horizontally, vertically and axial. Data were taken for various bearing conditions for speed of 400, 600, 800, 1000, and 1200 rpm. The extracted features covered mean, median, modus, root mean square, standard deviation, skewness, kurtosis, and beta kurtosis.

IOP Conf. Series: Materials Science and Engineering **851** (2020) 012063 doi:10.1088/1757-899X/851/1/012063

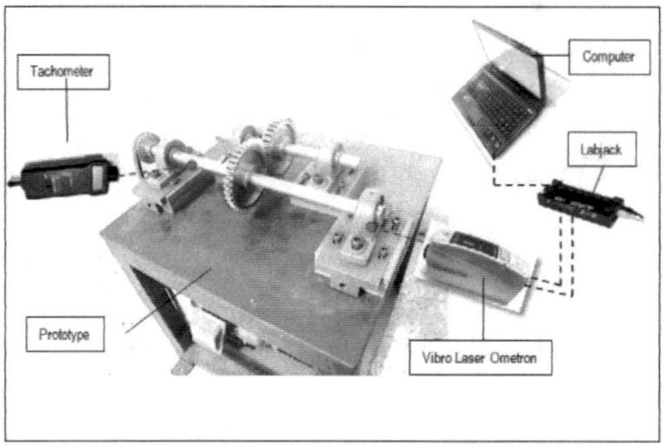

Figure 2. The assessment set up

3. Assessment results

Vibration measurements were performed in base tread, base plat and in bearings. Vibrations in four base treads and base plats have low number of harmonics. Vibration in bearing is much larger than the two other places. Figure 3 shown time domain vibration sample for broken bearing that reach amplitude up to 26 μm.

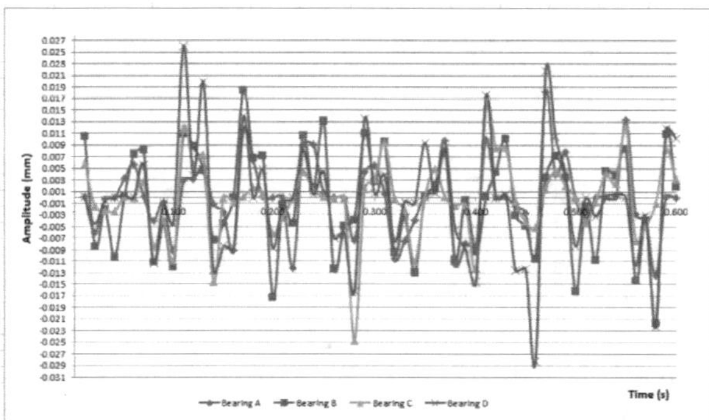

(a) Horizontal vibrations

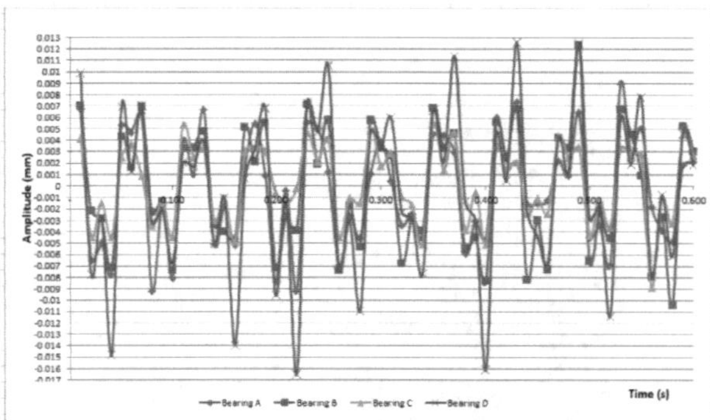

(b) Vertical vibrations

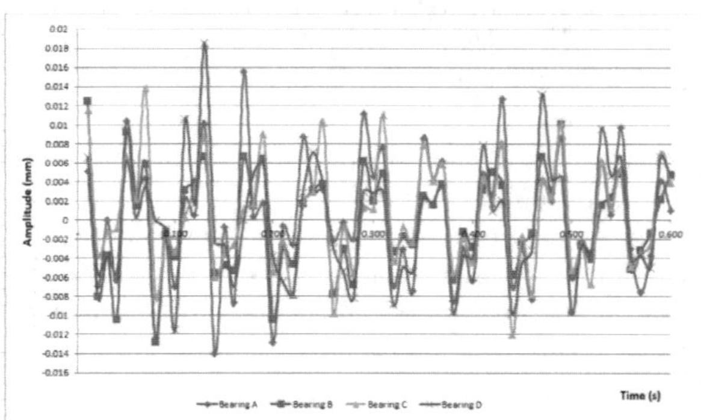

(c) Axial vibrations

Figure 3. Broken bearing vibrations

IOP Conf. Series: Materials Science and Engineering **851** (2020) 012063 doi:10.1088/1757-899X/851/1/012063

From the measured data, mean as the average value of data was calculated by using Equation 1. The root mean square was using Equation 2.

$$\mu_j = \frac{1}{n}\sum_{i=1}^{n} x_{ij} \tag{1}$$

$$RMS = \sqrt{\frac{1}{n}\sum_{n} x^2(t)} \tag{2}$$

Skewness (y) as the symmetric value around samples was calculated by using Equation 3. The sharpness of data distribution (kurtosis, K) and the forth moment of beta function (beta kurtosis, BK) were using Equation 4 and 5.

$$y = \frac{E(x-\mu)^3}{\sigma^3} \tag{3}$$

$$K = \frac{M_4}{\sigma^4} \tag{4}$$

$$BK = \frac{m_4}{\left(\sigma^2\right)^2} \tag{5}$$

In addition to the aforementioned features, additional vibration differentials were added. Displacement was calculated by using Equation 6 and acceleration used Equation 7. Velocity was obtained from measurement, while the angle speed (ω) was calculated based on rpm (n) (Equation 8).

$$X - \int \dot{X}(t) - A\sin\omega t \tag{6}$$

$$\ddot{X} = \frac{d\dot{X}}{dt} \tag{7}$$

$$\omega = 2\pi n/60 \tag{8}$$

The horizontal, vertical and axial vibrations generated by new bearings as in Figure 4 were made as training data for the support vector machine. The new bearing shows steady vibration with maximum amplitude of 3.2 μm.

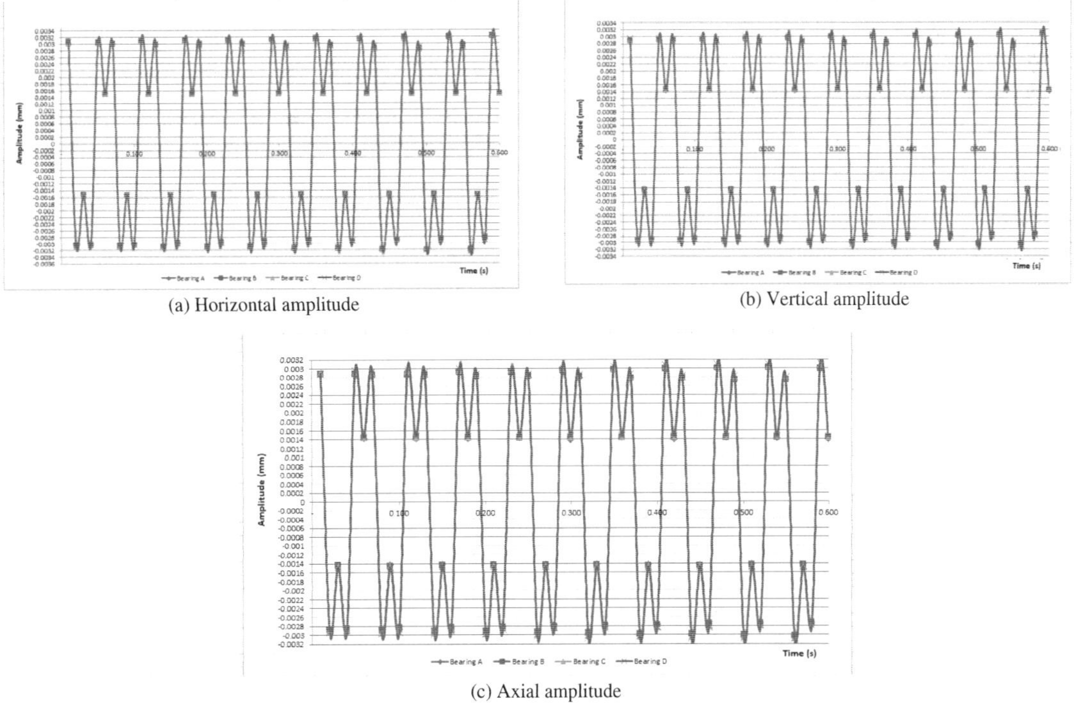

(a) Horizontal amplitude

(b) Vertical amplitude

(c) Axial amplitude

Figure 4. New bearing vibrations

The SVM was used to determine either bearing is new or broken. The predicted damages include the BPFI, BPFO, BSF and FTF. Figure 5 shows SVM accuracy in predicting the condition of bearing. The average accuracy is 93.125%. The lowest accuracy occurred when bearing experiences three types of damages. Bearing 5 has damages of BPFO, BSF and FTF, so that SVM predicted less accurate, only 61.25%.

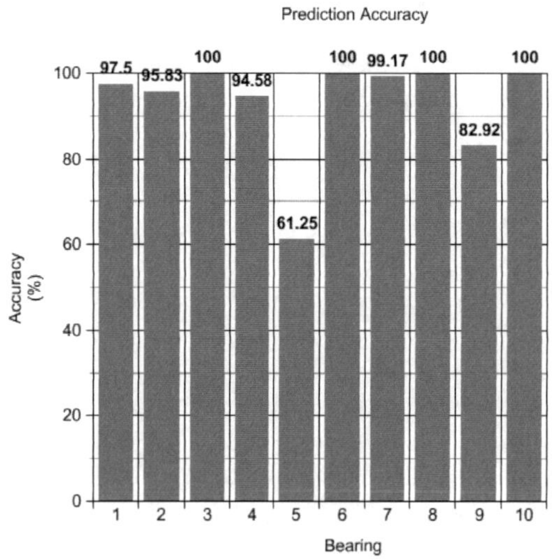

Figure 5. SVM prediction

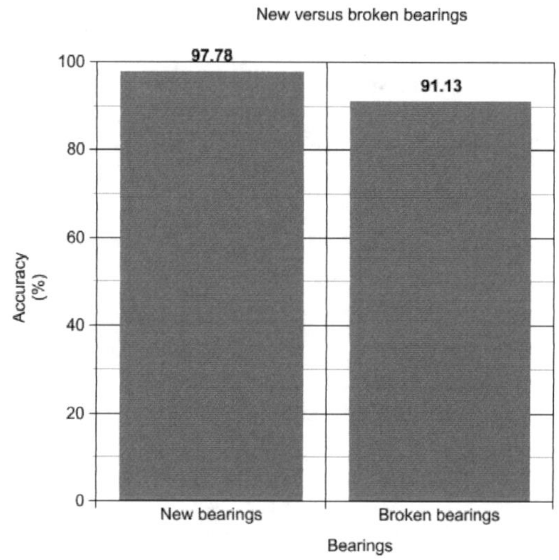

Figure 6. Bearing condition

The SVM performed best when predicting the new bearings as the features were likely uniform. SVM gained 97.78% accuracy. The broken bearing was rather difficult to predict as the positions of scratches or holes can be minor or major. It reached 6.65% lower accuracy, about 91.13% as shown in Figure 6.

4. Conclusions

This paper has discussed the use of laser vibrometer and SVM to detect damages on bearings: BPFI, BPFO, BSF, and FTF. Vibrometer along with labjack is able to detect vibration in order of μm through horizontal, vertical and axial direction of vibrations. SVM produces in average 93.125% accuracy. The lowest accuracy occurred when there were multiple types of damages. In order to enhance prediction, characteristics combination of damage types should be explored and new features should be added into classifier.

References

[1] Jung D, Zhang Z, and Winslett M. 2017. Vibration analysis for iot enabled predictive maintenance. *In IEEE 33rd International Conference on Data Engineering (ICDE)* 1271-1282

[2] Ameid T, Menacer A., Talhaoui H, and Harzelli I. 2017. Broken rotor bar fault diagnosis using fast Fourier transform applied to field-oriented control induction machine: simulation and experimental study. *The International Journal of Advanced Manufacturing Technology*, 92(1-4), 917-928.

[3] Singru P, Krishnakumar V, Natarajan D, and Raizada A. 2018. Bearing failure prediction using Wigner-Ville distribution, modified Poincare mapping and fast Fourier transform. *Journal of Vibroengineering*, 20(1), 127-137.

[4] Fuqing Y. 2011. Failure diagnostics using support vector machine (Doctoral dissertation, Luleå tekniska universitet).

[5] Suresh S, and Naidu V P S. 2019. Vibration Analysis of Heterogeneous Gearbox Faults using EMD Features and SVM Classifier. *In IOP Conference Series: Materials Science and Engineering*, 624 (1), 012032. IOP Publishing.

[6] Xie Z Q, Sun H E, Liu L, and Wu C. 2017. Fault Diagnosis of Rolling Bearing Based on CEEMDAN Sample Entropy and SVM. *Modular Machine Tool & Automatic Manufacturing Technique*, 3 (25).

[7] Chen Y, Peng G, Xie C, Zhang W, Li C, and Liu S. 2018. ACDIN: Bridging the gap between artificial and real bearing damages for bearing fault diagnosis. *Neurocomputing*, 294, 61-71.

IOP Conf. Series: Materials Science and Engineering **851** (2020) 012064 doi:10.1088/1757-899X/851/1/012064

Prediction of Regional Revenue and Expenditure Budget using Autoregressive Integrated Moving Average (ARIMA)

R F Rahmat[1], A H Syaputra[1], S Faza[2], and D Arisandi[1]

[1] Department of Information Technology, Faculty of Computer Science and Information Technology, Universitas Sumatera Utara, Medan, Indonesia

[2]Politeknik Negeri Medan, Medan, Indonesia

Email: romi.fadillah@usu.ac.id

Abstract. The Regional Revenue and Expenditure Budget (APBD in Indonesia) is a financial plan stipulated by Regional Regulations. The APBD of the North Sumatra Provincial Government is the reflection of development direction in the Province of North Sumatra since it would not be possible to execute the regional development process without being supported by adequate cash. Therefore, this study aims to predict the APBD allocation in 5 types of accounts, namely Regional Original Revenues, Balancing Funds, Other Legal Regional Revenues, Indirect Expenditures, and Direct Expenditures using time series data from 2002 to 2019. The proposed method in this study is Autoregressive Integrated Moving Average, which is a forecasting method using the time series data in the form of current and past data of the APBD. The results showed that the forecasting of the 2020 APBD value for the Regional Original Revenue account is IDR 6,989,217 Million, for the Balance Fund account is worth IDR 8045.259 Million, for Other Legitimate Income is at IDR 317,546 Million, for Indirect Shopping is worth IDR 9,308,607 Million and for Direct Shopping at IDR 5,495,597 Million.

1. Introduction

The Regional Revenue and Expenditure Budget (APBD in Indonesia) is the annual financial planning of the regional area. APBD consists of three groups, namely: Regional Revenue, Regional Expenditure, and Regional Financing. Regional Revenue is the regional government's right which is recognized as an income to the net worth, Regional Expenditure is a regional government obligation which is recognized as an expense in net worth, and Regional Financing is all revenue that needs to be paid back and/or expenditure to be received again, both on the ongoing fiscal year and in subsequent fiscal years [1].

The North Sumatra Provincial Government's APBD is a real picture of the development directions in North Sumatra. If Regional Revenue and Regional Expenditure are to increase proportionally each year, then it is a sign that the region is in the process of developing for the better. On the contrary, if the expenditure is greater than income and the numbers are worsening every year, it can be ascertained that the region is in bad condition. The reason is that it would be impossible to have a developed area with insufficient balance for development. Therefore, it is necessary to have prediction system to forecast the APBD in the coming year so that decision-makers can make more sensible policies related to the management of the upcoming APBD.

The Regional Revenue and Expenditure Budget (APBD) is the main determinant of regional development, the higher the income of an area, the more the amount of expenditure that can be spent for the infrastructure and other regional development activities. Therefore we need an approach to predict the value of regional budgets so that policymakers can have more precise objectives in planning the financial budgeting. The proposed method for this issue is the ARIMA method.

Numerous studies related to data prediction have been performed. One of which is Hafilah Hamimi (2014), by conducting research on data analysis of regional expenditure income using _6565 clustering k-means and its predictions. In the study, clustering data was performed using the k-means algorithm, and predictions were executed using multiple linear regression. The data used in this study were the Regional Budget data of Padang City in the period 2009-2013 (5 years) and was able to produce 3 clusters of 4 types of data used. The result of the predictions was that the Padang city area should be spending APBD worth of IDR 2,352,908.8 Million in the following year [2].

Subsequent research was conducted by Rahayu Mayang Sari (2015). The study discussed the prediction of regional budget revenue using the k-means algorithm. It has not much different from the previous study since it was also implemented k-means for the data clustering. The clustering process in this study resulted in 2 clusters each in the regional income and expenditure data. For the regional income, the first cluster of regional revenue obtained one substantial income from 13 data; the second one obtained 12 small income. As for the expenditure, the first 1616 clusters obtained six large expenditures from 10 types of expense, and the remaining 689 clusters obtained four small expenditures from 10 types of shopping. In addition, this study was also able to predict data in the following year, using only the previous year of data [3].

Another research was conducted by Liu Qingjie & Wang Huachun (2015). They performed a research to predict the development of income and financial expenditure in China using ARIMA. The ARIMA model processed 64 years of data (1950-2013), and the model (2,1,6) was obtained as the best model. From the generated model, the data prediction was created based on the income date in the last five years. The results indicated that there should be an increase in allocation from 9.93% to 9.97% on China's state finances in the following year [4].

CJ Ketaren performed a study to predict fuel needs using _6565 ARIMA. The data were the monthly data of PT. PERTAMINA UPMS -1 Medan from January 2011 to September 2015 with three types of data consisting Premium, Kerosene, and Solar. From these data, the result showed that the best type of premium model ARIMA was (0,1,1), for kerosene data was (2,1,0), and for solar was (0,1,1). Standard errors of the data were 0.10106 for premium, 0.1283 for kerosene, and 0.1109 for solar [5].

Another research was carried out by Almasarweh Muhammad & Wadi S. AL (2018). The authors conducted research to predict banking stock data in Jordan using _666 ARIMA. The data were acquired from the Amman stock market in the period 1993 to 2017. From this study, it was concluded that the best ARIMA model was (1,1,2) with a standard error of 1.4 [6]. Besides ARIMA there is other methods also suitable for this kind of problem such as DENFIS, or ANFIS [7]

2. Methodology

The proposed method in predicting the APBD consists of several processes. The processes carried out in this study were as follows: Data collection was carried out from the dashboard on the Ministry of Finance website and from the North Sumatra Province Financial and Asset Management Agency, then the data will go through the pre-processing using _666, after which the ARIMA model was formed. Based on the new model, a prediction was made for the following year's fund allocation. The system produced output in the form of a predicted value of the following year's financial planning. The general architecture of this study can be seen in Figure 1.

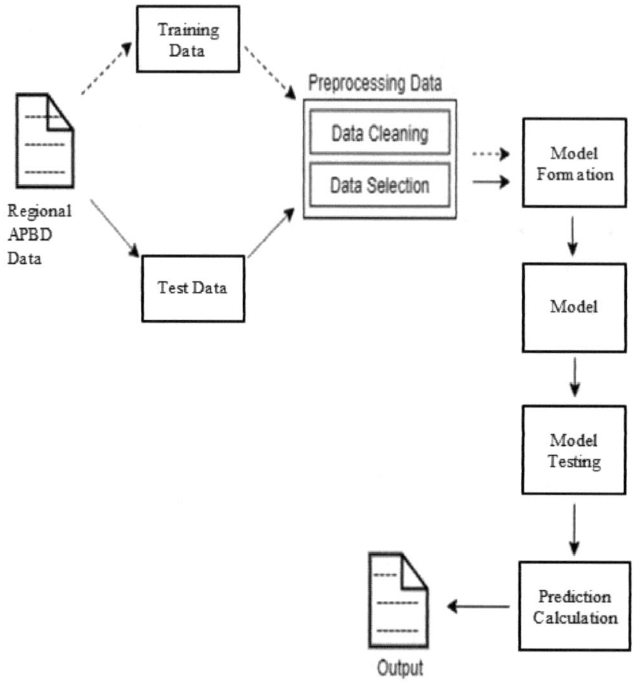

Figure 1. General architecture

2.1. Data Acquisition

The data were in the form of North Sumatra Province's APBD (Regional Revenue and Expenditure Budget) data from the period of 2002 to 2019. They were acquired from the Ministry of Finance data recap via their website and from the Regional Financial and Asset Management Agency of North Sumatra Province. The data used were in .csv format.

2.2. Pre-processing

Pre-processing is a stage in image processing to validate the data by performing data cleaning and selection.

Data cleaning is a process of checking data and deleting irrelevant and inconsistent data to increase the accuracy in the formation of data prediction models. The raw data can be seen in Figure 2.

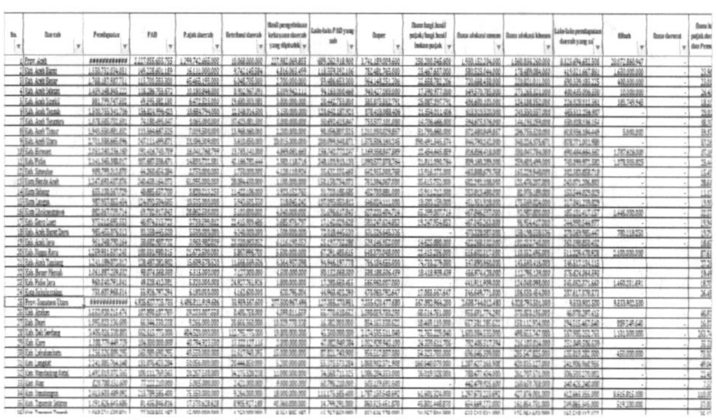

Figure 2. Data before the cleaning process

IOP Conf. Series: Materials Science and Engineering **851** (2020) 012064 doi:10.1088/1757-899X/851/1/012064

In this research, due to the process of forming the model and prediction performed at account level 2, then the field – field, which is a derivative of account level 2, will be deleted, such as field _749 regional income, regional retribution, and many more. The final result of this process in the form of recapitulation of North Sumatra Province APBD data is shown in Table 1.

Table 1. Result of data cleaning

Year	Local Original Revenue	Balancing Funds	Others Legitimate Income	Indirect Shopping	Direct Shopping
2002	440591	366644	17760	616423	355852
2003	621017	424955	0	517237	644795
2004	874268	485666	0	577225	863013
2005	1026891	475583	0	726859	919417
Etc …					

Data selection is the process of selecting data from the processed dataset. This process aims to ease the prediction process since the data have been sorted. The results of the data selection can be seen in Table 2.

Table 2. Result of data selection

Year	Local Original Revenue
2002	440591
2003	621017
2004	874268
2005	1026891
2006	1377138
2007	1502955
2008	1897497
2009	2104203
2010	2226498
2011	3181900
2012	4026427
2013	4809369
2014	4944502
2015	5257837
2016	4630468
2017	4925627
2018	5732425
2019	7583849

2.3. ARIMA model

IOP Conf. Series: Materials Science and Engineering **851** (2020) 012064 doi:10.1088/1757-899X/851/1/012064

ARIMA has parameters in the form of 3 integers, where these parameters will then become a model of ARIMA itself. Therefore the ARIMA model is denoted by the ARIMA notation of (p, d, q).

p is part of ARIMA's autoregressive, which allows the model to include past data patterns to predict future values. For example, in this study, PAD data from 2002 to 2009 continued to increase, then it was more likely the value of PAD in 2010 to also increasing. The general form of the $AR_{(p)}$ model is as follows:

$$X_t = \theta_1 X_{t-1} + \cdots + \theta_p X_{tp} + e_t$$

Where:

X_t	: variable value at the t-time
X_t , X_{t-1}..., X_{tp}	: past values of the determined time series at time of t, t-1, t-2, ..., tp
θ_1	: regression coefficient, i: 1, 2, 3, ..., p
e_t	: error value at t-time

d is the integrated part which will determine the number of differences. For example, if the number of values in the previous three years has a small difference, then d will assume the value in the following year will be the same as the previous year.

q is part of Moving Average, which will calculate the average movement in the data. The model of q can be declared as follows:

$$Xt = e_t + \theta_1 e_{t-1} + \theta_2 e_{t-2} + \theta_3 e_{t-3} + \ldots + \theta_q e_{tq}$$

Where:

X_t	: variable value at the t-time
e_t , e_{t-1}, e_{t-2}, ... e_{tq}	: error value at time t, t-1, t-2, .. ., tq
θ_1	: regression coefficient, i: 1, 2, 3, ..., p
q	: order of MA

One good indicator of a model ARIMA is when the model has a small value of _6767 AIC, so in this study, authors excluded all possible combinations of p, d, q and chose the combination with the smallest value of AIC .the sample of determining the value of AIC can be seen in Figure 3.

```
Output

SAR[Output] 0, 0)x(0, 0, 1, 12) - AIC:6787.3436240402125
SAR[...] 0, 0)x(0, 1, 1, 12) - AIC:1596.711172764114
SARIMAX(0, 0, 0)x(1, 0, 0, 12) - AIC:1058.9388921320026
SARIMAX(0, 0, 0)x(1, 0, 1, 12) - AIC:1056.2878315690562
SARIMAX(0, 0, 0)x(1, 1, 0, 12) - AIC:1361.6578978064144
SARIMAX(0, 0, 0)x(1, 1, 1, 12) - AIC:1044.7647912940095
...
...
...
SARIMAX(1, 1, 1)x(1, 0, 0, 12) - AIC:576.8647112294245
SARIMAX(1, 1, 1)x(1, 0, 1, 12) - AIC:327.9049123596742
SARIMAX(1, 1, 1)x(1, 1, 0, 12) - AIC:444.12436865161305
SARIMAX(1, 1, 1)x(1, 1, 1, 12) - AIC:277.7801413828764
```

Figure 3. List of p, d, q along with its AIC values

2.4. Model

The obtained model was then displayed in the form of a chart that is juxtaposed with the actual data graph to see the comparison of the generated data.

2.5. VAR Model Testing

In this study, the model testing was performed in 2 ways, one of which is to use the library calculation of python to get the value of standard error against the resulting prediction, as shown in Figure 4.

	coef	std err
ar.L1	0.9921	0.066
ma.L1	-0.9140	0.511
sigma2	2.242e+11	2.54e-12

Figure 4. Result of standard error

2.6. Prediction Calculation

At this stage, a prediction was calculated to determine the following year data for each type of account, as per the generated model. The prediction calculation used the following equation.

$$X_{-t+1} = X_{-t} + X_{-t-11}X_{-t-12} + e_{t+1} - \Theta_1 e_t - \Theta_1 e_{t-11} + \Theta_1\Theta_1 e_{t-12}$$

3. Result and Discussion

The model creation was performed in 1616 looping of the 16679 model of ARIMA 06689 (p, d, q) to find the smallest value of 06999 AIC. Table 3 shows the model for each account type along with the AIC value.

Table 3. Models and AIC

	Local Original Revenue	Balancing Funds	Others Legitimate Income	Indirect Shopping	Direct Shopping
ARIMA Model	1 , 3 , 0	3 , 3 , 1	0 , 2 , 1	3 , 3 , 2	1 , 3 , 1
AIC value	442,278	474,676	491,46	459,625	444,639

The tests were carried out based on the obtained model for each account type to predict the future value of each account. The test also calculated the standard error of each prediction result, where the smaller the standard error, the better the prediction will be. The generated prediction result, along with its standard error values for each account is as follows:

3.1. Test result of local revenue

From the model ARIMA (1, 3, 0) obtained a predictive value of IDR 10,102,217 Million, with a standard error _666 of 6.61e -15, or around 0.00000000000000661, as shown in Figure 5.

```
                          Statespace Model Results
==========================================================================================
Dep. Variable:      Pendapatan Asli Daerah   No. Observations:                   18
Model:                SARIMAX(1, 3, 0)     Log Likelihood                -219.364
Date:                 Wed, 18 Dec 2019     AIC                            442.729
Time:                         09:17:23     BIC                            444.145
Sample:                     01-01-2002     HQIC                           442.714
                          - 01-01-2019
Covariance Type:                   opg
==========================================================================================
                 coef    std err          z      P>|z|      [0.025      0.975]
------------------------------------------------------------------------------------------
ar.L1         -0.7086      0.141     -5.036      0.000      -0.984      -0.433
sigma2      3.152e+11   6.61e-15   4.77e+25      0.000    3.15e+11    3.15e+11
==========================================================================================
Ljung-Box (Q):                      nan   Jarque-Bera (JB):                   1.07
Prob(Q):                            nan   Prob(JB):                           0.58
Heteroskedasticity (H):           19.38   Skew:                               0.50
Prob(H) (two-sided):               0.01   Kurtosis:                           2.15
==========================================================================================
```

Figure 5. Standard error of local revenue

3.2. Test result of balanced fund

Based on the ARIMA model of (3, 3, 1), a prediction value of balance fund worth IDR 8,860,877 Million, with standard error _666 of 8.49e -12, or around 0.0000000000000849 as shown in Figure 6.

```
                          Statespace Model Results
==========================================================================================
Dep. Variable:        Dana Perimbangan   No. Observations:                   18
Model:                SARIMAX(3, 3, 1)   Log Likelihood                -232.338
Date:                 Wed, 18 Dec 2019   AIC                            474.676
Time:                         09:54:11   BIC                            478.216
Sample:                     01-01-2002   HQIC                           474.638
                          - 01-01-2019
Covariance Type:                   opg
==========================================================================================
                 coef    std err          z      P>|z|      [0.025      0.975]
------------------------------------------------------------------------------------------
ar.L1         -1.0008      1.529     -0.655      0.513      -3.997       1.996
ar.L2         -1.0007      3.241     -0.309      0.757      -7.353       5.352
ar.L3         -0.9999      3.355     -0.298      0.766      -7.576       5.576
ma.L1         -0.8177      1.792     -0.456      0.648      -4.330       2.695
sigma2      1.977e+12   8.49e-12   2.33e+23      0.000    1.98e+12    1.98e+12
==========================================================================================
Ljung-Box (Q):                      nan   Jarque-Bera (JB):                  78.88
Prob(Q):                            nan   Prob(JB):                           0.00
Heteroskedasticity (H):          559.79   Skew:                               3.25
Prob(H) (two-sided):               0.00   Kurtosis:                          12.16
==========================================================================================
```

Figure 6. Standard error of balance fund

3.3. Test result of other legitimate revenue

From the model ARIMA (3, 3, 1) obtained, a predictive value of IDR 10,102,217 Million with a standard error _666 of 6.61e -15, or around 0.00000000000000000661, as shown in Figure 7.

```
                          Statespace Model Results
==========================================================================================
Dep. Variable:   Lain-lain Pendapatan yang Sah   No. Observations:                   18
Model:                SARIMAX(0, 2, 1)   Log Likelihood                -243.730
Date:                 Wed, 18 Dec 2019   AIC                            491.460
Time:                         10:06:14   BIC                            493.005
Sample:                     01-01-2002   HQIC                           491.530
                          - 01-01-2019
Covariance Type:                   opg
==========================================================================================
                 coef    std err          z      P>|z|      [0.025      0.975]
------------------------------------------------------------------------------------------
ma.L1         -0.9802      0.194     -5.042      0.000      -1.361      -0.599
sigma2      9.937e+11   [illegible]   5.23e+24      0.000    9.94e+11    9.94e+11
==========================================================================================
Ljung-Box (Q):                      nan   Jarque-Bera (JB):                  27.40
Prob(Q):                            nan   Prob(JB):                           0.00
Heteroskedasticity (H):         7025.08   Skew:                              -1.79
Prob(H) (two-sided):               0.00   Kurtosis:                           8.33
==========================================================================================
```

Figure 7. Standard error of other legitimate revenue

3.4. Test result of indirect shopping

As per the ARIMA model of (3, 3, 2), the indirect shopping prediction was worth IDR 10,102,217 Million with the standard error _666 of 6.61e -15, or around 0.0000000000000000661 as shown in Figure 8.

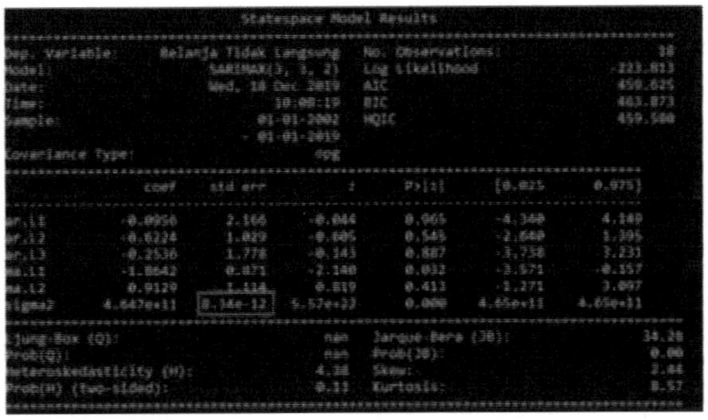

Figure 8. Standard error of indirect shopping

3.5. Test result of direct shopping

According to the ARIMA model of (3, 3, 1), the amount of direct shopping allocation should be at IDR 10,102,597 Million with the standard error _666 of 6.61e -12 or around 0.0000000000000000661 as shown in Figure 9.

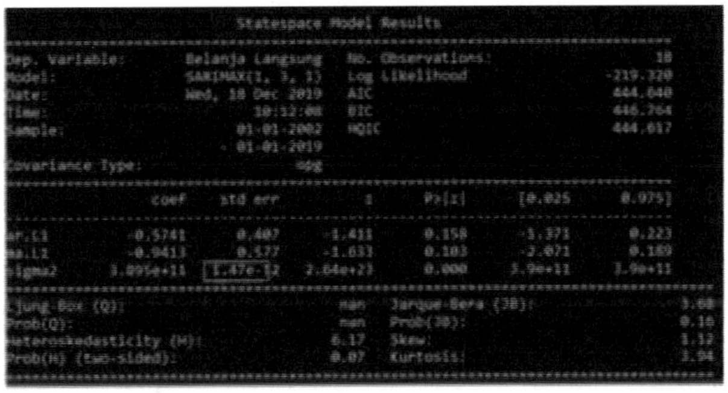

Figure 9. Standard error of direct shopping

4. Conclusion

Based on the test results, The ARIMA method was proven to be able to predict the future value of income and expenditure accounts in the North Sumatra Province Regional Budget. The prediction results were based on the ARIMA model with the smallest value AIC and could produce good predictive values with small standard errors.

References

[1] Government of the Republic of Indonesia. 2006 Minister of Home Affairs Regulation No. 13 of 2006 concerning Guidelines for Regional Financial Management.

[2] Hamimi H, 2014 Analysis of Regional Expenditure Budget Data Using Clustering K-Means and Forecasting. Padang: Universitas Negeri Padang.

[3] Sari R M 2015 Local Budget Revenue Prediction Data Using K-Means Algorithm. Riau: STMIK AMIK Riau.

[4] Liu Q and Huachun W, 2015 Research on the Forecast and Development of China's Public Fiscal Revenue Based on the ARIMA Model. *Theoretical Economics Letters*, Vol 5:482-493.

[5] Ketaren C J 2017 Prediction of Fuel Needs Using the ARIMA Method in Pertamina UPMS-1 Medan. Medan: Universitas Sumatera Utara.

[6] Almasarweh M and Alwadi S 2018 ARIMA Model in Predicting Banking Stock Market Data. *Modern Applied Science*.

[7] Rahmat R F, Nurmawan S, Sembiring S, Syahputra M F, Fadli 2018 Adaptive neuro-fuzzy inference system for forecasting rubber milk production *IOP Conference Series: Materials Science and Engineering* **308**(1),012014

IOP Conf. Series: Materials Science and Engineering **851** (2020) 012065 doi:10.1088/1757-899X/851/1/012065

Facial Recognition-Based Automatic Door Access System Using Extreme Learning Machine

R F Rahmat[1], E N Zai[1], I Fawwaz[2] and I Aulia[1]

[1] Department of Information Technology, Faculty of Computer Science and Information Technology, Universitas Sumatera Utara, Medan, Indonesia

[2] Universitas Prima Indonesia, Medan, Indonesia

Email: romi.fadillah@usu.ac.id

Abstract. Facial recognition is one of the best forms of security since it is biometric-based security that uses biological features of the face. One of the artificial neural networks that can be implemented in the face recognition case is Extreme Learning Machine with the assistance of Local Binary Pattern in obtaining the facial features. The testing process in this study used 150 face images for training data and 60 face images for test data. The test was carried out using several three parameters of the hidden neuron numbers, namely 10, 30, and 50, and also five parameters of the image condition, namely feature, expression, face direction, lighting, and webcam distance. The system achieved the highest accuracy of 90% using 50 hidden neurons.

1. Introduction

Security is crucial for daily activities, especially when it comes to authority or privacy. Conventional security systems require keys, passwords, RFID cards, or ID cards to have access. The disadvantages of these methods are they are either hard to remember, can be duplicated, or stolen by others. Therefore, we need a way to improve the security of certain rooms involving authority or privacy. One of the security forms that is difficult to imitate, open, or modify by others is to use the automatic characterization of biological characteristics that are possessed by each individual. This characterization is known as biometrics. These characteristics can be seen from physical characteristics, such as fingerprints, facial features, eye retina, and voice. Facial recognition is one of the biometric methods that is quite popular. Faces tend to difficult to emulate, modify, or steal compared to keys, passwords, or non-biometric security. Therefore, a face recognition system to access a door for a particular room is necessary in effort to increasing the security. Authors proposed Extreme Learning Machine as the method to allow the system to perform face identification.

Several previous studies related to facial recognition for door security have been conducted. Januzaj et al. used a haar-like feature algorithm to detect faces, and PCA to recognize faces. The research succeeded in making a tool that was able to carry out the training process and face identification directly, without being connected to other equipment. The results indicated that the accuracy could be improved by considering some factors such as background, lighting, and the amount of training data [1].

Another study was conducted by Lwin et al. The face detection and recognition processes were performed on a PC with Matlab. Microcontroller PIC 16F887 was used to control the door access system, which depends on the incoming data sent from the PC. The door will open automatically immediately once the face is successfully verified. After 2 seconds, the door will close automatically. The viola-jones method was implemented to detect faces in an image. This method has limits on head orientation. This method can only detect faces with foresight. Meanwhile, PCA method was applied to identify the face. The PCA method performed feature extraction on the face images and then used euclidian distances to recognize images [2].

Another research had also been carried out by Yang et al. to create an automatic door control system using a DSP-based multimedia module. In the study, the system was made to identify people through face detection, and then analyzed the trajectory path to determine whether the person has the intention to access the door or not. This was what controls the door. The system has a low error rate (up to 0%), a high correct activation rate of 99.6%, and a fast response time (within 2 seconds) in detecting a target, intention confirmation, and opening the door [3].

2. Methodology

In this study, the input data was acquired using a webcam connected to a PC; then, the data was processed for facial recognition. The connected microcontroller and the servo motor would receive a signal in the form of 1 or 0, depending on the output of the processing results on the PC, and the servo motor would move the door.

The webcam is a VGA camera with 0.3-megapixel resolution connected to the laptop with the capability of capturing images with a maximum dimension of 640 x 480. The microcontroller was a microcontroller type Arduino Uno (ATMega328), and the servo motor was the Tower Pro SG90 micro servo.

The face recognition process began with the collection of datasets for training and testing. The process continued to pre-processing to adjust the size of the entire image data to a size of 100x100, and convert those images to grayscale forms. After grayscale images obtained, the images would go through the feature extraction using LBP to get the characteristic value of each image. The feature extraction results (59 feature values) of the training image were stored as a comparison of the classification process using ELM with the feature extraction results from the test image. The output of the classification results in the recognition process would serve as the reference to control the door automatically.

This facial recognition program was created using Matlab r2016a. The training process (training) was conducted manually and simultaneously to all trained image data, while the testing process was performed directly through the program interface. The general architecture of this study can be seen in Figure 1.

2.1. Face Detection

At this stage, the input data was the captured image from the webcam. The aim is to detect whether there were faces in the input data.

Input : Image Data
Method : Viola-Jones
Output : Facial Image Data

2.2. Pre-processing

Pre-processing is a stage in image processing to produce a better image quality. In this research, there are two steps to this stage, which are resizing and grayscaling.

Resizing aims to obtain the same size images for the input data. The specified size for the image in this study is 100 x 100 pixels.

Input : Facial Image Data
Method : Bicubic interpolation
Output : Facial Image Data (100x100px)

Grayscaling is a process of converting RGB images to grayscaled ones.

Input : Facial Image Data (100x100px)

Method : Grayscaling
Output : Grayscaled Facial Image (100x100px)

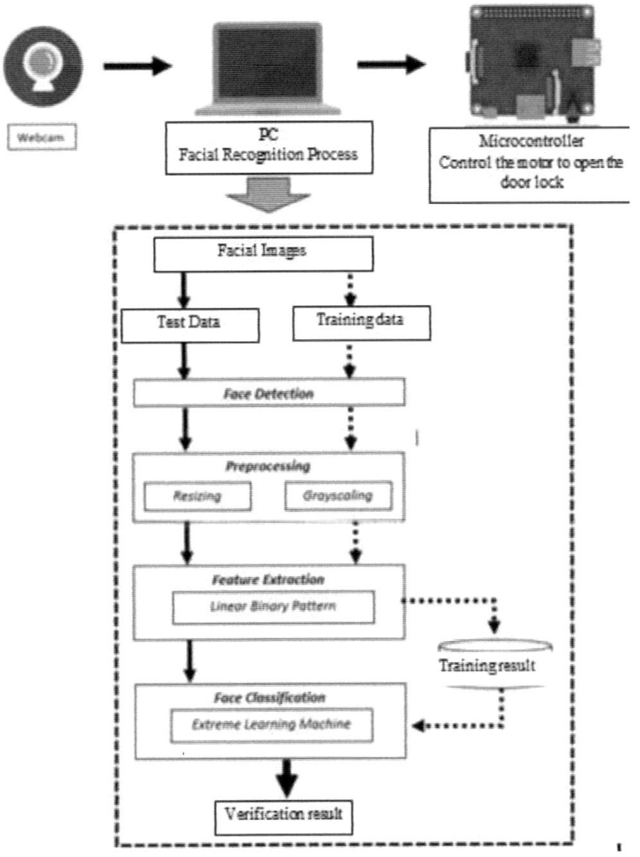

Figure 1. General architecture

2.3. Feature Extraction

At this stage, the data will be processed by extracting the features in the data. The proposed method was LBP (Local Binary Pattern). LBP is defined as a measure of invariant grayscale texture, derived from the general definition of texture in the surrounding area. LBP operators can be seen as a unitary approach with traditional statistical and structural models different from texture analysis. Simply stated, LBP is a binary code that describes local texture patterns. It is built with a limited environment of a gray value from its center [4]. The result of feature extraction per face image data is 59 feature values. These values will serve as the presentation of the image data.

Input : Grayscaled Facial Image (100x100px)
Method : LBP
Output : Extracted values of the faces (59 values per image).

2.4. Facial Recognition

The face recognition test will be carried out after feature extraction. The extraction results of the test data feature will be compared with the training data results using Extreme Learning Machine. Extreme Learning Machine (ELM) is supervised learning of neural network types a. ELM is one of the Feed-Forward Neural Networks, which has one single hidden layer. The ELM method is recommended to overcome the learning speed problem that has been happening in the implementation of other Feed-Forward Neural Network methods [5], a little bit similar approach with Deep Learning [8]. The activation function was sigmoid, while the numbers of hidden neurons were 10, 30, and 50.

Input : 59 extracted values
Method : LBP
Output : Facial recognition result (success or failed)

2.5. Dataset

The dataset of this study consisted of 210 face images with 150 images for training data and 60 for test data. The facial images were taken in varied conditions, namely normal face, face with certain expressions, face with sided, and tilted pose orientation. Training data and test data were acquired separately. The training data were obtained through a webcam. Then after being collected, the training data will be trained manually using ELM. The training results were the feature extraction values (in .mat format) and will be used as a comparison for the face classification. As for the test data, they were acquired in real-time via a webcam when the program runs. A total of 50 test data were the images from the trained facial images and 10 data for random facial images. Variations in the face conditions for the test data are normal, expressive faces, face-to-side orientation, lighting, and distance to the webcam.

Table 1. Allocation of training and test data

No	Dataset	Amount
1	Training	150
2	Test	60

Samples of the data training can be seen in Figure 2.

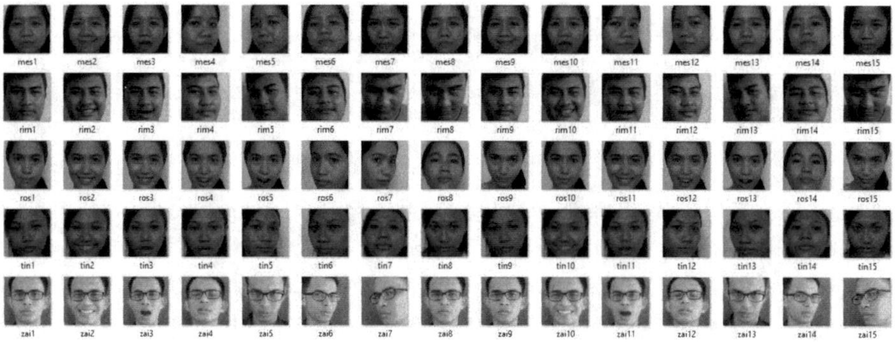

Figure 2. Sample of training data

3. Result and Discussion

System testing was performed to determine the ability of the built system. The system capability depended on the data training. The testing parameters were different hidden neurons of 10, 30, and 50, while the applied activation function was sigmoid. As for the image condition parameters, there were images with normal lighting (≥200 lux) and images with low lighting (≤20 lux), expressive faces, face orientation, and facial distance to the webcam (> 40 cm). The test data amounted to 60 facial images consisting of 12 facial images for each image condition. The output of the process can be seen in Figure 3.

Label	**Citra Awal**	**Deteksi & *Cropping***	***Grayscaling* & *Resizing***	**Ekstraksi Fitur (LBP)**		
Des				0,103443108499050	0,0460518114268780	0,115377098321915
				0,070320956408977	0,105028316378593	0,0371016114950180
				0,00366934365592897	0,0382384546101093	0,0445528291165829
				0,0291415285319090	0,137894570827484	0,0372977368533611
				0,00461532361805439	0,107396408915520	0,0439926646649838
				0,0421841554343700	0,414072036743164	0,0584534592926502
				0,00479727098718286	0,146608605980873	0,0735856294631958
				0,0376893989741802	0,193469166755676	0,0447964482009411
				0,00598277803510428	0,173712030053139	0,0447358004748821
				0,0366456471383572	0,303490161895752	0,0165262222290039
				0,0177617110311985	0,138052403926849	0,0260849595069885
				0,00828057527542114	0,357515066862106	0,0311577096581459
				0,0267836656421423	0,180523008108139	0,0278100240975618
				0,0165452212095261	0,149145051836967	0,0294877402484417
				0,0260453540831804	0,0952708274126053	0,0438336618244648
				0,0172573737800121	0,163291350007057	0,0233102999627590
				0,0407298915088177	0,0992223694920540	0,0253640394657850
				0,148148655891418	0,159174367785454	0,248092085123062
				0,0406701192259789	0,0947439819574356	0,312657088041306
				0,0724530220031738	0,260299801826477	

Figure 3. Result of data processing

The initial image captured by the webcam can be seen in Figure 3. Then, the images were examined to detect whether there is a face in the images, and cropping will be performed to separate the face from the background. The image was processed in the pre-processing stage to resize the images to a size of 100x100 pixels and to transform the RGB image into a grayscaled image. Furthermore, the image feature extraction was conducted to get the value of the image characteristics. There were 59 characteristic values of each facial image.

Table 2. System testing using 10 hidden neurons

	Normal	**Expressive**	**Face Orientation Direction**	**Bad lighting** $\leq$ **20** ***lux***	**Distance from *Webcam* (> 40 cm)**
Succeed	58	57	58	51	57
Failed	2	3	2	9	3
Accuracy (%)	**96,67**	**95**	**96,67**	**80,75**	**95**

Table 3. System testing using 30 hidden neurons

	Normal	**Expressive**	**Face Orientation Direction**	**Bad lighting** $\leq$ **20** ***lux***	**Distance from *Webcam* (> 40 cm)**
Succeed	60	56	60	57	58
Failed	-	4	-	3	2
Accuracy (%)	**100**	**93,3**	**100**	**95**	**96,67**

Table 4. System testing using 50 hidden neurons

	Normal	**Expressive**	**Face Orientation Direction**	**Bad lighting** $\leq$ **20** ***lux***	**Distance from *Webcam* (> 40 cm)**

Succeed	60	57	60	58	59
Failed	-	3	-	2	1
Accuracy (%)	**100**	**93,3**	**100**	**95**	**98,3**

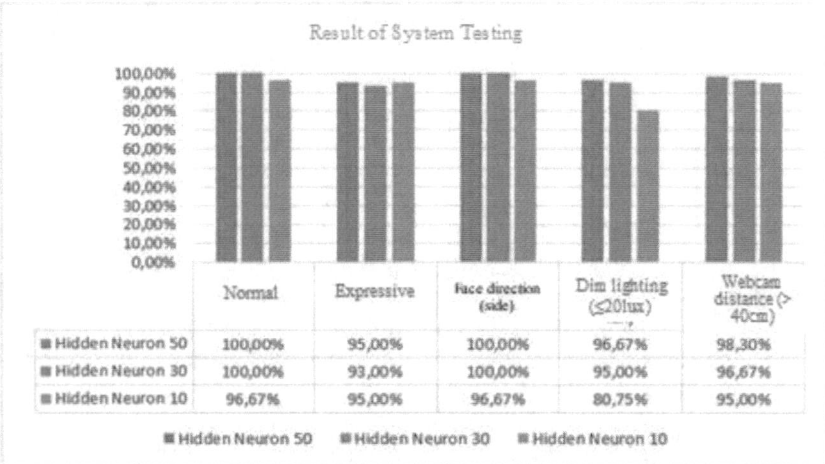

Figure 4. System test graph

The accuracy of the facial identification system was influenced by the condition of the test image, as shown in Table 2, Table 3, and Table 4, as well as n Figure 4. Changes in expression, the direction of facial poses, face distance to the webcam, and lighting affect the test results. Some misidentifications were found in images that were unknown or not included in the training data. However, the more the number of hidden neurons, the higher the percentage of recognition accuracy.

The performance of the system was reviewed using an evaluation standard where the assessment is based on True Positive (TP), False Positive (FP), True Negative (TN), False Negative (FN). True Positive (TP) is the number of faces that are correctly classified. TP shows precisely identified facial images in accordance with the database. FP shows the image that should be correctly identified but misidentified by the system. TN is the number of facial images that are not contained in the database and are not recognized. FN is the facial image that is not in the database but is identified as someone else's image. The values to be calculated for the comparison were the True Positive Rate (TPR), the False Positive Rate (FPR), and the Positive Predicted Value (PPV). The higher the value of TPR and PPV, the better the system performance and vice versa. Whereas FPR, if it is higher, then the system performance decreases and vice versa. The calculation results of the values of TPR, FPR, and PPV are shown in Table 5.

Table 5. The evaluation result of system performance

	Number of *Hidden Neuron*		
	10	**30**	**50**
TPR	80,77 %	89,65 %	94,91 %
FPR	100 %	42,86 %	25 %
PPV	84 %	94 %	96,43 %

IOP Conf. Series: Materials Science and Engineering **851** (2020) 012065 doi:10.1088/1757-899X/851/1/012065

The results of the overall system accuracy can be shown in Figure 5. It can be seen that the amount of hidden neurons affects facial recognition accuracy. In addition, these results were also influenced by the condition of the test and trained data.

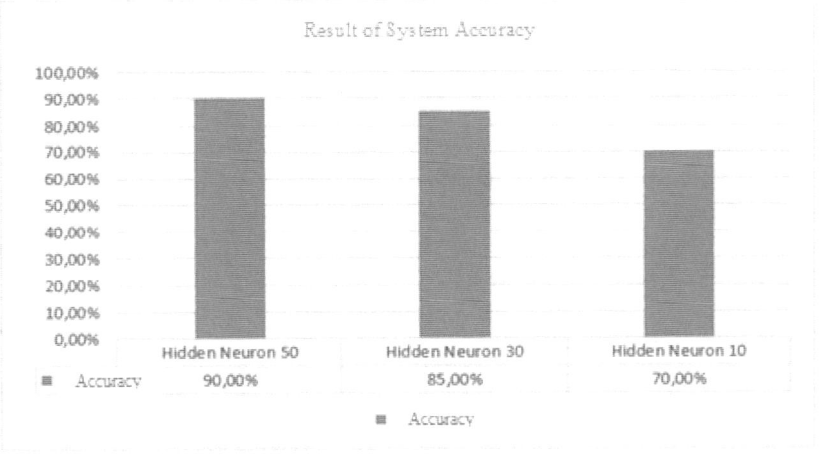

Figure 5. Result of system accuracy

After the face is recognized, the signal will be sent to the microcontroller to open the door. On the contrary, if the face fails to be recognized, then the door remains closed. The illustration of this process is shown in Figure 6.

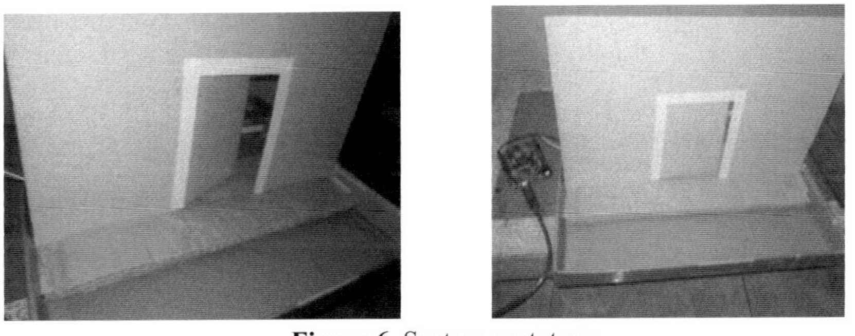

Figure 6. System prototype

4. Conclusion

The conclusion that can be drawn based on the test results of the automatic door access system using Extreme Learning Machine is that the door access system successfully opens the door if the system recognizes the face image. In contrast, the door will remain closed when the face image is not recognized or failed to be identified. So the system can be used as a security feature on a private room door. The Local Binary Pattern method served as the method for texture-based feature extraction worked well in facial feature extraction. These methods cannot distinguish facial images with many similarities, which results in the downfall of the percentage. Therefore, the system greatly depends on the ability of the classification method. As for the Extreme Learning Machine (ELM), the method can perform facial recognition through facial images very well. However, the accuracy is low when processing facial images which were not included in the database. The results showed the system accuracy is at 70% with 10 hidden neurons, 85% with 30 hidden neurons, and 90% with 50 hidden neurons.

Future research can be developed by adding the amount of training data to increase the system's ability to recognize faces and implementing a combination of other methods at the feature extraction stage to get a more accurate feature value.

References

[1] Januzaj Y, Luma A, and Ramaj V 2015 Real-Time Access Control Based on Face Recognition. *International Conference on Network Security and Computer Science (ICNSCCS-15).*

[2] Lwin H H, Khaing A S, Tun H M 2015 Automatic Door Access System Using Face Recognition. *International Journal of Scientific & Technology Research,* **4**(6).

[3] Yang J C, Lai C L, Sheu H T, Chen J J 2013 An Intelligent Automated Door Control System Based on a Smart Camera. Sensors (Basel, Switzerland).

[4] Ahonen T, Hadid A, and Pietik A M 2006. Face Description With Local Binary Patterns: Application to Face Recognition. *IEEE Transactions on Pattern Analysis and Machine Intelligence*, **28**(12): 2037-2041.

[5] Huang G B, Zhu Q Y, Siew C K 2006 Extreme Learning Machine: Theory and Applications. *Neurocomputing*, **70**: 489-501.

[6] Kasar M M, Bhattacharyya D, Kim T H 2016 Face Recognition Using Neural Network: A Review. *International Journal of Security and Its Applications,* Vol **10**(3): 81-100.

[7] Nazeer S A, Omar N, and Khalid M 2017 Face Recognition System using Artificial Neural Network Approach. *International Conference on Signal Processing, Communication, and Networking (ICSCN)*, 420-425.

[8] Rahmat R F, Dennis, Sitompul O S, Purnamawati S, Budiarto R 2019 Advertisement billboard detection and geotagging system with inductive transfer learning in deep convolutional neural network *Telkomnika* **17**(5) pp 2659-2666

IOP Conf. Series: Materials Science and Engineering **851** (2020) 012066 doi:10.1088/1757-899X/851/1/012066

Digitization of Batak Manuscripts Using Methods Learning Vector Quantization (LVQ)

M A Muchtar[1], I Jaya[1], Nandar Cholid Siregar[1], E B Nababan[1], Syahril Effendi[2], O S Sitompul[1], M Zarlis[2], U Andayani[1], T H Nasution[3], I Siregar[4], M F Syahputra[1]

[1]Department of Information Technology, Faculty of Computer Science and Information Technology, Universitas Sumatera Utara
[2]Department of Computer Science, Faculty of Computer Science and Information Technology, Universitas Sumatera Utara
[3] Department of Electrical Engineering, Faculty of Technic, Universitas Sumatera Utara
[4] Department of Industrial Engineering, Faculty of Technic, Universitas Sumatera Utara

Email : anggi.muchtar@usu.ac.id.

Abstract. Digitalization of the Batak manuscript itself is still done manually. [2][4] Dr. Uli Kozok has made a Batak script font that can run on Windows, Linux, and Apple OSX but there is no system that can digitize Batak manuscripts properly. There is no system that can digitize the Batak manuscript itself automatically. Therefore, we need a system that can digitize Batak manuscripts. This system was built by applying the LVQ Method. Batak manuscript digitization in .jpg format is used as an image input file for the recognition process. Then image pre-processing including grayscale, contrast, thresholding, and segmentation is done in order to facilitate the recognition of Batak documents and recognition as a result of document recognition. The application of LVQ method can digitize the Batak manuscript well and in accordance with what is desired. This is supported by an accuracy rate of 97.9%. Based on system testing, the absence of training data in the input document, the letters are unclear, the position of the letters is not appropriate and the background in the document is still noise which greatly affects the success rate of the system in recognizing Batak script documents.

1.Introduction

Batak script is one of the diversity of arts and culture in Indonesia. The culture and arts are dance, language, music and many more and the Batak tribe consists of several tribes namely Karo, Mandailing, Toba, Simalungun, and Pakpak. If seen from its history, the Batak script is an Indian writing family. The oldest Indian script is the Brahmi writings which revealed two groups namely North India and South India. Batak script can also be classified as an abugida (a type of phonetic writing whose language sounds can be accurately represented). There are two types of Batak alphabet sets, namely: "*ina ni surat*" (main letters) and "*anak ni surat*" (derivative letters). Indonesia is also an archipelago that has a diverse cultural heritage, one of which is manifested in the scripts or original writings of the regions included in the archipelago script. Batak script is a family of Batak scripts that needs full attention due to limited data and information. The digitalization of the Batak manuscript has so far been carried out manually. Therefore, the application of a technology is needed to be able to digitize Batak manuscripts.

Previous research has been conducted, [1] about character recognition between Myanmar and English characters. The system will change the Portable Document Format (.pdf) file to Machine

Editable Word Document (.doc). The introduction of Myanmar and English characters will use MICR (Myanmar Intelligent Character Recognition), which is one part of OCR. 92% success rate for the introduction of Myanmar and English scripts.

There is also research carried out, [5] on Kannada's Handwriting Character Recognition Using Hybrid Features. In this study, the hybrid feature extraction technique was used to recognize Kannada's handwriting. Kannada's handwritten characters are written in different font sizes and styles. The features taken from each image are 3600 samples that are used as training datasets to get good feature scores. The pre-processing stage has several steps to implement the hybrid feature method, namely correction, binarization, and noise removal. And the resulting output is in the form of documents containing parts of text then printed in the form of handwriting. Through this approach, researchers obtain an accuracy rate of 87.33%.

Subsequent research was conducted, [7] on the Introduction to Digit Handwriting Using LVQ. In this paper 10 (ten) structural features are used for handwriting recognition. Digit recognition system depends on the type of features used and the data used on this system as many as 50000 training data images and 5000 data testing. The success rate on this system reaches 95%.

The following research was conducted, [6] a comparison of the handwriting recognition of Arabic characters with Arabic characters that have been printed using the discrete method Hidden Markov Models (HMM). In this study, after the character pre-processing step, the segmentation process is continued using a recursive algorithm and feature extraction based on the structural properties of the text. The results of this study indicate an introduction rate of 93.40% for Arabic handwriting.

In addition, research has also been conducted, [3] on the Introduction of Arabic Handwriting Optical Characters using Artificial Neural Network. In this system Artificial Neural Network is used as a classification. Artificial Neural Networks are trained based on the Hopfield Algorithm. And the output of this system is able to recognize eight Arabic handwritten letters well. Through this approach, the accuracy rate on this system is 77.25% for Arabic handwriting.

Based on the description of the research that other researchers have done, the basic problem of digitizing the Batak manuscript itself is still done manually. There is no system that can digitize the Batak manuscript itself automatically. Therefore, we need a system that can digitize the Batak manuscript properly through the input image of the scanned .jpg image file and then convert it to font.docx.

2. Data and Methods Used

2.1.Data Used

The method proposed by the author for digitizing Batak manuscripts in this study uses LVQ. The image to be processed is in the form of an image with .jpg input which will then enter the image pre-processing process. In this process the image will undergo grayscale, contrast, and thresholding processes which will produce gray, black and white images to facilitate the next process. Then the segmentation process will be carried out, namely cropping, where each document will be cut in lines per line. The next process is the application of the LVQ method to digitize the Batak manuscripts, whether it is the Mandailing, Toba, Karo, Simalungun and Pakpak Batak scripts. The general architecture in digitizing Batak manuscripts in this study can be seen in figure 1.

2.2. Method Used

The general architecture in this study consists of several stages, namely:

2.2.1. Input

Training data is preliminary data that will be processed, while testing data is data that will be tested on the system and compared with previous data. At this stage, document files in .jpg format are included as testing data to be tested and processed by the system. The Batak script documents in this document file are script documents both from books, and bark with the font Dr. Uli Kozok.

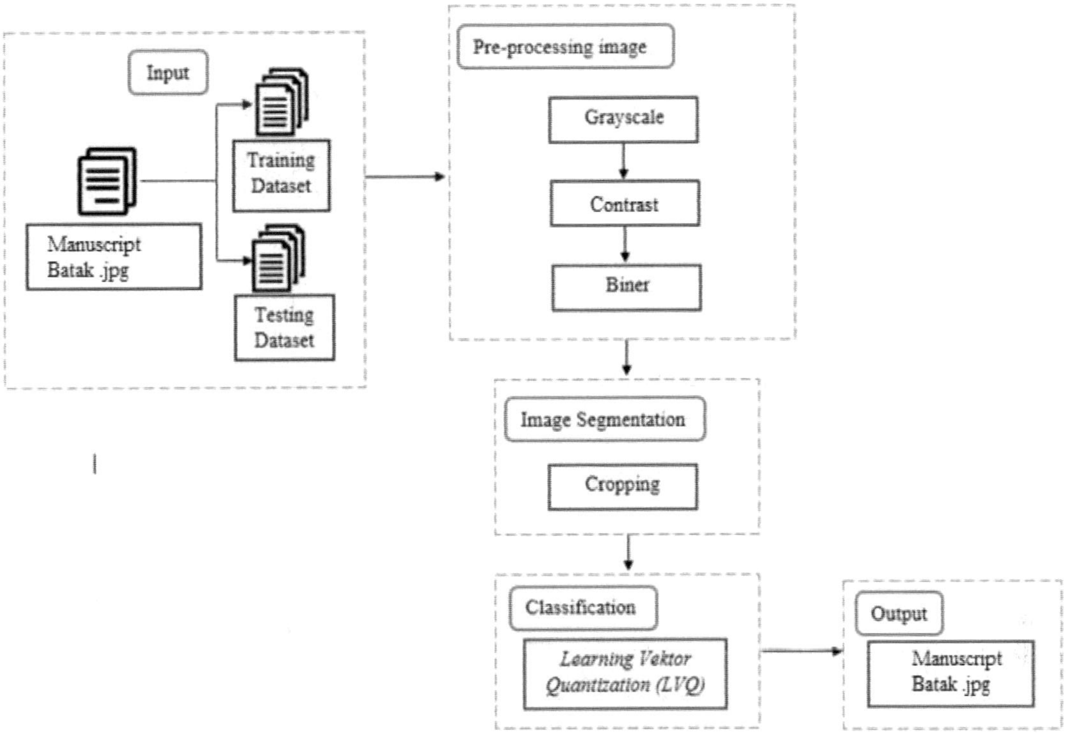

Figure 1. General architecture.

2.2.2. Pre-processing

Image pre-processing is the initial stage in this research. This stage aims to process the image that is input and converted into an image that will be further processed. The stages in image pre-processing are as follows:

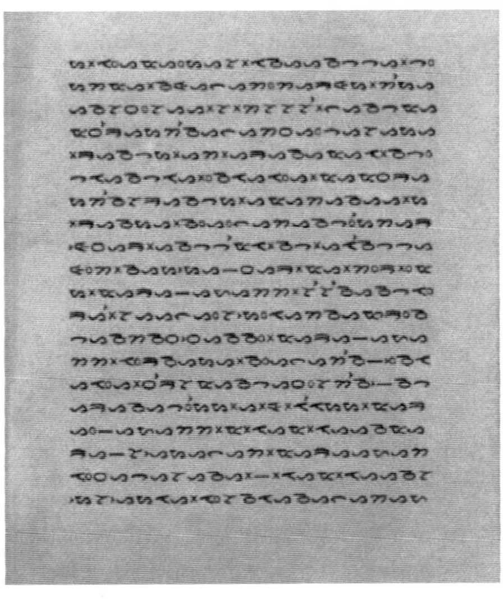

Figure 2. Greyscale process.

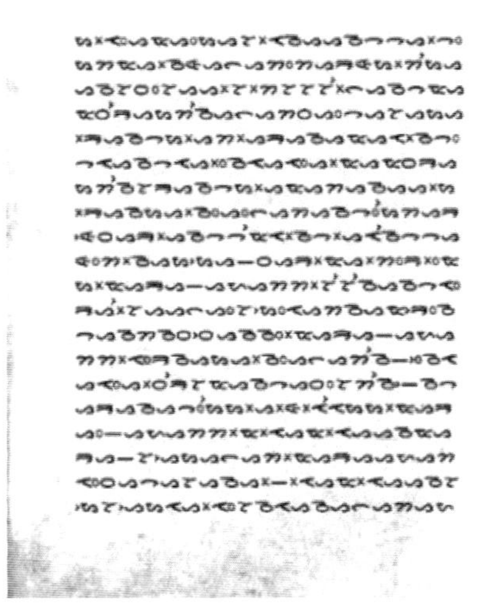

Figure 3. Contrast process.

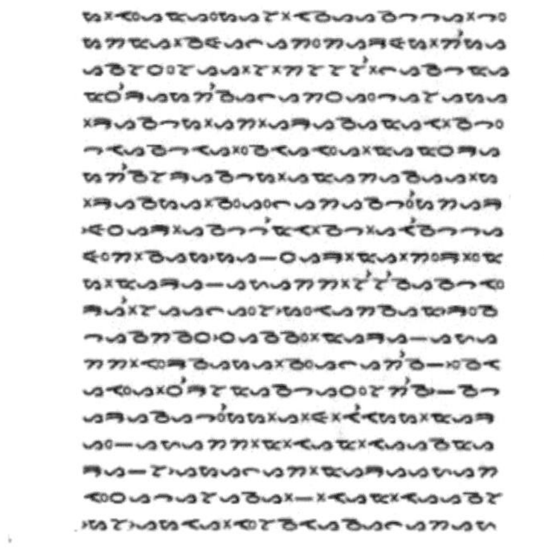

Figure 4. Threshold image process.

Figure 5. Cropping results image.

2.2.2.1 Grayscale
Menurut Basuki (2005), the initial process that is mostly done in image processing is changing the color image to grayish (grayscale). The grayscale process in this study can be seen in figure 2.

2.2.2.2 Contrast
At this stage a sharper contrast is carried out in order to facilitate the system to proceed to the next stage. The contrast process in this study can be seen in figure 3.

2.2.2.3 Thresholding
A thresholding image is an image where each pixel is only expressed by a value of two possibilities (i.e. values of 0 and 1). A value of 0 represents black and a value of 1 indicates white. Then the function of thresholding images at this stage is to simplify the process of digitizing Batak manuscripts, because documents will be more easily detected on images that contain less colour. The thresholding process in this study can be seen in figure 4.

2.2.3. Image Segmentation
The images used for input are Karo, Toba, Pakpak, Simalungun and Mandailing documents. The size for all images is 24 x 42 pixels, because it is to facilitate the user in processing the image of the Batak script document itself. The input that will be processed at this stage is the image that is scanned or photographed, and the system automatically captures the image to be processed to the next stage. The following is an example of the image when doing the cropping process can be seen in figure 5.

2.2.4 Learning Vector Quantization
LVQ is a method for learning in supervised competitive layers. A competitive layer will automatically learn to classify input vectors. The classes obtained as a result of this competitive layer only depend on the distance between the input vectors. If two input vectors approach the same, the competitive layer will put both input vectors into the same class[8].

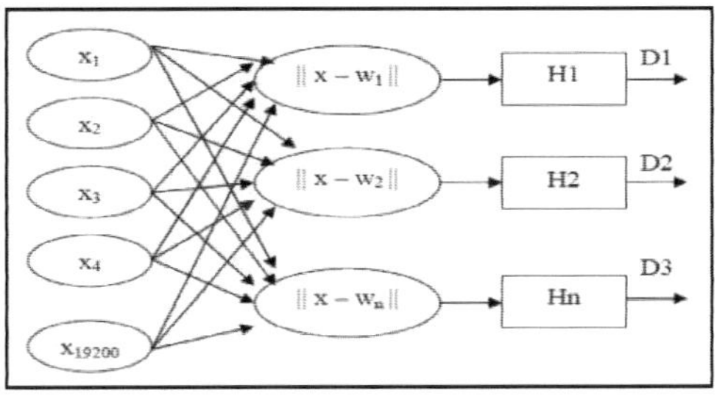

Figure 6. Network architecture LVQ

Explanation:
a) x1 until x19200 = input value
b) || x – w1 || until || x – wn || = weight distance
c) H1 until Hn = output layer
d) D1 until Dn = output value
e) n = number of character data (a number of class)

where:

- x1 until x19200 is a matrix element in each character pattern that will be used as an input value.
- || x – w1 || until || x – wn || is the calculation of the smallest distance between w1 and wn is the initialization data value.
- H1 until Hn are output layers.
- D1 until Dn is the final weight that will be used later in the testing process with new character data entered.

2.2.4.1 Learning Vector Quantization Learning Algorithm

The LVQ method will first make an introduction to the input patterns that will be presented in vector form so that the class can be searched. Because each output neuron represents a certain class or category, the input pattern can be recognized by the class based on the obtained neuron output. The LVQ method recognizes the input pattern based on the distance approach between two vectors, namely the vector of input units / neurons and the weight vector. The introduction of the LVQ method has two processes:

- Training process
- Testing process

In the initial training process, the input vector will experience a learning process that will be carried out through several epochs until the maximum epoch limit is reached. LVQ conducts learning in supervised competitive layers. A competitive layer will automatically learn to classify input vectors. The classes obtained as a result of this competitive layer only depend on the distance of the input vectors with the weight vectors of each class and the input vectors will enter the class that has the closest distance. The parameters used in this LVQ method are as follows:

1. Alfa (Learning rate) Alfa is defined as the level of learning. If the alpha is too large, then the algorithm will become unstable otherwise if the alpha is too small, the process will take too long. The alpha value is $0 < \alpha < 1$.
2. DecAlfa (Decreased Learning rate) That is a decrease in the level of learning.
3. MinAlfa (Minimum Learning rate) That is the minimum value of the level of learning that is still allowed.
4. Max Epoch (Maximum epoch); i.e. the maximum number of epochs or iterations that can be done during training. Iteration will be terminated if the value exceeds the maximum epoch.

IOP Conf. Series: Materials Science and Engineering **851** (2020) 012066 doi:10.1088/1757-899X/851/1/012066

2.3 Flowchart
Flowcharts are graphical descriptions of the steps and sequence of procedures of a program. Batak manuscript digitization system flowchart can be seen in figure 7.

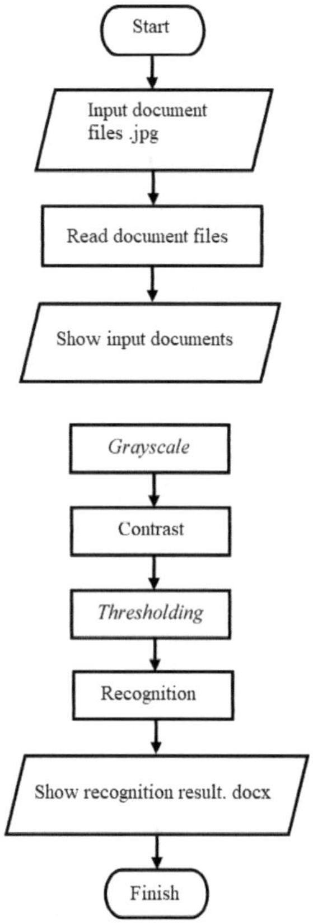

Figure 7. *Batak manuscript digitalization flowchart.*

2.4 Output
In this process, the Batak document file that has been done step by step will produce a document font with the .docx extension which has been structured according to the document input data, and then will be stored on the computer.

3. Result and Discussion
Batak manuscript digitization system is a system that can make it easier for users to process images by going through several image pre-processing for better results. This Batak Manuscript digitization application starts with the home view, where in this page has menus that have their respective functions. Before making an introduction, the user first conducts training data with the aim of facilitating the user in conducting an introduction system as shown in figure 8. When the user will digitize the Batak manuscript, then the next step to do is to select the Recognition menu on the home page. The following operational procedures that users must carry out in digitizing the Batak manuscript system as shown in figure 9.

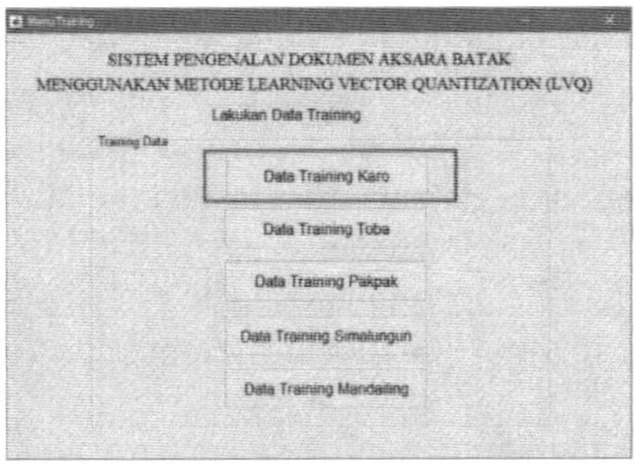

Figure 8. Select the data training.

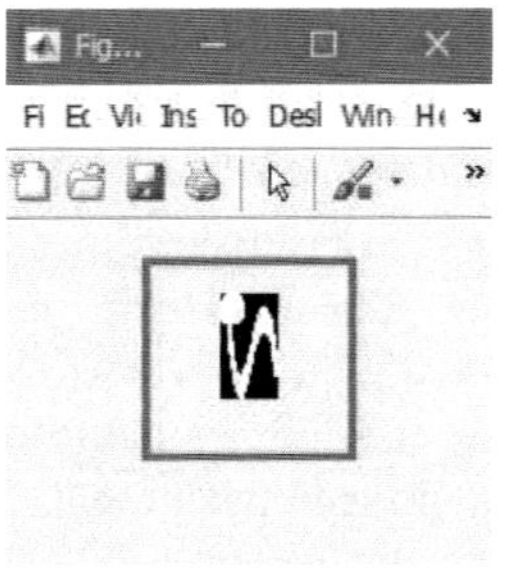

Figure 9. Batak manuscript.

Figure 10. Training successful message.

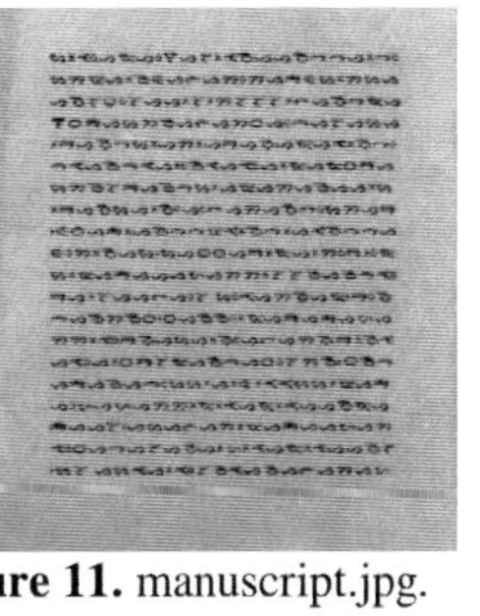

Figure 11. manuscript.jpg.

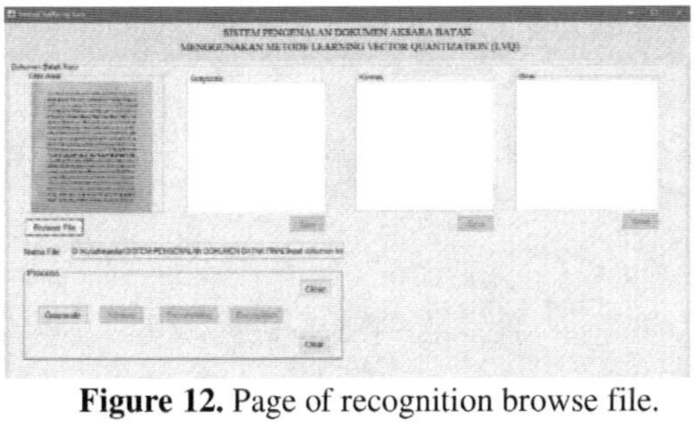

Figure 12. Page of recognition browse file.

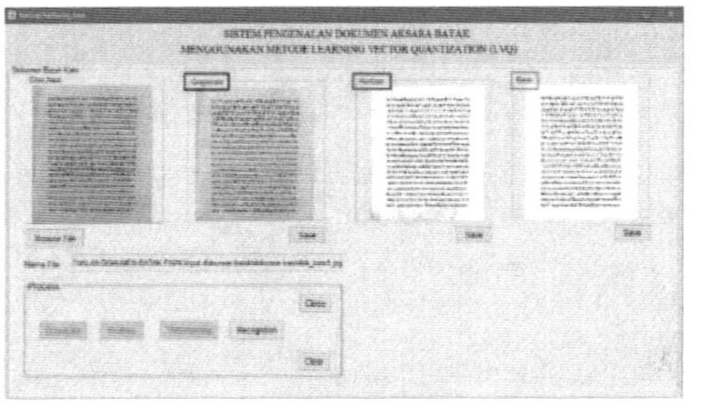

Figure 13. Pre-processing document image.

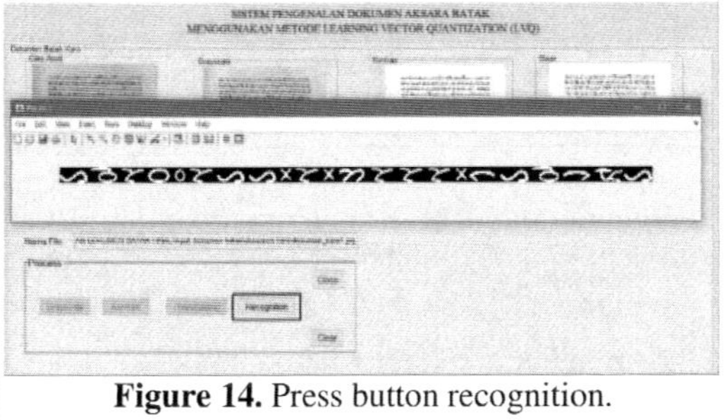

Figure 14. Press button recognition.

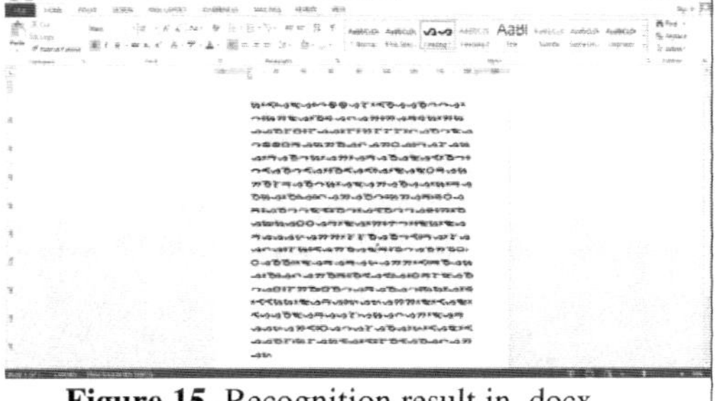

Figure 15. Recognition result in .docx format.

If the user has done training data as shown in figure 10, then the next step is Recognition. The user returns to the home page, then selects the Recognition menu, the system switches to Recognition and the user can choose what script they want to process as shown in figure 11. The recognition display can be seen in figure 12. After the user chooses Recognition precisely in the Karo character, then the recognition display, where the user will enter the Batak script document file that will be recognized by the system. Then the user can select the "Browse File" button which will direct the user to the storage of the document file and select the Batak script document to be recognized. At this stage, the author makes an example document that will be processed is document_karo1.jpg and click Open. The following dok_karo1 that will be processed to the next stage can be seen in figure 12. After the system displays the document file that you want to process, then the next step is pre-processing the image starting from grayscale, contrast, thresholding and recognition. Figure 13 can be seen the occurrence of image pre-processing. For grayscale processes marked in black, contrast is marked in red, and thresholding is marked in orange. After the user presses the recognition button, the system will automatically digitize the Batak manuscript by cropping rows by line until the operation completed message appears, can be seen in figure 14.

And this is the result of recognition that has been done by selecting the file dok_karo1. The following is the result of the dok_karo1 file recognition with the name "dok_karo.docx" can be seen in figure 15. After the recognition results are complete, then the dok_karo.docx file will automatically be saved in the file storage directory. System testing is done to find out whether the system that has been built is functioning properly and runs as desired. This test will explain how much the level of accuracy in the implementation of LVQ is applied to the Batak manuscript digitization system. This testing process uses data that has been obtained in previous studies. The test results on this system will use the formula:

$$Accuracy = \frac{Total\ True\ Testing\ Data}{Total\ All\ of\ Data}\ x\ 100\%$$

Testing the Batak manuscript digitization system and calculating the accuracy of the fit of the test results of each Batak script document can be seen in Table 1.

Table 1. System testing results.

No	input document	Output			Accuracy
		sucessfully recognised	Amount of manuscipts	Index error	
1	dok_karo1	225	228	3	98.6%
2	dok_karo2	208	211	3	98.5%
3	dok_karo3	216	216	0	100%
4	dok_karo4	217	219	2	99%
5	dok_karo5	204	211	7	96.6%
6	dok_pakpak1	208	212	4	98.1%
7	dok_pakpak2	216	219	3	98.6%
8	dok_pakpak3	220	220	0	100%
9	dok_pakpak4	213	217	4	98.1%
10	dok_pakpak5	214	214	0	100%
11	dok_toba1	210	210	0	100%
12	dok_toba2	213	213	0	100%
13	dok_toba3	216	216	0	100%
14	dok_toba4	208	213	5	97.6%
15	dok_simalungun1	217	220	0	98.6%
16	dok_simalungun2	219	221	2	99%
17	dok_simalungun3	225	228	3	98.6%
18	dok_simalungun4	223	227	4	98.2%
19	dok_simalungun5	200	216	16	92.5%
20	dok_mandailing1	226	231	5	97.8%
21	dok_mandailing2	229	234	5	97.8%
22	dok_mandailing3	223	229	6	97.3%
23	dok_mandailing4	226	232	6	97.4%
24	dok_mandailing5	229	233	4	98.2%
25	dok_mandailing6	224	227	3	98.6%
26	dok_karo6	110	124	14	88.7%
27	dok_toba5	45	60	15	75%
Jumlah		5584	5701	117	97.9%

IOP Conf. Series: Materials Science and Engineering **851** (2020) 012066 doi:10.1088/1757-899X/851/1/012066

As seen in Table 1, that dok_karo1 has an error index (unknown) of 3 letters. Examples of unknown letters can be seen in figure 16.

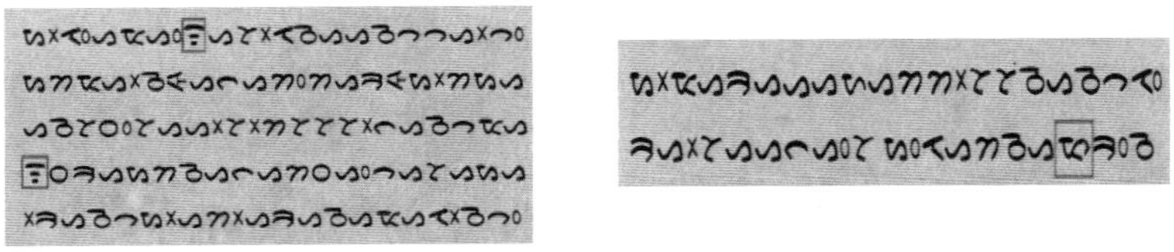

Figure 16. Example Batak script input dok_karo1.

In figure 16, there are 3 letters that are not recognized (error) that are marked in red, because the letters in the Batak script documents are not listed in the Karo training data and the letter positions are not appropriate.

So, there is no training data in the input document, the type of letters is unclear, the position of the letters is not appropriate and the background on the document greatly influences the success of the system in recognizing the Batak script documents.

4. Conclusions

Based on the results of research conducted, it can be concluded that the digitization system of the Batak manuscript uses the LVQ method as follows:

1. The application of the LVQ method can recognize Batak script documents properly and in accordance with what is desired. This is supported by an accuracy rate of 98%.
2. Based on system testing, the background to the document, the position of the letters, the type of letters is unclear and the absence of training data in the input document greatly influences the success rate of the system in recognizing Batak script documents.
3. The success rate of the system in digitizing book-shaped Batak manuscripts is 98%. While the document in the form of wood bark by 75%.
4. There are still documents recognition errors. For example, if there is no training data in the input document, the typeface is not clear, the letter position is not appropriate and the background is still noise, then one of the letters in the input document will be an error (not recognized).

Referensi

[1] Yadana and Cherry M 2010 *Mixed-Myanmar and English Character Recognition with Formatting. IJCSI International Journal of Computer Science* **7**(6).

[2] Https://ulikozok.com/aksara batak/sejarah-aksara batak/

[3]Hussien S R 2015. *Optical Character Recognition of Arabic Handwritten Characters using Neural Network.* International Conference on Computing, Control, Networking, Electronics and Embedded Systems Engineering.

[4]Kozok U 2017. Surat Batak . retrieved from : http://ulikozok.com

[5]Pasha S., Padma MC 2013. *Recognition of Handwritten Kannada Character using Hybrid Features.* Department of Information Science and Engineering, PES College of Engineering, Mandya - 571401, Karnataka, India

[6] Firoj P 2012. An Empirical Evaluation of Offline Arabic Handwriting And Printed Characters Recognition System. *IJCSI International Journal of Computer Science* **9**(6),

[7] Ravi, U B., Venkateswarlu Y, Kumar C A. 2014. *Handwritten Digit Recogniton Using K-Nearest Neighbour Classifier.* Aacharha Nagarjuna University, Assoc, India.

[8] Rahmat R F, Pulungan A F, Faza S, Budiarto R 2017 Image Classification of Ribbed Smoked Sheet using Learning Vector Quantization *Journal of Physics: Conference Series* **801**(1), 012050

IOP Conf. Series: Materials Science and Engineering **851** (2020) 012067 doi:10.1088/1757-899X/851/1/012067

Augmented Reality Technologies for Interior Design Planning using a Simultaneous Localization and Mapping Method

M F Syahputra[1], M R Aulia [1], Dedi Arisandy[1]

[1]Department of Information Technology, Faculty of Computer Science and Information Technology, Universitas Sumatera Utara, Medan, Indonesia.

| nca.fadly@usu.ac.id | mhd.rifki@gmail.com | dedi.arisandy@usu.ac.id

Abstract. Planning the placement of furniture in a room is an important planning process, with proper planning, the placement of furniture can be more efficient in the use of space, as well as from an aesthetic point of view. During this time the process can be done by using a sketch or other manual method. In this research, we use augmented reality technology to plan furniture placement in a room. We use the Markerless technique in the implementation of this AR system, and use the Simultaneous Localization and Mapping (SLAM) technique to determine the geospatial point where the furniture will be the desired layout. The results of this application ifound that good minimum distance in detecting SLAM on the ground is 60 cm and can only be seen up to a maximum position of 45 ° from the Z axis of the camera's initial position when tracking. This application runs on the Android mobile platform Introduction.

1. Introduction

Augmented Reality (AR) is a technology that aims to combine the real world with the virtual world. Development of Augmented Reality (AR) conducted by developers allows this technology to be applied in various fields, such as the world of entertainment, business, education, and so on. This is caused by the features presented by Augmented Reality where this technology is able to provide visual appearance in the form of interesting 2D and 3D objects [1].

Interior design is a trend that is in great demand to beautify the appearance of the room. The difficulty in changing the layout or adding to the collection of home furniture is to arrange it to look efficient, the room is not too narrow and so that the order of home furniture looks beautiful to look at. To change the layout so that it exactly requires repetitive settings so that it just wastes energy and time. Augmented Reality technology allows the work to be easy and efficient, so it will not take up the energy and time to display the 3D objects.

Today the use of smart phones is one of the most important needs for the community. Smart phones that have many functions can facilitate users in everyday life. In this research, the author will build a system or application of Augmented Reality in real-time and mobile-based and without using markers or called markerless augmented reality where it can facilitate a person to design the interior of the room more effectively and efficiently without having to experience difficulties in managing it manual and also use markers as markers to bring up 3D objects[2]. In this study the authors built an Android-based one

Interior design arrangement of a room requires a model that is drawn manually or using design software is considered less effective and efficient because it requires a long time in its creation and modelling and only presents a 2D-shaped object model that lacks visualization of the reality. Therefore, an application is needed by using Android-based Augmented Reality (AR) technology that can help users in designing a room to their liking more easily and can be done in real-time. Markerless Augmented Reality technology can make the implementation of Augmented Reality much more efficient, practical, attractive, and can be used anywhere, anytime without printing markers.

Research on Augmented Reality has been widely implemented in various fields. Such as education, medicine, marketing, culture, business and more. In general, Augmented Reality is implemented as a medium of recognition or learning. So that makes learning more interesting and interactive. Some of previous research: Interior Design in Augmented Reality Environment. This research makes Augmented Reality to display virtual furniture so that it becomes a learning media for interior design. Using Tangible AR and ARToolkit which still use markers and multi markers. [3]. Design and Implementation of Interior Design Applications Based on Augmented Reality Technology. This study uses Adobe Flex and ARToolkit Marker Generators by using markers to display visuals in 3D in accordance with the design [4]. Markerless Augmented Reality Android App For Interior Decoration. Make an Augmented Reality application based on markerless to design the interior, where each novice user can easily decorate his house. This application is based on Android and uses Metaio SDK and ARToolkit [5]. Setting Home Furniture Layout With Augmented Reality. The purpose of this study was to examine the state-of-the-art of technology, systems, and applications of augmented reality and also the application of Augmented Reality technology using markers to help arrange the layout of home furniture [1]. Visualization of Virtual House Model on Property Brochure Using Augmented Technology, which use augmented reality technologies to present 3d property brochure [6] with a finger interaction [7] has become its previous researches.

2. Methodology

2.1. General Architecture

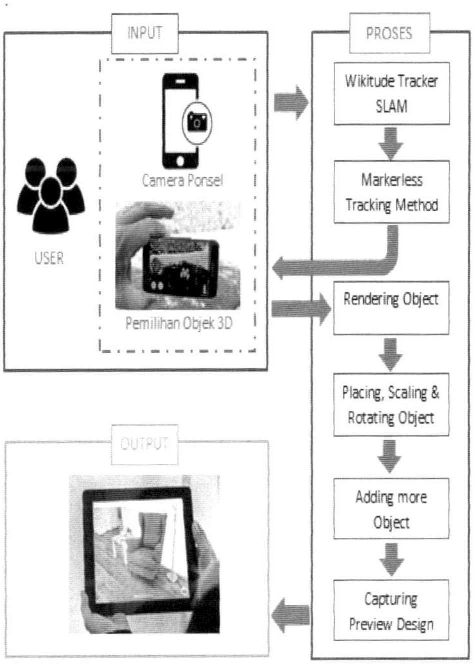

Figure 1. General Architecture

The following is an explanation of the stages in the general architecture:

2.1.1. Input

The user runs the Augmented Reality Decor AR application found on the smartphone. The user presses the Start button to start the application and will display the AR Camera display. The user selects the 3D object that he wants to display so that the display will appear selecting the object of the household furniture that will be displayed.

2.1.2. Process
The system identifies flat surfaces on the floor with AR Tracking SLAM to determine angles and track positions to display 3D objects. Previously 3D objects were created via 3D Studio Max, then exported to OBJ or FBX format then exported via Unity 3D. Then the system will adjust according to the selected 3D object. Then the system will render 3D objects. And displays 3D objects of household furniture to the smartphone screen. 3D objects will be placed in the lower left of the smartphone screen. and can be changed by shifting the object by sliding it by using a finger and can also enlarge and rotate the 3D object. Users can also add more than one desired object so that they can organize and design the layout of home furniture in the space.

2.1.3. Output
The resulting output is a 3D object that adapts to the floor surface. Users can also change their location by sliding the object by sliding it by using a finger and can also zoom in and rotate the 3D object. Users can also add more than one desired object so that they can organize and design the layout of home furniture in the room.

2.2. Markerless Tracking Method
Marker-less tracking is an Augmented Reality method where the tracking process is no longer only using markers as detection targets. With this method, the Augmented Reality process is no longer limited to markers, but visual images, 3D objects, GPS or faces that can be used as detection targets.

The difference between marker-based and marker-less is that the process of tracking the camera position and camera orientation is calculated with the markers that have been set. Whereas marker-less calculates the position and orientation of the camera and the real world without any specific provisions, using only natural features such as edges, corners, lines or 3D models. The marker-less method used in this study is the image tracking method where visual images are used as targets for the Augmented Reality application built. Examples of image tracking Augmented Reality applications can be seen in Figure 2.4. Can be seen after the teapot object appears above the marker pictured after tracking by the system.

Figure 2. Markerless Image Tracking

2.3. Activity Diagram

Activity diagrams or activity diagrams are diagrams that illustrate the various flow of activities in the system that are being designed, how each system flow begins, decisions that may occur, and how they end. Activity diagrams can also describe parallel processes that might occur in several executions. Start Activity diagrams explain the application system activity in running the camera to start the tracking process. The Start activity diagram can be seen in figure 3.

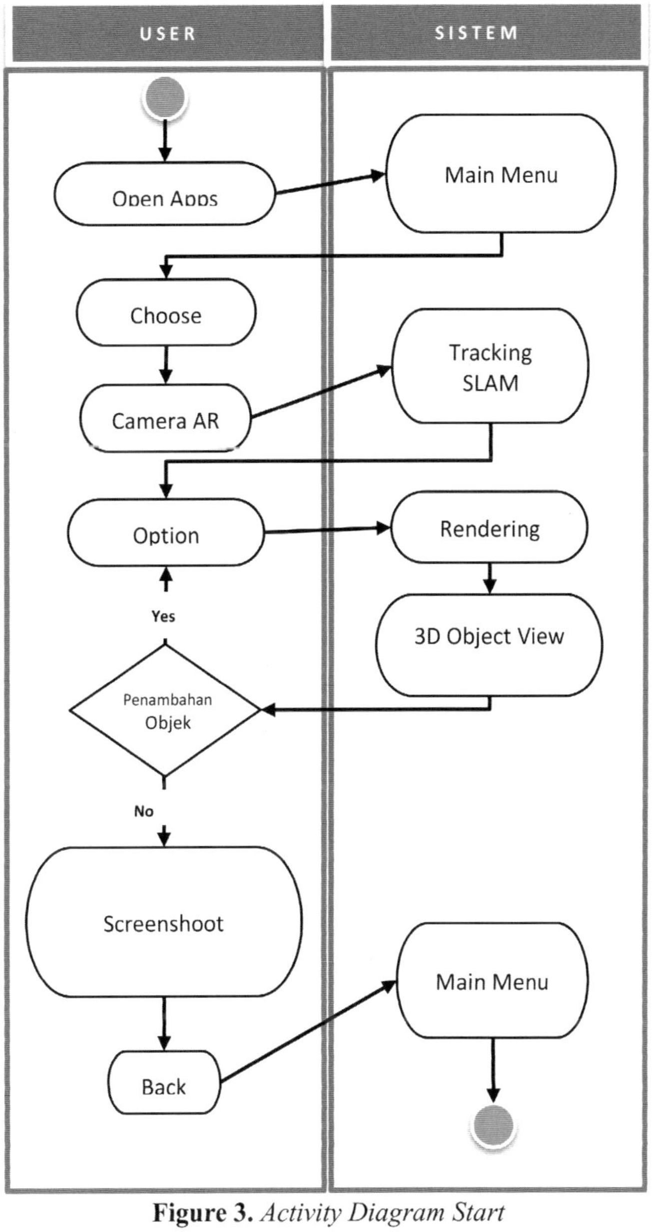

Figure 3. *Activity Diagram Start*

3. Result and discussion

3.1. Implementation

DecorAR application implementation is done by implementing the Wikitude SDK library and Simultaneous Localization and Mapping (SLAM) library into applications by using the developing Unity3D tool to build applications.

This research has a choice of 10 3D objects of furniture to be displayed, namely 3 cabinets, 2 tables, 2 chairs, 2 decorative lights, and 1 flower vase. Views that will be designed in this application, namely the main menu display page, Object menu page, Control Object page, Gallery menu page and About menu page.

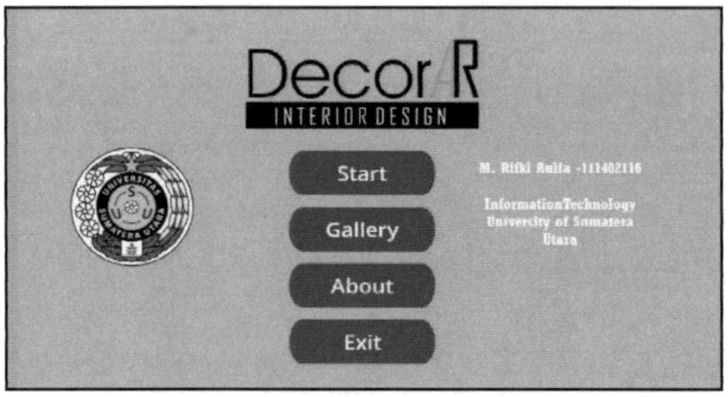

Figure 4. Main Menu Page Display

When starting to operate this application it will immediately show the main page. In the main page view there is a DecorAR application logo and 4 menus. The four menus are Start, Gallery, About, and Exit. Display of the main menu of the application can be seen in Figure 4.

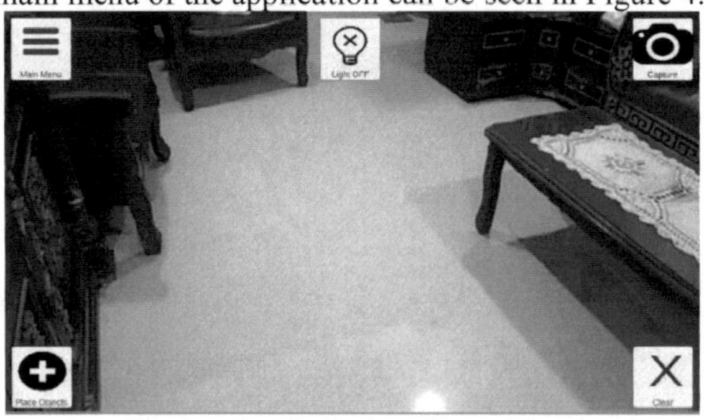

Figure 5. AR Camera Page View

In the Augmented Reality menu, there are 5 buttons that can be selected and each has its own function. Augmented Reality menu display on the application can be seen in Figure 5.

Navigation Description:

- Main Menu : Works to return to the initial menu.
- Capture : Functioning to take pictures from 3D object designs displayed.
- Place Object : Function to display a menu for selecting objects.
- Light button : Function to make the condition of the room glow or not glow
- Clear : Works to delete all 3D objects that have been raised on the AR Camera screen

Figure 6. Object Menu Display page

The function of this page is this is a page where users can select 3D objects that will be displayed on the AR Camera menu page. As in Figure 6, on this page there is a 3D object button that is ready to be selected to appear on AR Camera

Figure 7. Control Object Display page

This page contains control of 3D objects by changing positions, enlarging and rotating 3D objects of home furnishings so that they can design space more easily as desired. Users can use the available bar to rotate objects and use both fingers to zoom in and out on objects. The appearance of the Control Object page in the application can be seen in Figure 7.

3.2. Markerless testing

3.2.1. Testing slam detection

Markerless testing is tested by conducting SLAM testing to prove that users can use markerless methods to display 3D objects of household furniture. This test is carried out on different aspects that will be used to display 10 3D objects of home furnishings. The test is done by opening the DecorAR application, then activating AR Camera by pressing the Start button found on the application Main Menu display and pointing it to the desired place, then selecting the object of the household furniture that you want to display.

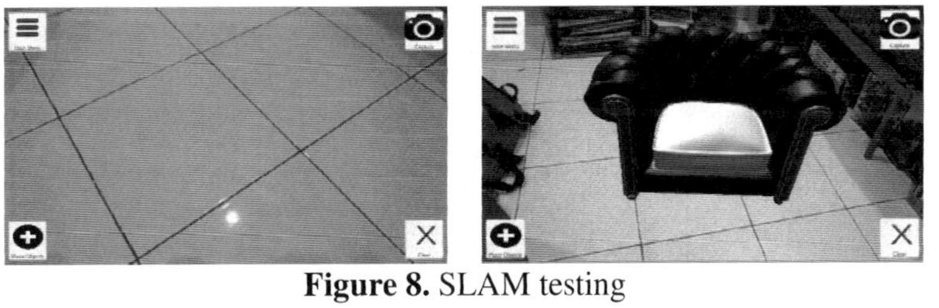

Figure 8. SLAM testing

In Figure 8 is a user-defined location to serve as a ground to place the 3D object. There is a grid in the form of a grid that is a checkerboard as a marker for ground detection. And the other figure in Figure 8, the object of household furniture that has been selected has been successfully raised. If the user exits the AR Camera page, the predetermined ground that is laying the 3D object will be removed from the system so that SLAM will detect it from the beginning again.

3.2.2. Testing slam detection based on distance

This experiment to find out from what distance SLAM is able to detect ground. This test is done using one 3D object. In the distance test, the closer the distance between the camera and the ground, the SLAM will not detect the ground properly so that 3D objects that appear will easily disappear because the ground position cannot be captured properly. There is an indicator on the grid contained in the ground that makes the user know if SLAM is functioning properly or not.

Figure 9. SLAM indicator for ground distance

In Figure 9 shows a green line indicator at the bottom of the AR Camera page which indicates that SLAM detects the ground properly. And the other figure 9 shows a red line indicator at the bottom of the AR Camera page which indicates that SLAM does not detect the ground properly. The results of the test distance can be seen in Table 1.

Table 1. SLAM distance test results to detect ground.

No	Distance	*Ground*	Trial Display (Object)	Result
1	10 cm	Not detect	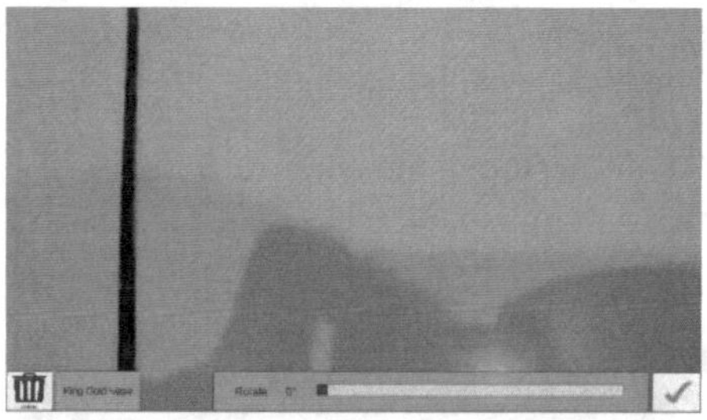	Object does not appear
2	50 cm	detect		Objects appear poorly so that objects can disappear easily because SLAM does not detect ground
3	90 cm	detect		Object appears well
4	120 cm	detect		Object appears well

5	150 cm	detect		Objects do not appear properly

The test results in table 1. give the results of the SLAM function to detect ground as a good object placement done at a minimum distance of 60 cm so that SLAM can work properly and a maximum distance that is good for using SLAM function is 120 cm.

3.2.3. Testing slam detection based on angle
Position detection testing is performed to determine which position SLAM can detect the ground by the system. The test was carried out using 1 3D object of household furniture by shifting the AR Camera towards a different one which is to the left or right based on the Z axis by using an angle to shift the position from the initial position. Position testing results can be seen in Table 2.

Table 2. Results of testing the position of the Object and Ground position.

No	Posisi	*Ground*	Tampilan Uji Coba (Objek)	Hasil Pengujian
1	0° from the starting position	detect		Object appears well
2	15° from the starting position	detect		Object appears well
3	30° from the starting position	detect		Object appears well
4	45° from the starting position	detect	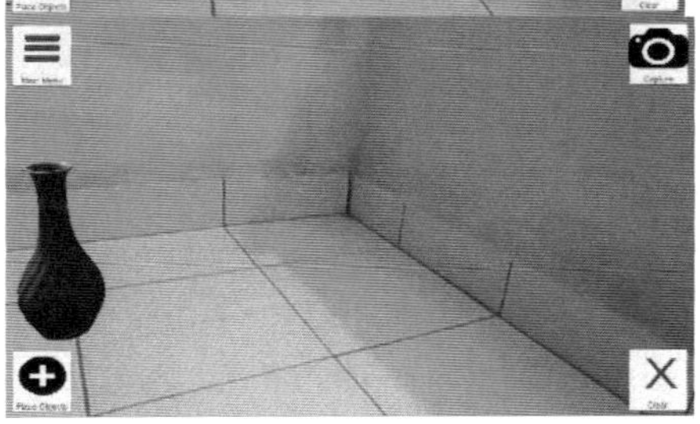	Object appears well

5	60° from the starting position	Not detect		Objects do not appear properly
6	90° from the starting position	Not detect		Object does not appear

The results in table 2. show the testing of the SLAM function that can be done by sliding the AR Camera to the right or left of the Z axis to detect the ground and the object that has been displayed only up to a maximum of 45 ° from the initial position of the AR Camera. If the AR Camera is shifted above 45 ° then the object or ground is not detected again so the object and ground disappear. However, if AR Camera is returned at the initial position then SLAM will detect the ground that has been locked during the tracking phase so that SLAM can reappear the ground and the object that has been displayed before.

4. Conclusion

After carrying out the implementation phase and testing the system, then obtained some conclusions found in this study, namely:

- The DecorAR application is able to display the AR Camera system and display 3D objects of home furniture.
- The DecorAR application is capable of displaying multiple 3D objects.
- A good minimum distance in detecting SLAM on the ground is 60 cm and can only be seen up to a maximum position of 45 ° from the Z axis of the camera's initial position when tracking.
- More interactive applications with objects that can be set by the user.

5. References

[1] Rosyid R. et al. 2011. Perancangan dan Implementasi Aplikasi Desain Interior Berbasis Teknologi Augmented Reality. Skripsi. Universitas Telkom Indonesia.

[2] Shedole A. et al. 2015. Markerless Augmented Reality For Interior Designing Using Android. *Asian Journal of Engineering and Technology Innovation* **3**(6): 10-14.

[3] Jani BY et al. 2015. Interior Design in Augmented Reality Environment. *International Journal of Advanced Research in Computer and Communication Engineering* **4**(3): 286-288.

[4] Chowanda A and Kanigoro B 2011. Pengaturan Tata Letak Perabotan Rumah Dengan Augmented Reality. *Konferensi Nasional Sistem dan Informatika*, pp. 202-207

[5] Renukdas P et al. 2013. Markerless Augmented Reality Android App For Interior Decoration. *International Journal of Engineering Research & Technology (IJERT)* **2**(4): 1367-1373

[6] Syahputra M F, Rinaldi, Muchtar M A, Andayani U, Dahlan A, Sriadhi S, Iskandar A, Laritmas, S 2019 Visualization of Virtual House Model on Property Brochure Using Augmented Technology. *Journal of Physics: Conference Series*.

[7] Syahputra M F, Siregar R K, Rahmat R F 2017 Finger recognition as interaction media in augmented reality for historical buildings in Matsum and Kesawan Regions of Medan city *Lecturer Notes in Computer Science* 10325, pp. 243-250

IOP Conf. Series: Materials Science and Engineering **851** (2020) 012068 doi:10.1088/1757-899X/851/1/012068

The Development of Mobile and Server Application for Traffic Data Collection Based on Location-Based Services (LBS)

D Gunawan, Priyanka, O S Sitompul
Department of Information Technology, Universitas Sumatera Utara, Medan, Indonesia
E-mail: danigunawan@usu.ac.id

Abstract. Traffic data includes various data such as the number of vehicles, vehicle velocity, or travel time that might lead to meaningful information for the traffic engineers. This research aims to collect the traffic data based on the Location Based Services (LBS) by utilizing the Global Positioning System (GPS) in a smartphone. The drivers bring the smartphone and collecting the traffic data while driving the vehicle. It is done in real-time by running the mobile application in background service. The data includes the date, timestamp, latitude, longitude, device ID, and vehicle velocity. These data are temporarily stored in the internal smartphone database, then synchronized to the database in the server application. According to the result, the large number of online taxi drivers and a longer time to collect the data should be considered to cover a wide area.

1. Introduction

Traffic data collection aims to obtain a real-world traffic situation. It includes various data, such as the number of vehicles, vehicle velocity, or travel time. The engineer might use the data to analyze traffic congestion or traffic engineering. The number of vehicles that are passing through a particular road segment over a certain period is an example of data processing results that can be performed. The result might lead to meaningful information that can be used by traffic engineers.

This research aims to propose an alternative to collect traffic data based on the vehicle location or position by utilizing the most common device, which is a mobile phone. Current mobile phones usually equipped with a feature called Global Positioning System (GPS). This research utilizes the GPS module in a mobile phone to obtain the current position of a driver who drives a vehicle. By obtaining the current location and velocity of the vehicle, we expect to obtain data that can be used to yield the travel time. Previously, research by Wu predicted travel time by utilizing support vector regression [1]. Furthermore, this data can also be used to yield traffic congestion information. As discussed by Dias [2], they found the relationship between congestion and traffic accidents.

Previous research [3] discussed the traffic data collection by designing a GPS device which embedded in a taxi vehicle. This research is done for a year. The research obtained the traffic data, which are used to yield the shortest route and estimated time based on traffic congestion conditions.

Another research discusses the classification of traffic congestion levels in the urban area by providing real-time location data every 30 seconds by using the tracking application [4]. The location data will be analyzed by using Kalman Filter to predict the level of traffic congestion in the urban area. Kalman filter also used by [5] to reduce noise in GPS data.

A different approach is used by the Medan city government while compiling the masterplan of a road network [6]. This masterplan is used to develop a transportation system among areas in Medan city. The government collected the traffic data by organizing the research form and gather the data from the institutes which have any relation with the transportation system and spatial planning. Those data are analyzed to obtain the weight of origin and destination matrix in the road network, which is the traffic weight in every road segment in a road network.

This paper consists of four sections. The introduction section discusses the background, the aim of this research, and previous studies. The next section discusses the research method to obtain the traffic data by using a mobile application, and transmission to the server application. The third section discusses the result of this research and then followed by the conclusion.

2. Research Method

2.1. General Architecture

As this research gathers the data based on Location-Based Services (LBS), we utilize smartphones that have a GPS feature. This research utilizes the existing feature in smartphones to collect driver positions and record them to the database. The mobile application records several data such as date, timestamp, latitude, longitude, device ID, and vehicle velocity in every vehicle position changes. The velocity will be recorded as a meter per second. According to Stipanic [7], the traffic congestion index in a road segment can be calculated using the vehicle velocity in that road segment.

The data updating occurs every two seconds. The mobile application has a background service feature to keep the data processing runs in the background without any interaction from the smartphone owner. The application also keeps running while the user opens another application. The background service feature allows the application to record the data in real-time. As shown in Figure 1, it stores all the collected data from the GPS satellite to the internal database. This research uses an internal database in the mobile application as temporary storage before it sends all the recorded data to the server. Saving the data to the internal storage aims to avoid data transmission failures due to loss of data communication because the mobile application and the web server are connected through the internet.

Figure 1 illustrates that the webserver listens to the request from the mobile application. The mobile application will synchronize its data to the server whenever the smartphone has an internet connection. The web server has a particular service called Application Program Interface (API) to handle the request from the mobile application.

Besides the date, timestamp, latitude, longitude, device ID, and vehicle velocity, this research will yield other data, namely road segment name for available coordinates. Because the GPS satellite does not provide this kind of data, then we perform a new process to obtain it. As the smartphone has limited resources, we decide to perform the process on the server-side. In order to yield the road segment name, we will convert the latitude and longitude by using the geocoding method. This research uses Google Maps API to perform the geocoding process. We use the following API code to obtain the road segment name:

```
https://maps.googleapis.com/maps/api/geocode/json?latlng=lat,lng&key=api_key&
sensor=false
```

Google Maps API will return requests in JSON format. Next, we decode the JSON to obtain the road name and store it to the traffic dataset.

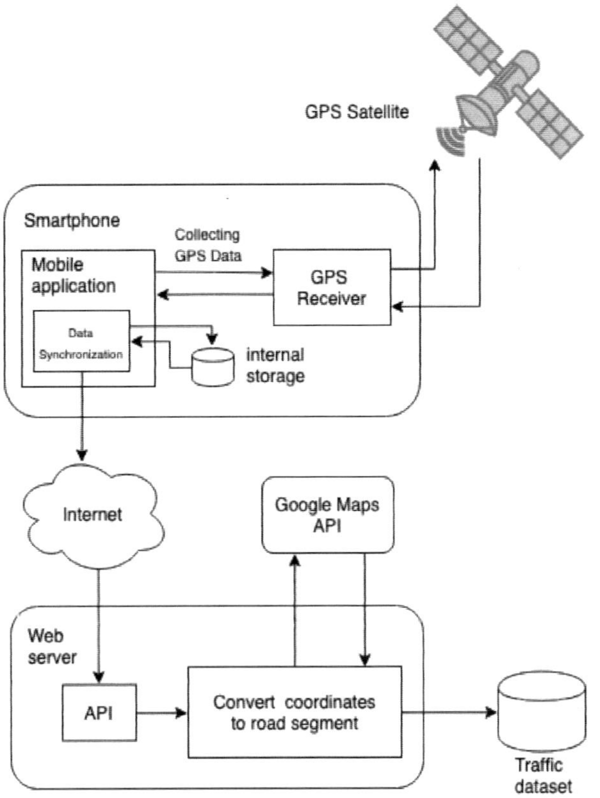

Figure 1. General Architecture

2.2. *Application Testing*

We test the mobile application to collect traffic data by installing it into 15 online taxi drivers' smartphones. Although other devices such as Personal Navigation Devices (PND), Personal Digital Assistant (PDA), or portable computer can access LBS as well [8], we decide to use a smartphone because it is the compulsory device for the online taxi drivers. We choose the online taxi drivers because they drive the car around the city more often. Therefore, the coverage area will be wider. The taxi drivers agreed and knew that the application would track their location. Therefore, they agree to keep the mobile application running while they are driving. We collect the traffic data in Medan city for one month.

3. Result and Discussion

The traffic data collection involved mobile and server applications. Generally, the mobile application will gather the data, then transmit it to the server. The mobile application is developed for Android OS (Version 6.0.1 or Marshmallow) and only requires a smartphone with GPS features. The server application is developed with PHP and hosted in Apache2 Web Server. We utilized MySQL version 10.1.10 to store the traffic dataset.

This research collects the data by using 15 online taxi drivers. The mobile application gathers the data while the drivers take guests to their destination. We collected the data for one month in Medan city. We collect the date, timestamp, latitude, longitude, device id, and the car velocity. This research obtains 50,000 locations, include the data where the velocity field shows zero values. From our observation, the zero values are caused where the drivers stopped or did

Table 1. The example of traffic data obtained from the smartphone

Date	Timestamp	Logitude	Latitude	Devide ID	Velocity
May 19, 2017	7:42:26 AM	98.659953	3.5673131	860153037496280	5.37
May 19, 2017	7:42:28 AM	98.659883	3.5673608	860153037496280	5.1
May 19, 2017	7:42:32 AM	98.659579	3.5673312	860153037496280	7.3
May 19, 2017	7:42:40 AM	98.659078	3.5673371	860153037496280	7.07
May 19, 2017	7:42:44 AM	98.658822	3.5673334	860153037496280	7.25
May 19, 2017	7:42:46 AM	98.658679	3.5673474	860153037496280	7.81
May 19, 2017	7:42:50 AM	98.658379	3.5673605	860153037496280	8.06
May 19, 2017	7:42:54 AM	98.658136	3.5673598	860153037496280	6.78
May 19, 2017	7:42:56 AM	98.658023	3.5673587	860153037496280	6.3
May 19, 2017	7:43:00 AM	98.6578	3.567369	860153037496280	5.15

Table 2. The example of geocoding result

Logitude	Latitude	Road Name
98.659953	3.5673131	Jl. DR. Mansyur No. 22, Merdeka, Medan Baru, Kota Medan, Sumatera Utara 20153, Indonesia
98.659883	3.5673608	Jl. DR. Mansyur No. 22, Merdeka, Medan Baru, Kota Medan, Sumatera Utara 20153, Indonesia
98.659579	3.5673312	Jl. DR. Mansyur No. 30, Merdeka, Medan Baru, Kota Medan, Sumatera Utara 20153, Indonesia
98.659078	3.5673371	Jl. DR. Mansyur No. 58, Merdeka, Medan Baru, Kota Medan, Sumatera Utara 20153, Indonesia
98.658822	3.5673334	Jl. DR. Mansyur No. 60A, Merdeka, Medan Baru, Kota Medan, Sumatera Utara 20153, Indonesia

not drive their car. For example, while the drivers were dropping the guest, waiting for the guest, having lunch, or taking a rest. These data should be eliminated because they do not reflect the traffic. We obtain 20,000 locations after eliminating the record, which contains zero values in the velocity field.

Table 1 shows an example of the data that was obtained by the mobile application. The mobile application stored them in the internal database instead of directly transmit them to the webserver. Storing the data to the internal database avoids the possibility of losing the data while transmitting them. The mobile application will transmit the data if the internet connection is available. It transmits the data to the webserver through the API. The server application will perform the geocoding process to obtain the road name by using Google Maps API, which returns the data in JSON format. The server application will combine the data from the mobile application and the geocoding result. Table 2 shows the result after the geocoding process. Next, the server application stores all the data to the traffic dataset.

According to the result, the data are scattered in various areas. Based on our observation, this occurs because the taxi drivers cannot determine the guest's origin, destination, and time. Besides, this causes the time gap in a similar road segment.

4. Conclusion

The smartphone can be used to obtain traffic data based on Location-Based Services (LBS). The mobile application has the ability to utilize the GPS feature in a smartphone to collect the traffic data. In order to avoid losing data while transmitting them, the data should be temporarily kept in the internal database. The mobile application will synchronize the data with the server application through the Internet. In the server application, we perform the new process, which called geocoding, to convert latitude and longitude to road name by utilizing Google Maps API. According to the result, we conclude that the mobile and server applications can be utilized as the system to gather and process the traffic data. In order to collect real-time traffic data, we can rely on online taxi drivers. However, we should consider a large number of online taxi drivers and a longer time to collect the data in order to cover a wide area.

References

[1] Chun-Hsin Wu, Jan-Ming Ho and Lee D T 2004 *IEEE Transactions on Intelligent Transportation Systems* **5** 276–281
[2] Dias B F, Miska M, Kuwahara M and Warita H 2009 Relationship between congestion and traffic accidents on expressways an investigation with bayesian belief networks *Proceeding of 40th Annual Meeting Of Infrastructure Planing (JSCE)* pp 93–110
[3] Lim S, Balakrishnan H, Gifford D, Madden S and Rus D 2010 *Stochastic Motion Planning and Applications to Traffic* (Berlin, Heidelberg: Springer Berlin Heidelberg) pp 483–500 ISBN 978-3-642-00312-7 URL `https://doi.org/10.1007/978-3-642-00312-7_30`
[4] Tao S, Manolopoulos V, Rodriguez S and Rusu A 2012 *Journal of Transportation Technologies* **02** 22–31 ISSN 2160-0473
[5] Nandhini C and Satheesbabu S 2015 *International Journal of Innovative Research in Computer and Communication Engineering* **03** 11398–11404 ISSN 2320-9801
[6] Medan P K 2013 Penyusunan master plan jaringan jalan kota medan
[7] Stipancic J, Miranda-Moreno L and Labbe A 2016 Measuring Congestion Using Large-Scale Smartphone-Collected GPS Data in an Urban Road Network *TAC 2016: Efficient Transportation - Managing the Demand - 2016 Conference and Exhibition of the Transportation Association of Canada*
[8] Kushwaha A and Kushwaha V 2011 Location Based Services using Android Mobile Operating System Tech. rep.

Combination of Rail Fence Cipher Algorithm and Least Significant Bit Technique to Secure the Image File

Dian Rachmawati[1*], Mohammad Andri Budiman[1*], and Akhmad Yusuf[1*]

[1]Departemen Ilmu Komputer, Fakultas Ilmu Komputer dan Teknologi Informasi, Universitas Sumatera Utara, Jl. Universitas No. 9-A, Medan 20155, Indonesia

*Email: dian.rachmawati@usu.ac.id, mandrib@usu.ac.id, akhmadyusufritonga@gmail.com

Abstract. Image is a visual representation of an object The process of sending information, especially images, can be done anywhere and with any intermediary, but not all information is public. Sometimes someone needs privacy or security from data or information that will be sent or received. One way to secure image files is to use cryptography. Cryptography is the study of how to keep data or messages safe when sent, from the sender to the recipient, without experiencing interference from third parties. In this study cryptography used is Rail Fence Cipher that was combined with steganography, so that the secret message could not be seen in plain view. Steganography serves to hide secret messages in other messages, so the secret is a very hidden existence. There are several methods in Steganography, one of which is Least Significant Bit. The concept of Least Significant Bit works by replacing the last bit of RGB image, where the message is converted first into bits. Based on experiments conducted on the Rail Fence Cipher algorithm, the time of encryption and decryption needed when the resolution of the image gets larger is linearly proportional to the value. The image does not show significant changes before and after inserting a secret message.

1. Introduction

Image is the visual form of information of an object. The delivery and distribution of images can already use a variety of media, so the security of the image becomes lower. One way to secure image files is to use cryptography and steganography.

In data transmission such as transferring an object especially image, aspect of confidentiality, and integrity are highly needed [1]. Cryptography is the science and art of keeping messages secure. Cryptography, in general, is science and art to maintain the confidentiality of news [2]. Rail Fence Cipher is one of the encryption techniques to disguise the writing by changing the position of the character with a diagonal shape down and up. To solve it, we must see the level of the writing, because this cipher is usually systematic [3].

Steganography is a technique of hiding confidential data into a medium so that hidden data is difficult to recognize by the human senses. Steganography requires two priorities, namely the container and secret data that's hidden. Digital steganography uses digital media as a container, for example, image, sound, text, and video. Confidential data that is hidden can also be in the form of image, audio, text, and video [4].

Least Significant Bit (LSB) is a method for inserting images or images into the cover image. Bits with the smallest weight of each pixel of an image are converted into bits of the message to be hidden [5].

To protect the data, both of cryptography and steganography can be done simultaneously. The science of cryptography use randomizes or disguise the data, whereas the science of steganography to hide data in a way [6].

2. Method

In this implementation, there are four main menus; they are encryption Rail Fence Cipher, Embedding LSB, Extraction LSB, and decryption Rail Fence Cipher. Rail Fence Cipher is one of transposition algorithm. A transposition cipher involves the rearranging of the letters in the plaintext to encrypt the message [7].

Steps for encryption Rail Fence Cipher [8]:

1. Input the image to be encrypted.
2. Take the RGB value of each pixel then sort it from the initial to the end pixel. In this test, I took 1x5 pixels of the image, as in Table 1 :

Table 1. The RGB value of the insert image

R= 169 G = 41 B = 41	R= 169 G = 41 B = 41	R= 169 G = 41 B = 41	R= 169 G = 41 B = 41	R= 169 G = 41 B = 41

3. The RGB value order of each pixel is used as a plain image value.
4. Key input. Key = 5
5. Create a table that has columns along with plain images and as many lines as keys. Then arrange the zigzag plain image in the table.

Table 2. Plain image values in a zigzag form

169								41						
	41						41		169					
		41				169				41				41
			169		41						41		41	
				41								169		

6. Read Table 2 from the first row to the last to get the cipher image. Cipher image = 169 41 41 41 169 41 169 41 41 169 41 41 41 41 169

Steps for embedding the secret image:

1. Input cipher image. Cipher image = 169 41 41 41 169 41 169 41 41 169 41 41 41 41 169
2. Change all RGB cipher image values in binary form. 10101001 00101001 00101001 00101001 10101001 00101001 10101001 00101001 00101001 10101001 00101001 00101001 00101001 00101001 00101001 00101001 00101001 00101001 00101001 10101001
3. Input container image or cover image. Take the RGB value from the cover image. In this test, I use 8x5 pixels as a container image.
4. Change each RGB value in the image cover into an 8-bit binary form.
5. Insert the value of the cipher image that has been converted into binary form to every last bit of the cover image.
6. In Table 3 the insertion process to convert plain image to stego image is done by replacing the last bits in the image of the container or the cover image with the RGB binary value of the cipher image.

Steps for extraction image:

1. In the message image extraction process, the system will take the last bits of the stego image that have been inserted in the previous process; the bits can be seen in Table 4.5.
2. Every 8 bits are arranged to get RGB to be used as a cipher image.

IOP Conf. Series: Materials Science and Engineering **851** (2020) 012069 doi:10.1088/1757-899X/851/1/012069

3. Cipher image is used to restore the image in the decryption process.
   Cipher image =
   10101001 00101001 00101001 00101001 10101001 00101001 10101001 00101001 00101001 10101001 00101001 00101001 00101001 00101001 10101001

Table 3. Embedding Process

00001011 00000000 10111001	00001010 00000001 10111000	00001010 00000001 10111000	00001010 00000001 10111000	00001011 00000000 10111000	0110100 0110001 0111001	11111101 11111010 00000011	0110100 0110001 0110100
00001010 00000000 10111001	00001010 00000001 10111000	00001010 00000001 10111001	00001010 00000001 10111000	0110100 0110001 0111001	11111101 11111010 00000010	0110100 0110001 0110100	00001010 00000010 10101011
00001011 00000000 10111001	00001010 00000001 10111000	00001010 00000001 10111000	0110100 0110001 0111001	11111101 11111010 00000010	0110101 0110001 0110100	00001011 00000010 10101011	00001010 00000010 10101011
00001011 00000000 10111001	00001010 00000001 10111000	0110100 0110001 0111001	11111100 11111011 00000010	0110101 0110001 0110100	00001011 00000010 10101010	00001011 00000010 10101011	00001010 00000010 10101011
00001010 00000000 10111001	0110100 0110001 0111001	11111100 11111011 00000010	0110100 0110001 0110100	00001011 00000010 10101010	00001011 00000011 10101010	00001011 00000010 10101011	00001010 00000010 10101011

Steps for decryption Rail Fence Cipher :

1. Input Cipher Image (in binary):
   10101001 00101001 00101001 00101001 10101001 00101001 10101001 00101001 00101001 10101001 00101001 00101001 00101001 00101001 00101001 00101001 00101001 00101001 10101001

   Cipher Image (in decimal):
   169 41 41 41 169 41 169 41 41 169 41 41 41 41 169
2. Key Input = 5
3. Create an empty table that has a cipher image length column and a key length row. Fill the table with an empty string. Then input an asterisk based on the position so that it forms a zigzag pattern.
4. Input the cipher image into a table that has a "*" starting from the first line to the last line, as illustrated in Table 4. Cipher image = 169 41 41 41 169 41 169 41 41 169 41 41 41 41 169

Table 4. Input Cipher Image

169								41						
	41						41		169					
		41				169				41				41
			169		41						41		41	
				41								169		

5. Read the RGB value in the table to return the plain image in the decryption process. The way of reading is based on position by following the Zig-Zag pattern. This process can be seen in Table 5.

Table 5. How to Read the result of decryption

169								41						
	41						41		169					
		41				169				41				41
			169		41						41		41	
				41								169		

6. In Table 5, the plain image value is read zigzag according to the arrow.
Plain Image = 169 41 41 169 41 41 169 41 41 169 41 41 169 41 41

3. Results and Discussions

The experiments were performed on Android 5.1 Lollipop with an Octa-core processor and 2GB RAM. The Integrated Development Environment (IDE) used for coding is Android Studio, and the programming language used is Java. The results of the experiments of each set are presented in Tables 6,7,8 and 9 as follows.

Table 6. Test results of the Image Encrypted with Key Variations

No.	Key Length	Plain Image Resolution	Time(Seconds)
1	5	50 x 50	0,892
2	10	50 x 50	0,767
3	15	50 x 50	0,729
4	20	50 x 50	0,792
5	25	50 x 50	0,767

In table 6, it can be seen that the time of the plain image encryption process using different keys is not always stable. The longer the key was used, it will not always take a long time to process or vice versa.

Table 7. Image testing results that are Encrypted with Plain Image Resolution Variations

No.	Key Length	Plain Image Resolution	Time(Seconds)
1	10	30 x 30	0,185
2	10	40 x 40	0,357
3	10	50 x 50	0,767
4	10	60 x 60	1,880
5	10	70 x 70	2,851

In table 7, the graph of the results of encryption testing with variations in plain image resolution shows the more significant the image resolution, so the encryption process longer, or vice versa, if the image resolution smaller, the encryption process time be faster.

Table 8. Test results of the Image Decrypted with Key Variations

No.	Key Length	Plain Image Resolution	Time(Seconds)
1	5	50 x 50	0,808
2	10	50 x 50	0,761
3	15	50 x 50	0,720
4	20	50 x 50	0,860
5	25	50 x 50	0,744

In table 8, it can be seen that the cipher image decryption process using different keys is not always stable. The longer the key is used, it is not always the processing time that takes a long time or vice versa.

Table 9. The results of testing the Image Decrypted with the Variation of Image Resolution

No.	Key Length	Plain Image Resolution	Time(Seconds)
1	10	30 x 30	0,122
2	10	40 x 40	0,305
3	10	50 x 50	0,761
4	10	60 x 60	1,527
5	10	70 x 70	2,613

In table 9, the results of decryption testing with variations in cipher image resolution shows the more significant the image resolution, so the decryption process longer or vice versa, if the image resolution smaller, the decryption process time be faster.

4. Conclusions

Based on the experiments conducted in this study, it can be concluded that the process of running time of encryption and decryption rail fence algorithm is the time needed when the resolution of the image gets bigger is relatively stable. Both of cryptography and steganography can be done simultaneously. In this research, it uses rail fence as algorithm in cryptography and least significant bit as algorithm in steganography.

5. Acknowledgments

The authors gratefully acknowledge that the present research is supported by Fund Dissemination IPTEKS Research Results for Lecturers / Researchers Universitas Sumatera Utara.

References

[1] Dian Rachmawati et al 2018 IOP Conf. Ser.: Mater. Sci. Eng. 300 012040
[2] Schneier, B. 1996. *Applied Cryptography: Protocols, Algorithms, and Source Code in C. 2nd Edition*. John Wiley & Sons, Inc: New Jersey
[3] Navaneethan, Ramkesh. Advanced Rail Fence Cipher Algorithm. School of Information Technology and Engineering, VIT University, Vellore. 2016
[4] Hidayat, Wildan. 2010. Perlindungan Pesan Rahasia Pada Citra Digital Menggunakan Metode Least Significant Bitsteganografi. Medan. Universitas Sumatera Utara.
[5] F Prashanti .G* , Sandhya Rani.K, Deepthi.S. 2013. LSB and MSB Based Steganography for Embedding Modified DES Encrypted Text. International Journal of Advanced Research in Computer Science and Software Engineering 3 (8): 788 – 799.
[6] D. Rachmawati and M. A. Budiman 2017 New approach toward data hiding by using affine cipher and least significant bit algorithm 2017 4th International Conference on Computer

Applications and Information Processing Technology (CAIPT) Kuta Bali 2017 pp 1-6. doi: 10.1109/CAIPT.2017.8320737
[7] Stallings, William, Cryptography and Network Security, 3rd edition, Pearson Education, 2003.
[8] Talbert, Robert, The Cycle Structure and Order of the Rail Fence Cipher, Cryptologia, Taylor & Francis, 2006

AUTHOR INDEX